Bayes Rules!

CHAPMAN & HALL/CRC
Texts in Statistical Science Series

Joseph K. Blitzstein, *Harvard University, USA*
Julian J. Faraway, *University of Bath, UK*
Martin Tanner, *Northwestern University, USA*
Jim Zidek, *University of British Columbia, Canada*

For more information about this series, please visit: https://www.routledge.com/Chapman--Hall-CRC-Texts-in-Statistical-Science/book-series/CHTEXSTASCI

Bayes Rules!
An Introduction to Applied Bayesian Modeling

Alicia A. Johnson
Miles Q. Ott
Mine Dogucu

CRC Press
Taylor & Francis Group
Boca Raton London New York

CRC Press is an imprint of the
Taylor & Francis Group, an **informa** business

A CHAPMAN & HALL BOOK

First edition published 2022
by CRC Press
6000 Broken Sound Parkway NW, Suite 300, Boca Raton, FL 33487-2742

and by CRC Press
4 Park Square, Milton Park, Abingdon, Oxon, OX14 4RN

CRC Press is an imprint of Taylor & Francis Group, LLC

© 2022 Alicia A. Johnson, Miles Q. Ott and Mine Dogucu

Library of Congress Cataloging-in-Publication Data

Names: Johnson, Alicia A., author. | Ott, Miles Q., editor. | Dogucu, Mine, editor.
Title: Bayes rules! : an introduction to Bayesian modeling with R / Alicia A. Johnson, Miles Ott, Mine Dogucu.
Description: Boca Raton : CRC Press, 2022. | Series: Chapman & Hall/CRC texts in statistical science | Includes bibliographical references and index.
Identifiers: LCCN 2021037969 | ISBN 9780367255398 (paperback) | ISBN 9781032191591 (hardback) | ISBN 9780429288340 (ebook)
Subjects: LCSH: Bayesian statistical decision theory. | R (Computer program language)
Classification: LCC QA279.5 .J64 2022 | DDC 519.5/42--dc23
LC record available at https://lccn.loc.gov/2021037969

ISBN: 978-1-032-19159-1 (hbk)
ISBN: 978-0-367-25539-8 (pbk)
ISBN: 978-0-429-28834-0 (ebk)

DOI: 10.1201/9780429288340

Publisher's note: This book has been prepared from camera-ready copy provided by the authors.

To the exclamation point. We couldn't talk about our friends, family, or Bayesian statistics without you.

Contents

Foreword

Even after decades of thinking about it, Bayes' Rule never ceases to amaze me. How can one simple formula have such a wide variety of applications? You will encounter a vibrant sample of such applications in this book, ranging from weather prediction to LGBTQ+ anti-discrimination laws, and from who calls soda "pop" (or calls pop "soda") to how to classify penguin species. Most importantly, careful study of this book will empower you to conduct thoughtful Bayesian analyses for the data and applications *you* care about.

Statistics and data science focus on using data to learn about the world and make predictions. The Bayesian approach gives a principled, powerful tool for obtaining probabilities and predictions about our unknown quantities of interest, given what we do know (the data). It gives easy-to-interpret results that directly quantify our uncertainties. Unfortunately, it is rarely taught in depth at the undergraduate level, perhaps out of concern that there would be too many scary-looking integrals to do or too much cryptic code to write.

Bayes Rules! shows that the Bayesian approach *is* in fact accessible to students and self-learners with basic statistics knowledge, even if they are not adept at calculus derivations or coding up fancy algorithms from scratch. The book achieves this with many reader-friendly features, such as clear explanations through words and pictures, quizzes to test your understanding, and the **bayesrules** R package that contains datasets and functions that facilitate trying out Bayesian methods.

Better yet, the accessibility is achieved through good pedagogy, not through giving a watered down, over-simplified look at the subject. For example, models called *hierarchical models* and an R package called **rstan** are introduced, with highly instructive examples showing how to apply these to interesting applications. Hierarchical models and **rstan** are among the state-of-the-art techniques used in modern Bayesian data analysis.

The Peter Parker principle from the Spider-Man comics says, "With great power comes great responsibility." Likewise, the great power of statistics and data science comes with the great responsibility to consider the benefits and risks to society, the privacy rights of the participants in a study, the biases in a dataset and whether a proposed algorithm amplifies those biases, and other ethical issues. *Bayes Rules!* emphasizes *fairness* and *ethics* rather than ignoring these crucial issues.

Given that you read *Bayes Rules!* (actively – make sure to try the self-quizzes and practice with some exercises!), the probability is high that you will strengthen your statistical problem-solving skills while experiencing the joy of Bayesian thinking.

—Joseph K. Blitzstein, Harvard University

Preface

Bayesian statistics?! Once an obscure term outside specialized industry and research circles, Bayesian methods are enjoying a renaissance. The title of this book speaks to what all the fuss is about: *Bayes Rules!* Bayesian methods provide a powerful alternative to the frequentist methods that are ingrained in the standard statistics curriculum. Though frequentist and Bayesian methods share a common goal – learning from data – the Bayesian approach to this goal is gaining popularity for many reasons: (1) Bayesian methods allow us to interpret new data in light of prior information, formally weaving both into a set of updated information; (2) relative to the confidence intervals and p-values utilized in frequentist analyses, Bayesian results are easier to interpret; (3) Bayesian methods can shine in settings where frequentist "likelihood" methods break down; and (4) the computational tools required for applying Bayesian techniques are increasingly accessible. Unfortunately, the popularity of Bayesian statistics has outpaced the curricular resources needed to support it. To this end, the primary goal of *Bayes Rules!* is to make modern Bayesian *thinking, modeling,* and *computing* accessible to a broader audience.

Audience

Bayes Rules! brings the power of Bayes to advanced undergraduate students and comparably trained practitioners. Accordingly, the book is neither written at the graduate level nor is it meant to be a first introduction to the field of statistics. At minimum, the book assumes that readers are familiar with the content covered in a typical undergraduate-level introductory statistics course. Readers will also, ideally, have some experience with undergraduate-level probability, calculus, and the R statistical software. *But wait!* Please don't go away if you don't check off all of these boxes. We provide all R code and enough probability review so that readers without this background will still be able to follow along so long as they are eager to pick up these tools on the fly. Further, though certain calculus *concepts* are important in Bayesian analysis (thus the book), calculus *derivations* are not. The latter are limited to the simple model settings in early chapters and easily skippable.

Navigating this book

Bayes Rules! empowers readers to weave Bayesian approaches into an everyday modern practice of statistics and data science. The overall spirit is applied: the book utilizes modern computing resources and a reproducible pipeline; the discussion emphasizes conceptual understanding; the material is motivated by data-driven inquiry; and the delivery blends traditional "content" with "activity." Despite the applied focus, readers will also develop the theoretical foundation required for a deeper understanding of Bayesian methodology and the flexibility to generalize it in new settings. The following are some tips and tricks for navigating the adventure.

Learn by doing. *Doing* Bayesian statistics requires practice. And software. *Bayes Rules!* utilizes R with the powerful **rstan** interface to the Stan language. All R code is included within the book, equipping readers with the tools to reproduce and generalize these computational techniques beyond the book.

Embrace a growth mindset. As you read the book and put Bayesian methodology into practice, you will make mistakes. Many mistakes. Making and learning from mistakes is simply part of learning. We hope that you persist through the struggle of learning so that you can contribute your unique insights, perspectives, and experiences to the Bayesian community.

Interpret Bayes in context. Bayesian statistics doesn't exist in a vacuum. It is a tool we can use to learn about the world around us, from data. Throughout, we encourage readers to reflect upon the potential implications of our Bayesian analyses (e.g., are they biased? fair?).

Practice, practice, practice. Each chapter concludes with a set of exercises. Some allow you to brush up on the mechanics of what you have learned, in bite-sized portions. Others encourage you to integrate and apply various concepts in realistic settings using R.

Getting set up

Once you're ready to dive into *Bayes Rules!*, take the following steps to get set up. First, download the most recent versions of the following software:

- R (https://www.r-project.org/)
- RStudio (https://rstudio.com/products/rstudio/download/)

Next, install the following packages within RStudio:

- Install the **rstan** package by carefully following the directions at https://github.com/stan-dev/rstan/wiki/RStan-Getting-Started.

- Install a set of data wrangling and Bayesian packages by typing the following in your console.

```
install.packages(c("bayesrules", "tidyverse", "janitor", "rstanarm",
                   "bayesplot", "tidybayes", "broom.mixed", "modelr",
                   "e1071", "forcats"),
                 dependencies = TRUE)
```

- The **bayesrules** package contains some datasets and functions we've built explicitly for the *Bayes Rules!* book. Several of the datasets in this package have been extracted from larger sources, using non-random methods to obtain subsets that match our pedagogical goals. These potentially biased datasets should *not* be used for rigorous research purposes. Should this be your goal, please see the relevant data help file to identify the original, complete data source.

Finally, check out `https://bayes-rules.github.io/`. This website contains more background information about the book as well as supporting resources that will continue to grow over time.

Accessibility and inclusion

We are dedicated to providing an inclusive and accessible Bayesian resource. We are continuing to learn, and shared our current efforts at `https://www.datapedagogy.com/posts/2020-07-24a-bayes-open-access`. All the figures in the online version book are supported by alternate text and should be able accessible to readers using a screen-reader.

Contact us

As you read and interact with Bayes Rules!, please feel free to drop us a note through the form posted on our website: `https://bayes-rules.github.io/posts/contact/`. We're especially curious to know if: you have any ideas for increasing the accessibility and inclusion of this book; you notice any errors or typos; and there's anything you'd like to see in future versions. When we revise the text, we will occasionally check in with this form to help guide that process.

Acknowledgments

First and foremost, we'd like to thank the students we've worked with at Augsburg University, Carleton College, Denison University, Macalester College, Smith College, and the University of California, Irvine. Their feedback, insights, and example inspired a high bar for the type of book we wanted to put out into the world. Beyond our students, we received valuable feedback from numerous colleagues in the Statistics and Data Science community. Special thanks to: James Albert, Virgilio Gómez Rubio, Amy Herring, David Hitchcock, Nick Horton, Yue Jiang, and Paul Roback. And to our supportive editor, David Grubbs. He certainly made this first adventure in book publishing more chill and enjoyable than we expected.

Beyond those above, Alicia would especially like to thank the following:

- John Kim, Martha Skold, the Johnsons, and Minneapolis friends for their support, even on the dreaded "writing brain" days.
- Galin Jones for his inviting enthusiasm about statistical computation.
- The STAT 454 students at Macalester College for their inspiring curiosity and the

STAT 454 teaching assistants for supporting their peers in learning about Bayes (Zuofu Huang, Sebastian Coll, Connie Zhang).

- Colleagues in the Department of Mathematics, Statistics, and Computer Science at Macalester College – you are a humane, fun, and reflective crew.

Miles would especially like to thank the following:

- Francesca Giardine, Sarah Glidden, and Elaona Lemoto for testing out half-formed exercises, gently pointing out errors, and giving excellent suggestions. It was an honor to get to work with you at this early stage of your statistics careers.
- The Smith College SDS 320 students from spring 2019 and the SDS 390 students from fall 2020 (especially Audrey Bretin, Dianne Caravela, Marlene Jackson, and Hannah Snell who provided helpful feedback on Chapter 8).
- His colleagues from Smith College SDS.
- The WSDS conference for being the starting point for many friendships and collaborations, including this book.
- His family, Bea Capistrant, Ethan Suniewick, Malkah Bird, Henry Schneiderman, Christopher Tradowsky, Ross Elfline, Jon Knapp, and Alex Callendar.

Mine would especially like to thank the following:

- Morteza Khakshoor, family, and friends who are far only in distance, for their love, support, and understanding.
- The late Binnaz Melin for being supportive of her career from a young age.
- Students of STATS 115 at UC Irvine who always are the best part of teaching Bayesian statistics.
- Her colleagues in the Department of Statistics at UC Irvine.
- All those in the statistics and R community who are supportive of, and kind towards, others, especially newcomers.

Finally, we would all like to thank each other for recognizing the importance of humor, empathy, and gratitude to effective collaboration. Deciding to tackle this project before we really knew one another was a pretty great gamble. It's been fun!

License

About the Authors

Alicia A. Johnson is an Associate Professor of Statistics at Macalester College in Saint Paul, Minnesota. She enjoys exploring and connecting students to Bayesian analysis, computational statistics, and the power of data in contributing to this shared world of ours.

Miles Q. Ott is a Senior Data Scientist at The Janssen Pharmaceutical Companies of Johnson & Johnson. Prior to his current position, he taught at Carleton College, Augsburg University, and Smith College. He is interested in biostatistics, LGBTQ+ health research, analysis of social network data, and statistics/data science education. He blogs at milesott.com and tweets on Twitter[1] about statistics, gardening, and his dogs.

Mine Dogucu is an Assistant Professor of Teaching in the Department of Statistics at University of California Irvine. She spends the majority of her time thinking about what to teach, how to teach it, and what tools to use while teaching. She likes intersectional feminism, cats, and R Ladies. She tweets on Twitter[2] about statistics and data science education.

---

[1]https://twitter.com/Miles_Ott
[2]https://twitter.com/MineDogucu

Unit I

Bayesian Foundations

1

The Big (Bayesian) Picture

> How can we live if we don't change?
> —Beyoncé. Lyric from "Satellites."

Everybody changes their mind. You likely even changed your mind in the last minute. Prior to ever opening it, you no doubt had some preconceptions about this book, whether formed from its title, its online reviews, or a conversation with a friend who has read it. And then you saw that the first chapter opened with a quote from Beyoncé, an unusual choice for a statistics book. Perhaps this made you think "This book is going to be even more fun than I realized!" Perhaps it served to do the opposite. No matter. The point we want to make is that we agree with Beyoncé – changing is simply part of life.

We continuously update our knowledge about the world as we accumulate lived experiences, or *collect data*. As children, it takes a few spills to understand that liquid doesn't stay in a glass. Or a couple attempts at conversation to understand that, unlike in cartoons, real dogs can't talk. Other knowledge is longer in the making. For example, suppose there's a new Italian restaurant in your town. It has a 5-star online rating *and* you love Italian food! Thus, prior to ever stepping foot in the restaurant, you anticipate that it will be quite delicious. On your first visit, you collect some edible data: your pasta dish arrives a soggy mess. Weighing the stellar online rating against your own terrible meal (which might have just been a fluke), you update your knowledge: this is a 3-star not 5-star restaurant. Willing to give the restaurant another chance, you make a second trip. On this visit, you're pleased with your Alfredo and increase the restaurant's rating to 4 stars. You continue to visit the restaurant, collecting edible data and updating your knowledge each time.

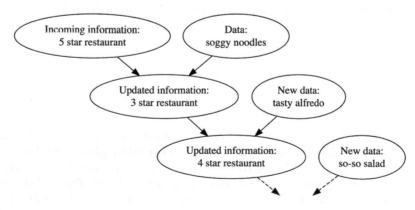

FIGURE 1.1: Your evolving knowledge about a restaurant.

DOI: 10.1201/9780429288340-1

Figure 1.1 captures the natural *Bayesian* knowledge-building process of acknowledging your preconceptions, using data to update your knowledge, and repeating. We can apply this same Bayesian process to rigorous research inquiries. If you're a political scientist, yours might be a study of demographic factors in voting patterns. If you're an environmental scientist, yours might be an analysis of the human role in climate change. You don't walk into such an inquiry without context – you carry a degree of incoming or **prior** information based on previous research and experience. Naturally, it's *in light of* this information that you interpret new **data**, weighing both in developing your updated or **posterior** information. You continue to refine this information as you gather new evidence (Figure 1.2).

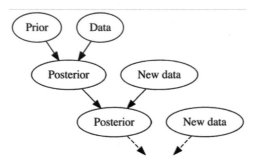

FIGURE 1.2: A Bayesian knowledge-building diagram.

The Bayesian philosophy represented in Figure 1.2 is the foundation of Bayesian statistics. Throughout this book, you will build the methodology and tools you need to *implement* this philosophy in a rigorous data analysis. This experience will require a sense of purpose and a map.

⊚ **Goals**

- Learn to think like a Bayesian.
- Explore the foundations of a Bayesian data analysis and how they contrast with the frequentist alternative.
- Learn a little bit about the history of the Bayesian philosophy.

1.1   Thinking like a Bayesian

Given our emphasis on how natural the Bayesian approach to knowledge-building is, you might be surprised to know that the alternative **frequentist** philosophy has traditionally dominated statistics. Before exploring their differences, it's important to note that **Bayesian and frequentist analyses share a common goal: to learn from data about the world around us.** Both Bayesian and frequentist analyses use data to fit models, make predictions, and evaluate hypotheses. When working with the same data, they will typically produce a similar set of conclusions. Moreover, though we'll often refer to people applying these philosophies as "Bayesians" and "frequentists," this is for brevity only. The distinction is not always so clear in practice – even the authors apply both philosophies in their work, and thus don't identify with a single label. Our goal here is not to "take sides," but to illustrate the key differences in the logic behind, approach to, and interpretation of Bayesian and frequentist analyses.

1.1.1 Quiz yourself

Before we elaborate upon the Bayesian vs frequentist philosophies, take a quick quiz to assess your current inclinations. In doing so, try to abandon any rules that you might have learned in the past and just go with your gut.

1. When flipping a fair coin, we say that "the probability of flipping Heads is 0.5." How do you interpret this probability?

    a) If I flip this coin over and over, roughly 50% will be Heads.
    b) Heads and Tails are equally plausible.
    c) Both a and b make sense.

2. An election is coming up and a pollster claims that candidate A has a 0.9 probability of winning. How do you interpret this probability?

    a) If we observe the election over and over, candidate A will win roughly 90% of the time.
    b) Candidate A is much more likely to win than to lose.
    c) The pollster's calculation is wrong. Candidate A will either win or lose, thus their probability of winning can only be 0 or 1.

3. Consider two claims. (1) Zuofu claims that he can predict the outcome of a coin flip. To test his claim, you flip a fair coin 10 times and he correctly predicts all 10. (2) Kavya claims that she can distinguish natural and artificial sweeteners. To test her claim, you give her 10 sweetener samples and she correctly identifies each. In light of these experiments, what do you conclude?

    a) You're more confident in Kavya's claim than Zuofu's claim.
    b) The evidence supporting Zuofu's claim is just as strong as the evidence supporting Kavya's claim.

4. Suppose that during a recent doctor's visit, you tested positive for a very rare disease. If you only get to ask the doctor one question, which would it be?

    a) What's the chance that I actually have the disease?
    b) If in fact I don't have the disease, what's the chance that I would've gotten this positive test result?

Next, tally up your quiz score using the scoring system below.[1] Totals from 4–5 indicate that your current thinking is fairly frequentist, whereas totals from 9–12 indicate alignment with the Bayesian philosophy. In between these extremes, totals from 6–8 indicate that you see strengths in both philosophies. Your current inclinations might be more frequentist than Bayesian or vice versa. These inclinations might change throughout your reading of this book. They might not. For now, we merely wish to highlight the key differences between the Bayesian and frequentist philosophies.

---

[1] 1. a = 1 point, b = 3 points, c = 2 points; 2. a = 1 point, b = 3 points, c = 1 point; 3. a = 3 points, b = 1 point; 4. a = 3 points, b = 1 point.

1.1.2   The meaning of probability

Probability theory is central to every statistical analysis. Yet, as illustrated by questions 1 and 2 in Section 1.1.1, Bayesians and frequentists differ on something as fundamental as the meaning of probability. For example, in question 1, a Bayesian and frequentist would both say that the probability of observing Heads on a fair coin flip is 1/2. The difference is in their interpretation.

Interpreting probability

- In the Bayesian philosophy, a probability measures the **relative plausibility** of an event.
- The frequentist philosophy is so named for its interpretation of probability as the **long-run relative frequency** of a repeatable event.

Thus, in the coin flip example, a Bayesian would conclude that Heads and Tails are equally likely. In contrast, a frequentist would conclude that if we flip the coin over and over and over, roughly 1/2 of these flips will be Heads. Let's try applying these same ideas to the question 2 setting in which a pollster declares that candidate A has a 0.9 probability of winning the upcoming election. This routine election calculation illustrates cracks within the frequentist interpretation. Since the election is a one-time event, the long-run relative frequency concept of observing the election over and over simply doesn't apply. A very strict frequentist interpretation might even conclude the pollster is just wrong. Since the candidate will either win or lose, their win probability must be either 1 or 0. A less extreme frequentist interpretation, though a bit awkward, is more reasonable: in long-run *hypothetical* repetitions of the election, i.e., elections with similar circumstances, candidate A would win roughly 90% of the time.

The election example is not rare. It's often the case that an event of interest is unrepeatable. Whether or not a politician wins an election, whether or not it rains tomorrow, and whether or not humans will live on Mars are all one-time events. Whereas the frequentist interpretation of probability can be awkward in these one-time settings, the more flexible Bayesian interpretation provides a path by which to express the uncertainty of these events. For example, a Bayesian would interpret "a 0.9 probability of winning" to mean that, based on election models, the relative plausibility of winning is high – the candidate is 9 times more likely to win than to lose.

1.1.3   The Bayesian balancing act

Inspired by an example by Berger (1984), question 3 in Section 1.1.1 presented you with two scenarios: (1) Zuofu claims that he can predict the outcome of a coin flip and (2) Kavya claims that she can distinguish between natural and artificial sweeteners. Let's agree here that the first claim is simply ridiculous but that the second is plausible (some people have sensitive palates!). Thus, imagine our surprise when, in testing their claims, both Zuofu and Kavya enjoyed a 10 out of 10 success rate: Zuofu correctly predicted the outcomes of 10 coin flips and Kavya correctly identified the source of 10 different sweeteners. What can we conclude from this data? Does it provide equal evidence in support of both Zuofu's and Kavya's claims? The Bayesian and frequentist philosophies view these questions through different lenses.

Let's begin by looking through the frequentist lens which, to oversimplify quite a bit, analyzes the data absent a consideration of our prior contextual understanding. Thus, in a frequentist analysis, "10 out of 10" is "10 out of 10" no matter if it's in the context of Zuofu's coins or Kavya's sweeteners. This means that a frequentist analysis would lead to *equally* confident conclusions that Zuofu can predict coin flips and Kavya can distinguish between natural and artificial sweeteners (at least on paper if not in our gut).[2] Given the absurdity of Zuofu's claim, this frequentist conclusion is a bit bizarre – it throws out all prior knowledge in favor of a mere 10 data points. We can't resist representing this conclusion with Figure 1.3, a frequentist complement to the Bayesian knowledge-building diagram in Figure 1.2, which solely consists of the data.

FIGURE 1.3: A frequentist knowledge-building diagram.

In contrast, a Bayesian analysis gives voice to our prior knowledge. Here, our experience on Earth suggests that Zuofu is probably overstating his abilities but that Kavya's claim is reasonable. Thus, after weighing their equivalent "10 out of 10" achievements against these different priors, our posterior understanding of Zuofu's and Kavya's claims differ. Since the data is consistent with our prior, we're even more certain that Kavya is a sweetener savant. However, given its inconsistency with our prior experience, we are chalking Zuofu's "psychic" achievement up to simple luck.

The idea of allowing one's prior experience to play a formal role in a statistical analysis might seem a bit goofy. In fact, a common critique of the Bayesian philosophy is that it's too subjective. We haven't done much to combat this critique yet. Favoring flavor over details, Figure 1.2 might even lead you to believe that Bayesian analysis involves a bit of subjective hocus pocus: combine your prior with some data and *poof*, out pops your posterior. In reality, the Bayesian philosophy provides a *formal* framework for such knowledge creation. This framework depends upon prior information, data, and the **balance** between them.

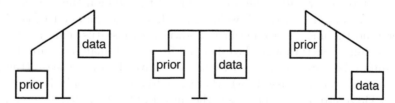

FIGURE 1.4: Bayesian analyses balance our prior experiences with new data. Depending upon the setting, the prior is given more weight than the data (left), the prior and data are given equal weight (middle), or the prior is given less weight than the data (right).

In building the posterior, the balance between the prior information and data is determined by the relative strength of each. For example, we had a very strong prior understanding that Zuofu isn't a psychic, yet very little data (10 coin flips) supporting his claim. Thus, like the left plot in Figure 1.4, the prior held more weight in our posterior understanding.

---

[2] *If* you have experience with frequentist statistics, you might be skeptical that these methods would produce such a silly conclusion. Yet in the frequentist null hypothesis significance testing framework, the hypothesis being tested in both Zuofu's and Kavya's settings is that their success rate exceeds 50%. Since their "10 out of 10" data is the same, the corresponding p-values (≈ 0.001) and resulting hypothesis test conclusions are also the same.

However, we're not stubborn. If Zuofu had correctly predicted the outcome of, say, *10,000* coin flips, the strength of this data would far surpass that of our prior, leading to a posterior conclusion that perhaps Zuofu is psychic after all (like the right plot in Figure 1.4)!

Allowing the posterior to balance out the prior and data is critical to the Bayesian knowledge-building process. When we have little data, our posterior can draw upon the power in our prior knowledge. As we collect more data, the prior loses its influence. Whether in science, policy-making, or life, this is how people tend to think (El-Gamal and Grether, 1995) and how progress is made. As they collect more and more data, two scientists will come to agreement on the human role in climate change, no matter their prior training and experience.[3] As they read more and more pages of this book, two readers will come to agreement on the power of Bayesian statistics. This logical and heartening idea is illustrated by Figure 1.5.

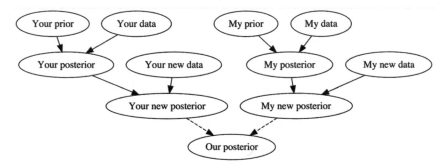

FIGURE 1.5: A two-person Bayesian knowledge-building diagram.

1.1.4 Asking questions

In question 4 of Section 1.1.1, you were asked to imagine that you tested positive for a rare disease and only got to ask the doctor one question: (a) what's the chance that I actually have the disease?, or (b) if in fact I do not have the disease, what's the chance that I would've gotten this positive test result? The authors are of the opinion that, though the answers to both questions would be *helpful*, we'd *rather* know the answer to (a). That is, we'd rather understand the uncertainty in our unknown disease status than in our observed test result. Unsurprising spoiler: though Bayesian and frequentist analyses share the goal of using your test results (the data) to assess whether you have the rare disease (the hypothesis), a Bayesian analysis would answer (a) whereas a frequentist analysis would answer (b). Specifically, a **Bayesian** analysis assesses the uncertainty of the hypothesis in light of the observed data, and a **frequentist** analysis assesses the uncertainty of the observed data in light of an assumed hypothesis.

Asking questions

- A Bayesian hypothesis test seeks to answer: In light of the observed data, what's the chance that the hypothesis is correct?
- A frequentist hypothesis test seeks to answer: If in fact the hypothesis is incorrect, what's the chance I'd have observed this, or even more extreme, data?

---

[3]There is an extreme exception to this rule. If someone assigns 0 prior weight to a given scenario, then no amount of data will change their mind. We explore this specific situation in Chapter 4.

For clarity, consider the scenario summarized in Table 1.1 where, in a population of 100, only four people have the disease. Among the 96 without the disease, nine test positive and thus get misleading test results. Among the four with the disease, three test positive and thus get accurate test results.

TABLE 1.1: Disease status and test outcomes for 100 people.

	test positive	test negative	total
disease	3	1	4
no disease	9	87	96
total	12	88	100

In this scenario, a *Bayesian* analysis would ask: Given my positive test result, what's the chance that I actually have the disease? Since only 3 of the 12 people that tested positive have the disease (Table 1.1), there's only a 25% chance that you have the disease. Thus, when we take into account the disease's rarity and the relatively high false positive rate, it's relatively unlikely that you actually have the disease. What a relief.

Recalling Section 1.1.2, you might anticipate that a frequentist approach to this analysis would differ. From the frequentist standpoint, since disease status isn't repeatable, the probability you have the disease is either 1 or 0 – you have it or you don't. To the contrary, medical testing (and data collection in general) *is* repeatable. You can get tested for the disease over and over and over. Thus, a *frequentist* analysis would ask: If I don't actually have the disease, what's the chance that I would've tested positive? Since only 9 of the 96 people without the disease tested positive, there's a roughly 10% (9/96) chance that you would've tested positive even if you didn't have the disease.

The 9/96 frequentist probability calculation is similar in spirit to a **p-value**. In general, p-values measure the chance of having observed data as or more extreme than ours if in fact our original hypothesis is incorrect. Though the p-value was prominent in the frequentist practice for decades, it's slowly being de-emphasized across the frequentist and Bayesian spectrum. Essentially, it's so commonly misinterpreted and misused (Goodman, 2011), that the American Statistical Association put out an official "public safety announcement" regarding its usage (Wasserstein, 2016). The *reason* the p-value is so commonly misinterpreted is simple – it's more *natural* to study the uncertainty of a yet-unproven hypothesis (whether you have the rare disease) than the uncertainty of data we have already observed (you tested positive for the rare disease).[4]

1.2 A quick history lesson

Given how natural Bayesian thinking is, you might be surprised to know that Bayes' momentum is relatively recent. Once an obscure term outside specialized industry and research circles, "Bayesian" has popped up on TV shows (e.g., *The Big Bang Theory*[5] and

---

[4] Authors' opinion.

[5] http://www.cracked.com/article_21544_6-tv-shows-that-put-insane-work-into-details-nobody-noticed_p2.html

Numb3rs[6]), has been popularized by various blogs, and regularly appears in the media (e.g., the *New York Times'* explanation of how to think like an epidemiologist (Roberts, 2020)).

Despite its recent rise in popularity, Bayesian statistics is rooted in the *eighteenth-century* work of Reverend Thomas Bayes, a statistician, minister, and philosopher. Though Bayes developed his philosophy during the 1740s, it wasn't until the late twentieth century that this work reached a broad audience. During the more than two centuries in between, the frequentist philosophy dominated statistical research and practice. The Bayesian philosophy not only fell out of popular favor during this time, it was stigmatized. As recently as 1990, Marilyn vos Savant was ridiculed by some readers of her *Parade* magazine column when she presented a Bayesian solution to the now classic *Monty Hall* probability puzzle. Apparently more than 10,000 readers wrote in to dispute and mock her solution.[7] Vos Savant's Bayesian solution was later proven to be indisputably correct. Yet with this level of scrutiny, it's no wonder that many researchers kept their Bayesian pursuits under wraps. Though neither proclaimed as much at the time, Alan Turing cracked Germany's Enigma code in World War II and John Tukey pioneered election-day predictions in the 1960s using Bayesian methods (McGrayne, 2012).

As the public existence of this book suggests, the stigma has largely eroded. Why? **(1) Advances in computing.** Bayesian applications require sophisticated computing resources that weren't broadly available until the 1990s. **(2) Departure from tradition.** Due to the fact that it's typically *what people learn*, frequentist methods are ingrained in practice. Due to the fact that it's often *what people use* in practice, frequentist methods are ingrained in the statistics curriculum, and thus are what people learn. This cycle is difficult to break. **(3) Reevaluation of "subjectivity."** Modern statistical practice is a child of the Enlightenment. A reflection of the Enlightenment ideals, frequentist methods were embraced as the superior, *objective* alternative to the *subjective* Bayesian philosophy. This subjective stigma is slowly fading for several reasons. First, the "subjective" label can be stamped on *all* statistical analyses, whether frequentist or Bayesian. Our prior knowledge naturally informs what we measure, why we measure it, and how we model it. Second, post-Enlightenment, "subjective" is no longer such a dirty word. After all, just as two bakers might use two different recipes to produce equally tasty bagels, two analysts might use two different techniques to produce equally informative analyses.

Figure 1.6 provides a sense of scale for the Bayesian timeline. Remember that Thomas Bayes started developing the Bayesian philosophy in the 1740s. It wasn't until 1953, more than 200 years later (!), that Arianna W. Rosenbluth wrote the first complete Markov chain Monte Carlo algorithm, making it possible to actually *do* Bayesian statistics. It was even later, in 1969, that David Blackwell introduced one of the first Bayesian textbooks, bringing these methods to a broader audience (Blackwell, 1969). The Bayesian community is now rapidly expanding around the globe,[8] with the Bayesian framework being used for everything from modeling COVID-19 rates in South Africa (Mbuvha and Marwala, 2020) to monitoring human rights violations (Lum et al., 2010). Not only are we, this book's authors, part of this Bayesian story now, we hope to have created a resource that invites you to be a part of this story too.

---

[6]http://pi.math.cornell.edu/~numb3rs/lipa/Episodes/

[7]https://priceonomics.com/the-time-everyone-corrected-the-worlds-smartest

[8]https://bayesian.org/chapters/australasian-chapter; or brazil; or chile; or east-asia; or india; or south-africa

FIGURE 1.6: In order from left to right: Portrait of Thomas Bayes (unknown author / public domain, Wikimedia Commons); photo of Arianna W. Rosenbluth (Marshall N. Rosenbluth, Wikimedia Commons); photo of David Blackwell at Berkeley, California (George M. Bergman, Wikimedia Commons); you! (insert a photo of yourself).

1.3 A look ahead

Throughout this book, you will learn how to bring your Bayesian thinking to life. The structure for this exploration is outlined below. Though the chapters are divided into broader units which each have a unique theme, there is a common thread throughout the book: building and analyzing Bayesian models for the behavior of some variable "Y."

1.3.1 Unit 1: Bayesian foundations

Motivating question

How can we incorporate Bayesian thinking into a formal model of some variable of interest, Y?

Unit 1 develops the *foundation* upon which to build our Bayesian analyses. You will explore the heart of every Bayesian model, *Bayes' Rule*, and put Bayes' Rule into action to build a few introductory but **fundamental Bayesian models**. These unique models are tied together in the broader **conjugate family**. Further, they are tailored toward variables Y of differing structures, and thus apply our Bayesian thinking in a wide variety of scenarios.

To begin, the **Beta-Binomial model** can help us determine the probability that it rains tomorrow in Australia using data on **binary categorical variable** Y, whether or not it rains for each of 1000 sampled days (Figure 1.7 (a)). The **Gamma-Poisson model** can help us explore the rate of bald eagle sightings in Ontario, Canada using data on variable Y, the **counts** of eagles seen in each of 37 one-week observation periods (Figure 1.7 (b)). Finally, the **Normal-Normal model** can provide insight into the average 3 p.m. temperature in Australia using data on the **bell-shaped** variable Y, temperatures on a sample of study days (Figure 1.7 (c)).[9]

---

[9]These plots use the `weather_perth`, `bald_eagles`, and `weather_australia` data in the `bayesrules` package. The weather datasets are subsets of the `weatherAUS` data in the `rattle` package. The eagles data was made available by Birds Canada (2018) and distributed by R for Data Science (2018).

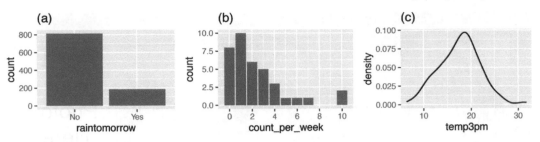

FIGURE 1.7: (a) Binomial output for the rain status of 1000 sampled days in Australia; (b) Poisson counts of bald eagles observed in 37 one-week observation periods; (c) Normally distributed 3 p.m. temperatures (in degrees Celsius) on 200 days in Australia.

1.3.2 Unit 2: Posterior simulation & analysis

Motivating questions

When our Bayesian models of Y become too complicated to mathematically *specify*, how can we *approximate* them? And once we've either specified or approximated a model, how can we *make meaning* of and draw formal conclusions from it?

We can use Bayes' Rule to mathematically *specify* each fundamental Bayesian model presented in Unit 1. Yet as we generalize these Bayesian modeling tools to broader settings, things get real complicated real fast. The result? We might not be able to actually specify the model. Never fear – data analysts are not known to throw up their hands in the face of the unknown. When we can't *know* something, we *approximate* it. In Unit 2 you will learn to use **Markov chain Monte Carlo** *simulation* techniques to *approximate* otherwise out-of-reach Bayesian models.

Once we've either specified or approximated a Bayesian model for a given scenario, we must also be able to *make meaning* of and draw formal conclusions from the results. **Posterior analysis** is the process of asking "what does this all mean?!" and revolves around three major elements: posterior estimation, hypothesis testing, and prediction.

1.3.3 Unit 3: Bayesian regression & classification

Motivating question

We're not always interested in the *lone* behavior of variable Y. Rather, we might want to understand the *relationship* between Y and a set of p potential **predictor variables** $(X_1, X_2, \ldots, X_p)$. How do we build a Bayesian model of this *relationship*? And how do we know whether ours is a *good* model?

Unit 3 is where things really keep staying fun. Prior to Unit 3, our motivating research questions all focus on a single variable Y. For example, in the Normal-Normal scenario we were interested in exploring Y, 3 p.m. temperatures in Australia. Yet once we have a grip on this **response variable** Y, we often have follow-up questions: Can we **model** and **predict** 3 p.m. temperatures based on *9 a.m.* temperatures (X_1) and precise location (X_2)? To this end, in Unit 3 we will survey Bayesian modeling tools that conventionally fall into two categories:

- Modeling and predicting a **quantitative** response variable Y is a **regression task**.
- Modeling and predicting a **categorical** response variable Y is a **classification task**.

Let's connect these terms with our three examples from Section 1.3.1. First, in the Australian temperature example, our sample data indicates that temperatures tend to be warmer in Wollongong and that the warmer it is at 9 a.m., the warmer it tends to be at 3 p.m. (Figure 1.8 (c)). Since the 3 p.m. temperature response variable is *quantitative*, modeling this relationship is a **regression** task. In fact, we can generalize the Unit 1 Normal-Normal model for the behavior in Y alone to build a **Normal regression model** of the relationship between Y and predictors X_1 and X_2. Similarly, we can extend our Unit 1 Gamma-Poisson analysis of the *quantitative* bird counts (Y) into a **Poisson regression model** that describes how these counts have increased over time (X_1) (Figure 1.8 (b)).

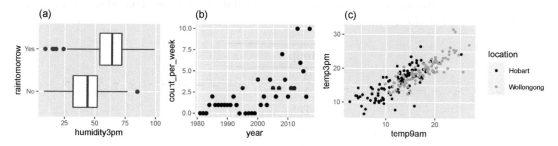

FIGURE 1.8: (a) Tomorrow's rain vs today's humidity; (b) the number of bald eagles over time; (c) 3 p.m. vs 9 a.m. temperatures in two different Australian cities.

Finally, consider our examination of the *categorical variable* Y, whether or not it rains tomorrow in Australia: *Yes* or *No*. Understanding *general* rain patterns in Y is nice. It's even nicer to be able to *predict* rain based on today's weather patterns. For example, in our sample data, rainy days tend to be preceded by higher humidity levels, X_1, than non-rainy days (Figure 1.8 (a)). Since Y is *categorical* in this setting, modeling and predicting the outcome of rain from the humidity level is a **classification task**. We will explore two approaches to classification in Unit 3. The first, **logistic regression**, is an extension of the Unit 1 Beta-Binomial model. The second, **Naive Bayes classification**, is a simplified extension of Bayes' Rule.

1.3.4 Unit 4: Hierarchical Bayesian models

Motivating question

Help! What if the structure of our data violates the assumptions of independence behind the Unit 3 regression and classification models? Specifically, suppose we have multiple observations per random "group" in our dataset. How do we tweak our Bayesian models to not only acknowledge, but *harness*, this structure?

The regression and classification models in Unit 3 operate under the assumption of **independence**. That is, they assume that our data on the response and predictor variables $(Y, X_1, X_2, \ldots, X_p)$ is a **random sample** – the observed values for any one subject in the sample are independent of those for any other subject. The structure of independent data is represented by the data table below:

observation	y	x
1	...	...
2	...	...
3	...	...

This assumption is often violated in practice. Thus, in Unit 4, we'll greatly expand the flexibility of our Bayesian modeling toolbox by accommodating hierarchical, or **grouped data**. For example, our data might consist of a sampled group of recording artists and data y and x on multiple individual songs within each artist; or a sampled group of labs and data from multiple individual experiments within each lab; or a sampled group of people on whom we make multiple individual observations of data over time. This idea is represented by the data table below:

group	y	x
A	...	...
A	...	...
B	...	...
B	...	...
B	...	...

The **hierarchical models** we'll explore in Unit 4 will both *accommodate* and *harness* this type of grouped data. By appropriately reflecting this data structure in our models, not only do we avoid doing the *wrong* thing, we can increase our power to detect the underlying trends in the relationships between Y and $(X_1, X_2, \ldots, X_p)$ that might otherwise be masked if we relied on the modeling techniques in Unit 3 alone.

1.4   Chapter summary

In Chapter 1, you learned how to think like a Bayesian. In the fundamental Bayesian knowledge-building process (Figure 1.2):

- we construct our **posterior knowledge** by balancing **information from our data** with our **prior knowledge**;
- as more data come in, we continue to **refine** this knowledge as the influence of our original prior fades into the background; thus,
- in light of more and more data, two analysts that start out with opposing knowledge will converge on the same posterior knowledge.

1.5   Exercises

In these first exercises, we hope that you make and learn from some mistakes as you incorporate the ideas you learned in this chapter into your way of thinking. Ultimately, we hope that you attain a greater understanding of these ideas than you would have had if you had never made a mistake at all.

Exercise 1.1 (Bayesian Chocolate Milk). In the fourth episode of the sixth season of the television show *Parks and Recreation*, Deputy Director of the Pawnee Parks and Rec department, Leslie Knope, is being subjected to an inquiry by Pawnee City Council member Jeremy Jamm due to an inappropriate tweet from the official Parks and Rec Twitter account. The following exchange between Jamm and Knope is an example of Bayesian thinking:

JJ: "When this sick depraved tweet first came to light, you said 'the account was probably hacked by some bored teenager'. Now you're saying it is an unfortunate mistake. Why do you keep flip-flopping?"

LK: "Well because I learned new information. When I was four, I thought chocolate milk came from brown cows. And then I 'flip-flopped' when I found that there was something called chocolate syrup."

JJ: "I don't think I'm out of line when I say this scandal makes Benghazi look like Whitewater."

  a)   Identify possible prior information for Leslie's chocolate milk story.
  b)   Identify the data that Leslie weighed against that incoming information in her chocolate milk story.
  c)   Identify the updated conclusion from the chocolate milk story.

Exercise 1.2 (Stats Tweets). In May 2020 the Twitter user @frenchpressplz tweeted[10] "Normalize changing your mind when presented with new information." We consider this a #BayesianTweet.

  a)   Write your own #BayesianTweet.
  b)   Write your own #FrequentistTweet.

Exercise 1.3 (When was the last time you changed your mind?). Think of a recent situation in which you changed your mind. As with the Italian restaurant example (Figure 1.1), make a diagram that includes your prior information, your new data that helped you change your mind, and your posterior conclusion.

Exercise 1.4 (When was the last time you changed someone else's mind?). Think of a recent situation in which you had a conversation in which you changed someone else's mind. As with the Italian restaurant example (Figure 1.1), make a diagram that includes the prior information, the new data that helped you change their mind, and the posterior conclusion.

Exercise 1.5 (Changing views on Bayes). When one of the book authors started their master's degree in biostatistics, they had never used Bayesian statistics before, and thus felt neutral about the topic. In their first semester, they used Bayes to learn about diagnostic tests for different diseases, saw how important Bayes was, and became very interested in the topic. In their second semester, their mathematical statistics course included a Bayesian exercise involving ant eggs which both disgusted them and felt unnecessarily difficult – they became disinterested in Bayesian statistics. In the first semester of their Biostatistics doctoral program, they took a required Bayes class with an excellent professor, and became

---

[10]https://twitter.com/frenchpressplz/status/1266424143207034880

exceptionally interested in the topic. Draw a Bayesian knowledge-building diagram that represents the author's evolving opinion about Bayesian statistics.

Exercise 1.6 (Applying for an internship). There are several data scientist openings at a much-ballyhooed company. Having read the job description, you know for a fact that you are qualified for the position: this is your *data*. Your goal is to ascertain whether you will actually be offered a position: this is your *hypothesis*.

a)   From the perspective of someone *using frequentist thinking*, what question is answered in testing the hypothesis that you'll be offered the position?

b)   Repeat part a from the perspective of someone *using Bayesian thinking*.

c)   Which question would you rather have the answer to: the frequentist or the Bayesian? Explain your reasoning.

Exercise 1.7 (You already know stuff).

a)   Identify a topic that you know about (e.g., a sport, a school subject, music).

b)   Identify a hypothesis about this subject.

c)   How would your current expertise inform your conclusion about this hypothesis?

d)   Which framework are you employing here, Bayesian or frequentist?

Exercise 1.8 (Benefits of Bayesian Statistics). Your friend just became interested in Bayesian statistics. Explain the following to them:

a)   Why is Bayesian statistics useful?

b)   What are the similarities in Bayesian and frequentist statistics?

2

Bayes' Rule

The Collins Dictionary named "fake news" the 2017 term of the year. And for good reason. Fake, misleading, and biased news has proliferated along with online news and social media platforms which allow users to post articles with little quality control. It's then increasingly important to help readers flag articles as "real" or "fake." In Chapter 2 you'll explore how the Bayesian philosophy from Chapter 1 can help us make this distinction. To this end, we'll examine a sample of 150 articles which were posted on Facebook and fact checked by five BuzzFeed journalists (Shu et al., 2017). Information about each article is stored in the `fake_news` dataset in the **bayesrules** package. To learn more about this dataset, type `?fake_news` in your console.

```
# Load packages
library(bayesrules)
library(tidyverse)
library(janitor)

# Import article data
data(fake_news)
```

> ❶ **Warning**
>
> The `fake_news` data contains the full text for *actual* news articles, both real and fake. As such, some of these articles contain disturbing language or topics. Though we believe it's important to provide our original resources, not metadata, you do *not* need to read the articles in order to do the analysis ahead.

The table below, constructed using the `tabyl()` function in the **janitor** package (Firke, 2021), illustrates that 40% of the articles in this particular collection are fake and 60% are real:

```
fake_news %>%
  tabyl(type) %>%
  adorn_totals("row")
  type    n percent
  fake   60     0.4
  real   90     0.6
 Total  150     1.0
```

Using this information alone, we *could* build a very simple news filter which uses the following rule: since most articles are real, we should read and believe all articles. This filter would certainly solve the problem of mistakenly disregarding real articles, but at the cost of reading

DOI: 10.1201/9780429288340-2

lots of fake news. It also only takes into account the overall rates of, not the typical *features* of, real and fake news. For example, suppose that the most recent article posted to a social media platform is titled: "The president has a funny secret!" Some features of this title probably set off some red flags. For example, the usage of an exclamation point might seem like an odd choice for a real news article. Our data backs up this instinct – in our article collection, 26.67% (16 of 60) of fake news titles but only 2.22% (2 of 90) of real news titles use an exclamation point:

```
# Tabulate exclamation usage and article type
fake_news %>%
  tabyl(title_has_excl, type) %>%
  adorn_totals("row")
 title_has_excl fake real
          FALSE   44   88
           TRUE   16    2
          Total   60   90
```

Thus, we have two pieces of contradictory information. Our **prior** information suggested that incoming articles are most likely real. However, the exclamation point **data** is more consistent with fake news. Thinking like Bayesians, we know that balancing both pieces of information is important in developing a **posterior** understanding of whether the article is fake (Figure 2.1).

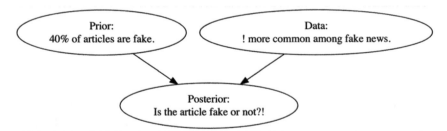

FIGURE 2.1: Bayesian knowledge-building diagram for whether or not the article is fake.

Put your own Bayesian thinking to use in a quick self-quiz of your current *intuition* about whether the most recent article is fake.

❓ Quiz Yourself!

What best describes your updated, posterior understanding about the article?

a.   The chance that this article is fake drops from 40% to 20%. The exclamation point in the title might simply reflect the author's enthusiasm.

b.   The chance that this article is fake jumps from 40% to roughly 90%. Though exclamation points are more common among fake articles, let's not forget that only 40% of articles are fake.

c.   The chance that this article is fake jumps from 40% to roughly 98%. Given that so few real articles use exclamation points, this article is most certainly fake.

The correct answer is given in the footnote below.[1] But if your intuition was incorrect, don't fret. By the end of Chapter 2, you will have learned how to support Bayesian thinking with rigorous Bayesian calculations using **Bayes' Rule**, the aptly named foundation of Bayesian statistics. And **heads up**: of any other chapter in this book, Chapter 2 introduces the most Bayesian concepts, notation, and vocabulary. No matter your level of previous probability experience, you'll want to take this chapter slowly. Further, our treatment focuses on the probability tools that are necessary to Bayesian analyses. For a broader probability introduction, we recommend that the interested reader visit Chapters 1 through 3 and Section 7.1 of Blitzstein and Hwang (2019).

◎ **Goals**

- **Explore foundational probability tools** such as marginal, conditional, and joint probability models and the *Binomial* model.
- **Conduct your first formal Bayesian analysis!** You will construct your first prior and data models and, from these, construct your first posterior models via Bayes' Rule.
- **Practice your Bayesian grammar.** Imagnie how dif ficult it would beto reed this bok if the authers didnt spellcheck or use proper grammar and! punctuation. In this spirit, you'll practice the formal notation and terminology central to Bayesian grammar.
- **Simulate Bayesian models.** Simulation is integral to building intuition for and supporting Bayesian analyses. You'll conduct your first simulation, using the R statistical software, in this chapter.

2.1   Building a Bayesian model for events

Our fake news analysis boils down to the study of two **variables**: an article's fake vs real status and its use of exclamation points. These features can *vary* from article to article. Some are fake, some aren't. Some use exclamation points, some don't. We can represent the *randomness* in these variables using **probability models**. In this section we will build a **prior probability model** for our prior understanding of whether the most recent article is fake; a model for interpreting the exclamation point **data**; and, eventually, a **posterior probability model** which summarizes the posterior plausibility that the article is fake.

2.1.1   Prior probability model

As a first step in our Bayesian analysis, we'll formalize our prior understanding of whether the new article is fake. Based on our `fake_news` data, which we'll assume is a fairly representative sample, we determined earlier that 40% of articles are fake and 60% are real. That is, before even reading the new article, there's a 0.4 **prior probability** that it's fake and a 0.6 prior

---

[1]Correct answer = b.

probability it's *not*. We can represent this information using mathematical notation. Letting B denote the event that an article is fake and B^c (read "B complement" or "B not") denote the event that it's *not* fake, we have

$$P(B) = 0.40 \quad \text{and} \quad P(B^c) = 0.60.$$

As a collection, $P(B)$ and $P(B^c)$ specify the simple **prior model** of fake news in Table 2.1. As a valid **probability model** must: (1) it accounts for all possible events (all articles must be fake or real); (2) it assigns prior probabilities to each event; and (3) these probabilities **sum to one**.

TABLE 2.1: Prior model of fake news.

event	B	B^c	Total
probability	0.4	0.6	1

2.1.2 Conditional probability & likelihood

In the second step of our Bayesian analysis, we'll summarize the insights from the data we collected on the new article. Specifically, we'll formalize our observation that the exclamation point **data** is more compatible with fake news than with real news. Recall that *if* an article is fake, *then* there's a roughly 26.67% chance it uses exclamation points in the title. In contrast, *if* an article is real, *then* there's only a roughly 2.22% chance it uses exclamation points. When stated this way, it's clear that the occurrence of exclamation points depends upon, or is *conditioned* upon, whether the article is fake. This dependence is specified by the following **conditional probabilities** of exclamation point usage (A) *given* an article's fake status (B or B^c):

$$P(A|B) = 0.2667 \quad \text{and} \quad P(A|B^c) = 0.0222.$$

Conditional vs unconditional probability

Let A and B be two events. The **unconditional probability** of A, $P(A)$, measures the probability of observing A, without any knowledge of B. In contrast, the **conditional probability** of A given B, $P(A|B)$, measures the probability of observing A in light of the information that B occurred.

Conditional probabilities are fundamental to Bayesian analyses, and thus a quick pause to absorb this concept is worth it. In general, comparing the conditional vs unconditional probabilities, $P(A|B)$ vs $P(A)$, reveals the extent to which information about B informs our understanding of A. In some cases, the certainty of an event A might *increase* in light of new data B. For example, if somebody practices the clarinet every day, then their probability of joining an orchestra's clarinet section is higher than that among the general population[2]:

$$P(\text{orchestra} \mid \text{practice}) > P(\text{orchestra}).$$

---

[2]We can't cite any rigorous research article here, but imagine what orchestras would sound like if this weren't true.

Conversely, the certainty of an event might *decrease* in light of new data. For example, if you're a fastidious hand washer, then you're less likely to get the flu:

$$P(\text{flu} \mid \text{wash hands}) < P(\text{flu}).$$

The *order* of conditioning is also important. Since they measure two different phenomena, it's typically the case that $P(A|B) \neq P(B|A)$. For instance, roughly 100% of puppies are adorable. Thus, if the next object you pass on the street is a puppy, $P(\text{adorable} \mid \text{puppy}) = 1$. However, the reverse is not true. Not every adorable object is a puppy, thus $P(\text{puppy} \mid \text{adorable}) < 1$.

Finally, information about B doesn't always change our understanding of A. For example, suppose your friend has a yellow pair of shoes and a blue pair of shoes, thus four shoes in total. They choose a shoe at random and don't show it to you. Without actually seeing the shoe, there's a 0.5 probability that it goes on the right foot: $P(\text{right foot}) = 2/4$. And even if they tell you that they happened to get one of the two yellow shoes, there's *still* a 0.5 probability that it goes on the right foot: $P(\text{right foot} \mid \text{yellow}) = 1/2$. That is, information about the shoe's color tells us nothing about which foot it fits – shoe color and foot are **independent**.

Independent events

Two events A and B are **independent** if and only if the occurrence of B doesn't tell us anything about the occurrence of A:

$$P(A|B) = P(A).$$

Let's reexamine our fake news example with these conditional concepts in place. The conditional probabilities we derived above, $P(A|B) = 0.2667$ and $P(A|B^c) = 0.0222$, indicate that a whopping 26.67% of fake articles versus a mere 2.22% of real articles use exclamation points. Since exclamation point usage is so much more **likely** among fake news than real news, this data provides some evidence that the article is fake. We should congratulate ourselves on this observation – we've evaluated the exclamation point data by flipping the conditional probabilities $P(A|B)$ and $P(A|B^c)$ on their heads. For example, on its face, the conditional probability $P(A|B)$ measures the uncertainty in event A given we know event B occurs. However, we find ourselves in the opposite situation. We *know* that the incoming article used exclamation points, A. What we *don't* know is whether or not the article is fake, B or B^c. Thus, in this case, we compared $P(A|B)$ and $P(A|B^c)$ to ascertain the relative **likelihoods** of observing data A under different scenarios of the *uncertain* article status. To help distinguish this application of conditional probability calculations from that when A is uncertain and B is known, we'll utilize the following **likelihood function** notation $L(\cdot|A)$:

$$L(B|A) = P(A|B) \quad \text{and} \quad L(B^c|A) = P(A|B^c).$$

We present a general definition below, but be patient with yourself here. The distinction is subtle, especially since people use the terms "likelihood" and "probability" interchangeably in casual conversation.

Probability vs likelihood

When B is known, the **conditional probability function** $P(\cdot|B)$ allows us to compare the probabilities of an unknown event, A or A^c, occurring with B:

$$P(A|B) \quad \text{vs} \quad P(A^c|B).$$

When A is known, the **likelihood function** $L(\cdot|A) = P(A|\cdot)$ allows us to evaluate the relative compatibility of data A with events B or B^c:

$$L(B|A) \quad \text{vs} \quad L(B^c|A).$$

Table 2.2 summarizes the information that we've amassed thus far, including the prior probabilities and likelihoods associated with the new article being fake or real, B or B^c. Notice that the prior probabilities add up to 1 but the likelihoods do not. Again, the **likelihood function is not a probability function**, but rather provides a framework to compare the relative compatibility of our exclamation point data with B and B^c. Thus, whereas the prior evidence suggested the article is most likely real ($P(B) < P(B^c)$), the data is more consistent with the article being fake ($L(B|A) > L(B^c|A)$).

TABLE 2.2: Prior probabilities and likelihoods of fake news.

event	B	B^c	Total
prior probability	0.4	0.6	1
likelihood	0.2667	0.0222	0.2889

2.1.3   Normalizing constants

Though the likelihood function in Table 2.2 nicely summarizes exclamation point usage in real vs fake news, the **marginal probability** of observing exclamation points across *all* news articles, $P(A)$, provides an important point of comparison. In our quest to calculate this **normalizing constant**,[3] we'll first use our prior model and likelihood function to fill in the table below. This table summarizes the possible **joint** occurrences of the fake news and exclamation point variables. We encourage you to take a crack at this before reading on, utilizing the information we've gathered on exclamation points.

	B	B^c	Total
A			
A^c			
Total	0.4	0.6	1

First, focus on the B column which splits fake articles into two groups: (1) those that are fake *and* use exclamation points, denoted $A \cap B$; and (2) those that are fake *and* don't use exclamation points, denoted $A^c \cap B$.[4] To determine the probabilities of these joint events, note that 40% of articles are fake and 26.67% of fake articles use exclamation points, $P(B) =$

---

[3]This term is mysterious now, but will make sense by the end of this chapter.
[4]We read "∩" as "and" or the "intersection" of two events.

0.4 and $P(A|B) = 0.2667$. It follows that across *all* articles, 26.67% of 40%, or 10.67%, are fake with exclamation points. That is, the **joint probability** of observing both A and B is

$$P(A \cap B) = P(A|B)P(B) = 0.2667 \cdot 0.4 = 0.1067.$$

Further, since 26.67% of fake articles use exclamation points, 73.33% do not. That is, the conditional probability that an article does *not* use exclamation points (A^c) given it's fake (B) is:

$$P(A^c|B) = 1 - P(A|B) = 1 - 0.2667 = 0.7333.$$

It follows that 73.33% of 40%, or 29.33%, of all articles are fake *without* exclamation points:

$$P(A^c \cap B) = P(A^c|B)P(B) = 0.7333 \cdot 0.4 = 0.2933.$$

In summary, the **total probability** of observing a fake article is the sum of its parts:

$$P(B) = P(A \cap B) + P(A^c \cap B) = 0.1067 + 0.2933 = 0.4.$$

We can similarly break down *real* articles into those that do and those that don't use exclamation points. Across all articles, only 1.33% (2.22% of 60%) are real and use exclamation points whereas 58.67% (97.78% of 60%) are real without exclamation points:

$$P(A \cap B^c) = P(A|B^c)P(B^c) = 0.0222 \cdot 0.6 = 0.0133$$
$$P(A^c \cap B^c) = P(A^c|B^c)P(B^c) = 0.9778 \cdot 0.6 = 0.5867.$$

Thus, the **total probability** of observing a real article is the sum of these two parts:

$$P(B^c) = P(A \cap B^c) + P(A^c \cap B^c) = 0.0133 + 0.5867 = 0.6.$$

In these calculations, we intuited a general formula for calculating joint probabilities $P(A \cap B)$ and, through rearranging, a formula for conditional probabilities $P(A|B)$.

Calculating joint and conditional probabilities

For events A and B, the joint probability of $A \cap B$ is calculated by weighting the conditional probability of A given B by the marginal probability of B:

$$P(A \cap B) = P(A|B)P(B). \tag{2.1}$$

Thus, when A and B are *independent*,

$$P(A \cap B) = P(A)P(B).$$

Dividing both sides of (2.1) by $P(B)$, and assuming $P(B) \neq 0$, reveals the definition of the conditional probability of A given B:

$$P(A|B) = \frac{P(A \cap B)}{P(B)}. \tag{2.2}$$

Thus, to evaluate the chance that A occurs in light of information B, we can consider the chance that they occur together, $P(A \cap B)$, relative to the chance that B occurs at all, $P(B)$.

Table 2.3 summarizes our new understanding of the *joint* behavior of our two article variables. The fact that the grand total of this table is one confirms that our calculations are reasonable. Table 2.3 also provides the point of comparison we sought: 12% of *all* news articles use exclamation points, $P(A) = 0.12$. So that we needn't always build similar marginal probabilities from scratch, let's consider the theory behind this calculation. As usual, we can start by recognizing the two ways that an article can use exclamation points: if it is fake $(A \cap B)$ and if it is not fake $(A \cap B^c)$. Thus, the **total probability** of observing A is the combined probability of these distinct parts:

$$P(A) = P(A \cap B) + P(A \cap B^c).$$

By (2.1), we can compute the two pieces of this puzzle using the information we have about exclamation point usage among fake and real news, $P(A|B)$ and $P(A|B^c)$, weighted by the prior probabilities of fake and real news, $P(B)$ and $P(B^c)$:

$$P(A) = P(A \cap B) + P(A \cap B^c) = P(A|B)P(B) + P(A|B^c)P(B^c). \qquad (2.3)$$

Finally, plugging in, we can confirm that roughly 12% of all articles use exclamation points: $P(A) = 0.2667 \cdot 0.4 + 0.0222 \cdot 0.6 = 0.12$. The formula we've built to calculate $P(A)$ here is a special case of the aptly named **Law of Total Probability (LTP)**.

TABLE 2.3: A joint probability model of the fake status and exclamation point usage across all articles.

	B	B^c	Total
A	0.1067	0.0133	0.12
A^c	0.2933	0.5867	0.88
Total	0.4	0.6	1

2.1.4   Posterior probability model via Bayes' Rule!

We're now in a position to answer the ultimate question: What's the probability that the latest article is fake? Formally speaking, we aim to calculate the **posterior probability** that the article is fake given that it uses exclamation points, $P(B|A)$. To build some intuition, let's revisit Table 2.3. Since our article uses exclamation points, we can zoom in on the 12% of articles that fall into the A row. Among these articles, proportionally 88.9% (0.1067 / 0.12) are fake and 11.1% (0.0133 / 0.12) are real. Though it might feel anti-climactic, this is the answer we were seeking: there's an 88.9% posterior chance that this latest article is fake.

Stepping back from the details, we've accomplished something big: we built **Bayes' Rule** from scratch! In short, Bayes' Rule provides the mechanism we need to put our Bayesian thinking into practice. It defines a posterior model for an event B from two pieces: the **prior** probability of B and the **likelihood** of observing data A if B were to occur.

Bayes' Rule for events

For events A and B, the posterior probability of B given A follows by combining (2.2) with (2.1) and recognizing that we can evaluate data A through the likelihood function, $L(B|A) = P(A|B)$ and $L(B^c|A) = P(A|B^c)$:

$$P(B|A) = \frac{P(A \cap B)}{P(A)} = \frac{P(B)L(B|A)}{P(A)} \tag{2.4}$$

where by the Law of Total Probability (2.3)

$$P(A) = P(B)L(B|A) + P(B^c)L(B^c|A). \tag{2.5}$$

More generally,

$$\text{posterior} = \frac{\text{prior} \cdot \text{likelihood}}{\text{normalizing constant}}.$$

To convince ourselves that Bayes' Rule works, let's directly apply it to our news analysis. Into (2.4), we can plug the prior information that 40% of articles are fake, the 26.67% likelihood that a fake article would use exclamation points, and the 12% marginal probability of observing exclamation points across all articles. The resulting posterior probability that the incoming article is fake is roughly 0.889, just as we calculated from Table 2.3:

$$P(B|A) = \frac{P(B)L(B|A)}{P(A)} = \frac{0.4 \cdot 0.2667}{0.12} = 0.889.$$

Table 2.4 summarizes our news analysis journey, from the prior to the posterior model. We started with a prior understanding that there's only a 40% chance that the incoming article would be fake. Yet upon observing the use of an exclamation point in the title *"The president has a funny secret!"*, a feature that's more common to fake news, our posterior understanding evolved quite a bit – the chance that the article is fake jumped to 88.9%.

TABLE 2.4: The prior and posterior models of fake news.

event	B	B^c	Total
prior probability	0.4	0.6	1
posterior probability	0.889	0.111	1

2.1.5   Posterior simulation

It's important to keep in mind that the probability models we built for our news analysis above are just that – *models*. They provide theoretical representations of what we observe in practice. To build intuition for the connection between the articles that might actually be posted to social media and their underlying models, let's run a **simulation**. First, define the possible article `type`, `real` or `fake`, and their corresponding prior probabilities:

```
# Define possible articles
article <- data.frame(type = c("real", "fake"))

# Define the prior model
prior <- c(0.6, 0.4)
```

To simulate the articles that might be posted to your social media, we can use the `sample_n()` function in the **dplyr** package (Wickham et al., 2021) to randomly sample rows from the `article` data frame. In doing so, we must specify the sample `size` and that the sample should be taken with replacement (`replace = TRUE`). Sampling with replacement ensures that we start with a fresh set of possibilities for each article – any article can either be fake or real. Finally, we set `weight = prior` to specify that there's a 60% chance an article is `real` and a 40% chance it's `fake`. To try this out, run the following code multiple times, each time simulating three articles.

```
# Simulate 3 articles
sample_n(article, size = 3, weight = prior, replace = TRUE)
```

Notice that you can get different results every time you run this code. That's because simulation, like articles, is random. Specifically, behind the R curtain is a **random number generator (RNG)** that's in charge of producing random samples. Every time we ask for a new sample, the RNG "starts" at a new place: the **random seed**. Starting at different seeds can thus produce different samples. This is a great thing in general – random samples *should* be random. However, within a single analysis, we want to be able to *reproduce* our random simulation results, i.e., we don't want the fine points of our results to change every time we re-run our code. To achieve this reproducibility, we can specify or *set* the seed by applying the `set.seed()` function to a positive integer (here 84735). Run the below code a few times and notice that the results are always the same – the first two articles are fake and the third is real:[5]

```
# Set the seed. Simulate 3 articles.
set.seed(84735)
sample_n(article, size = 3, weight = prior, replace = TRUE)
  type
1 fake
2 fake
3 real
```

> ❗ **Warning**
>
> We'll use `set.seed()` throughout the book so that readers can reproduce and follow our work. But it's important to remember that these results are still *random*. Reflecting the potential error and variability in simulation, different seeds would typically give different numerical results though similar conclusions.

---

[5] If you get different random samples than those printed here, it likely means that you are using a different version of R.

Now that we understand how to simulate a few articles, let's dream bigger: simulate 10,000 articles and store the results in `article_sim`.

```
# Simulate 10000 articles.
set.seed(84735)
article_sim <- sample_n(article, size = 10000,
                        weight = prior, replace = TRUE)
```

The composition of the 10,000 simulated articles is summarized by the bar plot below, constructed using the `ggplot()` function in the **ggplot2** package (Wickham, 2016):

```
ggplot(article_sim, aes(x = type)) +
  geom_bar()
```

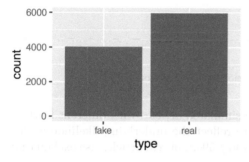

FIGURE 2.2: A bar plot of the fake vs real status of 10,000 simulated articles.

The table below provides a more thorough summary. Reflecting the model from which these 10,000 articles were generated, roughly (but not *exactly*) 40% are fake:

```
article_sim %>%
  tabyl(type) %>%
  adorn_totals("row")
  type      n percent
  fake   4031  0.4031
  real   5969  0.5969
 Total  10000  1.0000
```

Next, let's simulate the exclamation point usage among these 10,000 articles. The `data_model` variable specifies that there's a 26.67% chance that any fake article and a 2.22% chance that any real article uses exclamation points:

```
article_sim <- article_sim %>%
  mutate(data_model = case_when(type == "fake" ~ 0.2667,
                                type == "real" ~ 0.0222))

glimpse(article_sim)
Rows: 10,000
Columns: 2
```

```
$ type       <chr> "fake", "fake", "real", "fake", "f~
$ data_model <dbl> 0.2667, 0.2667, 0.0222, 0.2667, 0.~
```

From this `data_model`, we can simulate whether each article includes an exclamation point. This syntax is a bit more complicated. First, the `group_by()` statement specifies that the exclamation point simulation is to be performed separately for each of the 10,000 articles. Second, we use `sample()` to simulate the exclamation point data, `no` or `yes`, based on the `data_model` and store the results as `usage`. Note that `sample()` is similar to `sample_n()` but samples values from vectors instead of rows from data frames.

```
# Define whether there are exclamation points
data <- c("no", "yes")

# Simulate exclamation point usage
set.seed(3)
article_sim <- article_sim %>%
  group_by(1:n()) %>%
  mutate(usage = sample(data, size = 1,
                        prob = c(1 - data_model, data_model)))
```

The `article_sim` data frame now contains 10,000 simulated articles with different features, summarized in the table below. The patterns here reflect the underlying likelihoods that *roughly* 28% (1070 / 4031) of fake articles and 2% (136 / 5969) of real articles use exclamation points.

```
article_sim %>%
  tabyl(usage, type) %>%
  adorn_totals(c("col","row"))
 usage fake real Total
    no 2961 5833  8794
   yes 1070  136  1206
 Total 4031 5969 10000
```

Figure 2.3 provides a visual summary of these article characteristics. Whereas the left plot reflects the relative breakdown of exclamation point usage among real and fake news, the right plot frames this information within the normalizing context that only roughly 12% (1206 / 10000) of all articles use exclamation points.

```
ggplot(article_sim, aes(x = type, fill = usage)) +
  geom_bar(position = "fill")
ggplot(article_sim, aes(x = type)) +
  geom_bar()
```

Our 10,000 simulated articles now reflect the prior model of fake news, as well as the likelihood of exclamation point usage among fake vs real news. In turn, we can use them to approximate the posterior probability that the latest article is fake. To this end, we can filter out the simulated articles that match our data (i.e., those that use exclamation points) and examine the percentage of articles that are fake:

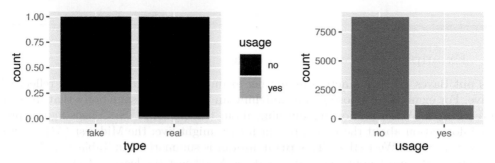

FIGURE 2.3: Bar plots of exclamation point usage, both within fake vs real news and overall.

```
article_sim %>%
  filter(usage == "yes") %>%
  tabyl(type) %>%
  adorn_totals("row")
 type    n percent
 fake 1070  0.8872
 real  136  0.1128
Total 1206  1.0000
```

Among the 1206 simulated articles that use exclamation points, roughly 88.7% are fake. This approximation is *quite* close to the actual posterior probability of 0.889. Of course, our posterior assessment of this article would change if we had seen different data, i.e., if the title *didn't* have exclamation points. Figure 2.4 reveals a simple rule: If an article uses exclamation points, it's most likely fake. Otherwise, it's most likely real (and we should read it!). NOTE: The same rule does *not* apply to this *real* book in which the liberal use of exclamation points simply conveys our enthusiasm!

```
ggplot(article_sim, aes(x = type)) +
  geom_bar() +
  facet_wrap(~ usage)
```

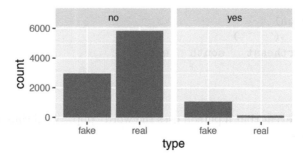

FIGURE 2.4: Bar plots of real vs fake news, broken down by exclamation point usage.

2.2   Example: Pop vs soda vs coke

Let's put Bayes' Rule into action in another example. Our word choices can reflect where we live. For example, suppose you're watching an interview of somebody that lives in the United States. Without knowing anything about this person, U.S. Census figures provide prior information about the region in which they might live: the Midwest (M), Northeast (N), South (S), or West (W).[6] This **prior model** is summarized in Table 2.5.[7] Notice that the South is the most populous region and the Northeast the least ($P(S) > P(N)$). Thus, based on population statistics alone, there's a 38% prior probability that the interviewee lives in the South:

$$P(S) = 0.38.$$

TABLE 2.5: Prior model of U.S. region.

region	M	N	S	W	Total
probability	0.21	0.17	0.38	0.24	1

But then, you see the person point to a fizzy cola drink and say "please pass my pop." Though the country is united in its love of fizzy drinks, it's divided in what they're called, with common regional terms including "pop," "soda," and "coke." This **data**, i.e., the person's use of "pop," provides further information about where they might live. To evaluate this data, we can examine the `pop_vs_soda` dataset in the **bayesrules** package (Dogucu et al., 2021) which includes 374250 responses to a volunteer survey conducted at popvssoda.com. To learn more about this dataset, type `?pop_vs_soda` in your console. Though the survey participants aren't directly representative of the regional populations (Table 2.5), we can use their responses to approximate the **likelihood** of people using the word `pop` in each `region`:

```
# Load the data
data(pop_vs_soda)

# Summarize pop use by region
pop_vs_soda %>%
  tabyl(pop, region) %>%
  adorn_percentages("col")
   pop midwest northeast    south    west
 FALSE  0.3553    0.7266  0.92078  0.7057
  TRUE  0.6447    0.2734  0.07922  0.2943
```

Letting A denote the event that a person uses the word "pop," we'll thus assume the following regional **likelihoods**:

$$L(M|A) = 0.6447, \quad L(N|A) = 0.2734, \quad L(S|A) = 0.0792, \quad L(W|A) = 0.2943$$

---

[6] https://www2.census.gov/geo/pdfs/maps-data/maps/reference/us_regdiv.pdf
[7] https://www.census.gov/popclock/data_tables.php?component=growth

For example, 64.47% of people in the Midwest but only 7.92% of people in the South use the term "pop." Comparatively then, the "pop" data is most likely if the interviewee lives in the Midwest and least likely if they live in the South, with the West and Northeast being in between these two extremes: $L(M|A) > L(W|A) > L(N|A) > L(S|A)$.

Weighing the **prior** information about regional populations with the **data** that the interviewee used the word "pop," what are we to think now? For example, considering the fact that 38% of people live in the South but that "pop" is relatively rare to that region, what's the **posterior probability** that the interviewee lives in the South? Per Bayes' Rule (2.4), we can calculate this probability by

$$P(S|A) = \frac{P(S)L(S|A)}{P(A)}. \tag{2.6}$$

We already have two of the three necessary pieces of the puzzle, the prior probability $P(S)$ and likelihood $L(S|A)$. Consider the third, the marginal probability that a person uses the term "pop" across the entire U.S., $P(A)$. By extending the Law of Total Probability (2.5), we can calculate $P(A)$ by combining the likelihoods of using "pop" in each region, while accounting for the regional populations. Accordingly, there's a 28.26% chance that a person in the U.S. uses the word "pop":

$$P(A) = L(M|A)P(M) + L(N|A)P(N) + L(S|A)P(S) + L(W|A)P(W)$$
$$= 0.6447 \cdot 0.21 + 0.2734 \cdot 0.17 + 0.0792 \cdot 0.38 + 0.2943 \cdot 0.24$$
$$\approx 0.2826.$$

Then plugging into (2.6), there's a roughly 10.65% posterior chance that the interviewee lives in the South:

$$P(S|A) = \frac{0.38 \cdot 0.0792}{0.2826} \approx 0.1065.$$

We can similarly update our understanding of the interviewee living in the Midwest, Northeast, or West. Table 2.6 summarizes the resulting posterior model of region alongside the original prior. Soak it in. Upon hearing the interviewee use "pop," we now think it's most likely that they live in the Midwest and least likely that they live in the South, despite the South being the most populous region.

TABLE 2.6: Prior and posterior models of U.S. region.

region	M	N	S	W	Total
prior probability	0.21	0.17	0.38	0.24	1
posterior probability	0.4791	0.1645	0.1065	0.2499	1

2.3 Building a Bayesian model for random variables

In our Bayesian analyses above, we constructed posterior models for *categorical* variables. In the fake news analysis, we examined the categorical status of an article: fake or real. In the pop vs soda example, we examined the categorical outcome of an interviewee's region:

Midwest, Northeast, South, or West. However, it's often the case in a Bayesian analysis that our outcomes of interest are *numerical*. Though some of the details will change, the same Bayes' Rule principles we built above generalize to the study of numerical **random variables**.

2.3.1 Prior probability model

In 1996, world chess champion (and human!) Gary Kasparov played a much anticipated six-game chess match against the IBM supercomputer Deep Blue. Of the six games, Kasparov won three, drew two, and lost one. Thus, Kasparov won the overall match, preserving the notion that machines don't perform as well as humans when it comes to chess. Yet Kasparov and Deep Blue were to meet again for a six-game match in 1997. Let π, read "pi" or "pie," denote Kasparov's chances of winning any particular game in the re-match.[8] Thus, π is a measure of his overall skill relative to Deep Blue. Given the complexity of chess, machines, and humans, π is *unknown* and can *vary* or *fluctuate* over time. Or, in short, π is a **random variable**.

As with the fake news analysis, our analysis of random variable π will start with a prior model which (1) identifies what values π can take, and (2) assigns a prior weight or probability to each, where (3) these probabilities sum to 1. Consider the prior model defined in Table 2.7. We'll get into how we might build such a prior in later chapters. For now, let's focus on interpreting and utilizing the given prior.

TABLE 2.7: Prior model of π, Kasparov's chance of beating Deep Blue.

π	0.2	0.5	0.8	Total
$f(\pi)$	0.10	0.25	0.65	1

The first thing you might notice is that this model greatly simplifies reality.[9] Though Kasparov's win probability π can *technically* be any number from zero to one, this prior assumes that π has a **discrete** set of possibilities: Kasparov's win probability is either 20%, 50%, or 80%. Next, examine the **probability mass function (pmf)** $f(\cdot)$ which specifies the prior probability of each possible π value. This pmf reflects the prior understanding that Kasparov learned from the 1996 match-up, and so will most likely improve in 1997. Specifically, this pmf places a 65% chance on Kasparov's win probability jumping to $\pi = 0.8$ and only a 10% chance on his win probability dropping to $\pi = 0.2$, i.e., $f(\pi = 0.8) = 0.65$ and $f(\pi = 0.2) = 0.10$.

Discrete probability model

Let Y be a discrete random variable. The probability model of Y is specified by a **probability mass function (pmf)** $f(y)$. This pmf defines the probability of any given outcome y,

$$f(y) = P(Y = y)$$

and has the following properties:

---

[8]Greek letters are conventionally used to denote our primary quantitative variables of interest.
[9]As we keep progressing with Bayes, we'll get the chance to make our models more nuanced and realistic.

- $0 \le f(y) \le 1$ for all y; and
- $\sum_{\text{all } y} f(y) = 1$, i.e., the probabilities of all possible outcomes y sum to 1.

2.3.2 The Binomial data model

In the second step of our Bayesian analysis, we'll collect and process data which can inform our understanding of π, Kasparov's skill level relative to that of Deep Blue. Here, our data Y is the number of the six games in the 1997 re-match that Kasparov wins.[10] Since the chess match outcome isn't predetermined, Y is a **random variable** that can take any value in $\{0, 1, ..., 6\}$. Further, Y inherently depends upon Kasparov's win probability π. If π were 0.80, Kasparov's victories Y would also tend to be high. If π were 0.20, Y would tend to be low. For our formal Bayesian analysis, we must *model* this dependence of Y on π. That is, we must develop a **conditional probability model** of how Y *depends upon* or is *conditioned upon* the value of π.

Conditional probability model of data Y

Let Y be a discrete random variable and π be a parameter upon which Y depends. Then the conditional probability model of Y given π is specified by conditional pmf $f(y|\pi)$. This pmf specifies the conditional probability of observing y given π,

$$f(y|\pi) = P(Y = y|\pi)$$

and has the following properties:

- $0 \le f(y|\pi) \le 1$ for all y; and
- $\sum_{\text{all } y} f(y|\pi) = 1$.

In modeling the dependence of Y on π in our chess example, we first make two assumptions about the chess match: (1) the outcome of any one game doesn't influence the outcome of another, i.e., games are **independent**; and (2) Kasparov has an equal probability, π, of winning any game in the match, i.e., his chances don't increase or decrease as the match goes on. This is a common framework in statistical analysis, one which can be represented by the **Binomial model**.

The Binomial model

Let random variable Y be the *number of successes* in a *fixed number of trials n*. Assume that the trials are *independent* and that the *probability of success* in each trial is π. Then the conditional dependence of Y on π can be modeled by the Binomial model with **parameters** n and π. In mathematical notation:

$$Y|\pi \sim \text{Bin}(n, \pi)$$

---

[10]Capital letters toward the end of the alphabet (e.g., X, Y, Z) are conventionally used to denote random variables related to our data.

where "~" can be read as "modeled by." Correspondingly, the Binomial model is specified by **conditional pmf**

$$f(y|\pi) = \binom{n}{y} \pi^y (1-\pi)^{n-y} \quad \text{for } y \in \{0, 1, 2, \ldots, n\} \tag{2.7}$$

where $\binom{n}{y} = \frac{n!}{y!(n-y)!}$.

We can now say that the dependence of Kasparov's victories Y in $n = 6$ games on his win probability π follows a Binomial model,

$$Y|\pi \sim \text{Bin}(6, \pi)$$

with conditional pmf

$$f(y|\pi) = \binom{6}{y} \pi^y (1-\pi)^{6-y} \quad \text{for } y \in \{0, 1, 2, 3, 4, 5, 6\}. \tag{2.8}$$

This pmf summarizes the conditional probability of observing any number of wins $Y = y$ for any given win probability π. For example, *if* Kasparov's underlying chance of beating Deep Blue were $\pi = 0.8$, then there's a roughly 26% chance he'd win all six games:

$$f(y = 6|\pi = 0.8) = \binom{6}{6} 0.8^6 (1-0.8)^{6-6} = 1 \cdot 0.8^6 \cdot 1 \approx 0.26.$$

And a near 0 chance he'd *lose* all six games:

$$f(y = 0|\pi = 0.8) = \binom{6}{0} 0.8^0 (1-0.8)^{6-0} = 1 \cdot 1 \cdot 0.2^6 \approx 0.000064.$$

Figure 2.5 plots the **conditional pmfs** $f(y|\pi)$, and thus the *random* outcomes of Y, under each possible value of Kasparov's win probability π. These plots confirm our intuition that Kasparov's victories Y would tend to be low if Kasparov's win probability π were low (far left) and high if π were high (far right).

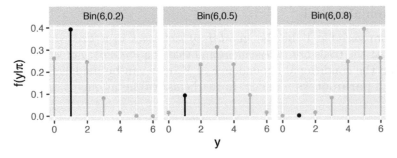

FIGURE 2.5: The pmf of a Bin(6, π) model is plotted for each possible value of $\pi \in \{0.2, 0.5, 0.8\}$. The masses marked by the black lines correspond to the eventual observed data, $Y = 1$ win.

2.3.3   The Binomial likelihood function

The Binomial provides a theoretical model of the data Y we *might* observe. In the end, Kasparov only won one of the six games against Deep Blue in 1997 ($Y = 1$). Thus, the next step in our Bayesian analysis is to determine how compatible this particular data is with the various possible π. Put another way, we want to evaluate the *likelihood* of Kasparov winning $Y = 1$ game under each possible π. It turns out that the answer is staring us straight in the face. Extracting only the masses in Figure 2.5 that correspond to our observed data, $Y = 1$, reveals the **likelihood function** of π (Figure 2.6).

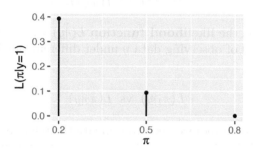

FIGURE 2.6: The likelihood function $L(\pi|y = 1)$ of observing $Y = 1$ win in six games for any win probability $\pi \in \{0.2, 0.5, 0.8\}$.

Just as the likelihood in our fake news example was obtained by flipping a conditional probability on its head, the formula for the likelihood function follows from evaluating the conditional pmf $f(y|\pi)$ in (2.8) at the observed data $Y = 1$. For $\pi \in \{0.2, 0.5, 0.8\}$,

$$L(\pi|y = 1) = f(y = 1|\pi) = \binom{6}{1} \pi^1 (1 - \pi)^{6-1} = 6\pi(1 - \pi)^5.$$

Table 2.8 summarizes the likelihood function evaluated at each possible value of π. For example, there's a low 0.0015 likelihood of Kasparov winning just one game if he were the superior player, i.e., $\pi = 0.8$:

$$L(\pi = 0.8|y = 1) = 6 \cdot 0.8 \cdot (1 - 0.8)^5 \approx 0.0015.$$

There are some not-to-miss details here. First, though it is equivalent in *formula* to the conditional pmf of Y, $f(y = 1|\pi)$, we use the $L(\pi|y - 1)$ notation to reiterate that the likelihood is a function of the unknown win probability π given the observed $Y = 1$ win data. In fact, the resulting likelihood formula depends only upon π. Further, the likelihood function does *not* sum to one across π, and thus is *not* a probability model. (Mental gymnastics!) Rather, it provides a mechanism by which to compare the compatibility of the observed data $Y = 1$ with different π.

TABLE 2.8: Likelihood function of π given Kasparov won 1 of 6 games.

π	0.2	0.5	0.8
$L(\pi|y = 1)$	0.3932	0.0938	0.0015

Putting this all together, the likelihood function summarized in Figure 2.6 and Table 2.8 illustrates that Kasparov's one game win is *most* consistent with him being the weaker player and *least* consistent with him being the better player: $L(\pi = 0.2|y = 1) > L(\pi = 0.5|y = 1)$

$> L(\pi = 0.8|y = 1)$. In fact, it's nearly impossible that Kasparov would have only won one game if his win probability against Deep Blue were as high as $\pi = 0.8$: $L(\pi = 0.8|y = 1) \approx 0$.

Probability mass functions vs likelihood functions

When π is known, the **conditional pmf** $f(\cdot|\pi)$ allows us to compare the probabilities of different possible values of data Y (e.g., y_1 or y_2) occurring with π:

$$f(y_1|\pi) \quad \text{vs} \quad f(y_2|\pi).$$

When $Y = y$ is known, the **likelihood function** $L(\cdot|y) = f(y|\cdot)$ allows us to compare the relative likelihood of observing data y under different possible values of π (e.g., π_1 or π_2):

$$L(\pi_1|y) \quad \text{vs} \quad L(\pi_2|y).$$

Thus, $L(\cdot|y)$ provides the tool we need to evaluate the relative compatibility of data $Y = y$ with various π values.

2.3.4   Normalizing constant

Consider where we are. In contrast to our prior model of π (Table 2.7), our 1997 chess match data provides evidence that Deep Blue is now the dominant player (Table 2.8). As Bayesians, we want to *balance* this prior and likelihood information. Our mechanism for doing so, Bayes' Rule, requires three pieces of information: the prior, likelihood, and a **normalizing constant**. We've taken care of the first two, now let's consider the third. To this end, we must determine the **total probability** that Kasparov would win $Y = 1$ game across all possible win probabilities π, $f(y = 1)$. As we did in our other examples, we can appeal to the Law of Total Probability (LTP) to calculate $f(y = 1)$. The idea is this. The overall probability of Kasparov's one win outcome ($Y = 1$) is the sum of its parts: the likelihood of observing $Y = 1$ with a win probability π that's either 0.2, 0.5, or 0.8 *weighted by* the prior probabilities of these π values. Taking a little leap from the LTP for events (2.5), this means that

$$f(y = 1) = \sum_{\pi \in \{0.2, 0.5, 0.8\}} L(\pi|y = 1) f(\pi)$$

or, expanding the summation Σ and plugging in the prior probabilities and likelihoods from Tables 2.7 and 2.8:

$$\begin{aligned} f(y = 1) &= L(\pi = 0.2|y = 1) f(\pi = 0.2) + L(\pi = 0.5|y = 1) f(\pi = 0.5) \\ &\quad + L(\pi = 0.8|y = 1) f(\pi = 0.8) \\ &\approx 0.3932 \cdot 0.10 + 0.0938 \cdot 0.25 + 0.0015 \cdot 0.65 \\ &\approx 0.0637. \end{aligned} \tag{2.9}$$

Thus, across all possible π, there's only a roughly 6% chance that Kasparov would have won only one game. It would, of course, be great if this all clicked. But if it doesn't, don't let this calculation discourage you from moving forward. We'll learn a magical shortcut in Section 2.3.6 that allows us to bypass this calculation.

2.3.5 Posterior probability model

Figure 2.7 summarizes what we know thus far and where we have yet to go. Heading into their 1997 re-match, our **prior** model suggested that Kasparov's win probability against Deep Blue was high (left plot). But! Then he only won one of six games, a result that is most **likely** when Kasparov's win probability is low (middle plot). Our updated, **posterior** model (right plot) of Kasparov's win probability will balance this prior and likelihood. Specifically, our formal calculations below will verify that Kasparov's chances of beating Deep Blue most likely dipped to $\pi = 0.20$ between 1996 to 1997. It's also relatively possible that his 1997 losing streak was a fluke, and that he's more evenly matched with Deep Blue ($\pi = 0.50$). In contrast, it's *highly* unlikely that Kasparov is still the superior player ($\pi = 0.80$).

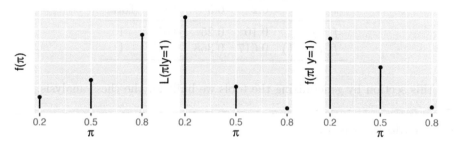

FIGURE 2.7: The prior (left), likelihood (middle), and posterior (right) models of π. The y-axis scales are omitted for ease of comparison.

The posterior model plotted in Figure 2.7 is specified by the **posterior pmf**

$$f(\pi|y = 1).$$

Conceptually, $f(\pi|y = 1)$ is the posterior probability of some win probability π given that Kasparov only won one of six games against Deep Blue. Thus, defining the posterior $f(\pi|y = 1)$ isn't much different than it was in our previous examples. Just as you might hope, Bayes' Rule still holds:

$$\text{posterior} = \frac{\text{prior} \cdot \text{likelihood}}{\text{normalizing constant}}.$$

In the chess setting, we can translate this as

$$f(\pi|y = 1) = \frac{f(\pi)L(\pi|y = 1)}{f(y = 1)} \quad \text{for } \pi \in \{0.2, 0.5, 0.8\}. \tag{2.10}$$

All that remains is a little "plug-and-chug": the prior $f(\pi)$ is defined by Table 2.7, the likelihood $L(\pi|y = 1)$ by Table 2.8, and the normalizing constant $f(y = 1)$ by (2.9). The posterior probabilities follow:

$$
\begin{aligned}
f(\pi = 0.2|y = 1) &= \frac{0.10 \cdot 0.3932}{0.0637} \approx 0.617 \\
f(\pi = 0.5|y = 1) &= \frac{0.25 \cdot 0.0938}{0.0637} \approx 0.368 \\
f(\pi = 0.8|y = 1) &= \frac{0.65 \cdot 0.0015}{0.0637} \approx 0.015
\end{aligned}
\tag{2.11}
$$

This posterior probability model is summarized in Table 2.9 along with the prior probability model for comparison. These details confirm the trends in and intuition behind Figure 2.7. Mainly, though we were fairly confident that Kasparov's performance would have improved from 1996 to 1997, after winning only one game, the chances of Kasparov being the dominant player ($\pi = 0.8$) dropped from 0.65 to 0.015. In fact, the scenario with the greatest posterior support is that Kasparov is the weaker player, with a win probability of only 0.2. Good news for machines. Bad news for humans.

TABLE 2.9: Prior and posterior probability models of π, Kasparov's chance of beating Deep Blue in chess.

π	0.2	0.5	0.8	Total
$f(\pi)$	0.10	0.25	0.65	1
$f(\pi\|y = 1)$	0.617	0.368	0.015	1

We close this section by generalizing the tools we built for the chess analysis.

Bayes' Rule for variables

For any variables π and Y, let $f(\pi)$ denote the prior pmf of π and $L(\pi|y)$ denote the likelihood function of π given observed data $Y = y$. Then the posterior pmf of π given data $Y = y$ is

$$f(\pi|y) = \frac{\text{prior} \cdot \text{likelihood}}{\text{normalizing constant}} = \frac{f(\pi)L(\pi|y)}{f(y)} \qquad (2.12)$$

where, by the Law of Total Probability, the overall probability of observing data $Y = y$ across all possible π is

$$f(y) = \sum_{\text{all } \pi} f(\pi)L(\pi|y). \qquad (2.13)$$

2.3.6   Posterior shortcut

We now make good on our promise that, moving forward, we needn't continue calculating the **normalizing constant**. To begin, notice in (2.11) that $f(y = 1) = 0.0637$ appears in the denominator of $f(\pi|y = 1)$ for each $\pi \in \{0.2, 0.5, 0.8\}$. This explains the term *normalizing constant* – its only purpose is to normalize the posterior probabilities so that they sum to one:

$$f(\pi = 0.2|y = 1) + f(\pi = 0.5|y = 1) + f(\pi = 0.8|y = 1) = 1.$$

Yet we needn't actually calculate $f(y = 1)$ to normalize the posterior probabilities. Instead, we can simply note that $f(y = 1)$ is some constant $1/c$, and thus replace (2.11) with

$$f(\pi = 0.2|y = 1) = c \cdot 0.10 \cdot 0.3932 \propto 0.039320$$
$$f(\pi = 0.5|y = 1) = c \cdot 0.25 \cdot 0.0938 \propto 0.023450$$
$$f(\pi = 0.8|y = 1) = c \cdot 0.65 \cdot 0.0015 \propto 0.000975$$

where $\propto$ denotes "proportional to." Though these unnormalized posterior probabilities don't add up to one,

$$0.039320 + 0.023450 + 0.000975 = 0.063745,$$

Figure 2.8 demonstrates that they preserve the *proportional* relationships of the normalized posterior probabilities.

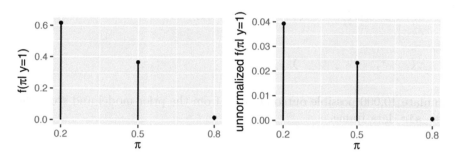

FIGURE 2.8: The normalized posterior pmf of π (left) and the unnormalized posterior pmf of π (right) with different y-axis scales.

Thus, to *normalize* these unnormalized probabilities while preserving their relative relationships, we can compare each to the whole. Specifically, we can divide each unnormalized probability by their sum. For example:

$$f(\pi = 0.2|y = 1) = \frac{0.039320}{0.039320 + 0.023450 + 0.000975} \approx 0.617.$$

Though we've just intuited this result, it also follows mathematically by combining (2.12) and (2.13):

$$f(\pi|y) = \frac{f(\pi)L(\pi|y)}{f(y)} = \frac{f(\pi)L(\pi|y)}{\sum_{\text{all } \pi} f(\pi)L(\pi|y)}.$$

We state the general form of this proportionality result below and will get plenty of practice with this concept in the coming chapters.

Proportionality

Since $f(y)$ is merely a normalizing constant which does not depend on π, the posterior pmf $f(\pi|y)$ is proportional to the product of $f(\pi)$ and $L(\pi|y)$:

$$f(\pi|y) = \frac{f(\pi)L(\pi|y)}{f(y)} \propto f(\pi)L(\pi|y).$$

That is,

$$\text{posterior} \propto \text{prior} \cdot \text{likelihood} .$$

The significance of this proportionality is that all the information we need to build the posterior model is held in the prior and likelihood.

2.3.7   Posterior simulation

We'll conclude this section with a simulation that provides insight into and supports our Bayesian analysis of Kasparov's chess skills. Ultimately, we'll simulate 10,000 scenarios of the six-game chess series. To begin, set up the possible values of win probability π and the corresponding prior model $f(\pi)$:

```r
# Define possible win probabilities
chess <- data.frame(pi = c(0.2, 0.5, 0.8))

# Define the prior model
prior <- c(0.10, 0.25, 0.65)
```

Next, simulate 10,000 possible outcomes of π from the prior model and store the results in the chess_sim data frame.

```r
# Simulate 10000 values of pi from the prior
set.seed(84735)
chess_sim <- sample_n(chess, size = 10000, weight = prior, replace = TRUE)
```

From each of the 10,000 prior plausible values pi, we can simulate six games and record Kasparov's number of wins, y. Since the dependence of y on pi follows a Binomial model, we can directly simulate y using the rbinom() function with size = 6 and prob = pi.

```r
# Simulate 10000 match outcomes
chess_sim <- chess_sim %>%
  mutate(y = rbinom(10000, size = 6, prob = pi))

# Check it out
chess_sim %>%
  head(3)
   pi y
1 0.5 3
2 0.5 3
3 0.8 4
```

The combined 10,000 simulated pi values closely approximate the prior model $f(\pi)$ (Table 2.7):

```r
# Summarize the prior
chess_sim %>%
  tabyl(pi) %>%
  adorn_totals("row")
   pi     n percent
  0.2  1017  0.1017
  0.5  2521  0.2521
  0.8  6462  0.6462
 Total 10000  1.0000
```

Further, the 10,000 simulated match outcomes y illuminate the dependence of these outcomes on Kasparov's win probability pi, closely mimicking the conditional pmfs $f(y|\pi)$ from Figure 2.5.

```
# Plot y by pi
ggplot(chess_sim, aes(x = y)) +
  stat_count(aes(y = ..prop..)) +
  facet_wrap(~ pi)
```

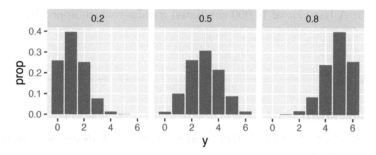

FIGURE 2.9: A bar plot of simulated win outcomes y under each possible win probability π.

Finally, let's focus on the simulated outcomes that match the observed data that Kasparov won one game. Among these simulations, the majority (60.4%) correspond to the scenario in which Kasparov's win probability π was 0.2 and very few (1.8%) correspond to the scenario in which π was 0.8. These observations very closely approximate the posterior model of π which we formally built above (Table 2.9).

```
# Focus on simulations with y = 1
win_one <- chess_sim %>%
  filter(y == 1)

# Summarize the posterior approximation
win_one %>%
  tabyl(pi) %>%
  adorn_totals("row")
    pi   n percent
   0.2 404 0.60389
   0.5 253 0.37818
   0.8  12 0.01794
 Total 669 1.00000

# Plot the posterior approximation
ggplot(win_one, aes(x = pi)) +
  geom_bar()
```

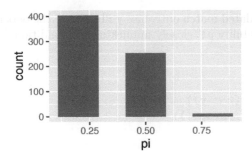

FIGURE 2.10: A bar plot of 10,000 simulated π values which approximates the posterior model.

2.4   Chapter summary

In Chapter 2, you learned Bayes' Rule and that *Bayes Rules!* Every Bayesian analysis consists of four common steps.

1. Construct a **prior model** for your variable of interest, π. The prior model specifies two important pieces of information: the possible values of π and the relative prior plausibility of each.

2. Summarize the dependence of data Y on π via a **conditional pmf** $f(y|\pi)$.

3. Upon *observing* data $Y = y$, define the **likelihood function** $L(\pi|y) = f(y|\pi)$ which encodes the relative likelihood of observing data $Y = y$ under different π values.

4. Build the **posterior model** of π via Bayes' Rule which balances the prior and likelihood:

$$\text{posterior} = \frac{\text{prior} \cdot \text{likelihood}}{\text{normalizing constant}} \propto \text{prior} \cdot \text{likelihood}.$$

More technically,

$$f(\pi|y) = \frac{f(\pi)L(\pi|y)}{f(y)} \propto f(\pi)L(\pi|y).$$

2.5   Exercises

2.5.1   Building up to Bayes' Rule

Exercise 2.1 (Comparing the prior and posterior). For each scenario below, you're given a pair of events, A and B. Explain what you believe to be the relationship between the posterior and prior probabilities of B: $P(B|A) > P(B)$ or $P(B|A) < P(B)$.

a) A = you just finished reading Lambda Literary Award-winning author Nicole Dennis-Benn's first novel, and you enjoyed it! B = you will also enjoy Benn's newest novel.

b) A = it's 0 degrees Fahrenheit in Minnesota on a January day. B = it will be 60 degrees tomorrow.

c) A = the authors only got 3 hours of sleep last night. B = the authors make several typos in their writing today.

d) A = your friend includes three hashtags in their tweet. B = the tweet gets retweeted.

Exercise 2.2 (Marginal, conditional, or joint?). Define the following events for a resident of a fictional town: A = drives 10 miles per hour above the speed limit, B = gets a speeding ticket, C = took statistics at the local college, D = has used R, E = likes the music of Prince, and F = is a Minnesotan. Several facts about these events are listed below. Specify each of these facts using probability notation, paying special attention to whether it's a marginal, conditional, or joint probability.

a) 73% of people that drive 10 miles per hour above the speed limit get a speeding ticket.

b) 20% of residents drive 10 miles per hour above the speed limit.

c) 15% of residents have used R.

d) 91% of statistics students at the local college have used R.

e) 38% of residents are Minnesotans that like the music of Prince.

f) 95% of the Minnesotan residents like the music of Prince.

Exercise 2.3 (Binomial practice). For each variable Y below, determine whether Y is Binomial. If yes, use notation to specify this model and its parameters. If not, explain why the Binomial model is not appropriate for Y.

a) At a certain hospital, an average of 6 babies are born each hour. Let Y be the number of babies born between 9 a.m. and 10 a.m. tomorrow.

b) Tulips planted in fall have a 90% chance of blooming in spring. You plant 27 tulips this year. Let Y be the number that bloom.

c) Each time they try out for the television show *Ru Paul's Drag Race*, Alaska has a 17% probability of succeeding. Let Y be the number of times Alaska has to try out until they're successful.

d) Y is the amount of time that Henry is late to your lunch date.

e) Y is the probability that your friends will throw you a surprise birthday party even though you said you hate being the center of attention and just want to go out to eat.

f) You invite 60 people to your "π day" party, none of whom know each other, and each of whom has an 80% chance of showing up. Let Y be the total number of guests at your party.

2.5.2 Practice Bayes' Rule for events

Exercise 2.4 (Vampires?). Edward is trying to prove to Bella that vampires exist. Bella thinks there is a 0.05 probability that vampires exist. She also believes that the probability that someone can sparkle like a diamond if vampires exist is 0.7, and the probability that someone can sparkle like a diamond if vampires don't exist is 0.03. Edward then goes into a meadow and shows Bella that he can sparkle like a diamond. Given that Edward sparkled like a diamond, what is the probability that vampires exist?

Exercise 2.5 (Sick trees). A local arboretum contains a variety of tree species, including elms, maples, and others. Unfortunately, 18% of all trees in the arboretum are infected with mold. Among the infected trees, 15% are elms, 80% are maples, and 5% are other species. Among the uninfected trees, 20% are elms, 10% are maples, and 70% are other species. In monitoring the spread of mold, an arboretum employee randomly selects a tree to test.

    a)   What's the prior probability that the selected tree has mold?
    b)   The tree happens to be a maple. What's the probability that the employee would have selected a maple?
    c)   What's the posterior probability that the selected maple tree has mold?
    d)   Compare the prior and posterior probability of the tree having mold. How did your understanding change in light of the fact that the tree is a maple?

Exercise 2.6 (Restaurant ratings). The probability that Sandra will like a restaurant is 0.7. Among the restaurants that she likes, 20% have five stars on Yelp, 50% have four stars, and 30% have fewer than four stars. What other information do we need if we want to find the posterior probability that Sandra likes a restaurant given that it has fewer than four stars on Yelp?

Exercise 2.7 (Dating app). Matt is on a dating app looking for love. Matt swipes right on 8% of the profiles he views. Of the people that Matt swipes right on, 40% are men, 30% are women, 20% are non-binary, and 10% identify in another way. Of the people that Matt does not swipe right on, 45% are men, 40% are women, 10% are non-binary, and 5% identify in some other way.

    a)   What's the probability that a randomly chosen person on this dating app is non-binary?
    b)   Given that Matt is looking at the profile of someone who is non-binary, what's the posterior probability that he swipes right?

Exercise 2.8 (Flight delays). For a certain airline, 30% of the flights depart in the morning, 30% depart in the afternoon, and 40% depart in the evening. Frustratingly, 15% of all flights are delayed. Of the delayed flights, 40% are morning flights, 50% are afternoon flights, and 10% are evening flights. Alicia and Mine are taking separate flights to attend a conference.

    a)   Mine is on a morning flight. What's the probability that her flight will be delayed?
    b)   Alicia's flight is not delayed. What's the probability that she's on a morning flight?

Exercise 2.9 (Good mood, bad mood). Your roommate has two moods, good or bad. In general, they're in a good mood 40% of the time. Yet you've noticed that their moods are related to how many text messages they receive the day before. If they're in a good mood today, there's a 5% chance they had 0 texts, an 84% chance they had between 1 and 45 texts, and an 11% chance they had more than 45 texts yesterday. If they're in a bad mood today, there's a 13% chance they had 0 texts, an 86% chance they had between 1 and 45 texts, and a 1% chance they had more than 45 texts yesterday.

    a)   Use the provided information to fill in the table above.

	good mood	bad mood	total
0 texts			
1-45 texts			
46+ texts			
Total			1

b)   Today's a new day. Without knowing anything about the previous day's text messages, what's the probability that your roommate is in a good mood? What part of the Bayes' Rule equation is this: the prior, likelihood, normalizing constant, or posterior?

c)   You surreptitiously took a peek at your roommate's phone (we are attempting to withhold judgment of this dastardly maneuver) and see that your roommate received 50 text messages yesterday. How likely are they to have received this many texts if they're in a good mood today? What part of the Bayes' Rule equation is this?

d)   What is the posterior probability that your roommate is in a good mood given that they received 50 text messages yesterday?

Exercise 2.10 (LGBTQ students: rural and urban). A recent study of 415,000 Californian public middle school and high school students found that 8.5% live in rural areas and 91.5% in urban areas.[11] Further, 10% of students in rural areas and 10.5% of students in urban areas identified as Lesbian, Gay, Bisexual, Transgender, or Queer (LGBTQ). Consider one student from the study.

a)   What's the probability they identify as LGBTQ?

b)   If they identify as LGBTQ, what's the probability that they live in a rural area?

c)   If they do not identify as LGBTQ, what's the probability that they live in a rural area?

2.5.3   Practice Bayes' Rule for random variables

Exercise 2.11 (Internship). Muhammad applies for six equally competitive data science internships. He has the following prior model for his chances of getting into any given internship, π:

π	0.3	0.4	0.5	Total
$f(\pi)$	0.25	0.60	0.15	1

a)   Let Y be the number of internship offers that Muhammad gets. Specify the model for the dependence of Y on π and the corresponding pmf, $f(y|\pi)$.

b)   Muhammad got some pretty amazing news. He was offered four of the six internships! How likely would this be if $\pi = 0.3$?

c)   Construct the posterior model of π in light of Muhammad's internship news.

---

[11]https://williamsinstitute.law.ucla.edu/wp-content/uploads/LGBTQ-Youth-in-CA-Public-Schools.pdf

Exercise 2.12 (Making mugs). Miles is learning how to make a mug in his ceramics class. A difficult part of the process is creating or "pulling" the handle. His prior model of π, the probability that one of his handles will actually be good enough for a mug, is below:

π	0.1	0.25	0.4	Total
$f(\pi)$	0.2	0.45	0.35	1

a) Miles has enough clay for 7 handles. Let Y be the number of handles that will be good enough for a mug. Specify the model for the dependence of Y on π and the corresponding pmf, $f(y|\pi)$.

b) Miles pulls 7 handles and only 1 of them is good enough for a mug. What is the posterior pmf of π, $f(\pi|y = 1)$?

c) Compare the posterior model to the prior model of π. How would you characterize the differences between them?

d) Miles' instructor Kris had a different prior for his ability to pull a handle (below). Find Kris's posterior $f(\pi|y = 1)$ and compare it to Miles'.

π	0.1	0.25	0.4	Total
$f(\pi)$	0.15	0.15	0.7	1

Exercise 2.13 (Lactose intolerant). Lactose intolerance is an inability to digest milk, often resulting in an upset stomach. Fatima wants to learn more about the proportion of adults who are lactose intolerant, π. Her prior model for π is:

π	0.4	0.5	0.6	0.7	Total
$f(\pi)$	0.1	0.2	0.44	0.26	1

a) Fatima surveys a random sample of 80 adults and 47 are lactose intolerant. Without doing any math, make a guess at the posterior model of π, and explain your reasoning.

b) Calculate the posterior model. How does this compare to your guess in part a?

c) If Fatima had instead collected a sample of 800 adults and 470 (keeping the sample proportion the same as above) are lactose intolerant, how does that change the posterior model?

Exercise 2.14 (Late bus). Li Qiang takes the 8:30am bus to work every morning. If the bus is late, Li Qiang will be late to work. To learn about the probability that her bus will be late (π), Li Qiang first surveys 20 other commuters: 3 think π is 0.15, 3 think π is 0.25, 8 think π is 0.5, 3 think π is 0.75, and 3 think π is 0.85.

a) Convert the information from the 20 surveyed commuters into a prior model for π.

b) Li Qiang wants to update that prior model with the data she collected: in 13 days, the 8:30am bus was late 3 times. Find the posterior model for π.

c)   Compare and comment on the prior and posterior models. What did Li Qiang learn about the bus?

Exercise 2.15 (Cuckoo birds). Cuckoo birds are *brood parasites*, meaning that they lay their eggs in the nests of other birds (hosts), so that the host birds will raise the cuckoo bird hatchlings. Lisa is an ornithologist studying the success rate, π, of cuckoo bird hatchlings that survive at least one week. She is taking over the project from a previous researcher who speculated in their notes the following prior model for π:

π	0.6	0.65	0.7	0.75	Total
$f(\pi)$	0.3	0.4	0.2	0.1	1

a)   If the previous researcher had been more sure that a hatchling would survive, how would the prior model be different?
b)   If the previous researcher had been less sure that a hatchling would survive, how would the prior model be different?
c)   Lisa collects some data. Among the 15 hatchlings she studied, 10 survived for at least one week. What is the posterior model for π?
d)   Lisa needs to explain the posterior model for π in a research paper for ornithologists, and can't assume they understand Bayesian statistics. Briefly summarize the posterior model in context.

Exercise 2.16 (Fake art). An article in *The Daily Beast* reports differing opinions on the proportion (π) of museum artworks that are fake or forged.[12]

a)   After reading the article, define your own prior model for π and provide evidence from the article to justify your choice.

b)   Compare your prior to that below. What's similar? Different?

π	0.2	0.4	0.6	Total
$f(\pi)$	0.25	0.5	0.25	1

c)   Suppose you randomly choose 10 artworks. Assuming the prior from part b, what is the minimum number of artworks that would need to be forged for $f(\pi = 0.6 | Y = y) > 0.4$?

2.5.4   Simulation exercises

Exercise 2.17 (Sick trees redux). Repeat Exercise 2.5 utilizing simulation to *approximate* the posterior probability that a randomly selected maple tree has mold. Specifically, simulate data for 10,000 trees and remember to set your random number seed.

Exercise 2.18 (Lactose intolerant redux). Repeat Exercise 2.13 utilizing simulation to *approximate* the posterior model of π corresponding to Fatima's survey data. Specifically, simulate data for 10,000 people and remember to set your random number seed.

[12]https://www.thedailybeast.com/are-over-half-the-works-on-the-art-market-really-fakes

Exercise 2.19 (Cuckoo birds redux). Repeat Exercise 2.15 utilizing simulation to *approximate* the posterior model of π.

Exercise 2.20 (Cat image recognition). Whether you like it or not, cats have taken over the internet.[13] Joining the craze, Zainab has written an algorithm to detect cat images. It correctly identifies 80% of cat images as cats, but falsely identifies 50% of non-cat images as cats. Zainab tests her algorithm with a new set of images, 8% of which are cats. What's the probability that an image is actually a cat if the algorithm identifies it as a cat? Answer this question by simulating data for 10,000 images.

Exercise 2.21 (Medical tests). A medical test is designed to detect a disease that about 3% of the population has. For 93% of those who have the disease, the test yields a positive result. In addition, the test falsely yields a positive result for 7% of those without the disease. What is the probability that a person has the disease given that they have tested positive? Answer this question by simulating data for 10,000 people.

---

[13]https://www.nytimes.com/2015/08/07/arts/design/how-cats-took-over-the-internet-at-the-museum-of-the-moving-image.html

3

The Beta-Binomial Bayesian Model

Every four years, Americans go to the polls to cast their vote for President of the United States. Consider the following scenario. "Michelle" has decided to run for president and you're her campaign manager for the state of Minnesota. As such, you've conducted 30 different polls throughout the election season. Though Michelle's support has hovered around 45%, she polled at around 35% in the dreariest days and around 55% in the best days on the campaign trail (Figure 3.1 (left)).

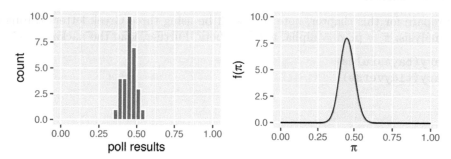

FIGURE 3.1: The results of 30 previous polls of Minnesotans' support of Michelle for president (left) and a corresponding continuous prior model for π, her current election support (right).

Elections are dynamic, thus Michelle's support is always in flux. Yet these past polls provide **prior** information about π, the proportion of Minnesotans that *currently* support Michelle. In fact, we can reorganize this information into a formal prior probability model of π. We worked a similar example in Section 2.3, in which context π was Kasparov's probability of beating Deep Blue at chess. In that case, we greatly over-simplified reality to fit within the framework of introductory Bayesian models. Mainly, we assumed that π could only be 0.2, 0.5, or 0.8, the corresponding chances of which were defined by a *discrete* probability model. However, in the reality of Michelle's election support and Kasparov's chess skill, π can be *any* value between 0 and 1. We can reflect this reality and conduct a more nuanced Bayesian analysis by constructing a **continuous prior probability model** of π. A reasonable prior is represented by the *curve* in Figure 3.1 (right). We'll examine continuous models in detail in Section 3.1. For now, simply notice that this curve preserves the overall information and variability in the past polls – Michelle's support π can be anywhere between 0 and 1, but is most likely around 0.45.

Incorporating this more nuanced, continuous view of Michelle's support π will require some new tools. BUT the *spirit* of the Bayesian analysis will remain the same. No matter if our parameter π is continuous or discrete, the posterior model of π will combine insights from the prior and data. Directly ahead, you will dig into the details and *build* Michelle's election model. You'll then generalize this work to the fundamental **Beta-Binomial** Bayesian model. The power of the Beta-Binomial lies in its broad applications. Michelle's election support

DOI: 10.1201/9780429288340-3

π isn't the only variable of interest that lives on $[0,1]$. You might also imagine Bayesian analyses in which we're interested in modeling the proportion of people that use public transit, the proportion of trains that are delayed, the proportion of people that prefer cats to dogs, and so on. The Beta-Binomial model provides the tools we need to study the proportion of interest, π, in each of these settings.

> ◎ **Goals**
>
> - **Utilize and tune continuous priors.** You will learn how to interpret and tune a continuous Beta prior model to reflect your prior information about π.
> - **Interpret and communicate features of prior and posterior models** using properties such as mean, mode, and variance.
> - **Construct the fundamental Beta-Binomial model** for proportion π.

Getting started

To prepare for this chapter, note that we'll be using three Greek letters throughout our analysis: π = pi, α = alpha, and β = beta. Further, load the packages below:

```
library(bayesrules)
library(tidyverse)
```

3.1   The Beta prior model

In building the Bayesian election model of Michelle's election support among Minnesotans, π, we begin as usual: with the prior. Our *continuous* prior probability model of π is specified by the **probability density function (pdf)** in Figure 3.1. Though it looks quite different, the role of this continuous pdf is the same as for the discrete probability mass function (pmf) $f(\pi)$ in Table 2.7: to specify all possible values of π and the relative plausibility of each. That is, $f(\pi)$ answers: What values can π take and which are more plausible than others? Further, in accounting for *all* possible outcomes of π, the pdf *integrates to* or has an area of 1, much like a discrete pmf *sums* to 1.

Continuous probability models

Let π be a continuous random variable with **probability density function** $f(\pi)$. Then $f(\pi)$ has the following properties:

- $f(\pi) \geq 0$;
- $\int_{\pi} f(\pi)d\pi = 1$, i.e., the area under $f(\pi)$ is 1; and
- $P(a < \pi < b) = \int_{a}^{b} f(\pi)d\pi$ when $a \leq b$, i.e., the area between any two possible values a and b corresponds to the probability of π being in this range.

Interpreting $f(\pi)$

It's possible that $f(\pi) > 1$, thus a continuous pdf *cannot* be interpreted as a probability. Rather, $f(\pi)$ can be used to *compare* the plausibility of two different values of π: the greater $f(\pi)$, the more plausible the corresponding value of π.

NOTE: Don't fret if integrals are new to you. You will not need to *perform* any integration to proceed with this book.

3.1.1 Beta foundations

The next step is to translate the *picture* of our prior in Figure 3.1 (right) into a formal probability model of π. That is, we must specify a formula for the pdf $f(\pi)$. In the world of probability, there are a variety of common "named" models, the pdfs and properties of which are well studied. Among these, it's natural to focus on the **Beta probability model** here. Like Michelle's support π, a Beta random variable is continuous and restricted to live on [0,1]. In this section, you'll explore the *properties* of the Beta model and how to **tune** the Beta to reflect our prior understanding of Michelle's support π. Let's begin with a general definition of the Beta probability model.

The Beta model

Let π be a random variable which can take any value between 0 and 1, i.e., $\pi \in [0,1]$. Then the variability in π might be well modeled by a Beta model with **shape hyperparameters** $\alpha > 0$ and $\beta > 0$:

$$\pi \sim \text{Beta}(\alpha, \beta).$$

The Beta model is specified by continuous pdf

$$f(\pi) = \frac{\Gamma(\alpha + \beta)}{\Gamma(\alpha)\Gamma(\beta)} \pi^{\alpha-1} (1 - \pi)^{\beta-1} \quad \text{for } \pi \in [0,1] \tag{3.1}$$

where $\Gamma(z) = \int_0^\infty x^{z-1} e^{-y} dx$ and $\Gamma(z + 1) = z\Gamma(z)$. Fun fact: when z is a positive integer, then $\Gamma(z)$ simplifies to $\Gamma(z) = (z - 1)!$.

Hyperparameter

A hyperparameter is a parameter used in a prior model.

This model is best understood by playing around. Figure 3.2 plots the Beta pdf $f(\pi)$ under a variety of **shape hyperparameters**, α and β. Check out the various shapes the Beta pdf can take. This flexibility means that we can *tune* the Beta to reflect our prior understanding of π by tweaking α and β. For example, notice that when we set $\alpha = \beta = 1$ (middle left plot), the Beta model is *flat* from 0 to 1. In this setting, the Beta model is equivalent to perhaps a more familiar model, the standard Uniform.

The standard Uniform model

When it's equally plausible for π to take on any value between 0 and 1, we can model π by the standard Uniform model

$$\pi \sim \text{Unif}(0,1)$$

with pdf $f(\pi) = 1$ for $\pi \in [0,1]$. The Unif(0,1) model is a special case of Beta(α,β) when $\alpha = \beta = 1$.

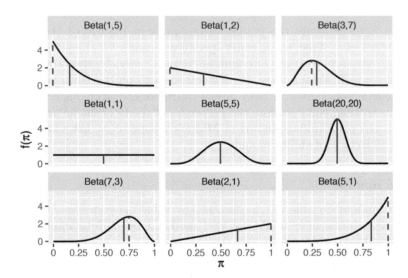

FIGURE 3.2: Beta(α,β) pdfs $f(\pi)$ under a variety of shape hyperparameters α and β (black curve). The mean and mode are represented by a blue solid line and dashed line, respectively.

Take a minute to see if you can identify some other patterns in how shape hyperparameters α and β reflect the **typical** values of π as well as the **variability** in π.[1]

---

❷ Quiz Yourself!

1. How would you describe the typical behavior of a Beta(α,β) variable π when $\alpha = \beta$?
    a) Right-skewed with π tending to be less than 0.5.
    b) Symmetric with π tending to be around 0.5.
    c) Left-skewed with π tending to be greater than 0.5.
2. Using the same options as above, how would you describe the typical behavior of a Beta(α,β) variable π when $\alpha > \beta$?
3. For which model is there greater variability in the plausible values of π, Beta(20,20) or Beta(5,5)?

We can support our observations of the behavior in π with numerical measurements. The **mean** (or "expected value") and **mode** of π provide **measures of central tendency**, or

---
[1]Answers: 1. b; 2. c; 3. Beta(5,5)

what's *typical*. Conceptually speaking, the mean captures the *average* value of π, whereas the mode captures the *most plausible* value of π, i.e., the value of π at which pdf $f(\pi)$ is maximized. These measures are represented by the solid and dashed vertical lines, respectively, in Figure 3.2. Notice that when α is less than β (top row), the Beta pdf is right skewed, thus the mean exceeds the mode of π and both are below 0.5. The opposite is true when α is greater than β (bottom row). When α and β are equal (center row), the Beta pdf is symmetric around a common mean and mode of 0.5. These trends reflect the formulas for the mean, denoted $E(\pi)$, and mode for a Beta(α, β) variable π:[2]

$$E(\pi) = \frac{\alpha}{\alpha + \beta}$$
$$\text{Mode}(\pi) = \frac{\alpha - 1}{\alpha + \beta - 2} \quad \text{when} \quad \alpha, \beta > 1. \tag{3.2}$$

For example, the central tendency of a Beta$(5, 5)$ variable π can be described by

$$E(\pi) = \frac{5}{5 + 5} = 0.5 \quad \text{and} \quad \text{Mode}(\pi) = \frac{5 - 1}{5 + 5 - 2} = 0.5.$$

Figure 3.2 also reveals patterns in the **variability** of π. For example, with values that tend to be closer to the mean of 0.5, the variability in π is smaller for the Beta$(20,20)$ model than for the Beta$(5,5)$ model. We can measure the variability of a Beta(α, β) random variable π by **variance**

$$\text{Var}(\pi) = \frac{\alpha\beta}{(\alpha + \beta)^2(\alpha + \beta + 1)}. \tag{3.3}$$

Roughly speaking, variance measures the typical *squared* distance of π values from the mean, $E(\pi)$. Since the variance thus has squared units, it's typically easier to work with the **standard deviation** which measures the typical *unsquared* distance of π values from $E(\pi)$:

$$\text{SD}(\pi) := \sqrt{\text{Var}(\pi)}.$$

For example, the values of a Beta$(5, 5)$ variable π tend to deviate from their mean of 0.5 by 0.151, whereas the values of a Beta$(20, 20)$ variable tend to deviate from 0.5 by only 0.078:

$$\sqrt{\frac{5 \cdot 5}{(5 + 5)^2(5 + 5 + 1)}} = 0.151 \quad \text{and} \quad \sqrt{\frac{20 \cdot 20}{(20 + 20)^2(20 + 20 + 1)}} = 0.078.$$

The formulas above don't magically pop out of nowhere. They are obtained by applying general definitions of mean, mode, and variance to the Beta pdf (3.1). We provide these definitions below, but you can skip them without consequence.

The theory behind measuring central tendency and variability

Let π be a continuous random variable with pdf $f(\pi)$. Consider two common measures of the **central tendency** in π. The **mean** or **expected value** of π captures the weighted average of π, where each possible π value is weighted by its corresponding pdf value:

---

[2]The mode when either $\alpha \leq 1$ or $\beta \leq 1$ is evident from a plot of the pdf.

$$E(\pi) = \int \pi \cdot f(\pi) d\pi.$$

The **mode** of π captures the *most plausible* value of π, i.e., the value of π for which the pdf is maximized:

$$\text{Mode}(\pi) = \text{argmax}_\pi f(\pi).$$

Next, consider two common measures of the **variability** in π. The **variance** in π roughly measures the typical or expected squared distance of possible π values from their mean:

$$\text{Var}(\pi) = E((\pi - E(\pi))^2) = \int (\pi - E(\pi))^2 \cdot f(\pi) d\pi.$$

The **standard deviation** in π roughly measures the typical or expected distance of possible π values from their mean:

$$\text{SD}(\pi) := \sqrt{\text{Var}(\pi)}.$$

NOTE: If π were *discrete* with pmf $f(\pi)$, we'd replace $\int$ with $\sum$.

3.1.2   Tuning the Beta prior

With a sense for how the Beta(α, β) model works, let's *tune* the shape hyperparameters α and β to reflect our prior information about Michelle's election support π. We saw in Figure 3.1 (left) that across 30 previous polls, Michelle's average support was around 45 percentage points, though she roughly polled as low as 25 and as high as 65 percentage points. Our Beta(α, β) prior should have similar patterns. For example, we want to pick α and β for which π tends to be around 0.45, $E(\pi) = \alpha/(\alpha + \beta) \approx 0.45$. Or, after some rearranging,

$$\alpha \approx \frac{9}{11}\beta.$$

In a trial and error process, we use `plot_beta()` in the **bayesrules** package to plot Beta models with α and β pairs that meet this constraint (e.g., Beta(9,11), Beta(27,33), Beta(45,55)). Among these, we find that the Beta(45,55) closely captures the typical outcomes *and* variability in the old polls:

```
# Plot the Beta(45, 55) prior
plot_beta(45, 55)
```

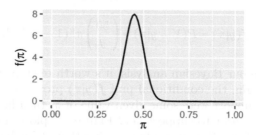

FIGURE 3.3: The Beta(45,55) probability density function.

Thus, a *reasonable* prior model for Michelle's election support is

$$\pi \sim \text{Beta}(45, 55)$$

with **prior pdf** $f(\pi)$ following from plugging 45 and 55 into (3.1),

$$f(\pi) = \frac{\Gamma(100)}{\Gamma(45)\Gamma(55)}\pi^{44}(1-\pi)^{54} \quad \text{for } \pi \in [0,1]. \tag{3.4}$$

By (3.2), this model specifies that Michelle's election support is most likely around 45 percentage points, with **prior mean** and **prior mode**

$$E(\pi) = \frac{45}{45+55} = 0.4500 \quad \text{and} \quad \text{Mode}(\pi) = \frac{45-1}{45+55-2} = 0.4490. \tag{3.5}$$

Further, by (3.3), the potential variability in π is described by a **prior standard deviation** of 0.05. That is, our other prior assumptions about Michelle's possible election support tend to deviate by 5 percentage points from the prior mean of 45%:

$$\begin{aligned} \text{Var}(\pi) &= \frac{45 \cdot 55}{(45+55)^2(45+55+1)} = 0.0025 \\ \text{SD}(\pi) &= \sqrt{0.0025} = 0.05. \end{aligned} \tag{3.6}$$

3.2 The Binomial data model & likelihood function

In the second step of our Bayesian analysis of Michelle's election support π, you're ready to collect some data. You plan to conduct a new poll of $n = 50$ Minnesotans and record Y, the number that support Michelle. The results depend upon, and thus will provide insight into, π – the greater Michelle's actual support, the greater Y will tend to be. To *model* the dependence of Y on π, we can make the following assumptions about the poll: 1) voters answer the poll *independently* of one another; and 2) the *probability* that any polled voter supports your candidate Michelle is π. It follows from our work in Section 2.3.2 that, *conditional* on π, Y is Binomial. Specifically,

$$Y|\pi \sim \text{Bin}(50, \pi)$$

with conditional pmf $f(y|\pi)$ defined for $y \in \{0, 1, 2, ..., 50\}$,

$$f(y|\pi) = P(Y = y|\pi) = \binom{50}{y} \pi^y (1-\pi)^{50-y}. \qquad (3.7)$$

Given its importance in our Bayesian analysis, it's worth re-emphasizing the details of the Binomial model. To begin, the conditional pmf $f(y|\pi)$ provides answers to a *hypothetical* question: *if* Michelle's support were some given value of π, then how many of the 50 polled voters $Y = y$ might we expect to support her? Figure 3.4 plots this pmf under a range of possible π. These plots formalize our understanding that *if* Michelle's support π were low (top row), the polling result Y is also likely to be low. *If* her support were high (bottom row), Y is also likely to be high.

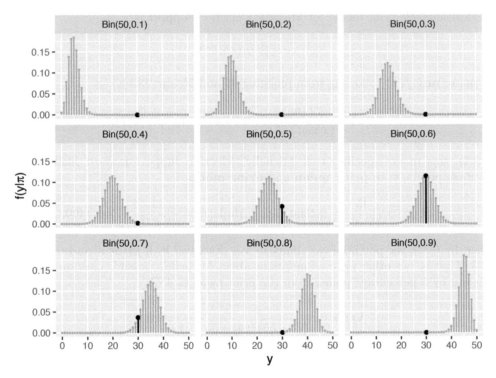

FIGURE 3.4: The Bin$(50, \pi)$ pmf $f(y|\pi)$ is plotted for values of $\pi \in \{0.1, 0.2, \ldots, 0.9\}$. The pmfs at the observed value of polling data $Y = 30$ are highlighted in black.

In *reality*, we ultimately observe that the poll was a huge success: $Y = 30$ of $n = 50$ (60%) polled voters support Michelle! This result is highlighted by the black lines among the pmfs in Figure 3.4. To focus on just these results that match the observed polling data, we extract and compare these black lines in a single plot (Figure 3.5). These represent the *likelihoods* of the observed polling data, $Y = 30$, at each potential level of Michelle's support π in $\{0.1, 0.2, \ldots, 0.9\}$. In fact, this discrete set of scenarios represents a small handful of points along the complete continuous **likelihood function** $L(\pi|y = 30)$ defined for any π between 0 and 1 (black curve).

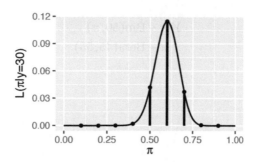

FIGURE 3.5: The likelihood function, $L(\pi|y = 30)$, of Michelle's election support π given the observed poll in which $Y = 30$ of $n = 50$ polled Minnesotans supported her. The vertical lines represent the likelihood evaluated at π in $\{0.1, 0.2, \ldots, 0.9\}$.

Recall that the likelihood function is defined by turning the Binomial pmf on its head. Treating $Y = 30$ as observed data and π as unknown, matching the reality of our situation, the Binomial likelihood function of π follows from plugging $y = 30$ into the Binomial pmf (3.7):

$$L(\pi|y = 30) = \binom{50}{30}\pi^{30}(1 - \pi)^{20} \quad \text{for } \pi \in [0, 1]. \tag{3.8}$$

For example, matching what we see in Figure 3.5, the chance that $Y = 30$ of 50 polled voters would support Michelle is 0.115 if her underlying support were $\pi = 0.6$:

$$L(\pi = 0.6|y = 30) = \binom{50}{30}0.6^{30}0.4^{20} \approx 0.115$$

but only 0.042 if her underlying support were $\pi = 0.5$:

$$L(\pi = 0.5|y = 30) = \binom{50}{30}0.5^{30}0.5^{20} \approx 0.042.$$

It's also important to remember here that $L(\pi|y = 30)$ is a *function of* π that provides insight into the relative compatibility of the observed polling data $Y = 30$ with different $\pi \in [0, 1]$. The fact that $L(\pi|y = 30)$ is *maximized* when $\pi = 0.6$ suggests that the 60% support for Michelle among *polled* voters is most likely when her *underlying* support is also at 60%. This makes sense! The further that a hypothetical π value is from 0.6, the less likely we would be to observe our poll result – $L(\pi|y = 30)$ effectively drops to 0 for π values under 0.3 and above 0.9. Thus, it's extremely unlikely that we would've observed a 60% support rate in the new poll if, in fact, Michelle's underlying support were as low as 30% or as high as 90%.

3.3 The Beta posterior model

We now have two pieces of our Bayesian model in place – the Beta prior model for Michelle's support π and the Binomial model for the dependence of polling data Y on π:

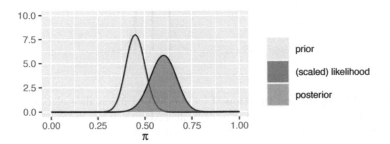

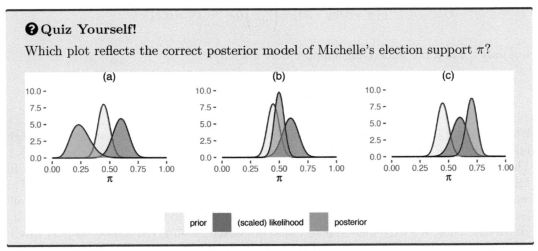

FIGURE 3.6: The prior model of π along with the (scaled) likelihood function of π given the new poll results in which $Y = 30$ of $n = 50$ polled Minnesotans support Michelle.

These pieces of the puzzle are shown together in Figure 3.6 where, *only for the purposes of visual comparison to the prior*, the likelihood function is scaled to integrate to 1.[3] The prior and data, as captured by the likelihood, don't completely agree. Constructed from old polls, the prior is a bit more pessimistic about Michelle's election support than the data obtained from the latest poll. Yet *both* insights are valuable to our analysis. Just as much as we shouldn't ignore the new poll in favor of the old, we also shouldn't throw out our bank of prior information in favor of the newest thing (also great life advice). Thinking like Bayesians, we can construct a **posterior model** of π which combines the information from the prior with that from the data.

> **❷ Quiz Yourself!**
>
> Which plot reflects the correct posterior model of Michelle's election support π?

Plot (b) is the only plot in which the posterior model of π strikes a *balance* between the relative pessimism of the prior and optimism of the data. You can reproduce this *correct* posterior using the `plot_beta_binomial()` function in the **bayesrules** package, plugging in the prior hyperparameters ($\alpha = 45, \beta = 55$) and data ($y = 30$ of $n = 50$ polled voters support Michelle):

---

[3]The scaled likelihood function is calculated by $L(\pi|y)/\int_0^1 L(\pi|y)d\pi$.

```
plot_beta_binomial(alpha = 45, beta = 55, y = 30, n = 50)
```

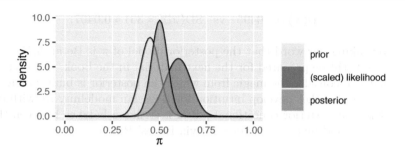

FIGURE 3.7: The prior pdf, scaled likelihood function, and posterior pdf of Michelle's election support π.

In its balancing act, the posterior here is slightly "closer" to the prior than to the likelihood. (We'll gain intuition for why this is the case in Chapter 4.) The posterior being *centered* at $\pi = 0.5$ suggests that Michelle's support is equally likely to be above or below the 50% threshold required to win Minnesota. Further, combining information from the prior and data, the *range* of posterior plausible values has narrowed: we can be fairly certain that Michelle's support is somewhere between 35% and 65%.

You might also recognize something new: like the prior, the posterior model of π is continuous and lives on [0,1]. That is, like the prior, the posterior appears to be a Beta(α, β) model where the shape parameters have been *updated* to combine information from the prior and data. This is indeed the case. Conditioned on the observed poll results ($Y = 30$), the posterior model of Michelle's election support is Beta(75, 75):

$$\pi|(Y = 30) \sim \text{Beta}(75, 75)$$

with a corresponding pdf which follows from (3.1):

$$f(\pi|y = 30) = \frac{\Gamma(150)}{\Gamma(75)\Gamma(75)}\pi^{74}(1 - \pi)^{74} \quad \text{for } \pi \in [0, 1]. \tag{3.9}$$

Before backing up this claim with some math, let's examine the evolution in your understanding of Michelle's election support π. The `summarize_beta_binomial()` function in the **bayesrules** package summarizes the typical values and variability in the prior and posterior models of π. These calculations follow directly from applying the prior and posterior Beta parameters into (3.2) and (3.3):

```
summarize_beta_binomial(alpha = 45, beta = 55, y = 30, n = 50)
      model alpha beta mean  mode      var      sd
1     prior    45   55 0.45 0.449 0.002450 0.04950
2 posterior    75   75 0.50 0.500 0.001656 0.04069
```

A comparison illuminates the polling data's influence on the posterior model. Mainly, after observing the poll in which 30 of 50 people supported Michelle, the expected value of her underlying support π nudged up from approximately 45% to 50%:

$$E(\pi) = 0.45 \quad \text{vs} \quad E(\pi|Y = 30) = 0.50.$$

Further, the variability within the model decreased, indicating a narrower range of posterior plausible π values in light of the polling data:

$$\text{SD}(\pi) \approx 0.0495 \quad \text{vs} \quad \text{SD}(\pi|Y = 30) \approx 0.0407.$$

If you're happy taking our word that the posterior model of π is Beta(75,75), you can skip to Section 3.4 and still be prepared for the next material in the book. However, we strongly recommend that you consider the magic from which the posterior is built. Going through the process can help you further develop intuition for Bayesian modeling. As with our previous Bayesian models, the posterior conditional pdf of π strikes a balance between the prior pdf $f(\pi)$ and the likelihood function $L(\pi|y = 30)$ via Bayes' Rule (2.12):

$$f(\pi|y = 30) = \frac{f(\pi)L(\pi|y = 30)}{f(y = 30)}.$$

Recall from Section 2.3.6 that $f(y = 30)$ is a **normalizing constant**, i.e., a constant across π which scales the posterior pdf $f(\pi|y = 30)$ to integrate to 1. **We don't need to calculate the normalizing constant in order to construct the posterior model.** Rather, we can simplify the posterior construction by utilizing the fact that the posterior pdf is *proportional to* the product of the prior pdf (3.4) and likelihood function (3.8):

$$\begin{aligned}
f(\pi|y = 30) &\propto f(\pi)L(\pi|y = 30) \\
&= \frac{\Gamma(100)}{\Gamma(45)\Gamma(55)}\pi^{44}(1 - \pi)^{54} \cdot \binom{50}{30}\pi^{30}(1 - \pi)^{20} \\
&= \left[\frac{\Gamma(100)}{\Gamma(45)\Gamma(55)}\binom{50}{30}\right] \cdot \pi^{74}(1 - \pi)^{74} \\
&\propto \pi^{74}(1 - \pi)^{74}.
\end{aligned}$$

In the third line of our calculation, we combined the constants and the elements that depend upon π into two different pieces. In the final line, we made a big simplification: we dropped all constants that don't depend upon π. We don't need these. Rather, it's the dependence of $f(\pi|y = 30)$ on π that we care about:

$$f(\pi|y = 30) = c\pi^{74}(1 - \pi)^{74} \propto \pi^{74}(1 - \pi)^{74}.$$

We *could* complete the definition of this posterior pdf by calculating the normalizing constant c for which the pdf integrates to 1:

$$1 = \int f(\pi|y = 30)d\pi = \int c \cdot \pi^{74}(1 - \pi)^{74}d\pi \quad \Rightarrow \quad c = \frac{1}{\int \pi^{74}(1 - \pi)^{74}d\pi}.$$

But again, we don't *need* to do this calculation. The pdf of π is defined by its structural dependence on π, that is, the **kernel** of the pdf. Notice here that $f(\pi|y = 30)$ has the *same* kernel as the normalized Beta(75,75) pdf in (3.9):

$$f(\pi|y = 30) = \frac{\Gamma(150)}{\Gamma(75)\Gamma(75)}\pi^{74}(1 - \pi)^{74} \propto \pi^{74}(1 - \pi)^{74}.$$

The fact that the unnormalized posterior pdf $f(\pi|y = 30)$ matches an unnormalized Beta(75,75) pdf verifies our claim that $\pi|(Y = 30) \sim$ Beta(75, 75). Magic. For extra practice in identifying the posterior model of π from an unnormalized posterior pdf or kernel, take the following quiz.[4]

---

❷ Quiz Yourself!

For each scenario below, identify the correct Beta posterior model of $\pi \in [0, 1]$ from its unnormalized pdf.

    a. $f(\pi|y) \propto \pi^{3-1}(1 - \pi)^{12-1}$
    b. $f(\pi|y) \propto \pi^{11}(1 - \pi)^2$
    c. $f(\pi|y) \propto 1$

Now, instead of identifying a model from a kernel, practice identifying the kernels of models.[5]

---

❷ Quiz Yourself!

Identify the kernels of each pdf below.

    1. $f(\pi|y) = ye^{-\pi y}$ for $\pi > 0$
        a. y
        b. $e^{-\pi}$
        c. $ye^{-\pi}$
        d. $e^{-\pi y}$
    2. $f(\pi|y) = \frac{2^y}{(y-1)!}\pi^{y-1}e^{-2\pi}$ for $\pi > 0$
        a. $\pi^{y-1}e^{-2\pi}$
        b. $\frac{2^y}{(y-1)!}$
        c. $e^{-2\pi}$
        d. π^{y-1}
    3. $f(\pi) = 3\pi^2$ for $\pi \in [0, 1]$

3.4 The Beta-Binomial model

In the previous section we developed the fundamental Beta-Binomial model for Michelle's election support π. In doing so, we assumed a specific Beta(45,55) prior and a specific polling result ($Y = 30$ of $n = 50$ polled voters supported your candidate) within a specific context. This was a special case of the more general Beta-Binomial model:

$$Y|\pi \sim \text{Bin}(n, \pi)$$
$$\pi \sim \text{Beta}(\alpha, \beta).$$

This general model has vast applications, applying to *any* setting having a parameter of interest π that lives on [0,1] with *any* tuning of a Beta prior and *any* data Y which is the number of "successes" in n fixed, independent trials, each having probability of success π.

---

[4] Answer: a. Beta(3,12); b. Beta(12,3); c. Beta(1,1) or, equivalently, Unif(0,1)
[5] Answers: 1. d; 2. a; 3. π^2

For example, π might be a coin's tendency toward Heads and data Y records the number of Heads observed in a series of n coin flips. Or π might be the proportion of adults that use social media and we learn about π by sampling n adults and recording the number Y that use social media. No matter the setting, upon observing $Y = y$ successes in n trials, the posterior of π can be described by a Beta model which reveals the influence of the prior (through α and β) and data (through y and n):

$$\pi|(Y = y) \sim \text{Beta}(\alpha + y, \beta + n - y). \tag{3.10}$$

Measures of posterior central tendency and variability follow from (3.2) and (3.3):

$$
\begin{aligned}
E(\pi|Y = y) &= \frac{\alpha + y}{\alpha + \beta + n} \\
\text{Mode}(\pi|Y = y) &= \frac{\alpha + y - 1}{\alpha + \beta + n - 2} \\
\text{Var}(\pi|Y = y) &= \frac{(\alpha + y)(\beta + n - y)}{(\alpha + \beta + n)^2(\alpha + \beta + n + 1)}.
\end{aligned}
\tag{3.11}
$$

Importantly, notice that the posterior follows a different parameterization of the same probability model as the prior – both the prior and posterior are Beta models with different tunings. In this case, we say that the Beta(α, β) model is a **conjugate prior** for the corresponding Bin(n, π) data model. Our work below will highlight that conjugacy simplifies the construction of the posterior, and thus can be a desirable property in Bayesian modeling.

Conjugate prior

We say that $f(\pi)$ is a conjugate prior for $L(\pi|y)$ if the posterior, $f(\pi|y) \propto f(\pi)L(\pi|y)$, is from the same model family as the prior.

The posterior construction for the general Beta-Binomial model is very similar to that of the election-specific model. First, the Beta prior pdf $f(\pi)$ is defined by (3.1) and the likelihood function $L(\pi|y)$ by (2.7), the conditional pmf of the Bin(n, π) model upon observing data $Y = y$. For $\pi \in [0, 1]$,

$$f(\pi) = \frac{\Gamma(\alpha + \beta)}{\Gamma(\alpha)\Gamma(\beta)}\pi^{\alpha-1}(1 - \pi)^{\beta-1} \quad \text{and} \quad L(\pi|y) = \binom{n}{y}\pi^y(1 - \pi)^{n-y}. \tag{3.12}$$

Putting these two pieces together, the posterior pdf follows from Bayes' Rule:

$$
\begin{aligned}
f(\pi|y) &\propto f(\pi)L(\pi|y) \\
&= \frac{\Gamma(\alpha + \beta)}{\Gamma(\alpha)\Gamma(\beta)}\pi^{\alpha-1}(1 - \pi)^{\beta-1} \cdot \binom{n}{y}\pi^y(1 - \pi)^{n-y} \\
&\propto \pi^{(\alpha+y)-1}(1 - \pi)^{(\beta+n-y)-1}.
\end{aligned}
$$

Again, we've dropped normalizing constants which don't depend upon π and are left with the *unnormalized* posterior pdf. Note that this shares the same structure as the normalized Beta($\alpha + y, \beta + n - y$) pdf,

$$f(\pi|y) = \frac{\Gamma(\alpha + \beta + n)}{\Gamma(\alpha + y)\Gamma(\beta + n - y)}\pi^{(\alpha+y)-1}(1 - \pi)^{(\beta+n-y)-1}.$$

Thus, we've verified our claim that the posterior model of π given an observed $Y = y$ successes in n trials is Beta$(\alpha + y, \beta + n - y)$.

3.5   Simulating the Beta-Binomial

Using Section 2.3.7 as a guide, let's *simulate* the posterior model of Michelle's support π. We begin by simulating 10,000 values of π from the Beta(45,55) prior using `rbeta()` and, subsequently, a potential Bin(50,π) poll result Y from each π using `rbinom()`:

```
set.seed(84735)
michelle_sim <- data.frame(pi = rbeta(10000, 45, 55)) %>%
  mutate(y = rbinom(10000, size = 50, prob = pi))
```

The resulting 10,000 pairs of π and y values are shown in Figure 3.8. In general, the greater Michelle's support, the better her poll results tend to be. Further, the highlighted pairs illustrate that the eventual observed poll result, $Y = 30$ of 50 polled voters supported Michelle, would most likely arise if her underlying support π were somewhere in the range from 0.4 to 0.6.

```
ggplot(michelle_sim, aes(x = pi, y = y)) +
  geom_point(aes(color = (y == 30)), size = 0.1)
```

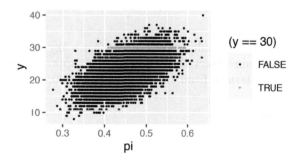

FIGURE 3.8: A scatterplot of 10,000 simulated pairs of Michelle's support π and polling outcome y.

When we zoom in closer on just those pairs that match our $Y = 30$ poll results, the behavior across the remaining set of π values well *approximates* the Beta(75,75) posterior model of π:

```r
# Keep only the simulated pairs that match our data
michelle_posterior <- michelle_sim %>%
  filter(y == 30)

# Plot the remaining pi values
ggplot(michelle_posterior, aes(x = pi)) +
  geom_density()
```

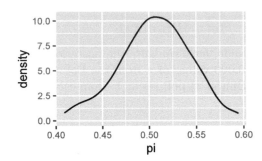

FIGURE 3.9: A density plot of simulated π values that produced polling outcomes in which $Y = 30$ voters supported Michelle.

As such, we can also use our simulated sample to approximate posterior features, such as the mean and standard deviation in Michelle's support. The results are quite similar to the theoretical values calculated above, $E(\pi|Y = 30) = 0.5$ and $\mathrm{SD}(\pi|Y = 30) = 0.0407$:

```r
michelle_posterior %>%
  summarize(mean(pi), sd(pi))
  mean(pi)  sd(pi)
1   0.5055 0.03732
```

In interpreting these simulation results, *"approximate"* is a key word. Since only 211 of our 10,000 simulations matched our observed $Y = 30$ data, this approximation might be *improved* by upping our original simulations from 10,000 to, say, 50,000:

```r
nrow(michelle_posterior)
[1] 211
```

3.6   Example: Milgram's behavioral study of obedience

In a 1963 issue of *The Journal of Abnormal and Social Psychology*, Stanley Milgram described a study in which he investigated the propensity of people to obey orders from authority figures, even when those orders may harm other people (Milgram, 1963). In the paper, Milgram describes the study as:

"consist[ing] of ordering a naive subject to administer electric shock to a victim. A simulated shock generator is used, with 30 clearly marked voltage levels that range from IS to 450 volts. The instrument bears verbal designations that range from Slight Shock *to* Danger: Severe Shock. *The responses of the victim, who is a trained confederate of the experimenter, are standardized. The orders to administer shocks are given to the naive subject in the context of a 'learning experiment' ostensibly set up to study the effects of punishment on memory. As the experiment proceeds the naive subject is commanded to administer increasingly more intense shocks to the victim, even to the point of reaching the level marked* Danger: Severe Shock.*"*

In other words, study participants were given the task of testing another participant (who was in truth a trained actor) on their ability to memorize facts. If the actor *didn't* remember a fact, the participant was ordered to administer a shock on the actor and to increase the shock level with every subsequent failure. Unbeknownst to the participant, the shocks were fake and the actor was only *pretending* to register pain from the shock.

3.6.1 A Bayesian analysis

We can translate Milgram's study into the Beta-Binomial framework. The parameter of interest here is π, the chance that a person would obey authority (in this case, administering the most severe shock), even if it meant bringing harm to others. Since Milgram passed away in 1984, we don't have the opportunity to ask him about his understanding of π *prior* to conducting the study. Thus, we'll diverge from the actual study here, and suppose that another psychologist helped carry out this work. Prior to collecting data, they indicated that a Beta(1,10) model accurately reflected their understanding about π, developed through previous work. Next, let Y be the number of the 40 study participants that would inflict the most severe shock. Assuming that each participant behaves independently of the others, we can model the dependence of Y on π using the Binomial. In summary, we have the following Beta-Binomial Bayesian model:

$$Y|\pi \sim \text{Bin}(40, \pi)$$
$$\pi \sim \text{Beta}(1, 10).$$

Before moving ahead with our analysis, let's examine the psychologist's prior model.

---

❷ Quiz Yourself!

What does the Beta(1,10) prior model in Figure 3.10 reveal about the psychologist's prior understanding of π?

   a)   They don't have an informed opinion.
   b)   They're fairly certain that a large proportion of people will do what authority tells them.
   c)   They're fairly certain that only a small proportion of people will do what authority tells them.

```
# Beta(1,10) prior
plot_beta(alpha = 1, beta = 10)
```

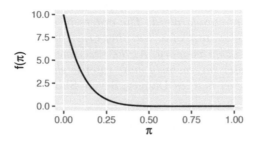

FIGURE 3.10: A Beta(1,10) prior model of π.

The correct answer to this quiz is c. The psychologist's prior is that π typically takes on values near 0 with low variability. Thus, the psychologist is fairly certain that very few people will just do whatever authority tells them. Of course, the psychologist's understanding will evolve upon seeing the results of Milgram's study.

---

❷ Quiz Yourself!

In the end, 26 of the 40 study participants inflicted what they understood to be the maximum shock. In light of this data, what's the psychologist's posterior model of π:

$$\pi|(Y = 26) \sim \text{Beta}(???, ???)$$

Plugging the prior hyperparameters ($\alpha = 1$, $\beta = 10$) and data ($y = 26$, $n = 40$) into (3.10) establishes the psychologist's posterior model of π:

$$\pi|(Y = 26) \sim \text{Beta}(27, 24).$$

This posterior is summarized and plotted below, contrasted with the prior pdf and scaled likelihood function. Note that the psychologist's understanding evolved quite a bit from their prior to their posterior. Though they started out with an understanding that fewer than ~25% of people would inflict the most severe shock, given the strong counterevidence in the study data, they now understand this figure to be somewhere between ~30% and ~70%.

```
summarize_beta_binomial(alpha = 1, beta = 10, y = 26, n = 40)
      model alpha beta    mean    mode      var      sd
1      prior     1   10 0.09091 0.0000 0.006887 0.08299
2 posterior    27   24 0.52941 0.5306 0.004791 0.06922


plot_beta_binomial(alpha = 1, beta = 10, y = 26, n = 40)
```

3.6.2   The role of ethics in statistics and data science

In working through the previous example, we hope you were a bit distracted by your inner voice – this experiment seems ethically dubious. You wouldn't be alone in this thinking. Stanley Milgram is a controversial historical figure. We chose the above example to not only

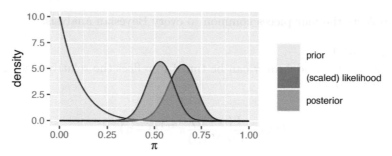

FIGURE 3.11: The Beta prior pdf, scaled Binomial likelihood function, and Beta posterior pdf for π, the proportion of subjects that would follow the given instructions.

practice building Beta-Binomial models, but to practice taking a critical eye to our work and the work of others. Every data collection, visualization, analysis, and communication engenders both harms and benefits to individuals and groups, both direct and indirect. As statisticians and data scientists, it is critical to always consider these harms and benefits. We encourage you to ask yourself the following questions each time you work with data:

- What are the study's potential **benefits** to society? To participants?
- What are the study's potential **risks** to society? To participants?
- What **ethical issues** might arise when generalizing observations on the study participants to a larger population?
- Who is **included** and **excluded** in this study? What are the corresponding risks and benefits? Are individuals in groups that have been historically (and currently) marginalized put at greater risk?
- Were the people who might be affected by your study *involved* in the study? If not, you may not be qualified to evaluate these questions.
- What's the personal story or experience of each subject represented by a row of data?

The importance of considering the context and implications for your statistical and data science work cannot be overstated. As statisticians and data scientists, we are responsible for considering these issues so as not to harm individuals and communities of people. Fortunately, there are many resources available to learn more. To name just a few: *Race After Technology* (Benjamin, 2019), *Data Feminism* (D'Ignazio and Klein, 2020), *Algorithms of Oppression* (Noble, 2018), *Datasheets for Datasets* (Gebru et al., 2018), *Model Cards for Model Reporting* (Mitchell et al., 2019), *Automating Inequality: How High-Tech Tools Profile, Police, and Punish the Poor* (Eubanks, 2018), *Closing the AI Accountability Gap* (Raji et al., 2020), and "Integrating Data Science Ethics Into an Undergraduate Major" (Baumer et al., 2020).

3.7 Chapter summary

In Chapter 3, you built the foundational Beta-Binomial model for π, an unknown **proportion** that can take any value between 0 and 1:

$$\begin{aligned} Y|\pi &\sim \text{Bin}(n, \pi) \\ \pi &\sim \text{Beta}(\alpha, \beta) \end{aligned} \quad \Rightarrow \quad \pi|(Y = y) \sim \text{Beta}(\alpha + y, \beta + n - y).$$

This model reflects the four pieces common to every Bayesian analysis:

1. **Prior model**
   The Beta prior model for π can be tuned to reflect the relative prior plausibility of each $\pi \in [0, 1]$.

$$f(\pi) = \frac{\Gamma(\alpha + \beta)}{\Gamma(\alpha)\Gamma(\beta)} \pi^{\alpha-1}(1 - \pi)^{\beta-1}.$$

2. **Data model**
   To learn about π, we collect **data** Y, the number of successes in n independent trials, each having probability of success π. The dependence of Y on π is summarized by the Binomial model $\text{Bin}(n, \pi)$.

3. **Likelihood function**
   Upon observing data $Y = y$ where $y \in \{0, 1, \ldots, n\}$, the likelihood function of π, obtained by plugging y into the Binomial pmf, provides a mechanism by which to compare the compatibility of the data with different π:

$$L(\pi|y) = \binom{n}{y} \pi^y(1 - \pi)^{n-y} \quad \text{for } \pi \in [0, 1].$$

4. **Posterior model**
   Via Bayes' Rule, the **conjugate** Beta prior combined with the Binomial data model produce a Beta posterior model for π. The updated Beta posterior parameters $(\alpha + y, \beta + n - y)$ reflect the influence of the prior (via α and β) and the observed data (via y and n).

$$f(\pi|y) \propto f(\pi)L(\pi|y) \propto \pi^{(\alpha+y)-1}(1 - \pi)^{(\beta+n-y)-1}.$$

3.8   Exercises

3.8.1   Practice: Beta prior models

Exercise 3.1 (Tune your Beta prior: Take I). In each situation below, tune a $\text{Beta}(\alpha, \beta)$ model that accurately reflects the given prior information. In many cases, there's no single "right" answer, but rather multiple "reasonable" answers.

a) Your friend applied to a job and tells you: "I think I have a 40% chance of getting the job, but I'm pretty unsure." When pressed further, they put their chances between 20% and 60%.

b) A scientist has created a new test for a rare disease. They expect that the test is accurate 80% of the time with a variance of 0.05.

c) Your Aunt Jo is a successful mushroom hunter. She boasts: "I expect to find enough mushrooms to feed myself and my co-workers at the auto-repair shop 90% of the time, but if I had to give you a likely range it would be between 85% and 100% of the time."

d)   Sal (who is a touch hyperbolic) just interviewed for a job, and doesn't know how to describe their chances of getting an offer. They say, "I couldn't read my interviewer's expression! I either really impressed them and they are absolutely going to hire me, or I made a terrible impression and they are burning my resumé as we speak."

Exercise 3.2 (Tune your Beta prior: Take II). As in Exercise 3.1, tune an appropriate Beta(α, β) prior model for each situation below.

a)   Your friend tells you "I think that I have a 80% chance of getting a full night of sleep tonight, and I am pretty certain." When pressed further, they put their chances between 70% and 90%.

b)   A scientist has created a new test for a rare disease. They expect that it's accurate 90% of the time with a variance of 0.08.

c)   Max loves to play the video game *Animal Crossing*. They tell you: "The probability that I play *Animal Crossing* in the morning is somewhere between 75% and 95%, but most likely around 85%."

d)   The bakery in Easthampton, Massachusetts often runs out of croissants on Sundays. Ben guesses that by 10 a.m., there is a 30% chance they have run out, but is pretty unsure about that guess.

Exercise 3.3 (It's OK to admit you don't know). You want to specify a Beta prior for a situation in which you have no idea about some parameter π. You think π is equally likely to be anywhere between 0 and 1.

a)   Specify and plot the appropriate Beta prior model.

b)   What is the mean of the Beta prior that you specified? Explain why that does or does not align with having no clue.

c)   What is the standard deviation of the Beta prior that you specified?

d)   Specify and plot an example of a Beta prior that has a **smaller standard deviation** than the one you specified.

e)   Specify and plot an example of a Beta prior that has a **larger standard deviation** than the one you specified.

Exercise 3.4 (Which Beta? Take I). Six Beta pdfs are plotted below. Match each to one of the following models: Beta(0.5,0.5), Beta(1,1), Beta(2,2), Beta(6,6), Beta(6,2), Beta(0.5,6).

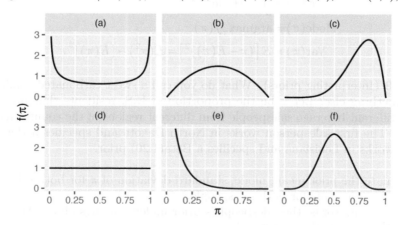

Exercise 3.5 (Which Beta? Take II). Six Beta pdfs are plotted below. Match each to one of the following models: Beta(1,0.3), Beta(2,1), Beta(3,3), Beta(6,3), Beta(4,2), Beta(5,6).

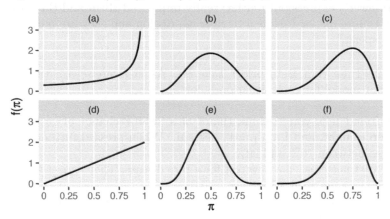

Exercise 3.6 (Beta properties). Examine the properties of the Beta models in Exercise 3.4.

a)   Which Beta model has the smallest mean? The biggest? Provide visual evidence and calculate the corresponding means.

b)   Which Beta model has the smallest mode? The biggest? Provide visual evidence and calculate the corresponding modes.

c)   Which Beta model has the smallest standard deviation? The biggest? Provide visual evidence and calculate the corresponding standard deviations.

Exercise 3.7 (Using R for Beta).

a)   Use `plot_beta()` to plot the six Beta models in Exercise 3.4.

b)   Use `summarize_beta()` to confirm your answers to Exercise 3.6.

Exercise 3.8 (Calculus challenge: establishing Beta features). Let π follow a Beta(α, β) model. Formulas for the mean, mode, and variance of π are given by (3.2) and (3.3). Confirm these properties by applying the following definitions of mean, mode, and variance directly to the Beta pdf $f(\pi)$, (3.1):

$$E(\pi) = \int \pi f(\pi) d\pi$$
$$\text{Mode}(\pi) = \text{argmax}_\pi f(\pi)$$
$$\text{Var}(\pi) = E\left[(\pi - E(\pi))^2\right] = E(\pi^2) - [E(\pi)]^2.$$

Exercise 3.9 (Interpreting priors). What do you call a sweet carbonated drink: pop, soda, coke, or something else? Let π be the proportion of U.S. residents that prefer the term "pop." Two different beverage salespeople from different regions of the country have different priors for π. The first salesperson works in North Dakota and specifies a Beta(8,2) prior. The second works in Louisiana and specifies a Beta(1,20) prior.

a)   Calculate the prior mean, mode, standard deviation of π for both salespeople.

b)   Plot the prior pdfs for both salespeople.

c)   Compare, in words, the salespeople's prior understandings about the proportion of U.S. residents that say "pop".

3.8.2 Practice: Beta-Binomial models

Exercise 3.10 (Different priors, different posteriors). Continuing Exercise 3.9, we poll 50 U.S. residents and 12 (24%) prefer the term "pop."

   a) Specify the unique posterior model of π for both salespeople. We encourage you to construct these posteriors from scratch.
   b) Plot the prior pdf, likelihood function, and posterior pdf for both salespeople.
   c) Compare the salespeople's posterior understanding of π.

Exercise 3.11 (Regular bike ridership). A university wants to know what proportion of students are regular bike riders, π, so that they can install an appropriate number of bike racks. Since the university is in sunny Southern California, staff think that π has a mean of 1 in 4 students, and a mode of 5/22.

   a) Specify and plot a Beta model that reflects the staff's prior ideas about π.
   b) Among 50 surveyed students, 15 are regular bike riders. What is the posterior model for π?
   c) What is the mean, mode, and standard deviation of the posterior model?
   d) Does the posterior model more closely reflect the prior information or the data? Explain your reasoning.

Exercise 3.12 (Same-sex marriage). A 2017 Pew Research survey found that 10.2% of LGBT adults in the U.S. were married to a same-sex spouse.[6] Now it's the 2020s, and Bayard guesses that π, the percent of LGBT adults in the U.S. who are married to a same-sex spouse, has most likely increased to about 15% but could reasonably range from 10% to 25%.

   a) Identify and plot a Beta model that reflects Bayard's prior ideas about π.
   b) Bayard wants to update his prior, so he randomly selects 90 US LGBT adults and 30 of them are married to a same-sex partner. What is the posterior model for π?
   c) Calculate the posterior mean, mode, and standard deviation of π.
   d) Does the posterior model more closely reflect the prior information or the data? Explain your reasoning.

Exercise 3.13 (Knowing someone who is transgender). A September 2016 Pew Research survey found that 30% of U.S. adults are aware that they know someone who is transgender.[7] It is now the 2020s, and Sylvia believes that the current percent of people who know someone who is transgender, π, has increased to somewhere between 35% and 60%.

   a) Identify and plot a Beta model that reflects Sylvia's prior ideas about π.
   b) Sylvia wants to update her prior, so she randomly selects 200 US adults and 80 of them are aware that they know someone who is transgender. Specify and plot the posterior model for π.

---

[6]https://news.gallup.com/poll/212702/lgbt-adults-married-sex-spouse.aspx?utm_source=alert&utm_medium=email&utm_content=morelink&utm_campaign=syndication
[7]https://www.pewforum.org/2016/09/28/5-vast-majority-of-americans-know-someone-who-is-gay-fewer-know-someone-who-is-transgender/

c) What is the mean, mode, and standard deviation of the posterior model?
d) Describe how the prior and posterior Beta models compare.

Exercise 3.14 (Summarizing the Beta-Binomial: Take I). Write the corresponding input code for the `summarize_beta_binomial()` output below.

	model	alpha	beta	mean	mode	var	sd
1	prior	2	3	0.4000	0.3333	0.040000	0.20000
2	posterior	11	24	0.3143	0.3030	0.005986	0.07737

Exercise 3.15 (Summarizing the Beta-Binomial: Take II). Write the corresponding input code for the `summarize_beta_binomial()` output below.

	model	alpha	beta	mean	mode	var	sd
1	prior	1	2	0.3333	0.0000	0.0555556	0.23570
2	posterior	100	3	0.9709	0.9802	0.0002719	0.01649

Exercise 3.16 (Plotting the Beta-Binomial: Take I). Below is output from `plot_beta_binomial()` function.

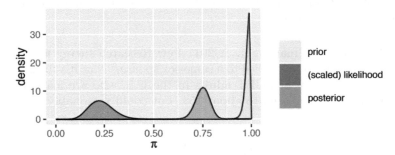

a) Describe and compare both the prior model and likelihood function in words.
b) Describe the posterior model in words. Does it more closely agree with the data (as reflected by the likelihood function) or the prior?
c) Provide the specific `plot_beta_binomial()` code you would use to produce a similar plot.

Exercise 3.17 (Plotting the Beta-Binomial: Take II). Repeat Exercise 3.16 for the `plot_beta_binomial()` output below.

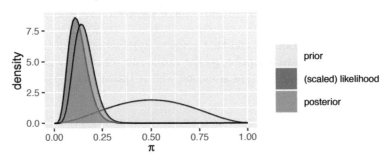

Exercise 3.18 (More Beta-Binomial).

a) Patrick has a Beta(3,3) prior for π, the probability that someone in their town attended a protest in June 2020. In their survey of 40 residents, 30 attended a protest. Summarize Patrick's analysis using `summarize_beta_binomial()` and `plot_beta_binomial()`.

b) Harold has the same prior as Patrick, but lives in a different town. In their survey, 15 out of 20 people attended a protest. Summarize Harold's analysis using `summarize_beta_binomial()` and `plot_beta_binomial()`.

c) How do Patrick and Harold's posterior models compare? Briefly explain what causes these similarities and differences.

4

Balance and Sequentiality in Bayesian Analyses

In Alison Bechdel's 1985 comic strip *The Rule*, a character states that they only see a movie if it satisfies the following three rules (Bechdel, 1986):

- the movie has to have at least two women in it;
- these two women talk to each other; and
- they talk about something besides a man.

These criteria constitute the **Bechdel test** for the representation of women in film. Thinking of movies you've watched, what percentage of *all* recent movies do you think pass the Bechdel test? Is it closer to 10%, 50%, 80%, or 100%?

Let π, a random value between 0 and 1, denote the unknown proportion of recent movies that pass the Bechdel test. Three friends – the feminist, the clueless, and the optimist – have some prior ideas about π. Reflecting upon movies that he has seen in the past, the feminist understands that the majority lack strong women characters. The clueless doesn't really recall the movies they've seen, and so are unsure whether passing the Bechdel test is common or uncommon. Lastly, the optimist thinks that the Bechdel test is a *really* low bar for the representation of women in film, and thus assumes almost all movies pass the test. All of this to say that three friends have three different prior models of π. No problem! We saw in Chapter 3 that a Beta prior model for π can be *tuned* to match one's prior understanding (Figure 3.2). Check your intuition for Beta prior tuning in the quiz below.[1]

> ❷ **Quiz Yourself!**
> Match each Beta prior in Figure 4.1 to the corresponding analyst: the feminist, the clueless, and the optimist.

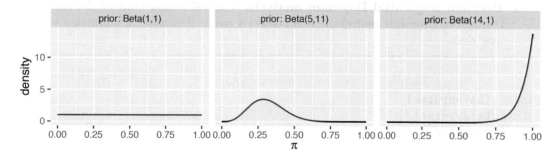

FIGURE 4.1: Three prior models for the proportion of films that pass the Bechdel test.

Placing the greatest prior plausibility on values of π that are less than 0.5, the Beta(5,11) prior reflects the feminist's understanding that the majority of movies fail the Bechdel test.

---

[1]Answer: Beta(1,1) = clueless prior. Beta(5,11) = feminist prior. Beta(14,1) = optimist prior.

DOI: 10.1201/9780429288340-4

In contrast, the Beta(14,1) places greater prior plausibility on values of π near 1, and thus matches the optimist's prior understanding. This leaves the Beta(1,1) or Unif(0,1) prior which, by placing equal plausibility on all values of π between 0 and 1, matches the clueless's figurative shoulder shrug – the only thing they know is that π is a proportion, and thus is *somewhere* between 0 and 1.

The three analysts agree to review a sample of n recent movies and record Y, the number that pass the Bechdel test. Recognizing Y as the number of "successes" in a fixed number of independent trials, they specify the dependence of Y on π using a Binomial model. Thus, each analyst has a unique Beta-Binomial model of π with differing prior hyperparameters α and β:

$$Y|\pi \sim \text{Bin}(n, \pi)$$
$$\pi \sim \text{Beta}(\alpha, \beta)^{\cdot}$$

By our work in Chapter 3, it follows that each analyst has a *unique* posterior model of π which depends upon their *unique* prior (through α and β) and the *common* observed data (through y and n)

$$\pi|(Y = y) \sim \text{Beta}(\alpha + y, \beta + n - y). \tag{4.1}$$

If you're thinking "Can *everyone* have their own prior?! Is this always going to be so subjective?!", you are asking the right questions! And the questions don't end there. To what *extent* might their different priors lead the analysts to three different posterior conclusions about the Bechdel test? How might this depend upon the sample size and outcomes of the movie data they collect? To what extent will the analysts' posterior understandings evolve as they collect more and more data? Will they *ever* come to agreement about the representation of women in film?! We will examine these fundamental questions throughout Chapter 4, continuing to build our capacity to think like Bayesians.

⊚ **Goals**

- **Explore the balanced influence of the prior and data on the posterior.** You will see how our choice of prior model, the features of our data, and the delicate balance between them can impact the posterior model.
- **Perform sequential Bayesian analysis.** You will explore one of the coolest features of Bayesian analysis: how a posterior model evolves as it's updated with new data.

```
# Load packages that will be used in this chapter
library(bayesrules)
library(tidyverse)
library(janitor)
```

4.1 Different priors, different posteriors

Reexamine Figure 4.1 which summarizes the prior models of π, the proportion of recent movies that pass the Bechdel test, tuned by the clueless, the feminist, and the optimist. Not only do the differing **prior means** reflect disagreement about whether π is closer to 0 or 1, the differing levels of **prior variability** reflect the fact that the analysts have different degrees of certainty in their prior information. Loosely speaking, the more **certain** the prior information, the *smaller* the prior variability. The more **vague** the prior information, the *greater* the prior variability. The priors of the optimist and the clueless represent these two extremes. With a Beta(14,1) prior which exhibits the smallest variability, the optimist is the most certain in their prior understanding of π (specifically, that almost all movies pass the Bechdel test). We refer to such priors as **informative**.

> **Informative prior**
>
> An **informative** prior reflects specific information about the unknown variable with high certainty, i.e., low variability.

With the largest prior variability, the clueless is the least certain about π. In fact, their Beta(1,1) prior assigns equal prior plausibility to each value of π between 0 and 1. This type of "shoulder shrug" prior model has an official name: it's a **vague prior**.

> **Vague prior**
>
> A **vague** or **diffuse** prior reflects little specific information about the unknown variable. A **flat** prior, which assigns equal prior plausibility to all possible values of the variable, is a special case.

The next natural question to ask is: how will their different priors influence the posterior conclusions of the feminist, the clueless, and the optimist? To answer this question, we need some data. Our analysts decide to review a random sample of $n = 20$ recent movies using data collected for the FiveThirtyEight article on the Bechdel test.[2] The `bayesrules` package includes a partial version of this dataset, named `bechdel`. A complete version is provided by the `fivethirtyeight` R package (Kim et al., 2020). Along with the `title` and `year` of each movie in this dataset, the `binary` variable records whether the film passed or failed the Bechdel test:

```
# Import data
data(bechdel, package = "bayesrules")

# Take a sample of 20 movies
set.seed(84735)
bechdel_20 <- bechdel %>%
  sample_n(20)
```

---

[2]https://fivethirtyeight.com/features/the-dollar-and-cents-case-against-hollywoods-exclusion-of-women/

```
bechdel_20 %>%
  head(3)
# A tibble: 3 x 3
   year title      binary
  <dbl> <chr>      <chr>
1  2005 King Kong  FAIL
2  1983 Flashdance PASS
3  2013 The Purge  FAIL
```

Among the 20 movies in this sample, only 9 (45%) passed the test:

```
bechdel_20 %>%
  tabyl(binary) %>%
  adorn_totals("row")
 binary  n percent
   FAIL 11    0.55
   PASS  9    0.45
  Total 20    1.00
```

Before going through any formal math, perform the following gut check of how you *expect* each analyst to react to this data. Answers are discussed below.

---

❷ Quiz Yourself!

The figure below displays our three analysts' unique priors along with the common scaled likelihood function which reflects the $Y = 9$ of $n = 20$ (45%) sampled movies that passed the Bechdel test. Whose posterior do you anticipate will look the *most* like the scaled likelihood? That is, whose posterior understanding of the Bechdel test pass rate will most agree with the observed 45% rate in the observed data? Whose do you anticipate will look the *least* like the scaled likelihood?

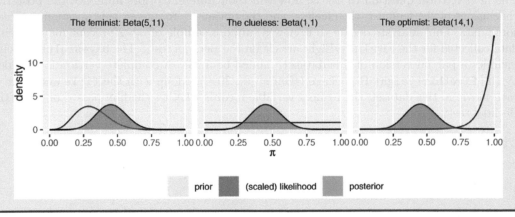

The three analysts' posterior models of π, which follow from applying (4.1) to their unique prior models and common movie data, are summarized in Table 4.1 and Figure 4.2. For example, the feminist's posterior parameters are calculated by $\alpha + y = 5 + 9 = 14$ and $\beta + n - y = 11 + 20 - 9 = 22$.

TABLE 4.1: The prior and posterior models for π, constructed in light of the data that $Y = 9$ of $n = 20$ sampled movies pass the Bechdel test.

Analyst	Prior	Posterior
the feminist	Beta(5,11)	Beta(14,22)
the clueless	Beta(1,1)	Beta(10,12)
the optimist	Beta(14,1)	Beta(23,12)

Were your instincts right? Recall that the optimist started with the most insistently optimistic prior about π – their prior model had a high mean with low variability. It's not very surprising then that *their* posterior model isn't as in sync with the data as the other analysts' posteriors. The dismal data in which only 45% of the 20 sampled movies passed the test wasn't enough to convince them that there's a problem in Hollywood – they still think that values of π above 0.5 are the most plausible. At the opposite extreme is the clueless who started with a flat, vague prior model of π. Absent any prior information, their posterior model directly reflects the insights gained from the observed movie data. In fact, their posterior is indistinguishable from the *scaled* likelihood function.

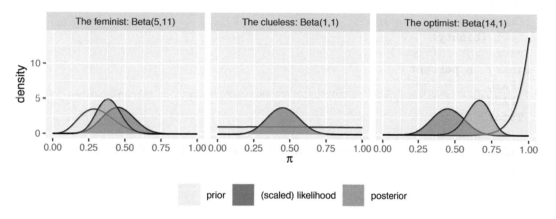

FIGURE 4.2: Posterior models of π, constructed in light of the sample in which $Y = 9$ of $n = 20$ movies passed the Bechdel.

---

❶ **Warning**

As a reminder, likelihood functions are *not* pdfs, and thus typically don't integrate to 1. As such, the clueless's actual (unscaled) likelihood is *not* equivalent to their posterior pdf. We're merely scaling the likelihood function here for simplifying the visual comparisons between the prior vs data evidence about π.

4.2   Different data, different posteriors

If you're concerned by the fact that our three analysts have differing posterior understandings
of π, the proportion of recent movies that pass the Bechdel, don't despair yet. Don't forget
the role that *data* plays in a Bayesian analysis. To examine these dynamics, consider three
new analysts – Morteza, Nadide, and Ursula – who all share the optimistic Beta(14,1) prior
for π *but* each have access to different data. Morteza reviews $n = 13$ movies from the year
1991, among which $Y = 6$ (about 46%) pass the Bechdel:

```
bechdel %>%
  filter(year == 1991) %>%
  tabyl(binary) %>%
  adorn_totals("row")
 binary  n percent
   FAIL  7  0.5385
   PASS  6  0.4615
  Total 13  1.0000
```

Nadide reviews $n = 63$ movies from 2000, among which $Y = 29$ (about 46%) pass the Bechdel:

```
bechdel %>%
  filter(year == 2000) %>%
  tabyl(binary) %>%
  adorn_totals("row")
 binary  n percent
   FAIL 34  0.5397
   PASS 29  0.4603
  Total 63  1.0000
```

Finally, Ursula reviews $n = 99$ movies from 2013, among which $Y = 46$ (about 46%) pass the
Bechdel:

```
bechdel %>%
  filter(year == 2013) %>%
  tabyl(binary) %>%
  adorn_totals("row")
 binary  n percent
   FAIL 53  0.5354
   PASS 46  0.4646
  Total 99  1.0000
```

What a coincidence! Though Morteza, Nadide, and Ursula have collected different data, each
observes a Bechdel pass rate of roughly 46%. Yet their sample sizes n differ – Morteza only
reviewed 13 movies whereas Ursula reviewed 99. Before doing any formal math, check your
intuition about how this different data will lead to different posteriors for the three analysts.
Answers are discussed below.

❓ Quiz Yourself!

The three analysts' common prior and unique Binomial likelihood functions (3.12), reflecting their different data, are displayed below. Whose posterior do you anticipate will be *most* in sync with their data, as visualized by the scaled likelihood? Whose posterior do you anticipate will be the *least* in sync with their data?

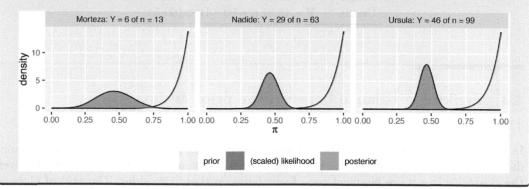

The three analysts' posterior models of π, which follow from applying (4.1) to their *common* Beta(14,1) prior model and *unique* movie data, are summarized in Figure 4.3 and Table 4.2. Was your intuition correct? First, notice that the larger the sample size n, the more "insistent" the likelihood function. For example, the likelihood function reflecting the 46% pass rate in Morteza's small sample of 13 movies is quite wide – his data are relatively plausible for any π between 15% and 75%. In contrast, reflecting the 46% pass rate in a much *larger* sample of 99 movies, Ursula's likelihood function is narrow – her data are implausible for π values outside the range from 35% to 55%. In turn, we see that the more insistent the likelihood, the more influence the data holds over the posterior. Morteza remains the least convinced by the low Bechdel pass rate observed in his small sample whereas Ursula is the most convinced. Her early prior optimism evolved into to a posterior understanding that π is likely only between 40% and 55%.

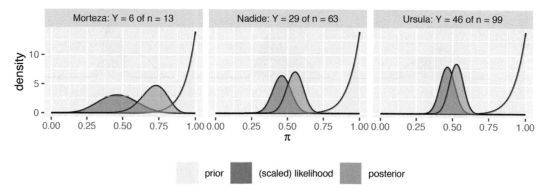

FIGURE 4.3: Posterior models of π, constructed from the same prior but different data, are plotted for each analyst.

TABLE 4.2: The prior and posterior models for π, constructed in light of a common Beta(14,1) prior and different data.

Analyst	Data	Posterior
Morteza	$Y = 6$ of $n = 13$	Beta(20,8)
Nadide	$Y = 29$ of $n = 63$	Beta(43,35)
Ursula	$Y = 46$ of $n = 99$	Beta(60,54)

4.3   Striking a balance between the prior & data

4.3.1   Connecting observations to concepts

In this chapter, we've observed the influence that different priors (Section 4.1) and different data (Section 4.2) can have on our posterior understanding of an unknown variable. However, the posterior is a more nuanced tug-of-war between these two sides. The grid of plots in Figure 4.4 illustrates the *balance* that the posterior model strikes between the prior and data. Each *row* corresponds to a unique prior model and each *column* to a unique set of data.

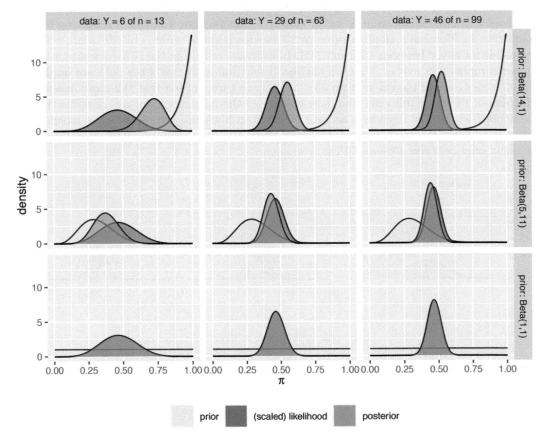

FIGURE 4.4: Posterior models of π constructed under different combinations of prior models and observed data.

Moving from left to right across the grid, the sample size increases from $n = 13$ to $n = 99$ movies while preserving the proportion of movies that pass the Bechdel test ($Y/n \approx 0.46$). The likelihood's insistence and, correspondingly, the data's influence over the posterior increase with sample size n. This also means that the influence of our prior understanding *diminishes* as we amass new data. Further, the *rate* at which the posterior balance tips in favor of the data depends upon the prior. Moving from top to bottom across the grid, the priors move from informative (Beta(14,1)) to vague (Beta(1,1)). Naturally, the more informative the prior, the greater its influence on the posterior.

Combining these observations, the last column in the grid delivers a very important Bayesian punchline: no matter the strength of and discrepancies among their prior understanding of π, three analysts will come to a common posterior understanding in light of strong data. This observation is a *relief*. If Bayesian models laughed in the face of more and more data, we'd have a problem.

Play around! To more deeply explore the roles the prior and data play in a posterior analysis, use the `plot_beta_binomial()` and `summarize_beta_binomial()` functions in the `bayesrules` package to visualize and summarize the Beta-Binomial posterior model of π under different combinations of Beta(α, β) prior models and observed data, Y successes in n trials:

```
# Plot the Beta-Binomial model
plot_beta_binomial(alpha = ___, beta = ___, y = ___, n = ___)

# Obtain numerical summaries of the Beta-Binomial model
summarize_beta_binomial(alpha = ___, beta = ___, y = ___, n = ___)
```

4.3.2 Connecting concepts to theory

The patterns we've observed in the posterior balance between the prior and data are intuitive. They're also supported by an elegant mathematical result. If you're interested in supporting your intuition with theory, read on. If you'd rather skip the technical details, you can continue on to Section 4.4 without major consequence.

Consider the general Beta-Binomial setting where π is the success rate of some event of interest with a Beta(α, β) prior. Then by (4.1), the posterior model of π upon observing $Y = y$ successes in n trials is Beta($\alpha + y, \beta + n - y$). It follows from (3.11) that the *central tendency* in our posterior understanding of π can be measured by the posterior mean,

$$E(\pi|Y = y) = \frac{\alpha + y}{\alpha + \beta + n}.$$

And with a little rearranging, we can isolate the influence of the prior and observed data on the posterior mean. The second step in this rearrangement might seem odd, but notice that we're just multiplying both fractions by 1 (e.g., n/n).

$$E(\pi|Y = y) = \frac{\alpha}{\alpha + \beta + n} + \frac{y}{\alpha + \beta + n}$$

$$= \frac{\alpha}{\alpha + \beta + n} \cdot \frac{\alpha + \beta}{\alpha + \beta} + \frac{y}{\alpha + \beta + n} \cdot \frac{n}{n}$$

$$= \frac{\alpha + \beta}{\alpha + \beta + n} \cdot \frac{\alpha}{\alpha + \beta} + \frac{n}{\alpha + \beta + n} \cdot \frac{y}{n}$$

$$= \frac{\alpha + \beta}{\alpha + \beta + n} \cdot E(\pi) + \frac{n}{\alpha + \beta + n} \cdot \frac{y}{n}.$$

We've now split the posterior mean into two pieces: a piece which depends upon the prior mean $E(\pi)$ (3.2) and a piece which depends upon the observed success rate in our sample trials, y/n. In fact, the posterior mean is a **weighted average** of the prior mean and sample success rate, their distinct weights summing to 1:

$$\frac{\alpha + \beta}{\alpha + \beta + n} + \frac{n}{\alpha + \beta + n} = 1.$$

For example, consider the posterior means for Morteza and Ursula, the settings for which are summarized in Table 4.2. With a shared Beta(14,1) prior for π, Morteza and Ursula share a prior mean of $E(\pi) = 14/15$. Yet their data differs. Morteza observed $Y = 6$ of $n = 13$ films pass the Bechdel test, and thus has a posterior mean of

$$E(\pi|Y = 6) = \frac{14 + 1}{14 + 1 + 13} \cdot E(\pi) + \frac{13}{14 + 1 + 13} \cdot \frac{y}{n}$$

$$= 0.5357 \cdot \frac{14}{15} + 0.4643 \cdot \frac{6}{13}$$

$$= 0.7143.$$

Ursula observed $Y = 46$ of $n = 99$ films pass the Bechdel test, and thus has a posterior mean of

$$E(\pi|Y = 46) = \frac{14 + 1}{14 + 1 + 99} \cdot E(\pi) + \frac{99}{14 + 1 + 99} \cdot \frac{y}{n}$$

$$= 0.1316 \cdot \frac{14}{15} + 0.8684 \cdot \frac{46}{99}$$

$$= 0.5263.$$

Again, though Morteza and Ursula have a common prior mean for π and observed similar Bechdel pass rates of roughly 46%, their posterior means differ due to their differing sample sizes n. Since Morteza observed only $n = 13$ films, his posterior mean put slightly more weight on the prior mean than on the observed Bechdel pass rate in his sample: 0.5357 vs 0.4643. In contrast, since Ursula observed a relatively large number of $n = 99$ films, her posterior mean put *much* less weight on the prior mean than on the observed Bechdel pass rate in her sample: 0.1316 vs 0.8684.

The implications of these results are mathemagical. In general, consider what happens to the posterior mean as we collect more and more data. As sample size n increases, the weight (hence influence) of the Beta(α, β) prior model approaches 0,

$$\frac{\alpha + \beta}{\alpha + \beta + n} \to 0 \quad \text{as } n \to \infty,$$

while the weight (hence influence) of the data approaches 1,

$$\frac{n}{\alpha + \beta + n} \to 1 \quad \text{as } n \to \infty.$$

Thus, the more data we have, the more the posterior mean will drift toward the trends exhibited in the data as opposed to the prior: as $n \to \infty$

$$E(\pi|Y = y) = \frac{\alpha + \beta}{\alpha + \beta + n} \cdot E(\pi) + \frac{n}{\alpha + \beta + n} \cdot \frac{y}{n} \quad \to \quad \frac{y}{n}.$$

The *rate* at which this drift occurs depends upon whether the prior tuning (i.e., α and β) is informative or vague. Thus, these mathematical results support the observations we made about the posterior's balance between the prior and data in Figure 4.4. And that's not all! In the exercises, you will show that we can write the posterior mode as the weighted average of the prior mode and observed sample success rate:

$$\text{Mode}(\pi|Y = y) = \frac{\alpha + \beta - 2}{\alpha + \beta + n - 2} \cdot \text{Mode}(\pi) + \frac{n}{\alpha + \beta + n - 2} \cdot \frac{y}{n}.$$

4.4 Sequential analysis: Evolving with data

In our discussions above, we examined the increasing influence of the data and diminishing influence of the prior on the posterior *as more and more data come in*. Consider the nuances of this concept. The phrase "as more and more data come in" evokes the idea that data collection, and thus the evolution in our posterior understanding, happens incrementally. For example, scientists' understanding of climate change has evolved over the span of *decades* as they gain new information. Presidential candidates' understanding of their chances of winning an election evolve over *months* as new poll results become available. Providing a formal framework for this evolution is one of the most powerful features of Bayesian statistics!

Let's revisit Milgram's behavioral study of obedience from Section 3.6. In this setting, π represents the proportion of people that will obey authority even if it means bringing harm to others. In Milgram's study, obeying authority meant delivering a severe electric shock to another participant (which, in fact, was a ruse). Prior to Milgram's experiments, our fictional psychologist expected that few people would obey authority in the face of harming another: $\pi \sim \text{Beta}(1, 10)$. They later observed that 26 of 40 study participants inflicted what they understood to be a severe shock.

Now, suppose that the psychologist collected this data *incrementally*, day by day, over a three-day period. Each day, they evaluated n subjects and recorded Y, the number that delivered the most severe shock (thus $Y|\pi \sim \text{Bin}(n, \pi)$). Among the $n = 10$ day-one participants, only $Y = 1$ delivered the most severe shock. Thus, by the end of day one, the psychologist's understanding of π had already evolved. It follows from (4.1) that[3]

$$\pi|(Y = 1) \sim \text{Beta}(2, 19).$$

Day two was much busier and the results grimmer: among $n = 20$ participants, $Y = 17$

---

[3]The posterior parameters are calculated by $\alpha + y = 1 + 1$ and $\beta + n - y = 10 + 10 - 1$.

delivered the most severe shock. Thus, by the end of day two, the psychologist's understanding of π had again evolved – π was likely larger than they had expected.

❓ Quiz Yourself!

What was the psychologist's posterior of π at the end of day two?

    a)   Beta(19,22)
    b)   Beta(18,13)

If your answer is "a," you are *correct!* On day two, the psychologist didn't simply forget what happened on day one and start afresh with the original Beta(1,10) prior. Rather, what they had learned by the end of day one, expressed by the Beta(2,19) posterior, provided a prior starting point on day two. Thus, by (4.1), the posterior model of π at the end of day two is Beta(19,22).[4] On day three, $Y = 8$ of $n = 10$ participants delivered the most severe shock, and thus their model of π evolved from a Beta(19,22) prior to a Beta(27,24) posterior.[5] The complete evolution from the psychologist's original Beta(1,10) prior to their Beta(27,24) posterior at the end of the three-day study is summarized in Table 4.3. Figure 4.5 displays this evolution in pictures, including the psychologist's big leap from day one to day two upon observing so many study participants deliver the most severe shock (17 of 20).

TABLE 4.3: A sequential Bayesian analysis of Milgram's data.

Day	Data	Model
0	NA	Beta(1,10)
1	$Y = 1$ of n = 10	Beta(2,19)
2	$Y = 17$ of n = 20	Beta(19,22)
3	$Y = 8$ of n = 10	Beta(27,24)

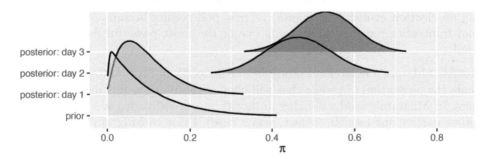

FIGURE 4.5: The sequential analysis of Milgram's data as summarized by Table 4.3.

The process we've just taken, incrementally updating the psychologist's posterior model of π, is referred to more generally as a **sequential Bayesian analysis** or **Bayesian learning**.

---

[4]The posterior parameters are calculated by $\alpha + y = 2 + 17$ and $\beta + n - y = 19 + 20 - 17$.
[5]The posterior parameters are calculated by $\alpha + y = 19 + 8$ and $\beta + n - y = 22 + 10 - 8$.

Sequential Bayesian analysis (aka Bayesian learning)

In a sequential Bayesian analysis, a posterior model is updated incrementally as more data come in. With each new piece of data, the previous posterior model reflecting our understanding prior to observing this data becomes the new prior model.

The ability to evolve as new data come in is one of the most powerful features of the Bayesian framework. These types of sequential analyses also uphold two fundamental and common sensical properties. First, the final posterior model is **data order invariant**, i.e., it isn't impacted by the *order* in which we observe the data. For example, suppose that the psychologist had observed Milgram's study data in the *reverse* order: $Y = 8$ of $n = 10$ on day one, $Y = 17$ of $n = 20$ on day two, and $Y = 1$ of $n = 10$ on day three. The resulting evolution in their understanding of π is summarized by Table 4.4 and Figure 4.6. In comparison to their analysis of the reverse data collection (Table 4.3), the psychologist's evolving understanding of π takes a different path. However, it still ends up in the same place – the Beta(27,24) posterior. These differing evolutions are highlighted by comparing Figure 4.6 to Figure 4.5.

TABLE 4.4: A sequential Bayesian analysis of Milgram's data, reversing the order in which the data was observed.

Day	Data	Model
0	NA	Beta(1,10)
1	$Y = 8$ of n = 10	Beta(9,12)
2	$Y = 17$ of n = 20	Beta(26,15)
3	$Y = 1$ of n = 10	Beta(27,24)

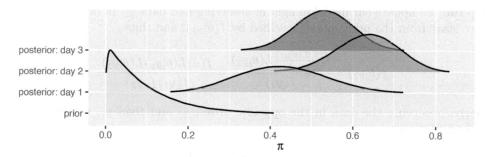

FIGURE 4.6: The sequential analysis of Milgram's data as summarized by Table 4.4.

The second fundamental feature of a sequential analysis is that the final posterior only depends upon the *cumulative* data. For example, in the *combined* three days of Milgram's experiment, there were $n = 10 + 20 + 10 = 40$ participants among whom $Y = 1 + 17 + 8 = 26$ delivered the most severe shock. In Section 3.6, we evaluated this data all at once, not incrementally. In doing so, we jumped straight from the psychologist's original Beta(1,10) prior model to the Beta(27,24) posterior model of π. That is, whether we evaluate the data incrementally or all in one go, we'll end up at the same place.

4.5   Proving data order invariance

In the previous section, you saw evidence of data order invariance in action. Here we'll *prove* that this feature is enjoyed by all Bayesian models. This section is fun but not a deal breaker to your future work.

Data order invariance

Let θ be any parameter of interest with prior pdf $f(\theta)$. Then a **sequential analysis** in which we *first* observe a data point y_1 and *then* a second data point y_2 will produce the same posterior model of θ as if we *first* observe y_2 and *then* y_1:

$$f(\theta|y_1, y_2) = f(\theta|y_2, y_1).$$

Similarly, the posterior model is invariant to whether we observe the data *all at once* or *sequentially*.

To prove the data order invariance property, let's first specify the structure of posterior pdf $f(\theta|y_1, y_2)$ which evolves by sequentially observing data y_1 followed by y_2. In step one of this evolution, we construct the posterior pdf from our original prior pdf, $f(\theta)$, and the likelihood function of θ given the *first* data point y_1, $L(\theta|y_1)$:

$$f(\theta|y_1) = \frac{\text{prior} \cdot \text{likelihood}}{\text{normalizing constant}} = \frac{f(\theta)L(\theta|y_1)}{f(y_1)}.$$

In step two, we update our model in light of observing *new* data y_2. In doing so, don't forget that we start from the prior model specified by $f(\theta|y_1)$, and thus

$$f(\theta|y_2) = \frac{\frac{f(\theta)L(\theta|y_1)}{f(y_1)}L(\theta|y_2)}{f(y_2)} = \frac{f(\theta)L(\theta|y_1)L(\theta|y_2)}{f(y_1)f(y_2)}.$$

Similarly, observing the data in the opposite order, y_2 and then y_1, would produce the equivalent posterior:

$$f(\theta|y_2, y_1) = \frac{f(\theta)L(\theta|y_2)L(\theta|y_1)}{f(y_2)f(y_1)}.$$

Finally, not only does the *order* of the data not influence the ultimate posterior model of θ, it doesn't matter whether we observe the data *all at once* or *sequentially*. To this end, suppose we start with the original $f(\theta)$ prior and observe data (y_1, y_2) together, not sequentially. Further, assume that these data points are unconditionally and conditionally **independent**, and thus

$$f(y_1, y_2) = f(y_1)f(y_2) \quad \text{and} \quad f(y_1, y_2|\theta) = f(y_1|\theta)f(y_2|\theta).$$

Then the posterior pdf resulting from this "data dump" is equivalent to that resulting from the sequential analyses above:

$$f(\theta|y_1, y_2) = \frac{f(\theta)L(\theta|y_1, y_2)}{f(y_1, y_2)}$$
$$= \frac{f(\theta)f(y_1, y_2|\theta)}{f(y_1)f(y_2)}$$
$$= \frac{f(\theta)L(\theta|y_1)L(\theta|y_2)}{f(y_1)f(y_2)}.$$

4.6 Don't be stubborn

Chapter 4 has highlighted some of the most compelling aspects of the Bayesian philosophy – it provides the framework and flexibility for our understanding to evolve over time. One of the only ways to *lose* this Bayesian benefit is by starting with an *extremely* stubborn prior model. A model so stubborn that it assigns a prior probability of *zero* to certain parameter values. Consider an example within the Milgram study setting where π is the proportion of people that will obey authority even if it means bringing harm to others. Suppose that a certain researcher has a stubborn belief in the good of humanity, insisting that π is equally likely to be anywhere between 0 and 0.25, and *surely* doesn't exceed 0.25. They express this prior understanding through a Uniform model on 0 to 0.25,

$$\pi \sim \text{Unif}(0, 0.25)$$

with pdf $f(\pi)$ exhibited in Figure 4.7 and specified by

$$f(\pi) = 4 \text{ for } \pi \in [0, 0.25].$$

Now, suppose this researcher was told that the first $Y = 8$ of $n = 10$ participants delivered the shock. This 80% figure runs counter to the stubborn researcher's belief. Check your intuition about how the researcher will update their posterior in light of this data.

> ❓**Quiz Yourself!**
>
> The stubborn researcher's prior pdf and likelihood function are illustrated in each plot of Figure 4.7. Which plot accurately depicts the researcher's corresponding posterior?

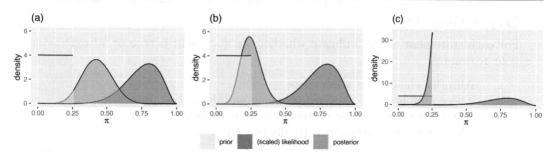

FIGURE 4.7: The stubborn researcher's prior and likelihood, with three potential corresponding posterior models.

As odd as it might seem, the posterior model in plot (c) corresponds to the stubborn researcher's updated understanding of π in light of the observed data. **A posterior model**

is defined on the same values for which the prior model is defined. That is, the *support* of the posterior model is inherited from the support of the prior model. Since the psychologist's prior model assigns zero probability to any value of π past 0.25, their posterior model must *also* assign zero probability to any value in that range. Mathematically, the posterior pdf $f(\pi|y = 8) = 0$ for any $\pi \notin [0, 0.25]$ and, for any $\pi \in [0, 0.25]$,

$$f(\pi|y = 8) \propto f(\pi)L(\pi|y = 8)$$
$$= 4 \cdot \binom{10}{8} \pi^8(1 - \pi)^2$$
$$\propto \pi^8(1 - \pi)^2.$$

The implications of this math are huge. No matter how much counterevidence the stubborn researcher collects, their posterior will *never* budge beyond the 0.25 cap, not even if they collect data on a billion subjects. Luckily, we have some good news for you: **this Bayesian bummer is completely preventable.**

Hot tip: How to avoid a regrettable prior model

Let π be some parameter of interest. No matter how much prior information you think you have about π or how informative you want to make your prior, be sure to assign non-0 plausibility to every *possible* value of π, even if this plausibility is *near* 0. For example, if π is a *proportion* which can technically range from 0 to 1, then your prior model should also be defined across this continuum.

4.7   A note on subjectivity

In Chapter 1, we alluded to a common critique about Bayesian statistics – it's too *subjective*. Specifically, some worry that "subjectively" tuning a prior model allows a Bayesian analyst to come to any conclusion that they want to. We can more rigorously push back against this critique in light of what we've learned in Chapter 4. Before we do, reconnect to and expand upon some concepts that you've explored throughout the book.

❓ Quiz Yourself!

For each statement below, indicate whether the statement is true or false. Provide your reasoning.

1.   All prior choices are informative.
2.   There may be good reasons for having an informative prior.
3.   Any prior choice can be overcome by enough data.
4.   The frequentist paradigm is totally objective.

Answers are provided in the footnotes.[6] Consider the main points. Throughout Chapter 4, you've confirmed that a Bayesian *can* indeed build a prior based on "subjective" experience.

---

[6] 1. False. Vague priors are typically uninformative. 2. True. We might have ample previous data or expertise from which to build our prior. 3. False. If you assign zero prior probability to a potential parameter value, no amount of data can change that! 4. False. Subjectivity always creeps in to both frequentist

Very seldom is this a bad thing, and quite often it's a great thing! In the best-case scenarios, a subjective prior can reflect a wealth of past experiences that *should* be incorporated into our analysis – it would be unfortunate *not* to. Even if a subjective prior runs counter to actual observed evidence, its influence over the posterior fades away as this evidence piles up. We've seen one worst-case scenario exception. And it was *preventable*. If a subjective prior is stubborn enough to assign zero probability on a possible parameter value, no amount of counterevidence will be enough to budge it.

Finally, though we encourage you to be critical in your application of Bayesian methods, please don't worry about them being any more subjective than frequentist methods. *No* human is capable of removing all subjectivity from an analysis. The life experiences and knowledge we carry with us inform everything from what research questions we ask to what data we collect. It's important to consider the potential implications of this subjectivity in *both* Bayesian and frequentist analyses.

4.8   Chapter summary

In Chapter 4 we explored the balance that a posterior model strikes between a prior model and the data. In general, we saw the following trends:

- **Prior influence**
  The less **vague** and more **informative** the prior, i.e., the greater our prior certainty, the more influence the prior has over the posterior.

- **Data influence**
  The more data we have, the more influence the data has over the posterior. Thus, if they have *ample* data, two researchers with different priors will have similar posteriors.

Further, we saw that in a **sequential Bayesian analysis**, we incrementally update our posterior model as more and more data come in. The final destination of this posterior is not impacted by the order in which we observe this data (i.e., the posterior is **data order invariant**) or whether we observe the data in one big dump or incrementally.

---

and Bayesian analyses. With the Bayesian paradigm, we can at least name and quantify aspects of this subjectivity.

4.9   Exercises

4.9.1   Review exercises

Exercise 4.1 (Match the prior to the description). Five different prior models for π are listed below. Label each with one of these descriptors: somewhat favoring $\pi < 0.5$, strongly favoring $\pi < 0.5$, centering π on 0.5, somewhat favoring $\pi > 0.5$, strongly favoring $\pi > 0.5$.

a)   Beta(1.8,1.8)
b)   Beta(3,2)
c)   Beta(1,10)
d)   Beta(1,3)
e)   Beta(17,2)

Exercise 4.2 (Match the plot to the code). Which arguments to the `plot_beta_binomial()` function generated the plot below?

a)   `alpha = 2, beta = 2, y = 8, n = 11`
b)   `alpha = 2, beta = 2, y = 3, n = 11`
c)   `alpha = 3, beta = 8, y = 2, n = 6`
d)   `alpha = 3, beta = 8, y = 4, n = 6`
e)   `alpha = 3, beta = 8, y = 2, n = 4`
f)   `alpha = 8, beta = 3, y = 2, n = 4`

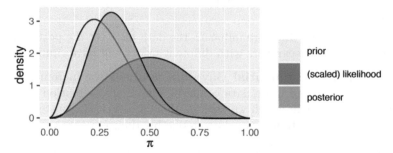

Exercise 4.3 (Choice of prior: gingko tree leaf drop). A ginkgo tree can grow into a majestic monument to the wonders of the natural world. One of the most notable things about ginkgo trees is that they shed all of their leaves at the same time, usually after the first frost. Randi thinks that the ginkgo tree in her local arboretum will drop all of its leaves next Monday. She asks 5 of her friends what they think about the probability (π) that this will happen. Identify some reasonable Beta priors to convey each of these beliefs.

a)   Ben says that it is really unlikely.
b)   Albert says that he is quite unsure and hates trees. He has no idea.
c)   Katie gives it some thought and, based on what happened last year, thinks that there is a very high chance.
d)   Daryl thinks that there is a decent chance, but he is somewhat unsure.
e)   Scott thinks it probably won't happen, but he's somewhat unsure.

4.9.2 Practice: Different priors, different posteriors

For all exercises in this section, consider the following story. The local ice cream shop is open until it runs out of ice cream for the day. It's 2 p.m. and Chad wants to pick up an ice cream cone. He asks his coworkers about the chance (π) that the shop is still open. Their Beta priors for π are below:

coworker	prior
Kimya	Beta(1, 2)
Fernando	Beta(0.5, 1)
Ciara	Beta(3, 10)
Taylor	Beta(2, 0.1)

Exercise 4.4 (Choice of prior). Visualize and summarize (in words) each coworker's prior understanding of Chad's chances to satisfy his ice cream craving.

Exercise 4.5 (Simulating the posterior). Chad peruses the shop's website. On 3 of the past 7 days, they were still open at 2 p.m.. Complete the following for each of Chad's coworkers:

- *simulate* their posterior model;
- create a histogram for the simulated posterior; and
- use the simulation to *approximate* the posterior mean value of π.

Exercise 4.6 (Identifying the posterior). Complete the following for each of Chad's coworkers:

- identify the *exact* posterior model of π;
- calculate the *exact* posterior mean of π; and
- compare these to the simulation results in the previous exercise.

4.9.3 Practice: Balancing the data & prior

Exercise 4.7 (What dominates the posterior?). In each situation below you will be given a Beta prior for π and some Binomial trial data. For each scenario, identify which of the following is true: the prior has more influence on the posterior, the data has more influence on the posterior, or the posterior is an equal compromise between the data and the prior.

a) Prior: $\pi \sim \text{Beta}(1, 4)$, data: $Y = 8$ successes in $n = 10$ trials
b) Prior: $\pi \sim \text{Beta}(20, 3)$, data: $Y = 0$ successes in $n = 1$ trial
c) Prior: $\pi \sim \text{Beta}(4, 2)$, data: $Y = 1$ success in $n = 3$ trials
d) Prior: $\pi \sim \text{Beta}(3, 10)$, data: $Y = 10$ successes in $n = 13$ trials
e) Prior: $\pi \sim \text{Beta}(20, 2)$, data: $Y = 10$ successes in $n = 200$ trials

Exercise 4.8 (Visualizing the evolution). For each scenario in Exercise 4.7, plot and compare the prior pdf, scaled likelihood function, and posterior pdf for π.

Exercise 4.9 (Different data: more or less sure). Let π denote the proportion of people that prefer dogs to cats. Suppose you express your prior understanding of π by a Beta(7, 2) model.

a) According to your prior, what are reasonable values for π?
b) If you observe a survey in which $Y = 19$ of $n = 20$ people prefer dogs, how would that change your understanding of π? Comment on both the evolution in your mean understanding and your level of certainty about π.
c) If instead, you observe that only $Y = 1$ of $n = 20$ people prefer dogs, how would that change your understanding about π?
d) If instead, you observe that $Y = 10$ of $n = 20$ people prefer dogs, how would that change your understanding about π?

Exercise 4.10 (What was the data?). In each situation below we give you a Beta prior and a Beta posterior. Further, we tell you that the data is Binomial, but we *don't* tell you the observed number of trials n or successes y in those trials. For each situation, identify n and y, and then utilize `plot_beta_binomial()` to sketch the prior pdf, scaled likelihood function, and posterior pdf.

a) Prior: Beta(0.5, 0.5), Posterior: Beta(8.5, 2.5)
b) Prior: Beta(0.5, 0.5), Posterior: Beta(3.5, 10.5)
c) Prior: Beta(10, 1), Posterior: Beta(12, 15)
d) Prior: Beta(8, 3), Posterior: Beta(15, 6)
e) Prior: Beta(2, 2), Posterior: Beta(5, 5)
f) Prior: Beta(1, 1), Posterior: Beta(30, 3)

Exercise 4.11 (Different data, uninformative prior). In each situation below we have the same prior on the probability of a success, $\pi \sim \text{Beta}(1,1)$, *but different data*. Identify the corresponding posterior model and utilize `plot_beta_binomial()` to sketch the prior pdf, likelihood function, and posterior pdf.

a) $Y = 10$ in $n = 13$ trials
b) $Y = 0$ in $n = 1$ trial
c) $Y = 100$ in $n = 130$ trials
d) $Y = 20$ in $n = 120$ trials
e) $Y = 234$ in $n = 468$ trials

Exercise 4.12 (Different data, informative prior). Repeat Exercise 4.11, this time assuming a $\pi \sim \text{Beta}(10,2)$ prior.

Exercise 4.13 (Bayesian bummer). Bayesian methods are great! But, like anything, we can screw it up. Suppose a politician specifies their prior understanding about their approval rating, π, by: $\pi \sim \text{Unif}(0.5,1)$ with pdf $f(\pi) = 2$ when $0.5 \leq \pi < 1$, and $f(\pi) = 0$ when $0 < \pi < 0.5$.

a) Sketch the prior pdf (by hand).
b) Describe the politician's prior understanding of π.
c) The politician's aides show them a poll in which 0 of 100 people approve of their job performance. Construct a formula for and sketch the politician's posterior pdf of π.

d) Describe the politician's posterior understanding of π. Use this to explain the mistake the politician made in specifying their prior.

Exercise 4.14 (Challenge: posterior mode).

a) In the Beta-Binomial setting, show that we can write the posterior mode of π as the weighted average of the prior mode and observed sample success rate:

$$\text{Mode}(\pi|Y = y) = \frac{\alpha + \beta - 2}{\alpha + \beta + n - 2} \cdot \text{Mode}(\pi) + \frac{n}{\alpha + \beta + n - 2} \cdot \frac{y}{n}.$$

b) To what value does the posterior mode converge as our sample size n increases? Support your answer with evidence.

4.9.4 Practice: Sequentiality

Exercise 4.15 (One at a time). Let π be the probability of success for some event of interest. You place a Beta(2, 3) prior on π, and are really impatient. Sequentially update your posterior for π with each new observation below.

a) First observation: Success
b) Second observation: Success
c) Third observation: Failure
d) Fourth observation: Success

Exercise 4.16 (Five at a time). Let π be the probability of success for some event of interest. You place a Beta(2, 3) prior on π, and are impatient, but you have been working on that aspect of your personality. So you sequentially update your posterior model of π after every five (!) new observations. For each set of five new observations, report the updated posterior model for π.

a) First set of observations: 3 successes
b) Second set of observations: 1 success
c) Third set of observations: 1 success
d) Fourth set of observations: 2 successes

Exercise 4.17 (Different data, different posteriors). A shoe company develops a new internet ad for their latest sneaker. Three employees share the same Beta(4, 3) prior model for π, the probability that a user will click on the ad when shown. However, the employees run three different studies, thus each has access to different data. The first employee tests the ad on 1 person – they do not click on the ad. The second tests 10 people, 3 of whom click on the ad. The third tests 100 people, 20 of whom click on the ad.

a) Sketch the prior pdf using `plot_beta()`. Describe the employees' prior understanding of the chance that a user will click on the ad.
b) Specify the unique posterior model of π for each of the three employees. We encourage you to construct these posteriors "from scratch", i.e., without relying on the Beta-Binomial posterior formula.
c) Plot the prior pdf, likelihood function, and posterior pdf for each employee.
d) Summarize and compare the employees' posterior models of π.

Exercise 4.18 (A sequential employee). The shoe company described in Exercise 4.17 brings in a fourth employee. They start with the same Beta(4, 3) prior for π as the first three employees but, not wanting to re-create work, don't collect their own data. Instead, in their first day on the job, the new employee convinces the first employee to share their data. On the second day they get access to the second employee's data and on the third day they get access to the third employee's data.

a)   Suppose the new employee updates their posterior model of π at the end of each day. What's their posterior at the end of day one? At the end of day two? At the end of day three?

b)   Sketch the new employee's prior and three (sequential) posteriors. In words, describe how their understanding of π evolved over their first three days on the job.

c)   Suppose instead that the new employee didn't update their posterior until the end of their third day on the job, after they'd gotten data from all three of the other employees. Specify their posterior model of π and compare this to the day three posterior from part (a).

Exercise 4.19 (Bechdel test). In this exercise we'll analyze π, the proportion of films that pass the Bechdel test, using the `bechdel` data. For each scenario below, specify the posterior model of π, and calculate the posterior mean and mode.

a)   John has a flat Beta(1, 1) prior and analyzes movies from the year 1980.

b)   The next day, John analyzes movies from the year 1990, while building off their analysis from the previous day.

c)   The third day, John analyzes movies from the year 2000, while again building off of their analyses from the previous two days.

d)   Jenna also starts her analysis with a Beta(1, 1) prior, but analyzes movies from 1980, 1990, 2000 all on day one.

Exercise 4.20 (Bayesian and frequentist: sequential edition). You learned in this chapter that we can use Bayes to sequentially update our understanding of a parameter of interest. How is this different from what the frequentist approach would be? How is it similar?

5

Conjugate Families

In the novel *Anna Karenina*, Tolstoy wrote "Happy families are all alike; every unhappy family is unhappy in its own way." In this chapter we will learn about conjugate families, which are all alike in the sense that they make the authors very happy. Read on to learn why.

◎ **Goals**

- **Practice building Bayesian models.** You will build Bayesian models by practicing how to recognize kernels and make use of proportionality.
- **Familiarize yourself with conjugacy.** You will learn about what makes a prior *conjugate* and why this is a helpful property. In brief, conjugate priors make it easier to build posterior models. Conjugate priors spark joy!

Getting started

To prepare for this chapter, note that we'll be using some new Greek letters throughout our analysis: λ = lambda, μ = mu or "mew", σ = sigma, τ = tau, and θ = theta. Further, load the packages below.

```
library(bayesrules)
library(tidyverse)
```

5.1 Revisiting choice of prior

How do we choose a prior? In Chapters 3 and 4 we used the *flexibility* of the Beta model to reflect our prior understanding of a proportion parameter $\pi \in [0, 1]$. There are other criteria to consider when choosing a prior model:

- **Computational ease**
  Especially if we don't have access to computing power, it is helpful if the posterior model is easy to build.
- **Interpretability**
  We've seen that posterior models are a compromise between the data and the prior model. A posterior model is interpretable, and thus more useful, when you can look at its formulation and *identify* the contribution of the data relative to that of the prior.

The Beta-Binomial has both of these criteria covered. Its calculation is easy. Once we know the Beta(α, β) prior hyperparameters and the observed data $Y = y$ for the Bin(n, π) model, the Beta($\alpha + y, \beta + n - y$) posterior model follows. This posterior reflects the influence of the

DOI: 10.1201/9780429288340-5

data, through the values of y and n, relative to the prior *hyperparameters* α and β. If α and β are large relative to the sample size n, then the posterior will not budge that much from the prior. However, if the sample size n is large relative to α and β, then the data will take over and be more influential in the posterior model. In fact, the Beta-Binomial belongs to a larger class of prior-data combinations called **conjugate families** that enjoy both computational ease and interpretable posteriors. Recall a general definition similar to that from Chapter 3.

Conjugate prior

Let the prior model for parameter θ have pdf $f(\theta)$ and the model of data Y conditioned on θ have likelihood function $L(\theta|y)$. If the resulting posterior model with pdf $f(\theta|y) \propto f(\theta)L(\theta|y)$ is of the same model family as the prior, then we say this is a **conjugate prior**.

To emphasize the utility (and fun!) of conjugate priors, it can be helpful to consider a *non*-conjugate prior. Let parameter π be a proportion between 0 and 1 and suppose we plan to collect data Y where, conditional on π, $Y|\pi \sim \text{Bin}(n, \pi)$. Instead of our conjugate Beta(α, β) prior for π, let's try out a **non-conjugate** prior with pdf $f(\pi)$, plotted in Figure 5.1:

$$f(\pi) = e - e^{\pi} \quad \text{for } \pi \in [0, 1]. \tag{5.1}$$

Though *not* a Beta pdf, $f(\pi)$ is indeed a *valid* pdf since $f(\pi)$ is non-negative on the support of π and the area under the pdf is 1, i.e., $\int_0^1 f(\pi) = 1$.

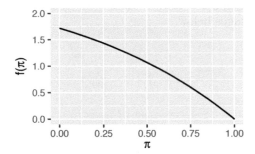

FIGURE 5.1: A non-conjugate prior for π.

Next, suppose we observe $Y = 10$ successes from $n = 50$ independent trials, all having the same probability of success π. The resulting Binomial likelihood function of π is:

$$L(\pi|y = 10) = \binom{50}{10} \pi^{10}(1 - \pi)^{40} \quad \text{for } \pi \in [0, 1].$$

Recall from Chapter 3 that, when we put the prior and the likelihood together, we are already on our path to finding the posterior model with pdf

$$f(\pi|y = 10) \propto f(\pi)L(\pi|y = 10) = (e - e^{\pi}) \cdot \binom{50}{10} \pi^{10}(1 - \pi)^{40}.$$

As we did in Chapter 3, we will drop all constants that do not depend on π since we are only specifying $f(\pi|y)$ up to a *proportionality constant*:

$$f(\pi|y) \propto (e - e^{\pi})\pi^{10}(1 - \pi)^{40}.$$

Notice here that our non-Beta prior didn't produce a neat and clean answer for the exact posterior model (fully specified and not up to a proportionality constant). We cannot squeeze this posterior pdf kernel into a Beta box or any other familiar model for that matter. That is, we *cannot* rewrite $(e - e^{\pi})\pi^{10}(1 - \pi)^{40}$ so that it shares the same structure as a Beta kernel, $\pi^{\blacksquare-1}(1 - \pi)^{\blacksquare-1}$. Instead we will need to integrate this kernel in order to complete the normalizing constant, and hence posterior specification:

$$f(\pi|y = 10) = \frac{(e - e^{\pi})\pi^{10}(1 - \pi)^{40}}{\int_0^1 (e - e^{\pi})\pi^{10}(1 - \pi)^{40}d\pi} \quad \text{for } \pi \in [0, 1]. \tag{5.2}$$

This is where we really start to feel the pain of not having a conjugate prior! Since this is a particularly unpleasant integral to evaluate, and we've been trying to avoid doing any integration altogether, we will leave ourselves with (5.2) as the final posterior. Yikes, what a mess. This *is* a valid posterior pdf – it's non-negative and integrates to 1 across $\pi \in [0, 1]$. But it doesn't have much else going for it. Consider a few characteristics about the posterior model that result from this particular non-conjugate prior (5.1):

- The calculation for this posterior was messy and unpleasant.
- It is difficult to derive any intuition about the balance between the prior information and the data we observed from this posterior model.
- It would be difficult to specify features such as the posterior mean, mode, and standard deviation, a process which would require even more integration.

This leaves us with the question: could we use the conjugate Beta prior and still capture the broader information of the messy non-conjugate prior (5.1)? If so, then we solve the problems of messy calculations and indecipherable posterior models.

❓ Quiz Yourself!

Which Beta model would most closely approximate the non-conjugate prior for π (Figure 5.1)?

    a.   Beta(3,1)
    b.   Beta(1,3)
    c.   Beta(2,1)
    d.   Beta(1,2)

The answer to this quiz is d. One way to find the answer is by comparing the non-conjugate prior in Figure 5.1 with plots of the possible Beta prior models. For example, the non-conjugate prior information is pretty well captured by the Beta(1,2) (Figure 5.2):

```
plot_beta(alpha = 1, beta = 2)
```

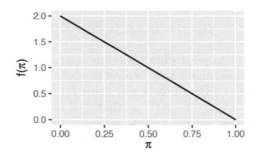

FIGURE 5.2: A conjugate Beta(1,2) prior model for π.

5.2 Gamma-Poisson conjugate family

> Stop callin' / Stop callin' / I don't wanna think anymore / I got my head and my
> heart on the dance floor
> —Lady Gaga featuring Beyoncé. Lyrics from the song "Telephone."

Last year, one of this book's authors got fed up with the number of fraud risk phone calls
they were receiving. They set out with a goal of modeling *rate* λ, the typical number of
fraud risk calls received per day. **Prior** to collecting any data, the author's guess was that
this rate was most likely around 5 calls per day, but could also reasonably range between 2
and 7 calls per day. To learn more, they planned to record the number of fraud risk phone
calls on each of n sampled days, $(Y_1, Y_2, \ldots, Y_n)$.

In moving forward with our investigation of λ, it's important to recognize that it will *not*
fit into the familiar Beta-Binomial framework. First, λ is *not* a proportion limited to be
between 0 and 1, but rather a rate parameter that can take on any *positive* value (e.g., we
can't receive -7 calls per day). Thus, a Beta prior for λ won't work. Further, each data point
Y_i is a *count* that can technically take on any non-negative integer in $\{0, 1, 2, \ldots\}$, and thus
is *not* limited by some number of trials n as is true for the Binomial. Not to fret. Our study
of λ will introduce us to a *new* conjugate family, the **Gamma-Poisson**.

5.2.1 The Poisson data model

The *spirit* of our analysis starts with a prior understanding of λ, the daily rate of fraud risk
phone calls. Yet before choosing a prior model structure and tuning this to match our prior
understanding, it's beneficial to identify a model for the dependence of our daily phone call
count data Y_i on the typical daily *rate* of such calls λ. Upon identifying a reasonable data
model, we can identify a prior model which can be tuned to match our prior understanding
while *also* mathematically complementing the data model's corresponding likelihood function.
Keeping in mind that each data point Y_i is a *random count* that can go from 0 to a really big
number, $Y \in \{0, 1, 2, \ldots\}$, the **Poisson model**, described in its general form below, makes a
reasonable candidate for modeling this data.

The Poisson model

Let discrete random variable Y be the *number of independent events* that occur in a fixed amount of time or space, where $\lambda > 0$ is the rate at which these events occur. Then the *dependence* of Y on **parameter** λ can be modeled by the Poisson. In mathematical notation:

$$Y|\lambda \sim \text{Pois}(\lambda).$$

Correspondingly, the Poisson model is specified by pmf

$$f(y|\lambda) = \frac{\lambda^y e^{-\lambda}}{y!} \quad \text{for } y \in \{0, 1, 2, \ldots\} \tag{5.3}$$

where $f(y|\lambda)$ sums to one across y, $\sum_{y=0}^{\infty} f(y|\lambda) = 1$. Further, a Poisson random variable Y has equal mean and variance,

$$E(Y|\lambda) = \text{Var}(Y|\lambda) = \lambda. \tag{5.4}$$

Figure 5.3 illustrates the Poisson pmf (5.3) under different rate parameters λ. In general, as the rate of events λ increases, the typical number of events increases, the variability increases, and the skew decreases. For example, when events occur at a rate of $\lambda = 1$, the model is heavily skewed toward observing a small number of events – we're most likely to observe 0 or 1 events, and rarely more than 3. In contrast, when events occur at a higher rate of $\lambda = 5$, the model is roughly symmetric and more variable – we're most likely to observe 4 or 5 events, though have a reasonable chance of observing anywhere between 1 and 10 events.

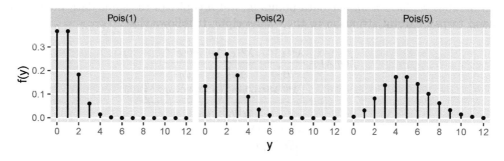

FIGURE 5.3: Poisson pmfs with different rate parameters.

Let $(Y_1, Y_2, \ldots, Y_n)$ denote the number of fraud risk calls we observed on each of the n days in our data collection period. We assume that the daily number of calls might differ from day to day and can be **independently modeled** by the Poisson. Thus, on each day i,

$$Y_i|\lambda \stackrel{ind}{\sim} \text{Pois}(\lambda)$$

with unique pmf

$$f(y_i|\lambda) = \frac{\lambda^{y_i} e^{-\lambda}}{y_i!} \quad \text{for } y_i \in \{0, 1, 2, \ldots\}.$$

Yet in weighing the evidence of the phone call data, we won't want to analyze each *individual* day. Rather, we'll need to process the *collective* or *joint* information in our n data points. This information is captured by the **joint probability mass function**.

Joint probability mass function

Let $(Y_1, Y_2, \ldots, Y_n)$ be an independent sample of random variables and $\vec{y} = (y_1, y_2, \ldots, y_n)$ be the corresponding vector of observed values. Further, let $f(y_i|\lambda)$ denote the pmf of an individual observed data point $Y_i = y_i$. Then by the assumption of independence, the following **joint pmf** specifies the randomness in and plausibility of the collective sample:

$$f(\vec{y}|\lambda) = \prod_{i=1}^{n} f(y_i|\lambda) = f(y_1|\lambda) \cdot f(y_2|\lambda) \cdot \cdots \cdot f(y_n|\lambda). \qquad (5.5)$$

Connecting concepts

This product is analogous to the joint probability of *independent events* being the product of the marginal probabilities, $P(A \cap B) = P(A)P(B)$.

The joint pmf for our fraud risk call sample follows by applying the general definition (5.5) to our Poisson pmfs. Letting the number of calls on each day i be $y_i \in \{0, 1, 2, \ldots\}$,

$$f(\vec{y}|\lambda) = \prod_{i=1}^{n} f(y_i|\lambda) = \prod_{i=1}^{n} \frac{\lambda^{y_i} e^{-\lambda}}{y_i!}. \qquad (5.6)$$

This looks like a mess, but it can be simplified. In this simplification, it's important to recognize that we have n *unique* data points y_i, *not* n copies of the same data point y. Thus, we need to pay careful attention to the i subscripts. It follows that

$$
\begin{aligned}
f(\vec{y}|\lambda) &= \frac{\lambda^{y_1} e^{-\lambda}}{y_1!} \cdot \frac{\lambda^{y_2} e^{-\lambda}}{y_2!} \cdots \frac{\lambda^{y_n} e^{-\lambda}}{y_n!} \\
&= \frac{\left[\lambda^{y_1} \lambda^{y_2} \cdots \lambda^{y_n}\right]\left[e^{-\lambda} e^{-\lambda} \cdots e^{-\lambda}\right]}{y_1! y_2! \cdots y_n!} \\
&= \frac{\lambda^{\sum y_i} e^{-n\lambda}}{\prod_{i=1}^{n} y_i!}
\end{aligned}
$$

where we've simplified the products in the final line by appealing to the properties below.

Simplifying products

Let (x, y, a, b) be a set of constants. Then we can utilize the following facts when simplifying products involving exponents:

$$x^a x^b = x^{a+b} \quad \text{and} \quad x^a y^a = (xy)^a \qquad (5.7)$$

Once we observe actual sample data, we can flip this joint pmf on its head to define the **likelihood function** of λ. The Poisson likelihood function is equivalent in formula to the joint pmf $f(\vec{y}|\lambda)$, yet is a function of λ which helps us assess the compatibility of different possible λ values with our observed *collection* of sample data $\vec{y}$:

$$L(\lambda|\vec{y}) = \frac{\lambda^{\sum y_i} e^{-n\lambda}}{\prod_{i=1}^{n} y_i!} \propto \lambda^{\sum y_i} e^{-n\lambda} \quad \text{for } \lambda > 0. \tag{5.8}$$

It is convenient to represent the likelihood function up to a proportionality constant here, especially since $\prod y_i!$ will be cumbersome to calculate when n is large, and what we really care about in the likelihood is λ. And when we express the likelihood up to a proportionality constant, note that the sum of the data points ($\sum y_i$) and the number of data points (n) is all the information that is required from the data. We don't need to know the value of each individual data point y_i. Taking this for a spin with real data points later in our analysis will provide some clarity.

5.2.2   Potential priors

The Poisson data model provides one of two key pieces for our Bayesian analysis of λ, the daily rate of fraud risk calls. The other key piece is a *prior model* for λ. Our original guess was that this rate is most likely around 5 calls per day, but could also reasonably range between 2 and 7 calls per day. In order to *tune* a prior to match these ideas about λ, we first have to identify a reasonable probability model structure.

Remember here that λ is a *positive* and *continuous* rate, meaning that λ does not have to be a whole number. Accordingly, a reasonable prior probability model will also have continuous and positive support, i.e., be defined on $\lambda > 0$. There are several named and studied probability models with this property, including the F, Weibull, and Gamma. We don't dig into all of these in this book. Rather, to make the λ posterior model construction more straightforward and convenient, we'll focus on identifying a **conjugate prior** model.

---

❷ Quiz Yourself!

Suppose we have a random sample of Poisson random variables $(Y_1, Y_2, \ldots, Y_n)$ with likelihood function $L(\lambda|\vec{y}) \propto \lambda^{\sum y_i} e^{-n\lambda}$ for $\lambda > 0$. What do you *think* would provide a convenient *conjugate* prior model for λ? Why?

   a.   A "Gamma" model with pdf $f(\lambda) \propto \lambda^{s-1} e^{-r\lambda}$
   b.   A "Weibull" model with pdf $f(\lambda) \propto \lambda^{s-1} e^{(-r\lambda)^s}$
   c.   A special case of the "F" model with pdf $f(\lambda) \propto \lambda^{\frac{s}{2}-1}(1+\lambda)^{-s}$

The answer is a. The Gamma model will provide a *conjugate* prior for λ when our data has a Poisson model. You might have guessed this from the section title (clever). You might also have guessed this from the shared features of the Poisson likelihood function $L(\lambda|\vec{y})$ (5.8) and the Gamma pdf $f(\lambda)$. Both are proportional to

$$\lambda^{\blacksquare} e^{-\blacksquare\lambda}$$

with differing $\blacksquare$. In fact, we'll prove that *combining* the prior and likelihood produces a posterior pdf with this same structure. That is, the posterior will be of the same Gamma model family as the prior. First, let's learn more about the Gamma model.

5.2.3   Gamma prior

The Gamma & Exponential models

Let λ be a continuous random variable which can take any positive value, i.e., $\lambda > 0$. Then the variability in λ might be well modeled by a Gamma model with **shape hyperparameter** $s > 0$ and **rate hyperparameter** $r > 0$:

$$\lambda \sim \text{Gamma}(s, r).$$

The Gamma model is specified by continuous pdf

$$f(\lambda) = \frac{r^s}{\Gamma(s)} \lambda^{s-1} e^{-r\lambda} \quad \text{for } \lambda > 0. \tag{5.9}$$

Further, the central tendency and variability in λ are measured by:

$$E(\lambda) = \frac{s}{r}$$

$$\text{Mode}(\lambda) = \frac{s-1}{r} \quad \text{for } s \geq 1 \tag{5.10}$$

$$\text{Var}(\lambda) = \frac{s}{r^2}.$$

The **Exponential model** is a special case of the Gamma with shape $s = 1$, Gamma$(1, r)$:

$$\lambda \sim \text{Exp}(r).$$

Notice that the Gamma model depends upon two hyperparameters, r and s. Assess your understanding of how these hyperparameters impact the Gamma model properties in the following quiz.[1]

❓ Quiz Yourself!

Figure 5.4 illustrates how different shape and rate hyperparameters impact the Gamma pdf (5.9). Based on these plots:

1. How would you describe the typical behavior of a Gamma(s, r) variable λ when $s > r$ (e.g., Gamma$(2,1)$)?
   a. Right-skewed with a mean greater than 1.
   b. Right-skewed with a mean less than 1.
   c. Symmetric with a mean around 1.
2. Using the same options as above, how would you describe the typical behavior of a Gamma(s, r) variable λ when $s < r$ (e.g., Gamma$(1,2)$)?
3. For which model is there greater variability in the plausible values of λ, Gamma$(20,20)$ or Gamma$(20,100)$?

---

[1] 1:a, 2:b, 3: Gamma(20,20)

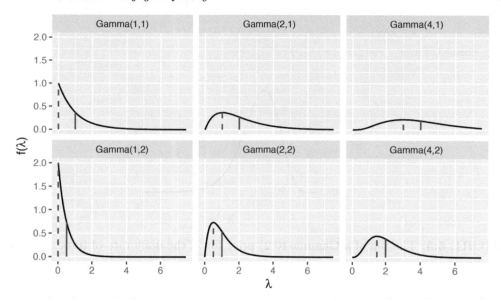

FIGURE 5.4: Gamma models with different hyperparameters. The dashed and solid vertical lines represent the modes and means, respectively.

In general, Figure 5.4 illustrates that $\lambda \sim \text{Gamma}(s,r)$ variables are positive and right skewed. Further, the general shape and rate of decrease in the skew are controlled by hyperparameters s and r. The quantitative measures of central tendency and variability in (5.10) provide some insight. For example, notice that the mean of λ, $E(\lambda) = s/r$, is greater than 1 when $s > r$ and less than 1 when $s < r$. Further, as s increases relative to r, the skew in λ decreases and the variability, $\text{Var}(\lambda) = s/r^2$, increases.

Now that we have some intuition for how the $\text{Gamma}(s,r)$ model works, we can *tune* it to reflect our prior information about the daily rate of fraud risk phone calls λ. Recall our earlier assumption that λ is about 5, and most likely somewhere between 2 and 7. Our $\text{Gamma}(s,r)$ prior should have similar patterns. For example, we want to pick s and r for which λ tends to be around 5,

$$E(\lambda) = \frac{s}{r} \approx 5.$$

This can be achieved by setting s to be 5 times r, $s = 5r$. Next we want to make sure that most values of our $\text{Gamma}(s,r)$ prior are between 2 and 7. Through some trial and error within these constraints, and plotting various Gamma models using `plot_gamma()` in the **bayesrules** package, we find that the Gamma(10,2) features closely match the central tendency *and* variability in our prior understanding (Figure 5.5). Thus, a *reasonable* prior model for the daily rate of fraud risk phone calls is

$$\lambda \sim \text{Gamma}(10, 2)$$

with **prior pdf** $f(\lambda)$ following from plugging $s = 10$ and $r = 2$ into (5.9).

$$f(\lambda) = \frac{2^{10}}{\Gamma(10)}\lambda^{10-1}e^{-2\lambda} \quad \text{for } \lambda > 0.$$

```
# Plot the Gamma(10, 2) prior
plot_gamma(shape = 10, rate = 2)
```

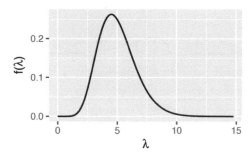

FIGURE 5.5: The pdf of a Gamma(10,2) prior for λ, the daily rate of fraud risk calls.

5.2.4   Gamma-Poisson conjugacy

As we discussed at the start of this chapter, conjugate families can come in handy. Fortunately for us, using a Gamma prior for a rate parameter λ and a Poisson model for corresponding count data Y is another example of a conjugate family. This means that, *spoiler*, the posterior model for λ will also have a Gamma model with *updated* parameters. We'll state and prove this in the general setting before applying the results to our phone call situation.

The Gamma-Poisson Bayesian model

Let $\lambda > 0$ be an unknown *rate* parameter and $(Y_1, Y_2, \ldots, Y_n)$ be an independent Pois(λ) sample. The Gamma-Poisson Bayesian model complements the Poisson structure of data Y with a Gamma prior on λ:

$$Y_i | \lambda \overset{ind}{\sim} \text{Pois}(\lambda)$$
$$\lambda \sim \text{Gamma}(s, r).$$

Upon observing data $\vec{y} = (y_1, y_2, \ldots, y_n)$, the posterior model of λ is also a Gamma with updated parameters:

$$\lambda | \vec{y} \sim \text{Gamma}\left(s + \sum y_i, \; r + n\right). \tag{5.11}$$

Let's prove this result. In general, recall that the posterior pdf of λ is proportional to the product of the prior pdf and likelihood function defined by (5.9) and (5.8), respectively:

$$f(\lambda | \vec{y}) \propto f(\lambda) L(\lambda | \vec{y}) = \frac{r^s}{\Gamma(s)} \lambda^{s-1} e^{-r\lambda} \cdot \frac{\lambda^{\sum y_i} e^{-n\lambda}}{\prod y_i!} \quad \text{for } \lambda > 0.$$

Next, remember that any non-λ multiplicative constant in the above equation can be "proportional-ed" out. Thus, boiling the prior pdf and likelihood function down to their kernels, we get

$$f(\lambda|\vec{y}) \propto \lambda^{s-1}e^{-r\lambda} \cdot \lambda^{\sum y_i}e^{-n\lambda}$$
$$= \lambda^{s+\sum y_i - 1}e^{-(r+n)\lambda}$$

where the final line follows by combining like terms. What we're left with here is the *kernel* of the posterior pdf. This particular kernel corresponds to the pdf of a Gamma model (5.9), with shape parameter $s + \sum y_i$ and rate parameter $r + n$. Thus, we've proven that

$$\lambda|\vec{y} \sim \text{Gamma}\left(s + \sum y_i, r + n\right).$$

Let's apply this result to our fraud risk calls. There we have a Gamma(10,2) prior for λ, the daily rate of calls. Further, on four separate days in the second week of August, we received $\vec{y} = (y_1, y_2, y_3, y_4) = (6, 2, 2, 1)$ such calls. Thus, we have a sample of $n = 4$ data points with a *total* of 11 fraud risk calls and an *average* of 2.75 phone calls per day:

$$\sum_{i=1}^{4} y_i = 6 + 2 + 2 + 1 = 11 \quad \text{and} \quad \bar{y} = \frac{\sum_{i=1}^{4} y_i}{4} = 2.75.$$

Plugging this data into (5.8), the resulting Poisson likelihood function of λ is

$$L(\lambda|\vec{y}) = \frac{\lambda^{11}e^{-4\lambda}}{6! \times 2! \times 2! \times 1!} \propto \lambda^{11}e^{-4\lambda} \quad \text{for } \lambda > 0.$$

We visualize a portion of $L(\lambda|\vec{y})$ for λ between 0 and 10 using the `plot_poisson_likelihood()` function in the **bayesrules** package. Here, `y` is the vector of data values and `lambda_upper_bound` is the maximum value of λ to view on the x-axis. (Why can't we visualize the whole likelihood? Because $\lambda \in (0, \infty)$ and this book would be pretty expensive if we had infinite pages.)

```
plot_poisson_likelihood(y = c(6, 2, 2, 1), lambda_upper_bound = 10)
```

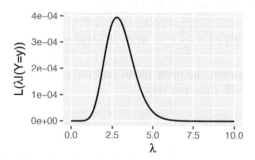

FIGURE 5.6: The likelihood function of λ, the daily rate of fraud risk calls, given a four-day sample of phone call data.

The punchline is this. Underlying *rates* λ of between one to five fraud risk calls per day are consistent with our phone call data. And across this spectrum, rates near 2.75 are the *most* compatible with this data. This makes sense. The Poisson data model assumes that λ is the underlying average daily phone call count, $E(Y_i|\lambda) = \lambda$. As such, we're *most* likely to observe a sample with an average daily phone call rate of $\bar{y} = 2.75$ when the underlying rate λ is also 2.75.

Combining these observations with our Gamma(10,2) prior model of λ, it follows from (5.11) that the posterior model of λ is a Gamma with an updated shape parameter of 21 ($s + \sum y_i = 10 + 11$) and rate parameter of 6 ($r + n = 2 + 4$):

$$\lambda|\vec{y} \sim \text{Gamma}(21, 6).$$

We can visualize the prior pdf, scaled likelihood function, and posterior pdf for λ all in a single plot with the `plot_gamma_poisson()` function in the **bayesrules** package. How magical. For this function to work, we must specify a few things: the prior `shape` and `rate` hyperparameters as well as the information from our data, the observed total number of phone calls `sum_y` ($\sum y_i$) and the sample size `n`:

```
plot_gamma_poisson(shape = 10, rate = 2, sum_y = 11, n = 4)
```

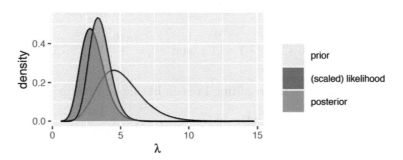

FIGURE 5.7: The Gamma-Poisson model of λ, the daily rate of fraud risk calls.

Our posterior notion about the daily rate of fraud calls is, of course, a *compromise* between our vague prior and the observed phone call data. Since our prior notion was quite variable in comparison to the strength in our sample data, the posterior model of λ is more in sync with the data. Specifically, utilizing the properties of the Gamma(10,2) prior and Gamma(21,6) posterior as defined by (5.10), notice that our posterior understanding of the typical daily rate of phone calls dropped from 5 to 3.5 per day:

$$E(\lambda) = \frac{10}{2} = 5 \quad \text{and} \quad E(\lambda|\vec{y}) = \frac{21}{6} = 3.5.$$

Though a compromise between the prior mean and data mean, this posterior mean is *closer* to the data mean of $\bar{y} = 2.75$ calls per day.

Hot tip

The posterior mean will always be between the prior mean and the data mean. If your posterior mean falls outside that range, it indicates that you made an error and should retrace some steps.

Further, with the additional information about λ from the data, the variability in our understanding of λ drops by more than half, from a standard deviation of 1.581 to 0.764 calls per day:

$$\text{SD}(\lambda) = \sqrt{\frac{10}{2^2}} = 1.581 \quad \text{and} \quad \text{SD}(\lambda|\vec{y}) = \sqrt{\frac{21}{6^2}} = 0.764.$$

The convenient `summarize_gamma_poisson()` function in the **bayesrules** package, which uses the same arguments as `plot_gamma_poisson()`, helps us contrast the prior and posterior models and confirms the results above:

```
summarize_gamma_poisson(shape = 10, rate = 2, sum_y = 11, n = 4)
      model shape rate mean  mode    var      sd
1     prior    10    2  5.0 4.500 2.5000 1.5811
2 posterior    21    6  3.5 3.333 0.5833 0.7638
```

5.3 Normal-Normal conjugate family

We now have two conjugate families in our toolkit: the Beta-Binomial and the Gamma-Poisson. But many more conjugate families exist! It's impossible to cover them all, but there is a third conjugate family that's especially helpful to know: the **Normal-Normal**. Consider a data story. As scientists learn more about brain health, the dangers of concussions (hence of activities in which participants sustain repeated concussions) are gaining greater attention (Bachynski, 2019). Among all people who have a history of concussions, we are interested in μ, the average volume (in cubic centimeters) of a specific part of the brain: the hippocampus. Though we don't have prior information about this group in particular, Wikipedia tells us that among the general population of human adults, both halves of the hippocampus have a volume between 3.0 and 3.5 cubic centimeters.[2] Thus, the *total* hippocampal volume of *both* sides of the brain is between 6 and 7 cm^3. Using this as a starting point, we'll assume that the mean hippocampal volume among people with a history of concussions, μ, is also somewhere between 6 and 7 cm^3, with an average of 6.5. We'll balance this prior understanding with data on the hippocampal volumes of $n = 25$ subjects, $(Y_1, Y_2, \ldots, Y_n)$, using the **Normal-Normal Bayesian model**.

5.3.1 The Normal data model

Again, the spirit of our Bayesian analysis starts with our prior understanding of μ. Yet the specification of an appropriate prior model structure for μ (which we can then tune) can be guided by first identifying a model for the dependence of our data Y_i upon μ. Since hippocampal volumes Y_i are measured on a continuous scale, there are many possible common models of the variability in Y_i from person to person: Beta, Exponential, Gamma, Normal, F, etc. From this list, we can immediately eliminate the Beta model – it assumes that $Y_i \in [0, 1]$, whereas hippocampal volumes tend to be around 6.5 cm^3. Among the remaining options, the Normal model is quite *reasonable* – biological measurements like hippocampal volume are often symmetrically or *Normally* distributed around some global average, here μ.

---

[2]https://en.wikipedia.org/wiki/Hippocampus

The Normal model

Let Y be a continuous random variable which can take any value between $-\infty$ and ∞, i.e., $Y \in (-\infty, \infty)$. Then the variability in Y might be well represented by a Normal model with **mean parameter** $\mu \in (-\infty, \infty)$ and **standard deviation parameter** $\sigma > 0$:

$$Y \sim N(\mu, \sigma^2).$$

The Normal model is specified by continuous pdf

$$f(y) = \frac{1}{\sqrt{2\pi\sigma^2}} \exp\left[-\frac{(y-\mu)^2}{2\sigma^2}\right] \quad \text{for } y \in (-\infty, \infty) \tag{5.12}$$

and has the following features:

$$E(Y) = \text{Mode}(Y) = \mu$$
$$\text{Var}(Y) = \sigma^2$$
$$\text{SD}(Y) = \sigma.$$

Further, σ provides a sense of scale for Y. Roughly 95% of Y values will be within 2 standard deviations of μ:

$$\mu \pm 2\sigma. \tag{5.13}$$

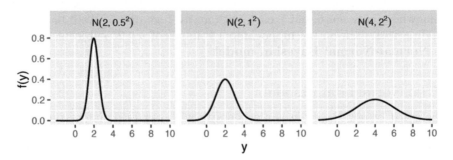

FIGURE 5.8: Normal pdfs with varying mean and standard deviation parameters.

Figure 5.8 illustrates the Normal model under a variety of mean and standard deviation parameter values, μ and σ. No matter the parameters, the Normal model is bell-shaped and symmetric around μ – thus as μ gets larger, the model shifts to the right along with it. Further, σ controls the variability of the Normal model – as σ gets larger, the model becomes more spread out. Finally, though a Normal variable Y can technically range from $-\infty$ to ∞, the Normal model assigns *negligible* plausibility to Y values that are more than 3 standard deviations σ from the mean μ. To play around some more, you can plot Normal models using the `plot_normal()` function from the **bayesrules** package.

Returning to our brain analysis, we can reasonably assume that the hippocampal volumes of our $n = 25$ subjects, $(Y_1, Y_2, \ldots, Y_n)$, are *independent* and Normally distributed around a mean volume μ with standard deviation σ. Further, to keep our focus on μ, we'll assume throughout our analysis that the standard deviation is *known* to be $\sigma = 0.5$ cm^3.[3] This choice of σ suggests that most people have hippocampal volumes within $2\sigma = 1$ cm^3 of the average. Thus, the dependence of Y_i on the *unknown* mean μ is:

$$Y_i|\mu \sim N(\mu, \sigma^2).$$

❗ **Warning**

Reasonable doesn't mean perfect. Though we'll later see that our hippocampal volume data does exhibit Normal behavior, the Normal model *technically* assumes that each subject's hippocampal volume can range from $-\infty$ to ∞. However, we're not too worried about this incorrect assumption here. Per our earlier discussion of Figure 5.8, the Normal model will put negligible weight on unreasonable values of hippocampal volume. In general, not letting *perfect* be the enemy of *good* will be a theme throughout this book (mainly because there is no perfect).

Accordingly, the **joint pdf** which describes the *collective* randomness in our $n = 25$ subjects' hippocampal volumes, $(Y_1, Y_2, \ldots, Y_n)$, is the product of the unique Normal pdfs $f(y_i|\mu)$ defined by (5.12),

$$f(\vec{y}|\mu) = \prod_{i=1}^{n} f(y_i|\mu) = \prod_{i=1}^{n} \frac{1}{\sqrt{2\pi\sigma^2}} \exp\left[-\frac{(y_i - \mu)^2}{2\sigma^2}\right].$$

Once we observe our sample data $\vec{y}$, we can flip the joint pdf on its head to obtain the Normal likelihood function of μ, $L(\mu|\vec{y}) = f(\vec{y}|\mu)$. Remembering that we're assuming σ is a known constant, we can simplify the likelihood up to a *proportionality* constant by dropping the terms that don't depend upon μ. Then for $\mu \in (-\infty, \infty)$,

$$L(\mu|\vec{y}) \propto \prod_{i=1}^{n} \exp\left[-\frac{(y_i - \mu)^2}{2\sigma^2}\right] = \exp\left[-\frac{\sum_{i=1}^{n}(y_i - \mu)^2}{2\sigma^2}\right].$$

Through a bit more rearranging (which we encourage you to verify if, like us, you enjoy algebra), we can make this even easier to digest by using the sample mean $\bar{y}$ and sample size n to summarize our data values:

$$L(\mu|\vec{y}) \propto \exp\left[-\frac{(\bar{y} - \mu)^2}{2\sigma^2/n}\right] \quad \text{for } \mu \in (-\infty, \infty). \tag{5.14}$$

Don't forget the whole point of this exercise! Specifying a model for the data along with its corresponding likelihood function provides the tools we'll need to assess the compatibility of our data $\vec{y}$ with different values of μ (once we actually collect that data).

5.3.2 Normal prior

With the likelihood in place, let's formalize a prior model for μ, the mean hippocampal volume among people that have a history of concussions. By the properties of the $Y_i|\mu \sim N(\mu, \sigma^2)$ data model, the Normal mean parameter μ can technically take any value between $-\infty$ and ∞. Thus, a Normal prior for μ, which is also defined for $\mu \in (-\infty, \infty)$, makes a reasonable choice. Specifically, we'll assume that μ itself is Normally distributed around some mean θ with standard deviation τ:

---

[3]You'll have a chance to relax this silly-ish assumption that we know σ but don't know μ in later chapters.

$$\mu \sim N(\theta, \tau^2),$$

where μ has prior pdf

$$f(\mu) = \frac{1}{\sqrt{2\pi\tau^2}} \exp\left[-\frac{(\mu - \theta)^2}{2\tau^2}\right] \quad \text{for } \mu \in (-\infty, \infty). \tag{5.15}$$

Not only does the Normal prior assumption that $\mu \in (-\infty, \infty)$ match the same assumption of the Normal data model, we'll prove below that this is a *conjugate* prior. You might anticipate this result from the fact that the likelihood function $L(\mu|\bar{y})$ (5.14) and prior pdf $f(\mu)$ (5.15) are both proportional to

$$\exp\left[-\frac{(\mu - \blacksquare)^2}{2\blacksquare^2}\right]$$

with different $\blacksquare$.

Using our understanding of a Normal model, we can now *tune* the prior hyperparameters θ and τ to reflect our prior understanding and uncertainty about the average hippocampal volume among people that have a history of concussions, μ. Based on our rigorous Wikipedia research that hippocampal volumes tend to be between 6 and 7 cm^3, we'll set the Normal prior mean θ to the midpoint, 6.5. Further, we'll set the Normal prior standard deviation to $\tau = 0.4$. In other words, by (5.13), we think there's a 95% chance that μ is somewhere between 5.7 and 7.3 cm^3 (6.5 ± 2 ∗ 0.4). This range is *wider*, and hence more conservative, than what Wikipedia indicated. Our uncertainty here reflects the fact that we didn't vet the Wikipedia sources, we aren't confident that the features for the typical adult translates to people with a history of concussions, and we generally aren't sure what's going on here (i.e., we're not brain experts). Putting this together, our tuned prior model for μ is:

$$\mu \sim N(6.5, 0.4^2).$$

```
plot_normal(mean = 6.5, sd = 0.4)
```

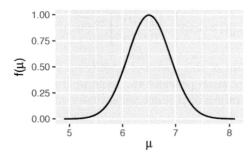

FIGURE 5.9: A Normal prior model for μ, with mean 6.5 and standard deviation 0.4.

5.3.3 Normal-Normal conjugacy

To obtain our posterior model of μ we must combine the information from our prior and our data. Again, we were clever to pick a Normal prior model – the Normal-Normal is another convenient conjugate family! Thus, the posterior model for μ will *also* be Normal with updated parameters that are informed by the prior and observed data.

The Normal-Normal Bayesian model

Let $\mu \in (-\infty, \infty)$ be an unknown *mean* parameter and $(Y_1, Y_2, \ldots, Y_n)$ be an independent $N(\mu, \sigma^2)$ sample where σ is assumed to be *known*. The Normal-Normal Bayesian model complements the Normal data structure with a Normal prior on μ:

$$Y_i | \mu \overset{ind}{\sim} N(\mu, \sigma^2)$$
$$\mu \sim N(\theta, \tau^2)$$

Upon observing data $\vec{y} = (y_1, y_2, \ldots, y_n)$ with mean $\bar{y}$, the posterior model of μ is also Normal with updated parameters:

$$\mu | \vec{y} \sim N\left(\theta \frac{\sigma^2}{n\tau^2 + \sigma^2} + \bar{y} \frac{n\tau^2}{n\tau^2 + \sigma^2}, \frac{\tau^2 \sigma^2}{n\tau^2 + \sigma^2} \right). \tag{5.16}$$

Whooo, that is a mouthful! We provide an optional proof of this result in Section 5.3.4. Even without that proof, we can observe the balance that the Normal posterior (5.16) strikes between the prior and the data. First, the posterior mean is a **weighted average** of the prior mean $E(\mu) = \theta$ and the sample mean $\bar{y}$. Second, the posterior variance is informed by the prior variability τ and variability in the data σ. Both are impacted by sample size n. First, as n increases, the posterior mean places less weight on the prior mean and more weight on sample mean $\bar{y}$:

$$\frac{\sigma^2}{n\tau^2 + \sigma^2} \to 0 \quad \text{and} \quad \frac{n\tau^2}{n\tau^2 + \sigma^2} \to 1.$$

Further, as n increases, the posterior variance decreases:

$$\frac{\tau^2 \sigma^2}{n\tau^2 + \sigma^2} \to 0.$$

That is, the more and more data we have, our posterior certainty about μ increases and becomes more in sync with the data.

Let's apply and examine this result in our analysis of μ, the average hippocampal volume among people that have a history of concussions. We've already built our prior model of μ, $\mu \sim N(6.5, 0.4^2)$. Next, consider some data. The `football` data in **bayesrules**, a subset of the **FootballBrain** data in the **Lock5Data** package (Lock et al., 2016), includes results for a cross-sectional study of hippocampal volumes among 75 subjects (Singh et al., 2014): 25 collegiate football players with a history of concussions (`fb_concuss`), 25 collegiate football players that do not have a history of concussions (`fb_no_concuss`), and 25 control subjects. For our analysis, we'll focus on the $n = 25$ subjects with a history of concussions (`fb_concuss`):

```
# Load the data
data(football)
concussion_subjects <- football %>%
  filter(group == "fb_concuss")
```

These subjects have an average hippocampal volume of $\bar{y} = 5.735$ cm^3:

```
concussion_subjects %>%
  summarize(mean(volume))
  mean(volume)
1       5.735
```

Further, the hippocampal volumes appear to vary *normally* from subject to subject, ranging from roughly 4.5 to 7 cm^3. That is, our assumed Normal data model about individual hippocampal volumes, $Y_i|\mu \sim N(\mu, \sigma^2)$ with an assumed standard deviation of $\sigma = 0.5$, seems *reasonable*:

```
ggplot(concussion_subjects, aes(x = volume)) +
  geom_density()
```

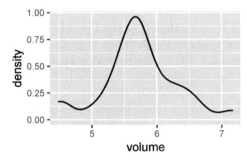

FIGURE 5.10: A density plot of the hippocampal volumes (in cubic centimeters) among 25 subjects that have experienced concussions.

Plugging this information from the data ($n = 25$, $\bar{y} = 5.735$, and $\sigma = 0.5$) into (5.14) defines the Normal likelihood function of μ:

$$L(\mu|\vec{y}) \propto \exp\left[-\frac{(5.735 - \mu)^2}{2(0.5^2/25)}\right] \quad \text{for } \mu \in (-\infty, \infty).$$

We plot this likelihood function using `plot_normal_likelihood()`, providing our observed `volume` data and data standard deviation $\sigma = 0.5$ (Figure 5.11). This likelihood illustrates the compatibility of our observed hippocampal data with different μ values. To this end, the hippocampal patterns observed in our data would most likely have arisen if the mean hippocampal volume across *all* people with a history of concussions, μ, were between 5.3 and 6.1 cm^3. Further, we're *most* likely to have observed a mean volume of $\bar{y} = 5.735$ among our 25 sample subjects if the underlying population mean μ were also 5.735.

```
plot_normal_likelihood(y = concussion_subjects$volume, sigma = 0.5)
```

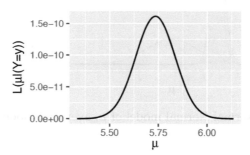

FIGURE 5.11: The Normal likelihood function for mean hippocampal volume μ.

We now have all necessary pieces to plug into (5.16), and hence to specify the posterior model of μ:

- our Normal prior model of μ had mean $\theta = 6.5$ and standard deviation $\tau = 0.4$;
- our $n = 25$ sample subjects had a sample mean volume $\bar{y} = 5.735$;
- we assumed a known standard deviation among individual hippocampal volumes of $\sigma = 0.5$.

It follows that the posterior model of μ is:

$$\mu|\vec{y} \sim N\left(6.5 \cdot \frac{0.5^2}{25 \cdot 0.4^2 + 0.5^2} + 5.735 \cdot \frac{25 \cdot 0.4^2}{25 \cdot 0.4^2 + 0.5^2}, \frac{0.4^2 \cdot 0.5^2}{25 \cdot 0.4^2 + 0.5^2}\right).$$

Further simplified,

$$\mu|\vec{y} \sim N\left(5.78, 0.009^2\right)$$

where the posterior mean places roughly 94% of its weight on the data mean ($\bar{y} = 5.375$) and only 6% of its weight on the prior mean ($E(\mu) = 6.5$):

$$E(\mu|\vec{y}) = 6.5 \cdot 0.0588 + 5.735 \cdot 0.9412 = 5.78.$$

Bringing all of these pieces together, we plot and summarize our Normal-Normal analysis of μ using `plot_normal_normal()` and `summarize_normal_normal()` in the **bayesrules** package. Though a compromise between the prior and data, our posterior understanding of μ is more heavily influenced by the latter. In light of our data, we are much more *certain* about the mean hippocampal volume among people with a history of concussions, and believe that this figure is somewhere in the range from 5.586 to 5.974 cm^3 ($5.78 \pm 2 * 0.097$).

```
plot_normal_normal(mean = 6.5, sd = 0.4, sigma = 0.5,
                   y_bar = 5.735, n = 25)
```

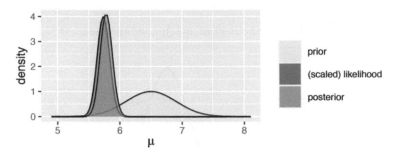

FIGURE 5.12: The Normal-Normal model of μ, average hippocampal volume.

```
summarize_normal_normal(mean = 6.5, sd = 0.4, sigma = 0.5,
                        y_bar = 5.735, n = 25)
      model mean mode      var       sd
1      prior 6.50 6.50 0.160000 0.40000
2 posterior 5.78 5.78 0.009412 0.09701
```

5.3.4 Optional: Proving Normal-Normal conjugacy

For completeness sake, we prove here that the Normal-Normal model produces posterior model (5.16). If derivations are not your thing, that's totally fine. Feel free to skip ahead to the next section and know that you can use this Normal-Normal model with a light heart. Let's get right into it. The posterior pdf of μ is proportional to the product of the Normal prior pdf (5.15) and the likelihood function (5.14). For $\mu \in (-\infty, \infty)$:

$$f(\mu|\vec{y}) \propto f(\mu)L(\mu|\vec{y}) \propto \exp\left[\frac{-(\mu-\theta)^2}{2\tau^2}\right] \cdot \exp\left[-\frac{(\bar{y}-\mu)^2}{2\sigma^2/n}\right].$$

Next, we can expand the squares in the exponents and sweep under the rug of proportionality both the θ^2 in the numerator of the first exponent and the $\bar{y}^2$ in the numerator of the second exponent:

$$f(\mu|\vec{y}) \propto \exp\left[\frac{-\mu^2 + 2\mu\theta - \theta^2}{2\tau^2}\right] \exp\left[\frac{-\mu^2 + 2\mu\bar{y} - \bar{y}^2}{2\sigma^2/n}\right]$$

$$\propto \exp\left[\frac{-\mu^2 + 2\mu\theta}{2\tau^2}\right] \exp\left[\frac{-\mu^2 + 2\mu\bar{y}}{2\sigma^2/n}\right].$$

Next, we give the exponents common denominators and combine them into a single exponent:

$$f(\mu|\vec{y}) \propto \exp\left[\frac{(-\mu^2 + 2\mu\theta)\sigma^2/n}{2\tau^2\sigma^2/n}\right] \exp\left[\frac{(-\mu^2 + 2\mu\bar{y})\tau^2}{2\tau^2\sigma^2/n}\right]$$

$$\propto \exp\left[\frac{(-\mu^2 + 2\mu\theta)\sigma^2 + (-\mu^2 + 2\mu\bar{y})n\tau^2}{2\tau^2\sigma^2}\right].$$

Now let's combine μ terms and rearrange so that μ^2 is by itself:

$$f(\mu|\vec{y}) \propto \exp\left[\frac{-\mu^2(n\tau^2 + \sigma^2) + 2\mu(\theta\sigma^2 + \bar{y}n\tau^2)}{2\tau^2\sigma^2}\right]$$

$$\propto \exp\left[\frac{-\mu^2 + 2\mu\left(\frac{\theta\sigma^2 + \bar{y}n\tau^2}{n\tau^2 + \sigma^2}\right)}{2(\tau^2\sigma^2)/(n\tau^2 + \sigma^2)}\right].$$

This may seem like too much to deal with, but if you look closely, you can see that we can bring back some constants which do not depend upon μ to complete the square in the numerator:

$$f(\mu|\vec{y}) \propto \exp\left[\frac{-\left(\mu - \frac{\theta\sigma^2 + \bar{y}n\tau^2}{n\tau^2 + \sigma^2}\right)^2}{2(\tau^2\sigma^2)/(n\tau^2 + \sigma^2)}\right].$$

This may still *seem* messy, but once we complete the square, we actually have the kernel of a Normal pdf for μ, $\exp\left[-\frac{(\mu - \blacksquare)^2}{2\blacksquare^2}\right]$. By identifying the missing pieces $\blacksquare$, we can thus conclude that

$$\mu|\vec{y} \sim N\left(\frac{\theta\sigma^2 + \bar{y}n\tau^2}{n\tau^2 + \sigma^2}, \ \frac{\tau^2\sigma^2}{n\tau^2 + \sigma^2}\right)$$

where we can reorganize the posterior mean as a weighted average of the prior mean μ and data mean $\bar{y}$:

$$\frac{\theta\sigma^2 + \bar{y}n\tau^2}{n\tau^2 + \sigma^2} = \theta\frac{\sigma^2}{n\tau^2 + \sigma^2} + \bar{y}\frac{n\tau^2}{n\tau^2 + \sigma^2}.$$

5.4 Why no simulation in this chapter?

As you may have gathered, we love some simulations. There is something reassuring about the reality check that a simulation can provide. Yet we are at a crossroads. The Gamma-Poisson and Normal-Normal models we've studied here are tough to simulate using the techniques we've learned thus far. Letting θ represent some parameter of interest, recall the steps we've used for past simulations:

1. Simulate, say, 10000 values of θ from the prior model.
2. Simulate a set of sample data Y from each simulated θ value.
3. Filter out only those of the 10000 simulated sets of (θ, Y) for which the simulated Y data matches the data we actually observed.
4. Use the remaining θ values to approximate the posterior of θ.

The issue with extending this simulation technique to the Gamma-Poisson and Normal-Normal examples in this chapter comes with step 3. In both of our examples, we had a sample size greater than one, $(Y_1, Y_2, \ldots, Y_n)$. Further, in the Normal-Normal example, our data values Y_i are *continuous*. In both of these scenarios, it's very likely that *no* simulated sets of (θ, Y) will perfectly match our observed sample data. This is true even if we carry out millions of simulations in step 3. Spoiler alert! There are other ways to approximate the posterior model which we will learn in Unit 2.

5.5   Critiques of conjugate family models

Before we end the chapter, we want to acknowledge that conjugate family models also have drawbacks. In particular:

- A conjugate prior model isn't always flexible enough to fit your prior understanding. For example, a Normal model is always unimodal and symmetric around the mean μ. So if your prior understanding is not symmetric or is not unimodal, then the Normal prior might not be the best tool for the job.
- Conjugate family models do not always allow you to have an entirely flat prior. While we can tune a flat Beta prior by setting $\alpha = \beta = 1$, neither the Normal nor Gamma priors (or any proper models with infinite support) can be tuned to be *totally* flat. The best we can do is tune the priors to have very high variance, so that they're *almost* flat.

5.6   Chapter summary

In Chapter 5, you learned about conjugacy and applied it to a few different situations. Our main takeaways for this chapter are:

- Using conjugate priors allows us to have easy-to-derive and readily interpretable posterior models.
- The Beta-Binomial, Gamma-Poisson, and Normal-Normal conjugate families allow us to analyze data Y in different scenarios. The Beta-Binomial is convenient when our data Y is the number of successes in a set of n trials, the Gamma-Poisson when Y is a count with no upper limit, and the Normal-Normal when Y is continuous.
- We can use several functions from the **bayesrules** package to explore these conjugate families: `plot_poisson_likelihood()`, `plot_gamma()`, `plot_gamma_poisson()`, `summarize_gamma_poisson()`, `plot_normal()`, `plot_normal_normal()`, and `summarize_normal_normal()`.

We hope that you now appreciate the utility of conjugate priors!

5.7   Exercises

5.7.1   Practice: Gamma-Poisson

Exercise 5.1 (Tuning a Gamma prior). For each situation below, tune an appropriate Gamma(s, r) prior model for λ.

a)   The most common value of λ is 4, and the mean is 7.
b)   The most common value of λ is 10 and the mean is 12.
c)   The most common value of λ is 5, and the variance is 3.
d)   The most common value of λ is 14, and the variance is 6.
e)   The mean of λ is 4 and the variance is 12.
f)   The mean of λ is 22 and the variance is 3.

Exercise 5.2 (Poisson likelihood functions). For each situation below, we observe the outcomes for a random sample of Poisson variables, $Y_i | \lambda \overset{ind}{\sim} \text{Pois}(\lambda)$. Specify and plot the likelihood function of λ.

a) $(y_1, y_2, y_3) = (3, 7, 19)$
b) $(y_1, y_2, y_3, y_4) = (12, 12, 12, 0)$
c) $y_1 = 12$
d) $(y_1, y_2, y_3, y_4, y_5) = (16, 10, 17, 11, 11)$

Exercise 5.3 (Finding the Gamma-Poisson posterior). Assume a prior model of $\lambda \sim$ Gamma(24, 2). Specify the posterior models of λ corresponding to the data in each scenario of Exercise 5.2.

Exercise 5.4 (Finding the Gamma-Poisson posterior with a different prior). Assume a prior model of $\lambda \sim$ Gamma(2, 2). Specify the posterior models of λ corresponding to the data in each scenario of Exercise 5.2.

Exercise 5.5 (Text messages). Let random variable λ represent the rate of text messages people receive in an hour. At first, you believe that the typical number of messages per hour is 5 with a standard deviation of 0.25 messages.

a)   Tune and plot an appropriate Gamma(s, r) prior model for λ.
b)   What is the prior probability that the rate of text messages per hour is larger than 10? Hint: learn about `pgamma()`.

Exercise 5.6 (Text messages with data). Continuing with Exercise 5.5, you collect data from six friends. They received 7, 3, 8, 9, 10, 12 text messages in the previous hour.

a)   Plot the resulting likelihood function of λ.
b)   Plot the prior pdf, likelihood function, and the posterior pdf of λ.
c)   Use `summarize_gamma_poisson()` to calculate descriptive statistics for the prior and the posterior models of λ.
d)   Comment on how your understanding about λ changed from the prior (in the previous exercise) to the posterior based on the data you collected from your friends.

Exercise 5.7 (World Cup). Let λ be the average number of goals scored in a Women's World Cup game. We'll analyze λ by the following Gamma-Poisson model where data Y_i is the observed number of goals scored in a sample of World Cup games:

$$Y_i | \lambda \overset{ind}{\sim} \text{Pois}(\lambda)$$
$$\lambda \sim \text{Gamma}(1, 0.25)$$

a) Plot and summarize our prior understanding of λ.

b) Why is the Poisson model a reasonable choice for our data Y_i?

c) The `wwc_2019_matches` data in the **fivethirtyeight** package includes the number of goals scored by the two teams in each 2019 Women's World Cup match. Define, plot, and discuss the total number of goals scored per game:

```
library(fivethirtyeight)
data("wwc_2019_matches")
wwc_2019_matches <- wwc_2019_matches %>%
  mutate(total_goals = score1 + score2)
```

d) Identify the posterior model of λ and verify your answer using `summarize_gamma_poisson()`.

e) Plot the prior pdf, likelihood function, and posterior pdf of λ. Describe the evolution in your understanding of λ from the prior to the posterior.

5.7.2 Practice: Normal-Normal

Exercise 5.8 (Normal likelihood functions). In each situation below, we observe the outcomes for a Normal random sample, $Y_i | \mu \overset{ind}{\sim} N(\mu, \sigma^2)$ with known σ. Specify and plot the corresponding likelihood function of μ.

a) $(y_1, y_2, y_3) = (-4.3, 0.7, -19.4)$ and $\sigma = 10$
b) $(y_1, y_2, y_3, y_4) = (-12, 1.2, -4.5, 0.6)$ and $\sigma = 6$
c) $(y_1, y_2) = (12.4, 6.1)$ and $\sigma = 5$
d) $(y_1, y_2, y_3, y_4, y_5) = (1.6, 0.09, 1.7, 1.1, 1.1)$ and $\sigma = 0.6$

Exercise 5.9 (Investing in stock). You just bought stock in *FancyTech*. Let random variable μ be the average dollar amount that your *FancyTech* stock goes up or down in a one-day period. At first, you believe that μ is 7.2 dollars with a standard deviation of 2.6 dollars.

a) Tune and plot an appropriate Normal prior model for μ.
b) According to your plot, does it seem plausible that the *FancyTech* stock would increase by an average of 7.6 dollars in a day?
c) Does it seem plausible that the *FancyTech* stock would increase by an average of 4 dollars in a day?
d) What is the prior probability that, on average, the stock price goes *down*? Hint: `pnorm()`.
e) What is the prior probability that, on average, your stock price goes up by more than 8 dollars per day?

Exercise 5.10 (Investing in stock with data). Continuing with Exercise 5.9, it's reasonable to assume that the daily changes in *FancyTech* stock value are Normally distributed around an *unknown* mean of μ with a *known* standard deviation of $\sigma = 2$ dollars. On a random sample of 4 days, you observe changes in stock value of -0.7, 1.2, 4.5, and -4 dollars.

a) Plot the corresponding likelihood function of μ.

b) Plot the prior pdf, likelihood function, and the posterior pdf for μ.

c) Use `summarize_normal_normal()` to calculate descriptive statistics for the prior and the posterior models.

d) Comment on how your understanding about μ evolved from the prior (in the previous exercise) to the posterior based on the observed data.

e) What is the posterior probability that, on average, the stock price goes *down*? Hint: `pnorm()`.

f) What is the posterior probability that, on average, your stock price goes up by more than 8 dollars per day?

Exercise 5.11 (Normal-Normal calculation). Prof. Abebe and Prof. Morales both recently finished their PhDs and are teaching their first statistics classes at Bayesian University. Their colleagues told them that the *average* final exam score across all students, μ, varies Normally from year to year with a mean of 80 points and a standard deviation of 4. Further, individual students' scores Y vary Normally around μ with a known standard deviation of 3 points.

a) Prof. Abebe conducts the final exam and observes that his 32 students scored an average of 86 points. Calculate the posterior mean and variance of μ using the data from Prof. Abebe's class.

b) Prof. Morales conducts the final exam and observes that her 32 students scored an average of 82 points. Calculate the posterior mean and variance of μ using the data from Prof. Morales' class.

c) Next, use Prof. Abebe and Prof. Morales' *combined* exams to calculate the posterior mean and variance of μ.

Exercise 5.12 (Control brains). In this chapter we used the Normal-Normal model to analyze the mean hippocampal volume among people who have sustained concussions. In this exercise we will explore the mean volume, μ, for people who *have not* been diagnosed with a concussion. As before, we'll assume that the hippocampal volumes of individuals in this group, Y, vary Normally around μ with a standard deviation of $\sigma = 0.5$ cubic centimeters. We'll also start with the same prior as before: μ has a Normal prior with mean $\theta = 6.5$ and standard deviation $\tau = 0.4$.

a) Use the `football` data to calculate the sample mean hippocampal volume and sample size of the *control* subjects who have *not* been diagnosed with a concussion.

b) Identify the posterior model of μ and verify your answer using `summarize_normal_normal()`.

c) Plot the prior pdf, likelihood function, and posterior pdf of μ. Describe the evolution in your understanding of μ from the prior to the posterior.

Exercise 5.13 (Australia). Let μ be the average 3 p.m. temperature in Perth, Australia. Not knowing much about Australian weather, your friend's prior understanding is that the average temperature is likely around 30 degrees Celsius, though might be anywhere between 10 and 50 degrees Celsius. To learn about μ, they plan to analyze 1000 days of temperature data. Letting Y_i denote the 3 p.m. temperature on day i, they'll assume that daily temperatures vary Normally around μ with a standard deviation of 5 degrees: $Y_i | \mu \sim N(\mu, 5^2)$.

a) Tune and plot a Normal prior for μ that reflects your friend's understanding.

b)   The `weather_perth` data in the **bayesrules** package includes 1000 daily obser-
      vations of 3 p.m. temperatures in Perth (`temp3pm`). Plot this data and discuss
      whether it's reasonable to assume a Normal model for the temperature data.

c)   Identify the posterior model of μ and verify your answer using
      `summarize_normal_normal()`.

d)   Plot the prior pdf, likelihood function, and posterior pdf of μ. Describe the
      evolution in your understanding of μ from the prior to the posterior.

Exercise 5.14 (Normal-Normal Simulation).

a)   Your friend Alex has read Chapter 4 of this book, but not Chapter 5. Explain to
      Alex why it's difficult to simulate a Normal-Normal posterior using the simulation
      methods we have learned thus far.

b)   To prove your point, try (and fail) to simulate the posterior of μ for the following
      model upon observing a single data point $Y = 1.1$:

$$Y|\mu \sim N(\mu, 1^2)$$
$$\mu \sim N(0, 1^2)$$

5.7.3   General practice exercises

Exercise 5.15 (Which model?). Below are kernels for Normal, Poisson, Gamma, Beta, and
Binomial models. Identify the appropriate model with specific parameter values.

a) $f(\theta) \propto 0.3^\theta 0.7^{16-\theta}$ for $\theta \in \{0, 1, 2, \ldots, 16\}$
b) $f(\theta) \propto 1/\theta!$ for $\theta \in \{0, 1, 2, \ldots, \infty\}$
c) $f(\theta) \propto \theta^4 (1 - \theta)^7$ for $\theta \in [0, 1]$
d) $f(\theta) \propto e^{-\theta^2}$ for $\theta \in (-\infty, \infty)$

Exercise 5.16 (Which model: Back for more!). Below are kernels for Normal, Poisson,
Gamma, Beta, and Binomial models. Identify the appropriate model with specific parameter
values.

a) $f(\theta) \propto e^{-2\theta}\theta^{15}$ for $\theta > 0$
b) $f(\theta) \propto e^{\frac{-(\theta-12)^2}{18}}$ for $\theta \in (-\infty, \infty)$
c) $f(\theta) \propto 0.3^\theta/\theta!$ for $\theta \in \{0, 1, 2, \ldots, \infty\}$

Exercise 5.17 (Weighted average). In Section 4.3.2, we expressed the Beta-Binomial
posterior mean of proportion π as a weighted average of the prior mean and the observed
sample proportion. In Section 5.3.3, we expressed the Normal-Normal posterior mean of
μ as a weighted average of the prior mean θ and the observed sample mean $\bar{y}$. Similarly,
express the Gamma-Poisson posterior mean of rate λ, $E(\lambda|\bar{y})$, as a weighted average of the
prior mean $E(\lambda)$ and sample mean $\bar{y}$.

Exercise 5.18 (Counting insects). A biologist hires you to study the density of a certain
insect in the local region. Their prior idea about θ, the number of insects per m^2 area, is well
captured by a Gamma model with an expected value of 0.5 insects per m^2 and a standard
deviation of 0.25 insects per m^2. You then go out to inspect 20 separate m^2 areas of a nearby
field. You count 3, 2, 5, 1, 2 insects in the first five square meters, and then do not encounter
the insect in the next $15m^2$.

a) What model, Normal or Poisson, should you use to model the dependence of your insect count data on the underlying insect density θ? Explain why.

b) Plot the prior pdf, likelihood function, and posterior pdf of insect density θ. Comment on whether the posterior model is more in sync with the data or prior. Explain why this makes sense.

c) What is the posterior mean and standard deviation of the insect density?

d) Describe your posterior conclusions in context, in a way that a biologist would find helpful.

Exercise 5.19 (The Beta-Geometric model). Consider the following *new* Bayesian model:

$$Y|\theta \sim \text{Geometric}(\theta)$$
$$\theta \sim \text{Beta}(\alpha, \beta)$$

where the Geometric model has pmf $f(y|\theta) = \theta(1-\theta)^{y-1}$ for $y \in \{1, 2, \ldots\}$.

a. Derive the posterior model for θ given observed data $Y = y$. If possible, identify the *name* of the posterior model and its parameters.

b. Is the Beta model a conjugate prior for the Geometric data model?

Unit II

Posterior Simulation & Analysis

6

Approximating the Posterior

Welcome to Unit 2!

Unit 2 serves as a critical bridge to applying the fundamental concepts from Unit 1 in the more sophisticated model settings of Unit 3 and beyond. In Unit 1, we learned to think like Bayesians and to build some fundamental Bayesian models in this spirit. Further, by cranking these models through Bayes' Rule, we were able to mathematically *specify* the corresponding posteriors. Those days are over. Though merely hypothetical for now, *some* day (starting in Chapter 9) the models we'll be interested in analyzing will get too complicated to mathematically specify. Never fear – data analysts are not known to throw up their hands in the face of the unknown. When we can't *know* or specify something, we *approximate* it. In Unit 2 we'll explore **Markov chain Monte Carlo simulation** techniques for *approximating* otherwise out-of-reach posterior models.

No matter whether we're able to specify or must approximate a posterior model, we must then be able to understand and apply the results. To this end, we learned how to *describe* our posterior understanding using model features such as central tendency and variability in Unit 1. Yet in practice, we typically want to perform a deeper **posterior analysis**. This process of asking "what does it all mean?!" revolves around three major elements that we'll explore in Unit 2: posterior estimation, hypothesis testing, and prediction.

Learning requires the occasional leap. You've already taken a few. From Chapter 2 to Chapter 3, you took the leap from using simple discrete priors to using continuous Beta priors for a proportion π. From Chapter 3 to Chapter 5, you took the leap from engineering the Beta-Binomial model to a family of Bayesian models that can be applied in a wider variety of settings. With each leap, your Bayesian toolkit became more flexible and powerful, but at the cost of the underlying math becoming a bit more complicated. As you continue to generalize your Bayesian methods in more sophisticated settings, this complexity will continue to grow.

Consider Michelle's run for president. In Chapter 3 you built a model of Michelle's support in Minnesota based on polling data in that state. You could *continue* to refine your analysis of Michelle's chances of becoming president. To begin, you could model Michelle's support in *each* of the fifty states and Washington, D.C. Better yet, this model might incorporate data on past state-level voting trends and demographics. The trade-off is that increasing your model's flexibility also makes it more complicated. Whereas your Minnesota-only model depended upon only one parameter π, Michelle's level of support in that state, the new model depends upon *dozens* of parameters. Here, let $\theta = (\theta_1, \theta_2, \ldots, \theta_k)$ denote a generic set of $k \geq 1$ parameters upon which a Bayesian model depends. In your growing Bayesian model of Michelle's election chances, θ includes 51 parameters that represent her support in each state as well as multiple parameters that define the *relationships* between Michelle's support among voters, state-level demographics, and past voting trends. Our posterior analysis thus relies on building the posterior pdf of θ given a set of observed data y on state-level polls, demographics, and voting trends,

$$f(\theta|y) = \frac{f(\theta)L(\theta|y)}{f(y)} \propto f(\theta)L(\theta|y).$$

Though this formula looks familiar, complexity lurks beneath. Since θ is a vector with k elements, $f(\theta)$ and $L(\theta|y)$ are complicated *multivariate* functions. And, if we value our time, we can forget about calculating the normalizing constant $f(y)$ across all possible θ, an intractable multiple integral (or multiple sum) for which a closed form solution might not exist:

$$f(y) = \int_{\theta_1} \int_{\theta_2} \cdots \int_{\theta_k} f(\theta)L(\theta|y)d\theta_k \cdots d\theta_2 d\theta_1.$$

Fortunately, when a Bayesian posterior is either impossible or prohibitively difficult to specify, we're not out of luck. We must simply change our strategy: instead of *specifying* the posterior, we can **approximate the posterior via simulation**. We'll explore two simulation techniques: **grid approximation** and **Markov chain Monte Carlo (MCMC)**. When done well, both techniques produce a **sample** of N θ values,

$$\left\{\theta^{(1)}, \theta^{(2)}, \ldots, \theta^{(N)}\right\},$$

with properties that reflect those of the posterior model for θ. In Chapter 6, we'll explore these simulation techniques in the familiar Beta-Binomial and Gamma-Poisson model contexts. Though these models don't *require* simulation (we can and did specify their posteriors in Unit 1), exploring simulation in these familiar settings will help us build intuition for the process and give us peace of mind that it actually works when we eventually *do* need it.

◎ **Goals**

- Implement and examine the limitations of using **grid approximation** to simulate a posterior model.
- Explore the fundamental properties of **MCMC** posterior simulation techniques and how these differ from grid approximation.
- Implement MCMC simulation in R.
- Learn several **Markov chain diagnostics** for examining the *quality* of an MCMC posterior simulation.

But first, load some packages that we'll be utilizing throughout the remainder of this chapter (and book). Among these, **rstan** is quite unique, thus be sure to revisit the Preface for directions on installing this package.

```
# Load packages
library(tidyverse)
library(janitor)
library(rstan)
library(bayesplot)
```

6.1   Grid approximation

Imagine there's an image that you can't view in its entirety – you only observe snippets along a grid that sweeps from left to right across the image. The finer the grid, the clearer the image. And if the grid is fine enough, the result is an excellent approximation of the complete image:

This is the big picture idea behind Bayesian grid approximation, in which case the target "image" is posterior pdf $f(\theta|y)$. We needn't observe $f(\theta|y)$ at *every* possible θ to get a sense of its structure. Rather, we can evaluate $f(\theta|y)$ at a *finite, discrete grid* of possible θ values. Subsequently, we can take random samples from this *discretized* pdf to approximate the full posterior pdf $f(\theta|y)$. We formalize these ideas here and apply them below.

Grid approximation

Grid approximation produces a sample of N **independent** θ values, $\{\theta^{(1)}, \theta^{(2)}, \ldots, \theta^{(N)}\}$, from a **discretized approximation** of posterior pdf $f(\theta|y)$. This algorithm evolves in four steps:

1. Define a discrete grid of possible θ values.
2. Evaluate the prior pdf $f(\theta)$ and likelihood function $L(\theta|y)$ at each θ grid value.
3. Obtain a discrete approximation of the posterior pdf $f(\theta|y)$ by: (a) calculating the product $f(\theta)L(\theta|y)$ at each θ grid value; and then (b) *normalizing* the products so that they sum to 1 across all θ.
4. Randomly sample N θ grid values with respect to their corresponding normalized posterior probabilities.

6.1.1   A Beta-Binomial example

To bring the grid approximation technique to life, let's explore the following generic Beta-Binomial model (one without a corresponding data story):

$$Y|\pi \sim \text{Bin}(10, \pi)$$
$$\pi \sim \text{Beta}(2, 2).$$

(6.1)

We can interpret Y here as the number of successes in 10 independent trials. Each trial has probability of success π where our prior understanding about π is captured by a Beta$(2, 2)$ model. Suppose we observe $Y = 9$ successes. Then by our work in Chapter 3, we know that the updated posterior model of π is Beta with parameters 11 ($Y + \alpha = 9 + 2$) and 3 ($n - Y + \beta = 10 - 9 + 2$):

$$\pi|(Y = 9) \sim \text{Beta}(11, 3).$$

We're now going to ask you to *forget* that we were able to specify this posterior. Instead, we'll try to *approximate* the posterior using grid approximation. As **Step 1**, we need to split the continuum of possible π values on 0 to 1 into a finite grid. We'll start with a course grid of only 6 π values, $\pi \in \{0, 0.2, 0.4, 0.6, 0.8, 1\}$:

```
# Step 1: Define a grid of 6 pi values
grid_data <- data.frame(pi_grid = seq(from = 0, to = 1, length = 6))
```

In **Step 2** we use `dbeta()` and `dbinom()`, respectively, to evaluate the Beta(2, 2) prior pdf and Bin(10, π) likelihood function with $Y = 9$ at each π in `pi_grid`:

```
# Step 2: Evaluate the prior & likelihood at each pi
grid_data <- grid_data %>%
  mutate(prior = dbeta(pi_grid, 2, 2),
         likelihood = dbinom(9, 10, pi_grid))
```

In **Step 3**, we calculate the product of the `likelihood` and `prior` at each grid value. This provides an *unnormalized* discrete approximation of the posterior pdf that doesn't sum to one. We subsequently **normalize** this approximation by dividing each unnormalized posterior value by their collective sum:

```
# Step 3: Approximate the posterior
grid_data <- grid_data %>%
  mutate(unnormalized = likelihood * prior,
         posterior = unnormalized / sum(unnormalized))

# Confirm that the posterior approximation sums to 1
grid_data %>%
  summarize(sum(unnormalized), sum(posterior))
  sum(unnormalized) sum(posterior)
1             0.318              1
```

The resulting discretized posterior pdf, rounded to 2 decimal places and plotted below, provides a very rigid glimpse into the *actual* posterior pdf. It places a roughly 99% chance on π being either 0.6 or 0.8, a 1% chance on π being 0.4, and a near 0% chance on the other 3 π grid values:

```
# Examine the grid approximated posterior
round(grid_data, 2)
  pi_grid prior likelihood unnormalized posterior
1     0.0  0.00       0.00         0.00      0.00
2     0.2  0.96       0.00         0.00      0.00
3     0.4  1.44       0.00         0.00      0.01
4     0.6  1.44       0.04         0.06      0.18
5     0.8  0.96       0.27         0.26      0.81
6     1.0  0.00       0.00         0.00      0.00
```

```
# Plot the grid approximated posterior
ggplot(grid_data, aes(x = pi_grid, y = posterior)) +
  geom_point() +
  geom_segment(aes(x = pi_grid, xend = pi_grid, y = 0, yend = posterior))
```

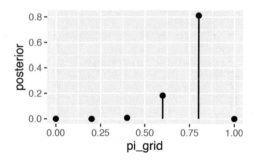

FIGURE 6.1: The discretized posterior pdf of π at only 6 grid values.

You might anticipate what will happen when we use this approximation to simulate samples from the posterior in **Step 4**. Each sample draw has only 6 possible outcomes and is highly likely to be 0.6 or 0.8. Let's try it: use `sample_n()` to take a sample of `size = 10000` values from the 6-length `grid_data`, with replacement, and using the discretized `posterior` probabilities as sample weights.

```
# Set the seed
set.seed(84735)
```

```
# Step 4: sample from the discretized posterior
post_sample <- sample_n(grid_data, size = 10000,
                        weight = posterior, replace = TRUE)
```

As expected, most of our 10,000 sample values of π were 0.6 or 0.8, few were 0.4, and *none* were below 0.4 or above 0.8:

```
# A table of the 10000 sample values
post_sample %>%
  tabyl(pi_grid) %>%
  adorn_totals("row")
 pi_grid      n percent
```

0.4	69	0.0069
0.6	1885	0.1885
0.8	8046	0.8046
Total	10000	1.0000

This is an *extremely* oversimplified approximation of the true Beta(11, 3) posterior. If you need any more convincing, Figure 6.2 superimposes the true posterior pdf $f(\pi|y)$ on a scaled histogram of the 10,000 sample values. If this *were* a good approximation, the histogram would mimic the shape, location, and spread of the smooth pdf. It does not.

```
# Histogram of the grid simulation with posterior pdf
ggplot(post_sample, aes(x = pi_grid)) +
  geom_histogram(aes(y = ..density..), color = "white") +
  stat_function(fun = dbeta, args = list(11, 3)) +
  lims(x = c(0, 1))
```

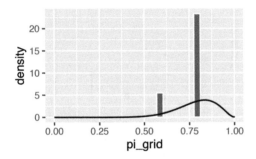

FIGURE 6.2: A grid approximation of the posterior pdf of π using only 6 grid values. The actual pdf is represented by the curve.

Remember the rainbow image and how we got a more complete picture by viewing snippets along a finer grid? Similarly, instead of chopping up the 0-to-1 continuum of possible π values into a grid of only 6 values, let's try a more reasonable grid of **101** values: $\pi \in \{0, 0.01, 0.02, \ldots, 0.99, 1\}$. The first 3 grid approximation steps using this refined grid are performed below:

```
# Step 1: Define a grid of 101 pi values
grid_data  <- data.frame(pi_grid = seq(from = 0, to = 1, length = 101))

# Step 2: Evaluate the prior & likelihood at each pi
grid_data <- grid_data %>%
  mutate(prior = dbeta(pi_grid, 2, 2),
         likelihood = dbinom(9, 10, pi_grid))


# Step 3: Approximate the posterior
grid_data <- grid_data %>%
  mutate(unnormalized = likelihood * prior,
         posterior = unnormalized / sum(unnormalized))
```

The resulting discretized posterior pdf is quite smooth, especially in comparison to the rigid approximation when we only used 6 grid values:

```
ggplot(grid_data, aes(x = pi_grid, y = posterior)) +
  geom_point() +
  geom_segment(aes(x = pi_grid, xend = pi_grid, y = 0, yend = posterior))
```

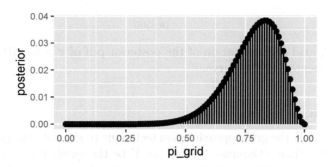

FIGURE 6.3: The discretized posterior pdf of π at 101 grid values.

Finally, in step 4 of the grid approximation, we sample 10,000 draws from the discretized posterior pdf.

```
# Set the seed
set.seed(84735)

# Step 4: sample from the discretized posterior
post_sample <- sample_n(grid_data, size = 10000,
                         weight = posterior, replace = TRUE)
```

Figure 6.4 displays a histogram of the resulting sample values. The shape, location, and spread of these values mimic those of the true target posterior pdf $f(\pi|y)$, represented by the smooth curve. Exciting! Take a step back to appreciate what we've just accomplished. We've taken an independent *sample* from a *discretized approximation* of the posterior pdf $f(\pi|y)$ that's defined only on a *grid* of π values. The algorithm is pretty intuitive and didn't require advanced programming skills to implement. Most of all, the result was a sample that behaves much like a random sample taken directly from $f(\pi|y)$ itself. Thus, grid approximation can be a lifeline in scenarios in which the posterior is intractable, providing an approximation that's nearly indistinguishable from the "real thing."

```
ggplot(post_sample, aes(x = pi_grid)) +
  geom_histogram(aes(y = ..density..), color = "white", binwidth = 0.05) +
  stat_function(fun = dbeta, args = list(11, 3)) +
  lims(x = c(0, 1))
```

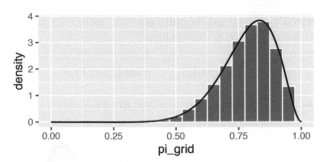

FIGURE 6.4: A grid approximation of the posterior pdf of π using 101 grid values. The actual pdf is represented by the curve.

6.1.2   A Gamma-Poisson example

For practice, let's apply the grid approximation technique to simulate the posterior of another Bayesian model, this time a Gamma-Poisson. Let Y be the number of events that occur in a one-hour period, where events occur at an average rate of λ per hour. Further, suppose we collect two data points (Y_1, Y_2) and place a Gamma$(3, 1)$ prior on λ:

$$
\begin{aligned}
Y_i|\lambda &\stackrel{ind}{\sim} \text{Pois}(\lambda) \\
\lambda &\sim \text{Gamma}(3, 1).
\end{aligned}
\tag{6.2}
$$

If we observe $Y_1 = 2$ events in the first one-hour observation period and $Y_2 = 8$ in the next, then by our work in Chapter 5, the updated posterior model of λ is Gamma with parameters 13 $(s + \sum Y = 3 + 10)$ and 3 $(r + n = 1 + 2)$:

$$
\lambda|((Y_1, Y_2) = (2, 3)) \sim \text{Gamma}(13, 3).
$$

To *simulate* this posterior using grid approximation, **Step 1** is to specify a discrete grid of reasonable λ values. Unlike the π parameter in the Beta-Binomial which is restricted to the finite 0-to-1 interval, the Poisson parameter λ can technically take on any non-negative real value ($\lambda \in [0, \infty)$). However, in a plot of the Gamma prior pdf and Poisson likelihood function it appears that, though possible, values of λ beyond 15 are implausible (Figure 6.5). With this in mind, we'll set up a discrete grid of λ values between 0 and 15, essentially truncating the posterior's tail. Before completing the corresponding grid approximation together, we encourage you to challenge yourself by first trying the quiz below.

```
plot_gamma_poisson(s = 3, r = 1, sum_y = 10, n = 2, posterior = FALSE)
```

❓ Quiz Yourself!

Fill in the code below to construct a grid approximation of the Gamma-Poisson posterior corresponding to (6.2). In doing so, use a grid of 501 λ values between 0 and 15.

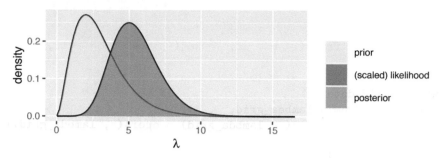

FIGURE 6.5: A Gamma prior pdf and scaled Poisson likelihood function for λ.

```
# Step 1: Define a grid of 501 lambda values
grid_data   <- data.frame(
  lambda_grid = seq(from = ___, to = ___, length = ___))

# Step 2: Evaluate the prior & likelihood at each lambda
grid_data <- grid_data %>%
  ___(prior = dgamma(___, ___, ___),
      likelihood = dpois(___, ___) * dpois(___, ___))

# Step 3: Approximate the posterior
grid_data <- grid_data %>%
  ___(unnormalized = ___,
      posterior = ___)

# Set the seed
set.seed(84735)

# Step 4: sample from the discretized posterior
post_sample <- sample_n(___, size = ___,
                        weight = ___, replace = ___)
```

Check out the complete code below. Much of this is the same as it was for the Beta-Binomial model. There are two key differences. First, we use **dgamma()** and dpois() instead of **dbeta()** and **dbinom()** to evaluate the prior pdf and likelihood function of λ. Second, since we have a sample of two data points $(Y_1, Y_2) = (2, 8)$, the Poisson likelihood function of λ must be calculated by the product of the marginal Poisson pdfs, $L(\lambda|y_1, y_2) = f(y_1, y_2|\lambda) = f(y_1|\lambda)f(y_2|\lambda)$.

```
# Step 1: Define a grid of 501 lambda values
grid_data    <- data.frame(lambda_grid = seq(from = 0, to = 15, length = 501))

# Step 2: Evaluate the prior & likelihood at each lambda
grid_data <- grid_data %>%
  mutate(prior = dgamma(lambda_grid, 3, 1),
         likelihood = dpois(2, lambda_grid) * dpois(8, lambda_grid))

# Step 3: Approximate the posterior
grid_data <- grid_data %>%
  mutate(unnormalized = likelihood * prior,
         posterior = unnormalized / sum(unnormalized))

# Set the seed
set.seed(84735)

# Step 4: sample from the discretized posterior
post_sample <- sample_n(grid_data, size = 10000,
                        weight = posterior, replace = TRUE)
```

Most importantly, grid approximation again produced a decent approximation of our target posterior:

```
# Histogram of the grid simulation with posterior pdf
ggplot(post_sample, aes(x = lambda_grid)) +
  geom_histogram(aes(y = ..density..), color = "white") +
  stat_function(fun = dgamma, args = list(13, 3)) +
  lims(x = c(0, 15))
```

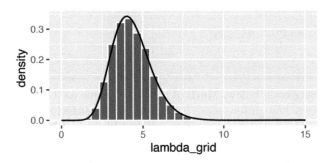

FIGURE 6.6: A grid approximation of the posterior pdf of λ using 101 grid values. The actual pdf is represented by the curve.

6.1.3   Limitations

Some say that all good things must come to an end. Though we don't agree with this saying in general, it happens to be true in the case of grid approximation. Limitations in the grid approximation method quickly present themselves as our models get more complicated. For example, by the end of Unit 4 we'll be working with models that have *lots* of model parameters $\theta = (\theta_1, \theta_2, \ldots, \theta_k)$. In such settings, grid approximation suffers from the **curse**

of dimensionality. Let's return to our image approximation for some intuition. Above, we assumed that we could only see snippets of the image along a grid that sweeps from left to right along the x-axis. This is analogous to using grid approximation to simulate a model with *one* parameter. Suppose instead that our model has *two* parameters. Or, in the case of the image approximation, we can only see snippets along a grid that sweeps from left to right along the x-axis *and* from top to bottom along the y-axis:

When we chop *both* the x- and y-axes into grids, there are bigger *gaps* in the image approximation. To achieve a more refined approximation, we need a finer grid than when we only chopped the x-axis into a grid. Analogously, when using grid approximation to simulate multivariate posteriors, we need to divide the multidimensional sample space of $\theta = (\theta_1, \theta_2, \ldots, \theta_k)$ into a very, very fine grid in order to prevent big gaps in our approximation. In practice, this might not be feasible. When evaluated on finer and finer grids, the grid approximation method becomes computationally expensive. You can't merely start the simulation, get up for a cup of coffee, come back, and *poof* the simulation is done. You might have to start the simulation and go off to a month-long meditation retreat (to practice the patience you'll need for grid approximation). MCMC methods provide a more flexible alternative.

6.2 Markov chains via rstan

MCMC methods and their coinage hold some historical significance. The *Markov chain* component of MCMC is named for the Russian mathematician Andrey Markov (1856–1922). The etymology of the *Monte Carlo* component is more dubious. As part of their top secret nuclear weapons project in the 1940s, Stanislav Ulam, John von Neumann, and their collaborators at the Los Alamos National Laboratory used Markov chains to simulate and better understand neutron travel (Eckhardt, 1987). The Los Alamos team referred to their work by the code name "Monte Carlo," a choice said to be inspired by the opulent Monte Carlo casino in the French Riviera. What inspired the link between random simulation methods and casinos? We can't help you there.

Today, "Markov chain Monte Carlo" refers to the application of Markov chains to simulate probability models using methods pioneered by the Monte Carlo project. In contrast to grid approximation, MCMC simulation methods scale up for more complicated Bayesian models. But this flexibility doesn't come for free. Like grid approximation samples, **MCMC samples are *not* taken directly from the posterior pdf** $f(\theta|y)$. Yet unlike grid approximation samples, **MCMC samples aren't even independent** – each subsequent sample value depends directly upon the previous value. This feature is evoked by the "chain" terminology. Specifically, let $\{\theta^{(1)}, \theta^{(2)}, \ldots, \theta^{(N)}\}$ be an N-length MCMC sample, or **Markov chain**. In constructing this chain, $\theta^{(2)}$ is drawn from some model that depends upon $\theta^{(1)}$, $\theta^{(3)}$ is drawn from some model that depends upon $\theta^{(2)}$, $\theta^{(4)}$ is drawn from some model that depends upon $\theta^{(3)}$, and so on and so on the chain grows. In general, the $(i+1)$st chain value $\theta^{(i+1)}$ is drawn from a model that depends on data y and the previous chain value $\theta^{(i)}$ with conditional pdf

FIGURE 6.7: A billboard at a uranium processing plant in Oak Ridge, TN, a sister site of Los Alamos. Image: `https://commons.wikimedia.org/wiki/File:Oak_Ridge_Wise_Mon keys.jpg`

$$f\left(\theta^{(i+1)} \mid \theta^{(i)}, y\right).$$

There are a couple of things to note about this dependence among chain values. First, by the **Markov property**, $\theta^{(i+1)}$ depends upon the preceding chain values only through the most recent value $\theta^{(i)}$:

$$f\left(\theta^{(i+1)} \mid \theta^{(1)}, \theta^{(2)}, \ldots, \theta^{(i)}, y\right) = f\left(\theta^{(i+1)} \mid \theta^{(i)}, y\right).$$

For example, since $\theta^{(i+i)}$ depends on $\theta^{(i)}$ which depends on $\theta^{(i-1)}$, it's also true that $\theta^{(i+i)}$ depends on $\theta^{(i-1)}$. However, if we know $\theta^{(i)}$, then $\theta^{(i-1)}$ is of no consequence to $\theta^{(i+1)}$ – the only information we need to simulate $\theta^{(i+1)}$ is the value of $\theta^{(i)}$. Further, each chain value can be drawn from a different model, and *none* of these models are the target posterior. That is, the pdf from which a Markov chain value is simulated is not equivalent to the posterior pdf:

$$f\left(\theta^{(i+1)} \mid \theta^{(i)}, y\right) \neq f\left(\theta^{(i+1)} \mid y\right).$$

Markov chain Monte Carlo

MCMC simulation produces a *chain* of N **dependent** θ values, $\left\{\theta^{(1)}, \theta^{(2)}, \ldots, \theta^{(N)}\right\}$, which are **not** drawn from the posterior pdf $f(\theta|y)$.

It likely seems strange to approximate the posterior using a dependent sample that's not even taken from the posterior. But mathemagically, with reasonable MCMC *algorithms*, it can work. The rub is that these algorithms have a steeper learning curve than the grid approximation technique. We provide a glimpse into the details in Chapter 7. Though a review of Chapter 7 and a firm grasp of these details would be ideal, there's a growing number of MCMC computing resources that can do the heavy lifting for us. In this book, we'll conduct MCMC simulation using the **rstan** package (Guo et al., 2020) which combines the power of R with the Stan engine. You'll get into the nitty gritty of **rstan** in Chapter 6, building up the code line by line and thereby gaining familiarity with the key elements of an **rstan** simulation. Starting in Chapter 9, you will utilize the complementary **rstanarm** package, which provides shortcuts for simulating a broad framework of Bayesian *applied regression models* (**arm**).

6.2.1   A Beta-Binomial example

There are two essential steps to all **rstan** analyses: (1) **define the Bayesian model structure** in **rstan** notation and (2) **simulate the posterior**. Let's examine these steps in the context of the Beta-Binomial model defined by (6.1), starting with **step 1**. In defining the structure of this model, we must specify the three aspects upon which it depends:

- data
  Data Y is the observed number of successes in 10 trials. Since **rstan** isn't a mind reader, we must specify that Y is an *integer* between 0 and 10.

- parameters
  The model depends upon parameter π, or pi in **rstan** notation. We must specify that π can be any *real* number between 0 and 1.

- model
  The model is defined by the Bin$(10, \pi)$ model for data Y and the Beta$(2, 2)$ prior for π. We specify these using binomial() and beta().

Below we translate this Beta-Binomial structure into **rstan** syntax and store it as the *character string* bb_model. We encourage you to pause and examine the code, noting how it matches up with the three aspects above.

```
# STEP 1: DEFINE the model
bb_model <- "
  data {
    int<lower = 0, upper = 10> Y;
  }
  parameters {
    real<lower = 0, upper = 1> pi;
  }
  model {
    Y ~ binomial(10, pi);
    pi ~ beta(2, 2);
  }
"
```

In **step 2**, we *simulate* the posterior using the stan() function. *Very* loosely speaking, stan() designs and runs an MCMC algorithm to produce an approximate sample from the Beta-Binomial posterior.

> **❶ Warning**
>
> Since `stan()` has to do the double duty of identifying an appropriate MCMC algorithm
> for simulating the given model, and then applying this algorithm to our data, the
> simulation will be quite slow for each new model.

```
# STEP 2: SIMULATE the posterior
bb_sim <- stan(model_code = bb_model, data = list(Y = 9),
               chains = 4, iter = 5000*2, seed = 84735)
```

Note that `stan()` requires two types of arguments. First, we must specify the **model
information** by:

- `model_code` = the character string defining the model (here `bb_model`)
- `data` = a list of the observed data (here `Y = 9`).

Second, we must specify the desired **Markov chain information** using three additional
arguments:

- The `chains` argument specifies how many **parallel Markov chains** to run. We run
  four chains here, thus obtain four distinct samples of π values. We discuss this choice in
  Section 6.3.

- The `iter` argument specifies the desired number of **iterations** in, or *length* of, each
  Markov chain. By default, the first half of these iterations are thrown out as "burn-in" or
  "warm-up" samples (see below for details). The second half are kept as the final Markov
  chain sample.

- To set the random number generating seed for an **rstan** simulation, we utilize the `seed`
  argument *within* the `stan()` function.

The result, stored in `bb_sim`, is a `stanfit` object. This object includes four parallel Markov
chains run for 10,000 iterations each. After tossing out the first 5,000 iterations of all four
chains, we end up with four separate Markov chain samples of size 5,000, or a combined
Markov chain sample size of 20,000.

> **Burn-in**
>
> If you've ever made a batch of pancakes or crêpes, you know that the first pancake
> is always the worst – the pan isn't yet at the perfect temperature, you haven't yet
> figured out how much batter to use, and you need more time to practice your flipping
> technique. MCMC chains are similar. Without direct knowledge of the posterior it's
> trying to simulate, the Markov chain might start out sampling unreasonable values of
> our parameter of interest, say π. Eventually though, it *learns* and starts producing
> values that mimic a random sample from the posterior. And just as we might need to
> toss out the first pancake, we might want to toss the Markov chain values produced
> during this learning period – keeping them in our sample might lead to a poor posterior
> approximation. As such, "burn-in" is the practice of discarding the first portion of
> Markov chain values.

The first four π values for each of the four parallel chains are extracted and shown here:

```
as.array(bb_sim, pars = "pi") %>%
  head(4)
, , parameters = pi

          chains
iterations chain:1 chain:2 chain:3 chain:4
      [1,]  0.9403  0.8777  0.3803  0.6649
      [2,]  0.9301  0.9802  0.8186  0.6501
      [3,]  0.9012  0.9383  0.8458  0.7001
      [4,]  0.9224  0.9540  0.7336  0.5902
```

It's important to remember that these **Markov chain values are NOT a random sample from the posterior and are NOT independent**. Rather, each of the four parallel chains forms a dependent 5,000 length Markov *chain* of π values, $\left(\pi^{(1)}, \pi^{(2)}, \ldots, \pi^{(5000)}\right)$. For example, in iteration 1, `chain:1` starts at a value of roughly $\pi^{(1)} = 0.9403$. The value at iteration 2, $\pi^{(2)}$, *depends upon* $\pi^{(1)}$. In this case the chain moves from 0.9403 to 0.9301. Similarly, the chain moves from 0.9301 to 0.9012, from 0.9012 to 0.9224, and so on. Thus, the chain traverses the **sample space** or range of posterior plausible π values. A Markov chain **trace plot** illustrates this traversal, plotting the π value (y-axis) in each iteration (x-axis). We use the `mcmc_trace()` function in the **bayesplot** package (Gabry et al., 2019) to construct the trace plots of all four Markov chains:

```
mcmc_trace(bb_sim, pars = "pi", size = 0.1)
```

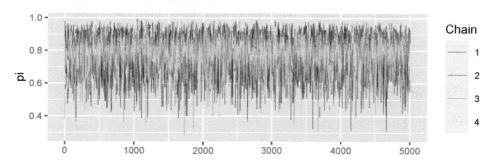

FIGURE 6.8: Trace plots of the four parallel Markov chains of π.

Figure 6.9 zooms in on the trace plot of chain 1. In the first 20 iterations (left), the chain largely explores values between 0.57 and 0.94. After 200 iterations (right), the Markov chain has started to explore new territory, traversing a slightly wider range of values between 0.49 and 0.96. Both trace plots also exhibit evidence of the slight dependence among the Markov chain values, places where the chain tends to float up for multiple iterations and then down for multiple iterations.

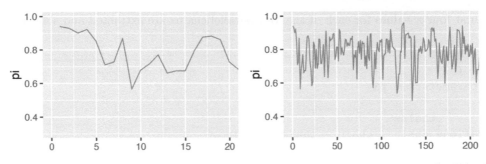

FIGURE 6.9: A trace plot of the first 20 iterations (left) and 200 iterations (right) of the first Beta-Binomial Markov chain.

Marking the *sequence* of the chain values, the trace plots in Figure 6.9 illuminate the Markov chains' **longitudinal** behavior. We also want to examine the **distribution** of the values these chains visit along their journey, ignoring the order of these visits. The histogram and density plot in Figure 6.10 provide a snapshot of this distribution for the combined 20,000 chain values, 5,000 from each of the four separate chains. Notice the important punchline here: the distribution of the Markov chain values is an excellent *approximation* of the target Beta(11, 3) posterior model of π (superimposed in black). That's a relief – that was the whole point.

> ❗**Warning**
>
> Like some other plotting functions in the **bayesplot** package, the `mcmc_hist()` and `mcmc_dens()` functions don't automatically include axis labels and scales. As we're new to these plots, we add labels and scales here using `yaxis_text(TRUE)` and `ylab()`. As we become more and more comfortable with these plots, we'll fall back on the defaults.

```
# Histogram of the Markov chain values
mcmc_hist(bb_sim, pars = "pi") +
  yaxis_text(TRUE) +
  ylab("count")

# Density plot of the Markov chain values
mcmc_dens(bb_sim, pars = "pi") +
  yaxis_text(TRUE) +
  ylab("density")
```

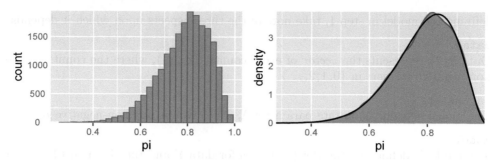

FIGURE 6.10: A histogram (left) and density plot (right) of the combined 20,000 Markov chain π values from the 4 parallel chains. The target pdf is superimposed in black.

6.2.2 A Gamma-Poisson example

For more practice with **rstan** and MCMC simulation, let's use these tools to approximate the Gamma-Poisson posterior corresponding to (6.2) upon observing data $(Y_1, Y_2) = (2, 8)$. Recall that this involves two steps. In **step 1**, we define the Gamma-Poisson model structure, specifying the three aspects upon which it depends: the **data**, **parameters**, and **model**. In **step 2**, we simulate the posterior. We encourage you to challenge yourself by trying this on your own first. Your code will require two terms you haven't yet seen, but you might guess how to use: **poisson()** and **gamma()**. Further, your code will incorporate a *vector* of (Y_1, Y_2) variables and observations as opposed to a single variable Y.

> **❓ Quiz Yourself!**
>
> Fill in the code below to construct an MCMC approximation of the Gamma-Poisson posterior corresponding to (6.2). In doing so, run four parallel chains for 10,000 iterations each (resulting in a sample size of 5,000 per chain).
>
> ```
> # STEP 1: DEFINE the model
> gp_model <- "
>     data {
>         int<___> Y[2];
>     }
>     parameters {
>         real<___> lambda;
>     }
>     model {
>         Y ~ ___(___);
>         lambda ~ ___(___, ___);
>     }
> "
>
> # STEP 2: SIMULATE the posterior
> gp_sim <- ___(model_code = ___, data = list(___),
>               chains = ___, iter = ___, seed = 84735)
> ```

In defining the model in step 1, take note of the three aspects upon which it depends.

- `data`
  Data `Y[2]` represents the *vector* of event counts, (Y_1, Y_2), where the counts can be any non-negative *integers* in $\{0, 1, 2, \ldots\}$.

- `parameters`
  The model depends upon rate parameter λ, which can be any non-negative *real* number.

- `model`
  The model is defined by the Pois(λ) model for data Y and the Gamma(3,1) prior for λ. We specify these using `poisson()` and `gamma()`.

In step 2, we must feed in the vector of both `data` points, `Y = c(2,8)`. It follows that:

```
# STEP 1: DEFINE the model
gp_model <- "
  data {
    int<lower = 0> Y[2];
  }
  parameters {
    real<lower = 0> lambda;
  }
  model {
    Y ~ poisson(lambda);
    lambda ~ gamma(3, 1);
  }
"

# STEP 2: SIMULATE the posterior
gp_sim <- stan(model_code = gp_model, data = list(Y = c(2,8)),
               chains = 4, iter = 5000*2, seed = 84735)
```

Figure 6.11 illustrates the simulation results. First, the trace plots (top) illustrate the paths these chains take through the λ sample space. To this end, note that the λ chain values travel around the rough range from 0 to 10, but mostly remain below 7.5. Further, the histogram and density plot (bottom) provide a better look into the overall distribution of λ Markov chain values. The punchline: this distribution provides an excellent approximation of the Gamma(13,3) posterior (superimposed in black).

```
# Trace plots of the 4 Markov chains
mcmc_trace(gp_sim, pars = "lambda", size = 0.1)

# Histogram of the Markov chain values
mcmc_hist(gp_sim, pars = "lambda") +
  yaxis_text(TRUE) +
  ylab("count")

# Density plot of the Markov chain values
mcmc_dens(gp_sim, pars = "lambda") +
  yaxis_text(TRUE) +
  ylab("density")
```

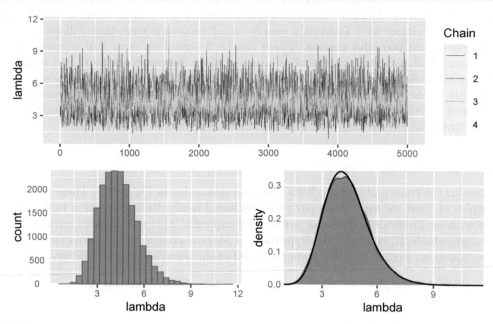

FIGURE 6.11: Trace plots of the four parallel Markov chains of λ (top). A histogram and density plot of the 20,000 combined Markov chain values of λ (bottom). The black curve represents the actual posterior pdf of λ.

6.3 Markov chain diagnostics

Simulation is fantastic. Even when we're able to derive a Bayesian posterior, we can use posterior simulation to verify our work and provide some intuition. More importantly, when we're *not* able to derive a posterior, simulated samples provide a crucial *approximation*. In the MCMC examples above, we saw that the Markov chains traverse the sample space of our parameter (π or λ) and, in the end, *mimic* a random sample that eventually *converges* to the posterior. "Approximation" and "converges" are key words here – **simulations aren't perfect.** This begs the following questions:

- What does a *good* Markov chain look like?
- How can we tell if our Markov chain sample produces a *reasonable* approximation of the posterior?
- How *big* should our Markov chain sample size be?

Answering these questions is both an art and science. There are no one-size-fits-all magic formulas that provide definitive answers here. Rather, it's through experience that you get a feel for what "good" Markov chains look like and what you can do to fix a "bad" Markov chain. In this section, we'll focus on a couple visual diagnostic tools that will get us started: **trace plots** and **parallel chains**. These can be followed up with and supplemented by a few numerical diagnostics: **effective sample size**, **autocorrelation**, and **R-hat** ($\hat{R}$). Utilizing these diagnostics should be done *holistically*. Since no single visual or numerical diagnostic is one-size-fits-all, they provide a fuller picture of Markov chain quality when considered together. Further, other excellent diagnostics exist. We focus here on those that are common and easy to implement in the software packages we'll be using.

6.3.1 Examining trace plots

Reexamine the trace plots for the Beta-Binomial simulation in Figure 6.8. These are textbook examples of what we *want* trace plots to look like. Mainly, they look like a bunch of white noise with no discernible trends or notable phenomena. This nothingness implies that the chains are *stable*. In contrast, the hypothetical Markov chains for the same Beta-Binomial model shown in Figure 6.12 illustrate potential chain behavior that should give us pause.

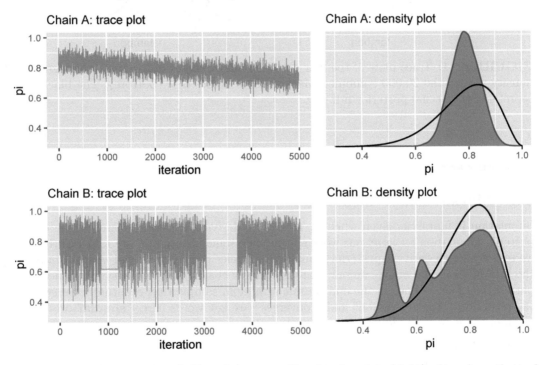

FIGURE 6.12: Trace plots (left) and corresponding density plots (right) of two hypothetical Markov chains. These provide examples of what "bad" Markov chains might look like. The superimposed black lines (right) represent the target Beta(11,3) posterior pdf.

First, consider the trace plots in Figure 6.12 (left). The downward trend in Chain A indicates that it has not yet stabilized after 5,000 iterations – it has not yet "found" or does not yet know how to explore the range of posterior plausible π values. The downward trend also hints at strong correlation among the chain values – they don't look like independent noise. All of this to say that Chain A is **mixing slowly**. This is bad. Though Markov chains are inherently dependent, the more they *behave* like fast mixing (noisy) independent samples, the smaller the error in the resulting posterior approximation (roughly speaking). Chain B exhibits a different problem. As evidenced by the two completely flat lines in the trace plot, it tends to get *stuck* when it visits smaller values of π.

The density plots in Figure 6.12 (right) confirm that both of these goofy-looking chains result in a serious issue: they produce poor approximations of the Beta(11,3) posterior (superimposed in black), and thus misleading posterior conclusions. Consider Chain A. Since it's mixing so slowly, it has only explored π values in the rough range from 0.6 to 0.9 in its first 5,000 iterations. As a result, its posterior approximation overestimates the plausibility of π values in this range while completely underestimating the plausibility of values outside this range. Next, consider Chain B. In getting stuck, Chain B over-samples some values in the left tail of the posterior π values. This phenomenon produces the erroneous spikes in the posterior approximation.

In practice, we run `rstan` simulations when we can't specify, and thus want to approximate the posterior. This means that we won't have the privilege of being able to compare our simulation results to the "real" posterior. This is why diagnostics are so important. If we see bad trace plots like those in Figure 6.12, there are some immediate steps we can take:

1.   Check the model. Are the assumed prior and data models appropriate?
2.   Run the chain for more iterations. Some undesirable short-term chain trends might iron out in the long term.

We'll get practice with the more nuanced Step 1 throughout the book. Step 2 is easy, though it requires extra computation time.

6.3.2   Comparing parallel chains

Recall that our `stan()` simulation for the Beta-Binomial model produced four parallel Markov chains. Not only do we want to see stability in each individual chain (as discussed above), we want to see consistency *across* the four chains. Mainly, though we expect different chains take different paths, they should exhibit similar features and produce similar posterior approximations. For example, the trace plots for the four parallel chains in Figure 6.8 appear similar in their randomness. Further, in the Figure 6.13 density plots, we observe that these four chains produce nearly indistinguishable posterior approximations. This provides evidence that our simulation is stable and sufficiently long – running the chains for more iterations likely wouldn't produce drastically different or improved posterior approximations.

```
# Density plots of individual chains
mcmc_dens_overlay(bb_sim, pars = "pi") +
  ylab("density")
```

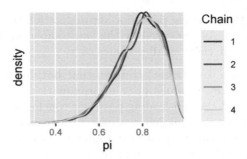

FIGURE 6.13: Density plot of the four parallel Markov chains for π.

For a point of comparison, let's implement a *shorter* Markov chain simulation for the same model. Instead of running four parallel chains for 10,000 iterations and a resulting sample size of 5,000 each, run four parallel chains for only 100 iterations and a resulting sample size of 50 each:

```
# STEP 2: SIMULATE the posterior
bb_sim_short <- stan(model_code = bb_model, data = list(Y = 9),
                     chains = 4, iter = 50*2, seed = 84735)
```

The trace plots and corresponding density plots of the short Markov chains are shown below. Though the chains' trace plots exhibit similar random behavior, their corresponding density plots differ, hence they produce discrepant posterior approximations. In the face of such instability and confusion about which of these four approximations is the most accurate, it would be a mistake to stop our simulation after only 100 iterations.

```
# Trace plots of short chains
mcmc_trace(bb_sim_short, pars = "pi")

# Density plots of individual short chains
mcmc_dens_overlay(bb_sim_short, pars = "pi")
```

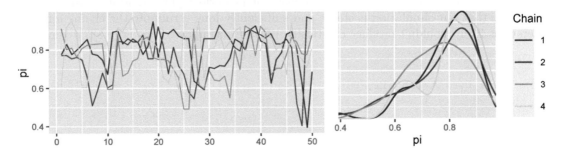

FIGURE 6.14: Trace plots and density plots of the four short parallel Markov chains for π, each of length 50.

6.3.3   Calculating effective sample size & autocorrelation

Recall from Section 6.3.1 that the more a dependent Markov chain *behaves* like an independent sample, the smaller the error in the resulting posterior approximation (loosely speaking).

Though trace plots provide some visual insight into this behavior, supplementary numerical assessments can provide more nuanced information. To begin, recall that our Markov chain simulation `bb_sim` produced a combined 20,000 *dependent* values of π, sampled from models *other* than the posterior. We then used this chain to approximate the posterior model of π. Knowing that the error in this approximation is likely larger than if we had used 20,000 *independent* sample values drawn directly *from the posterior* begs the following question: Relatively, how many independent sample values would it take to produce an equivalently accurate posterior approximation? The **effective sample size ratio** provides an answer.

Effective sample size ratio

Let N denote the *actual* sample size or length of a *dependent* Markov chain. The **effective sample size** of this chain, N_{eff}, quantifies the number of *independent* samples it would take to produce an equivalently accurate posterior approximation. The greater the N_{eff} the better, yet it's typically true that the accuracy of a Markov chain approximation is only as good as that of a smaller independent sample. That is, it's typically true that $N_{eff} < N$, thus the **effective sample size ratio** is less than 1:

$$\frac{N_{eff}}{N}$$

There's no magic rule for interpreting this ratio, and it should be utilized alongside other diagnostics such as the trace plot. That said, we might be suspicious of a Markov chain for which the effective sample size ratio is less than 0.1, i.e., the effective sample size N_{eff} is less than 10% of the actual sample size N.

A quick call to the `neff_ratio()` function in the **bayesplot** package provides the estimated effective sample size ratio for our Markov chain sample of `pi` values. Here, the accuracy in using our 20,000 length Markov chain to approximate the posterior of π is roughly as great as if we had used only 34% as many *independent* values. Put another way, our 20,000 Markov chain values are about as useful as only 6800 independent samples ($0.34 \cdot 20000$). Since this ratio is above 0.1, we're not going to stress.

```
# Calculate the effective sample size ratio
neff_ratio(bb_sim, pars = c("pi"))
[1] 0.3361
```

Autocorrelation provides another metric by which to evaluate whether our Markov chain sufficiently mimics the behavior of an independent sample. *Strong* autocorrelation or dependence is a bad thing – it goes hand in hand with small effective sample size ratios, and thus provides a warning sign that our resulting posterior approximations might be unreliable. We saw some evidence of this in Chain A of Figure 6.12. Now, by the simple construction of a Markov chain, there's inherently going to be *some* autocorrelation among the chain values – one chain value $(\pi^{(i)})$ depends upon the previous chain value $(\pi^{(i-1)})$, which depends upon the one before that $(\pi^{(i-2)})$, which depends upon the one before that $(\pi^{(i-3)})$, and so on. This chain of dependencies also means that each chain value depends in some degree on all previous chain values. For example, since $\pi^{(i)}$ is dependent on $\pi^{(i-1)}$ which is dependent on $\pi^{(i-2)}$, $\pi^{(i)}$ is *also* dependent on $\pi^{(i-2)}$. Yet this dependence, or autocorrelation, fades. It's like Tobler's first law of geography: *everything is related to everything else, but near things are more related than distant things.* Thus, it's typically the case that a chain value $\pi^{(i)}$ is more strongly related to the previous value $(\pi^{(i-1)})$ than to a chain value 100 steps back $(\pi^{(i-100)})$.

Autocorrelation

Lag 1 autocorrelation measures the correlation between pairs of Markov chain values that are one "step" apart (e.g., $\pi^{(i)}$ and $\pi^{(i-1)}$). **Lag 2** autocorrelation measures the correlation between pairs of Markov chain values that are two "steps" apart (e.g., $\pi^{(i)}$ and $\pi^{(i-2)}$). And so on.

Let's apply these concepts to our `bb_sim` analysis. Check out the trace plot and **autocorrelation plot** of our simulation results in Figure 6.15. (For simplicity, we show the results for only one of our four parallel chains.)

```
mcmc_trace(bb_sim, pars = "pi")
mcmc_acf(bb_sim, pars = "pi")
```

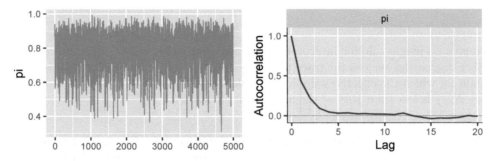

FIGURE 6.15: A trace plot (left) and autocorrelation plot (right) for a single Markov chain from the `bb_sim` analysis.

Again, notice that there are no obvious patterns in the trace plot. This provides one visual clue that, though the chain values are inherently dependent, this dependence is relatively weak and limited to small lags or values that are just a few steps apart. This observation is supported by the autocorrelation plot which marks the autocorrelation (y-axis) at lags 0 through 20 (x-axis). The lag 0 autocorrelation is naturally 1 – it measures the correlation between a Markov chain value and *itself*. From there, the lag 1 autocorrelation is roughly 0.5, indicating moderate correlation among chain values that are only 1 step apart. But then the autocorrelation quickly drops off and is effectively 0 by lag 5. That is, there's very little correlation between Markov chain values that are more than a few steps apart. This is all good news. It's more confirmation that our Markov chain is **mixing quickly**, i.e., quickly moving around the range of posterior plausible π values, and thus at least mimicking an independent sample.

Presuming you've never seen an autocorrelation plot before, we imagine that it's not very obvious that the plot in Figure 6.15 is a "good" one. For contrast, consider the results for an unhealthy Markov chain (Figure 6.16). The trace plot exhibits strong trends, and hence autocorrelation, in the Markov chain values. This observation is echoed and further formalized by the autocorrelation plot. The slow decrease in the autocorrelation curve indicates that the dependence between chain values does *not* quickly fade away. In fact, there's a roughly 0.9 correlation between Markov chain values that are a full *20 steps apart!*

Since its chain values are so strongly tied to the previous values, this chain is **slow mixing** –
it would take a long time for it to adequately explore the full range of the posterior. Thus,
just as with the slow mixing Chain A in Figure 6.12, we should be wary about using this
chain to approximate the posterior. Let's tie these ideas together.

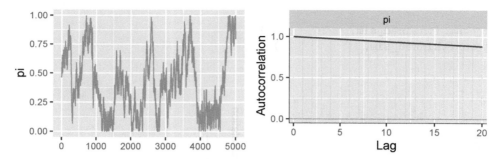

FIGURE 6.16: A trace plot (left) and autocorrelation plot (right) for a slow mixing Markov
chain of π.

Fast vs slow mixing Markov chains

Fast mixing chains exhibit behavior similar to that of an independent sample: the
chains move "quickly" around the range of posterior plausible values, the autocorrelation among the chain values drops off quickly, and the effective sample size ratio is
reasonably large. **Slow mixing** chains do *not* enjoy the features of an independent
sample: the chains move "slowly" around the range of posterior plausible values, the
autocorrelation among the chain values drops off very slowly, and the effective sample
size ratio is small.

So what do we do if ours is a dreaded slow mixing chain? The first and, in our opinion best,
possible solution is the tried and true "run a longer chain." Even a slow mixing chain can
eventually produce a good posterior approximation if we run it for enough iterations. Another
common approach is to **thin** the Markov chain. For example, from our original chain of 5000 π
values, we might keep every *second* value and toss out the rest: $\left\{\pi^{(2)}, \pi^{(4)}, \pi^{(6)}, \ldots, \pi^{(5000)}\right\}$.
Or, we might keep every *ten* chain values: $\left\{\pi^{(10)}, \pi^{(20)}, \pi^{(30)}, \ldots, \pi^{(5000)}\right\}$. By discarding the
draws in between, we remove the strong correlations at low lags. For example, $\pi^{(20)}$ is less
correlated with the previous value in the thinned chain ($\pi^{(10)}$) than with the previous value
in the original chain ($\pi^{(19)}$).

For illustration only, we thin out our original `bb_sim` chains to just every tenth value
using the `thin` argument in `stan()`. As a result, the autocorrelation drops off a bit earlier,
after 1 lag instead of 5 (Figure 6.17). BUT, since this was *already* a quick mixing simulation
with quickly decreasing autocorrelation and a relatively high effective sample size ratio, this
minor improvement in autocorrelation **isn't worth the information we lost**. Instead of a
sample size of 5000 chain values, we now only have 500 values with which to approximate
the posterior.

```
# Simulate a thinned MCMC sample
thinned_sim <- stan(model_code = bb_model, data = list(Y = 9),
                    chains = 4, iter = 5000*2, seed = 84735, thin = 10)

# Check out the results
mcmc_trace(thinned_sim, pars = "pi")
mcmc_acf(thinned_sim, pars = "pi")
```

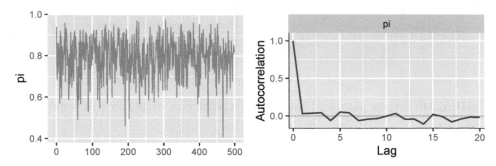

FIGURE 6.17: A trace plot (left) and autocorrelation plot (right) for a single Markov chain from the `bb_sim` analysis, thinned to every tenth value.

We similarly thin our slow mixing chain down to every tenth value (Figure 6.18). The resulting chain still exhibits slow mixing trends in the trace plot, but the autocorrelation drops more quickly than the pre-thinned chain. This is good, but is it worth losing 90% of our original sample values? We're not so sure.

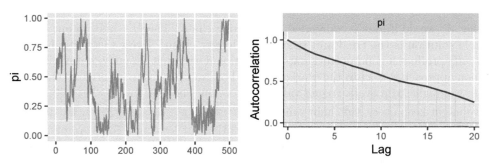

FIGURE 6.18: A trace plot (left) and autocorrelation plot (right) for a slow mixing Markov chain of π, thinned to every tenth value.

❶ Warning

There is a careful line to walk when deciding whether or not to thin a Markov chain. The benefits of reduced autocorrelation don't necessarily outweigh the loss of precious chain values. That is, 5000 Markov chain values with stronger autocorrelation might produce a better posterior approximation than 500 chain values with weaker autocorrelation. The effectiveness of thinning also depends in part on the algorithm used to construct the Markov chain. For example, the **rstan** and **rstanarm** packages used throughout this book employ an efficient Hamiltonian Monte Carlo algorithm. As such, in the current `stan()` help file, the package authors advise *against* thinning unless your simulation hogs up too much memory on your machine.

6.3.4 Calculating R-hat

Just as effective sample size and autocorrelation provide numerical supplements to the visual trace plot diagnostic in assessing the degree to which a Markov chain behaves like an independent sample, the split-$\hat{R}$ metric (or "R-hat" for short) provides a numerical supplement to the visual comparison of parallel chains. Recall from Section 6.3.2 that, not only do we want each *individual* Markov chain in our simulation to be stable, we want there to be consistency *across* the parallel chains. R-hat addresses this consistency by comparing the variability in sampled π values across all chains *combined* to the variability *within* each individual chain. Before presenting a more formal definition, check your intuition.[1]

> ❷ **Quiz Yourself!**
>
> Figure 6.19 provides simulation results for `bb_sim` (top row) along with a *bad* hypothetical alternative (bottom row). Based on the patterns in these plots, what do you think is a marker of a "good" Markov chain simulation?
>
> a. The variability in π values *within* any individual chain is **less than** the variability in π values across all chains *combined*.
> b. The variability in π values *within* any individual chain is **comparable to** the variability in π values across all chains *combined*.

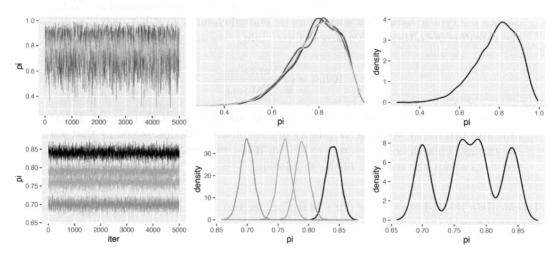

FIGURE 6.19: Simulation results for `bb_sim` (top row) and a hypothetical alternative (bottom row). Included are trace plots of the four parallel chains (left), density plots for each individual chain (middle), and a density plot of the combined chains (right).

---

[1]Answer: b

To answer this quiz, let's dig into Figure 6.19. Based on what we learned in Section 6.3.2, we can see that `bb_sim` is superior to the alternative – its parallel chains exhibit similar features and produce similar posterior approximations. In particular, the variability in π values is nearly identical *within* each chain (top middle plot). As a consequence, the variability in π values across all chains *combined* (top right plot) is similar to that of the individual chains. In contrast, notice that the four parallel chains in the alternative simulation produce conflicting posterior approximations (bottom middle plot), and hence an unstable and poor posterior approximation when we combine these chains (bottom right plot). As a consequence, the range and variability in π values across all chains *combined* are much larger than the range and variability in π values *within* any individual chain.

Bringing this analysis together, we've intuited the importance of the relationship between the variability in values across all chains *combined* and *within* the individual parallel chains. Specifically:

- in a "good" Markov chain simulation, the variability across all parallel chains combined will be roughly comparable to the variability within any individual chain;

- in a "bad" Markov chain simulation, the variability across all parallel chains combined might exceed the typical variability within each chain.

We can *quantify* the relationship between the combined chain variability and within-chain variability using R-hat. We provide an intuitive definition of R-hat here. For a more detailed definition, see Vehtari et al. (2021). And to learn more about the exciting connection between effective sample size and R-hat, see Vats and Knudson (2018)!

R-hat

Consider a Markov chain simulation of parameter θ which utilizes four parallel chains. Let $\text{Var}_{\text{combined}}$ denote the variability in θ across all four chains combined and $\text{Var}_{\text{within}}$ denote the typical variability within any individual chain. The R-hat metric calculates the ratio between these two sources of variability:

$$\text{R-hat} \approx \sqrt{\frac{\text{Var}_{\text{combined}}}{\text{Var}_{\text{within}}}}.$$

Ideally, R-hat ≈ 1, reflecting stability across the parallel chains. In contrast, R-hat > 1 indicates instability, with the variability in the combined chains exceeding that within the chains. Though no golden rule exists, an R-hat ratio greater than 1.05 raises some red flags about the stability of the simulation.

To calculate the R-hat ratio for our simulation, we can apply the `rhat()` function from the `bayesplot` package:

```
rhat(bb_sim, pars = "pi")
[1] 1
```

Reflecting our observation that the variability across and within our four parallel chains is comparable, `bb_sim` has an R-hat value that's effectively equal to 1. In contrast, the bad hypothetical simulation exhibited in Figure 6.19 has an R-hat value of 5.35. That is, the variance across all chain values combined is more than 5 *times* the typical variance within each chain. This *well* exceeds the 1.05 red flag marker, providing ample evidence that the hypothetical parallel chains do *not* produce consistent posterior approximations, thus the simulation is unstable.

6.4 Chapter summary

As our Bayesian models get more sophisticated, their posteriors will become too difficult, if not impossible, to specify. In Chapter 6, you learned two **simulation techniques** for **approximating the posterior** in such scenarios: grid approximation and Markov chain Monte Carlo. Both techniques produce a **sample** of N θ values,

$$\left\{ \theta^{(1)}, \theta^{(2)}, \ldots, \theta^{(N)} \right\}.$$

However, the properties of these samples differ:

- **Grid approximation** takes an **independent** sample of $\theta^{(i)}$ from a *discretized approximation* of posterior pdf $f(\theta|y)$. This approach is nicely straightforward but breaks down in more complicated model settings.
- **Markov chain Monte Carlo** offers a more flexible alternative. Though MCMC produces a **dependent** sample of $\theta^{(i)}$ which are **not** drawn from the posterior pdf $f(\theta|y)$, this sample will mimic the posterior model so long as the chain length N is big enough.

Finally, you learned some MCMC **diagnostics** for checking the resulting simulation quality. In short, we can visually examine **trace plots** and density plots of multiple **parallel chains** for **stability** and **mixing** in our simulation:

```
# Diagnostic plots
mcmc_trace(my_sim, pars = "___")
mcmc_dens_overlay(my_sim, pars = "___")
```

And we can supplement these visual diagnostics with numerical diagnostics such as **effective sample size ratio**, **autocorrelation**, and **R-hat**:

```
neff_ratio(my_sim, pars = "___")
mcmc_acf(my_sim, pars = "___")
rhat(my_sim, pars = "___")
```

6.5   Exercises

6.5.1   Conceptual exercises

Exercise 6.1 (Steps for grid approximation).

a)   Identify the steps for the grid approximation of a posterior model.
b)   *Which* step(s) would you change to make the approximation more accurate? *How* would you change them?

Exercise 6.2 (Trace plot diagnostics). For each MCMC simulation scenario described below, sketch by hand what a single chain trace plot might look like for each simulation.

a)   The chain is mixing too slowly.
b)   The chain has high correlation.
c)   The chain has a tendency to get "stuck."
d)   The chain has no problems!

Exercise 6.3 (MCMC woes). For each MCMC simulation scenario described below, describe how the scenario could impact the posterior approximation.

a)   The chain is mixing too slowly.
b)   The chain has high correlation.
c)   The chain has a tendency to get "stuck."

Exercise 6.4 (MCMC simulation: thank you for being a friend). Your friend missed class this week and they are allergic to reading textbooks (a common affliction). Since you are a true friend, you decide to help them out and answer their following questions:

a)   Why is it important to look at MCMC diagnostics?
b)   Why are MCMC simulations helpful?
c)   What are the benefits of using RStan?
d)   What don't *you* understand about the chapter?

6.5.2   Practice: Grid approximation

Exercise 6.5 (Beta-Binomial grid approximation). Consider the Beta-Binomial model for π with $Y|\pi \sim \text{Bin}(n, \pi)$ and $\pi \sim \text{Beta}(3, 8)$. Suppose that in $n = 10$ independent trials, you observe $Y = 2$ successes.

a)   Utilize grid approximation with grid values $\pi \in \{0, 0.25, 0.5, 0.75, 1\}$ to approximate the posterior model of π.
b)   Repeat part a using a grid of 201 equally spaced values between 0 and 1.

Exercise 6.6 (Gamma-Poisson grid approximation). Consider the Gamma-Poisson model for λ with $Y_i|\lambda \sim \text{Pois}(\lambda)$ and $\lambda \sim \text{Gamma}(20, 5)$. Suppose you observe $n = 3$ independent data points $(Y_1, Y_2, Y_3) = (0, 1, 0)$.

a)   Utilize grid approximation with grid values $\lambda \in \{0, 1, 2, \ldots, 8\}$ to approximate the posterior model of λ.

b)   Repeat part a using a grid of 201 equally spaced values between 0 and 8.

Exercise 6.7 (Normal-Normal grid approximation). Consider the Normal-Normal model for μ with $Y_i|\mu \sim N(\mu, 1.3^2)$ and $\mu \sim N(10, 1.2^2)$. Suppose that on $n = 4$ independent observations, you observe data $(Y_1, Y_2, Y_3, Y_4) = (7.1, 8.9, 8.4, 8.6)$.

a)   Utilize grid approximation with grid values $\mu \in \{5, 6, 7, \ldots, 15\}$ to approximate the posterior model of μ.

b)   Repeat part a using a grid of 201 equally spaced values between 5 and 15.

Exercise 6.8 (The Curse of Dimensionality). As we note in this chapter, grid approximation suffers from **the curse of dimensionality**.

a)   Describe a situation in which we would want to have inference for multiple parameters (i.e., high-dimensional Bayesian models).

b)   In your own words, explain how dimensionality can affect grid approximation and why this is a curse.

6.5.3   Practice: MCMC

Exercise 6.9 (Comparing MCMC to Grid Approximation).

a)   What drawback(s) do MCMC and grid approximation share?

b)   What advantage(s) do MCMC and grid approximation share?

c)   What is an advantage of grid approximation over MCMC?

d)   What is an advantage of MCMC over grid approximation?

Exercise 6.10 (Is it a Markov Chain?). Below are examples of "chains" $\{\theta^{(1)}, \theta^{(2)}, \ldots, \theta^{(N)}\}$, for different probability parameters θ. For each example, determine whether the given chain is a *Markov* chain. Explain.

a)   You go out to eat N nights in a row and $\theta^{(i)}$ is the probability you go to a Thai restaurant on day i.

b)   You play the lottery N days in a row and $\theta^{(i)}$ is the probability you win the lottery on day i.

c)   You play your roommate in chess for N games in a row and $\theta^{(i)}$ is the probability you win game i against your roommate.

Exercise 6.11 (MCMC with RStan: Step 1). Use the given information to define the Bayesian model structure using the correct RStan syntax. You don't need to run the code, just provide the syntax.

a) $Y|\pi \sim \text{Bin}(20, \pi)$ with $\pi \sim \text{Beta}(1, 1)$.

b) $Y|\lambda \sim \text{Pois}(\lambda)$ with $\lambda \sim \text{Gamma}(4, 2)$.

c) $Y|\mu \sim N(\mu, 1^2)$ with $\mu \sim N(0, 10^2)$.

Exercise 6.12 (MCMC with RStan: Steps 1 and 2). Use the given information to (1) define the Bayesian model structure, and (2) simulate the posterior using the correct RStan syntax. You don't need to run the code, just provide the syntax.

a) $Y|\pi \sim \text{Bin}(20, \pi)$ and $\pi \sim \text{Beta}(1, 1)$ with $Y = 12$.
b) $Y|\lambda \sim \text{Pois}(\lambda)$ and $\lambda \sim \text{Gamma}(4, 2)$ with $Y = 3$.
c) $Y|\mu \sim \text{N}(\mu, 1^2)$ and $\mu \sim \text{N}(0, 10^2)$ with $Y = 12.2$.

Exercise 6.13 (MCMC with RStan: Beta-Binomial). Consider the Beta-Binomial model for π with $Y|\pi \sim \text{Bin}(n, \pi)$ and $\pi \sim \text{Beta}(3, 8)$. Suppose that in $n = 10$ independent trials, you observe $Y = 2$ successes.

a)   Simulate the posterior model of π with RStan using 3 chains and 12000 iterations per chain.
b)   Produce trace plots for each of the three chains.
c)   What is the range of values on the trace plot x-axis? Why is the maximum value of this range not 12000?
d)   Create a density plot of the values for each of the three chains.
e)   Hearkening back to Chapter 5, specify the posterior model of π. How does your MCMC approximation compare?

Exercise 6.14 (MCMC with RStan: once more with feeling). Repeat Exercise 6.13 for the Beta-Binomial model with $Y|\pi \sim \text{Bin}(n, \pi)$ and $\pi \sim \text{Beta}(4, 3)$, where you observe $Y = 4$ successes in $n = 12$ independent trials.

Exercise 6.15 (MCMC with RStan: Gamma-Poisson). Consider the Gamma-Poisson model for λ with $Y_i|\lambda \sim \text{Pois}(\lambda)$ and $\lambda \sim \text{Gamma}(20, 5)$. Suppose that you observe $n = 3$ independent data points $(Y_1, Y_2, Y_3) = (0, 1, 0)$.

a)   Simulate the posterior model of λ with RStan using 4 chains and 10000 iterations per chain.
b)   Produce trace and density plots for all four chains.
c)   From the density plots, what seems to be the most posterior plausible value of λ?
d)   Hearkening back to Chapter 5, specify the posterior model of λ. How does your MCMC approximation compare?

Exercise 6.16 (MCMC with RStan: Gamma-Poisson again). Repeat exercise 6.15 using a $\lambda \sim \text{Gamma}(5, 5)$ prior model.

Exercise 6.17 (MCMC with RStan: Normal-Normal). Repeat exercise 6.15 for the Normal-Normal model of μ with $Y_i|\mu \sim N(\mu, 1.3^2)$ and $\mu \sim N(10, 1.2^2)$. Suppose that on $n = 4$ independent observations, you observe data $(Y_1, Y_2, Y_3, Y_4) = (7.1, 8.9, 8.4, 8.6)$.

Exercise 6.18 (MCMC with RStan: Normal-Normal part deux). Repeat exercise 6.15 for the Normal-Normal model of μ with $Y_i|\mu \sim N(\mu, 8^2)$ and $\mu \sim N(-14, 2^2)$. Suppose that on $n = 5$ independent observations, you observe data $(Y_1, Y_2, Y_3, Y_4, Y_5) = (-10.1, 5.5, 0.1, -1.4, 11.5)$.

7

MCMC under the Hood

In Chapter 6, we discussed the need for MCMC posterior simulation in advanced Bayesian analyses and how to achieve these ends using the **rstan** package. You now have everything you need to analyze more sophisticated Bayesian models in Unit 3 and beyond. Mainly, you can drive from point a to point b without a working understanding of the engine. However, whether now or later, it will be beneficial to your deeper understanding of Bayesian methodology (and fun) to learn about how MCMC methods actually *work*. We'll take a peak under the hood in Chapter 7. If learning about computation isn't your thing at this particular point in time, you'll still be able to do everything that comes after this chapter. You do you.

Bayesian MCMC simulation is a rich field, spanning various algorithms that share a common goal: approximate the posterior model. For example, the **rstan** package utilized throughout this book employs a **Hamiltonian Monte Carlo** algorithm. The alternative **rjags** package for Bayesian modeling employs a **Gibbs sampling** algorithm. Though the details differ, both algorithms are variations on the fundamental **Metropolis-Hastings** algorithm. Thus, instead of studying all variations, which would require a book itself, we'll focus our attention on the Metropolis-Hastings in Chapter 7. You will both build intuition for this algorithm and learn how to *implement* it, *from scratch*, in R. Though this implementation will require computer programming skills that are otherwise outside the scope of this book (e.g., writing **functions** and **for loops**), we'll keep the focus on the concepts so that all can follow along.

```
# Load packages you'll need for this chapter
library(tidyverse)
```

> ◎ **Goals**
> - Build a strong **conceptual understanding** of how Markov chain algorithms work.
> - Explore the foundational **Metropolis-Hastings** algorithm.
> - **Implement** the Metropolis-Hastings algorithm in the Normal-Normal and Beta-Binomial settings.

7.1 The big idea

Consider the following Normal-Normal model with numerical outcome Y that varies Normally around an unknown mean μ with a standard deviation of 0.75:

DOI: 10.1201/9780429288340-7

$$Y|\mu \sim N(\mu, 0.75^2)$$
$$\mu \sim N(0, 1^2) \tag{7.1}$$

The corresponding likelihood function $L(\mu|y)$ and prior pdf $f(\mu)$ for $y \in (-\infty, \infty)$ and $\mu \in (-\infty, \infty)$ are:

$$L(\mu|y) = \frac{1}{\sqrt{2\pi \cdot 0.75^2}} \exp\left[-\frac{(y-\mu)^2}{2 \cdot 0.75^2}\right] \quad \text{and} \quad f(\mu) = \frac{1}{\sqrt{2\pi}} \exp\left[-\frac{\mu^2}{2}\right]. \tag{7.2}$$

Suppose we observe an outcome $Y = 6.25$. Then by our Normal-Normal work in Chapter 5 (5.16), the posterior model of μ is Normal with mean 4 and standard deviation 0.6:

$$\mu|(Y = 6.25) \sim N(4, 0.6^2).$$

As we saw in Chapter 6, if we *weren't* able to specify this posterior model of μ (just pretend), we could *approximate* it using MCMC simulation. To get a sense for how this works, consider the results of a potential $N = 5000$ iteration MCMC simulation (Figure 7.1). It helps to think of the illustrated Markov chain $\{\mu^{(1)}, \mu^{(2)}, \ldots, \mu^{(N)}\}$ as a **tour** around the range of posterior plausible values of μ and yourself as the **tour manager**. The trace plot (left) illustrates the tour route or sequence of tour stops, $\mu^{(i)}$. The histogram (right) illustrates the relative amount of time you spent in each μ region throughout the tour.

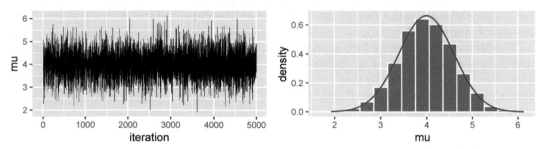

FIGURE 7.1: A trace plot (left) and histogram (right) of a 5,000 iteration MCMC simulation of the $N(4, 0.6^2)$ posterior. The posterior pdf is superimposed in blue (right).

As tour manager, it's your job to ensure that the density of tour stops in each μ region is proportional to its posterior plausibility. That is, the chain should spend more time touring values of μ between 2 and 6, where the Normal posterior pdf is greatest, and less time visiting μ values less than 2 or greater than 6, where the posterior drops off. This consideration is crucial to producing a collection of tour stops that accurately approximate the posterior, as does the tour here (Figure 7.1 right).

As tour manager, you can automate the tour route using the **Metropolis-Hastings algorithm**. This algorithm iterates through a two-step process. Assuming the Markov chain is at location $\mu^{(i)} = \mu$ at iteration or "tour stop" i, the next tour stop $\mu^{(i+1)}$ is selected as follows:

- **Step 1**: Propose a random location, μ', for the next tour stop.
- **Step 2**: Decide whether to go to the proposed location ($\mu^{(i+1)} = \mu'$) or to stay at the current location for another iteration ($\mu^{(i+1)} = \mu$).

This might seem easy. For example, if there were no constraints on your tour plan, you could simply draw proposed tour stops μ' from the $N(4, 0.6^2)$ posterior pdf $f(\mu'|y = 6.25)$ (Step 1) and then go there (Step 2). This special case of the Metropolis-Hastings algorithm has a special name – **Monte Carlo**.

Monte Carlo algorithm

To construct an *independent* Monte Carlo sample *directly* from posterior pdf $f(\mu|y)$, $\{\mu^{(1)}, \mu^{(2)}, ..., \mu^{(N)}\}$, select each tour stop $\mu^{(i)} = \mu$ as follows:

- **Step 1: Propose a location.**
  Draw a location μ from the posterior model with pdf $f(\mu|y)$.

- **Step 2: Go there.**

The Monte Carlo algorithm is so convenient! We can implement this algorithm by simply using `rnorm()` to sample directly from the $N(4, 0.6^2)$ posterior. The result is a nice independent sample from the posterior which, in turn, produces an accurate posterior approximation:

```
set.seed(84375)
mc_tour <- data.frame(mu = rnorm(5000, mean = 4, sd = 0.6))
ggplot(mc_tour, aes(x = mu)) +
  geom_histogram(aes(y = ..density..), color = "white", bins = 15) +
  stat_function(fun = dnorm, args = list(4, 0.6), color = "blue")
```

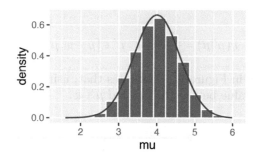

FIGURE 7.2: A histogram of a Monte Carlo sample from the posterior pdf of μ. The actual pdf is superimposed in blue.

BUT there's a glitch. Remember that we only need MCMC to approximate a Bayesian posterior when that posterior is too complicated to specify. And if a posterior is too complicated to specify, it's typically too complicated to directly sample or draw from as we did in our Monte Carlo tour above. This is where the more general Metropolis-Hastings MCMC algorithm comes in. Metropolis-Hastings relies on the fact that, even if we don't know the posterior model, we *do* know that the posterior pdf is proportional to the product of the *known* prior pdf and likelihood function (7.2):

$$f(\mu|y = 6.25) \propto f(\mu)L(\mu|y = 6.25).$$

This *unnormalized* pdf is drawn in Figure 7.3. Importantly, though it's not properly scaled to integrate to 1, this unnormalized pdf preserves the shape, central tendency, and variability of the actual posterior.

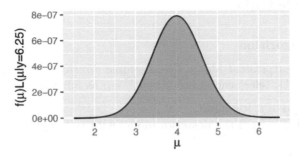

FIGURE 7.3: The *unnormalized* $N(4, 0.6^2)$ posterior pdf.

Further, **Step 1** of the Metropolis-Hastings algorithm relies on the fact that, even when we don't know and thus can't sample from the posterior model, we *can* propose Markov chain tour stops by sampling from a *different* and more convenient model. As one of many options here, we'll utilize a **Uniform proposal model** with half-width w. Specifically, let $\mu^{(i)} = \mu$ denote the current tour location. *Conditioned* on this current location, we propose the next location by taking a random draw μ' from the Uniform model which is centered at the current location μ and ranges from $\mu - w$ to $\mu + w$:

$$\mu'|\mu \sim \text{Unif}(\mu - w, \mu + w)$$

with pdf

$$q(\mu'|\mu) = \frac{1}{2w} \quad \text{for} \quad \mu' \in [\mu - w, \mu + w].$$

The Uniform pdf plotted in Figure 7.4 illustrates that, using this method, proposals μ' are equally likely to be any value between $\mu - w$ and $\mu + w$.

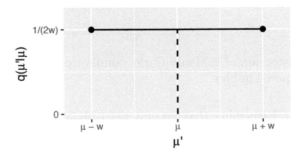

FIGURE 7.4: The Uniform proposal model.

Figure 7.5 illustrates this idea in a specific scenario. Suppose we're utilizing a Uniform half-width of $w = 1$ and that the Markov chain tour is at location $\mu = 3$. *Conditioned* on this current location, we'll then propose the next location by taking a random draw from a $\text{Unif}(3 - w, 3 + w) = \text{Unif}(2, 4)$ model. Thus, the chosen half-width $w = 1$ plays an important role here, defining the *neighborhood* of potential proposals. Specifically, the proposed next stop is equally likely to be anywhere within the restricted neighborhood stretching from 2 to 4, around the chain's current location of 3.

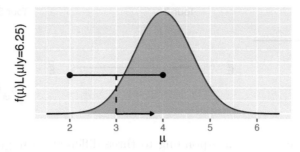

FIGURE 7.5: A visual representation of Step 1 in the Metropolis-Hastings algorithm for the $N(4, 0.6^2)$ posterior. A Unif(2,4) proposal model centered at the current chain location of 3 (black curve) is drawn against the unnormalized posterior pdf.

The whole idea behind Step 1 might seem goofy. How can proposals drawn from a *Uniform* model produce a decent approximation of the *Normal* posterior model?!? Well, they're only *proposals*. As with any other proposal in life, they can thankfully be *rejected* or *accepted*. Mainly, if a proposed location μ' is "bad," we can reject it. When we do, the chain sticks around at its current location μ for at least another iteration.

Step 2 of the Metropolis-Hastings algorithm provides a formal process for deciding whether to accept or reject a proposal. Let's first check in with our intuition about how this process *should* work. Revisiting Figure 7.5, suppose that our random Unif(2, 4) draw proposes that the chain move from its current location of 3 to 3.8. Does this proposal seem *desirable* to you? Well, sure. Notice that the (unnormalized) posterior plausibility of 3.8 is greater than that of 3. Thus, we *want* our Markov chain tour to spend more time exploring values of μ around 3.8 than around 3. Accepting the proposal gives us the chance to do so. In contrast, if our random Unif(2, 4) draw proposed that the chain move from 3 to 2.1, a location with very *low* posterior plausibility, we might be more hesitant. Consider three possible rules for *automating* Step 2 in the following quiz.[1]

> ❓ **Quiz Yourself!**
>
> Suppose we start our Metropolis-Hastings Markov chain tour at location $\mu^{(1)} = 3$ and utilize a Uniform proposal model in Step 1 of the algorithm. Consider three possible rules to follow in Step 2, deciding whether or not to accept a proposal:
> - Rule 1: *Never* accept the proposed location.
> - Rule 2: *Always* accept the proposed location.
> - Rule 3: Only accept the proposed location *if* its (unnormalized) posterior plausibility is greater than that of the current location.
>
> Each rule was used to generate one of the Markov chain tours in Figure 7.6. Match each rule to the correct tour.

The quiz above presented you with three *poor* options for determining whether to accept or reject proposed tour stops in Step 2 of the Metropolis-Hastings algorithm. **Rule 1** presents one extreme: *never* accept a proposal. This is a terrible idea. It results in the Markov chain remaining at the same location at every iteration (Tour 2), which would certainly produce a silly posterior approximation. **Rule 2** presents the opposite extreme: *always* accept a proposal. This results in a Markov chain which is not at all discerning in where it travels

---

[1]Answers: Rule 1 produces Tour 2, Rule 2 produces Tour 3, and Rule 3 produces Tour 1.

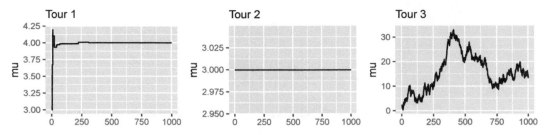

FIGURE 7.6: Trace plots corresponding to three different strategies for step 2 of the Metropolis-Hastings algorithm.

(Tour 3), completely ignoring the information we have from the unnormalized posterior model regarding the plausibility of a proposal. For example, Tour 3 spends the majority of its time exploring posterior values μ above 6, which we know from Figure 7.3 to be *implausible*.

Rule 3 might seem like a reasonable balance between the two extremes: it neither always rejects nor always accepts proposals. However, it's still problematic. Since this rule only accepts a proposed stop if its posterior plausibility is greater than that at the current location, it ends up producing a Markov chain similar to that of Tour 1 above. Though this chain floats toward values near $\mu = 4$, where the (unnormalized) posterior pdf is greatest, it then gets stuck there *forever*.

Putting all of this together, we're closer to understanding how to make the Metropolis-Hastings algorithm work. Upon proposing a next tour stop (Step 1), the process for rejecting or accepting this proposal (Step 2) must embrace the idea that the chain should spend more time exploring areas of high posterior plausibility *but* shouldn't get stuck there forever:

- **Step 1**: Propose a location, μ', for the next tour stop by taking a draw from a proposal model.
- **Step 2**: Decide whether to go to the proposed location ($\mu^{(i+1)} = \mu'$) or to stay at the current location for another iteration ($\mu^{(i+1)} = \mu$) as follows.
    - If the (unnormalized) posterior plausibility of the proposed location μ' is *greater* than that of the current location μ, $f(\mu')L(\mu'|y) > f(\mu)L(\mu|y)$, *definitely* go there.
    - Otherwise, *maybe* go there.

There are more details to fill in here (e.g., what does it mean to "maybe" accept a proposal?!). We'll do that next.

7.2   The Metropolis-Hastings algorithm

The Metropolis-Hastings algorithm for constructing a Markov chain tour $\left\{\mu^{(1)}, \mu^{(2)}, ..., \mu^{(N)}\right\}$ is formalized here. We'll break down the details below, exploring how to implement this algorithm in the context of our Normal-Normal Bayesian model.

Metropolis-Hastings algorithm

Conditioned on data y, let parameter μ have posterior pdf $f(\mu|y) \propto f(\mu)L(\mu|y)$. A Metropolis-Hastings Markov chain for $f(\mu|y)$, $\{\mu^{(1)}, \mu^{(2)}, ..., \mu^{(N)}\}$, evolves as follows. Let $\mu^{(i)} = \mu$ be the chain's location at iteration $i \in \{1, 2, ..., N-1\}$ and identify the next location $\mu^{(i+1)}$ through a two-step process:

- **Step 1: Propose a new location.**
  Conditioned on the current location μ, draw a location μ' from a proposal model with pdf $q(\mu'|\mu)$.

- **Step 2: Decide whether or not to go there.**
  - Calculate the **acceptance probability**, i.e., the probability of accepting the proposal μ':

$$\alpha = \min\left\{1, \ \frac{f(\mu')L(\mu'|y)}{f(\mu)L(\mu|y)} \frac{q(\mu|\mu')}{q(\mu'|\mu)}\right\}. \tag{7.3}$$

  - Figuratively, flip a weighted coin. If it's Heads, with probability α, go to the proposed location μ'. If it's Tails, with probability $1-\alpha$, stay at μ:

$$\mu^{(i+1)} = \begin{cases} \mu' & \text{with probability } \alpha \\ \mu & \text{with probability } 1-\alpha. \end{cases}$$

Though the notation and details are new, the algorithm above matches the concepts we developed in the previous section. First, recall that for our Normal-Normal simulation, we utilized a $\mu'|\mu \sim \text{Unif}(\mu-w, \mu+w)$ proposal model in Step 1. In fact, this is a bit lazy. Though our Normal-Normal posterior model is defined for $\mu \in (-\infty, \infty)$, the Uniform proposal model lives on a truncated neighborhood around the current chain location. However, utilizing a Uniform proposal model *simplifies* the Metropolis-Hastings algorithm by the fact that it's **symmetric**. This symmetry exhibits itself in the plot of the Uniform pdf (Figure 7.4), as well as numerically – the conditional pdf of μ' given μ is equivalent to that of μ given μ':

$$q(\mu'|\mu) = q(\mu|\mu') = \begin{cases} \frac{1}{2w} & \text{when } \mu \text{ and } \mu' \text{ are within } w \text{ units of each other} \\ 0 & \text{otherwise} \end{cases}.$$

This symmetry means that the chance of proposing a chain move from μ to μ' is the same as proposing a move from μ' to μ. For example, the Uniform model with a half-width of 1 is equally likely to propose a move from $\mu = 3$ to $\mu' = 3.8$ as a move from $\mu = 3.8$ to $\mu' = 3$. We refer to this special case of the Metropolis-Hastings algorithm as simply the **Metropolis algorithm**.

Metropolis algorithm

The Metropolis algorithm is a special case of the Metropolis-Hastings in which the proposal model is *symmetric*. That is, the chance of proposing a move to μ' from μ is equal to that of proposing a move to μ from μ': $q(\mu'|\mu) = q(\mu|\mu')$. Thus, the acceptance probability (7.3) simplifies to

$$\alpha = \min\left\{1, \frac{f(\mu')L(\mu'|y)}{f(\mu)L(\mu|y)}\right\}. \qquad (7.4)$$

Inspecting (7.4) reveals the nuances of Step 2, determining whether to accept or reject the proposed location drawn in Step 1. First, notice that we can rewrite the acceptance probability α by dividing both the numerator and denominator by $f(y)$:

$$\alpha = \min\left\{1, \frac{f(\mu')L(\mu'|y)/f(y)}{f(\mu)L(\mu|y)/f(y)}\right\} = \min\left\{1, \frac{f(\mu'|y)}{f(\mu|y)}\right\}. \qquad (7.5)$$

This rewrite emphasizes that, though we can't calculate the posterior pdfs of μ' and μ, $f(\mu'|y)$ and $f(\mu|y)$, their *ratio* is equivalent to that of the *unnormalized* posterior pdfs (which we *can* calculate). Thus, the probability of accepting a move from a current location μ to a proposed location μ' comes down to a comparison of their posterior plausibility: $f(\mu'|y)$ versus $f(\mu|y)$. There are two possible scenarios here:

- Scenario 1: $f(\mu'|y) \geq f(\mu|y)$
  When the posterior plausibility of μ' is *at least* as great as that of μ, $\alpha = 1$. Thus, we'll *definitely* move there.

- Scenario 2: $f(\mu'|y) < f(\mu|y)$
  If the posterior plausibility of μ' is *less* than that of μ, then

$$\alpha = \frac{f(\mu'|y)}{f(\mu|y)} < 1.$$

Thus, we *might* move there. Further, α approaches 1 as $f(\mu'|y)$ nears $f(\mu|y)$. That is, the probability of accepting the proposal increases with the plausibility of μ' relative to μ.

Scenario 1 is straightforward. We'll always jump at the chance to move our tour to a more plausible posterior region. To wrap our minds around Scenario 2, a little R simulation is helpful. For example, suppose our Markov tour is currently at location "3":

```
current <- 3
```

Further, suppose we're utilizing a Uniform proposal model with half-width $w = 1$. Then to determine the next tour stop, we first propose a location by taking a random draw from the Unif(current - 1, current + 1) model (Step 1):

```
set.seed(8)
proposal <- runif(1, min = current - 1, max = current + 1)
proposal
[1] 2.933
```

Revisiting Figure 7.3, we observe that the (unnormalized) posterior plausibility of the proposed tour location (2.93) is *slightly* less than that of the current location (3). We can calculate the *unnormalized* posterior plausibility of these two μ locations, $f(\mu)L(\mu|y = 6.25)$, using `dnorm()` to evaluate both pieces of the product:

- dnorm(..., 0, 1) calculates the $N(0, 1^2)$ prior pdf $f(\mu)$ at a given μ value;
- dnorm(6.25, ..., 0.75) calculates the Normal likelihood function $L(\mu|y = 6.25)$ with $Y = 6.25$ and standard deviation 0.75 at an unknown mean μ.

```
proposal_plaus <- dnorm(proposal, 0, 1) * dnorm(6.25, proposal, 0.75)
proposal_plaus
[1] 1.625e-07
current_plaus  <- dnorm(current, 0, 1) * dnorm(6.25, current, 0.75)
current_plaus
[1] 1.972e-07
```

It follows that, though not certain, the probability α of accepting and subsequently moving to the proposed location is relatively high:

```
alpha <- min(1, proposal_plaus / current_plaus)
alpha
[1] 0.824
```

To make the final determination, we set up a weighted coin which accepts the proposal with probability α (0.824) and rejects the proposal with probability $1 - \alpha$ (0.176). In a random flip of this coin using the `sample()` function, we *accept* the proposal, meaning that the `next_stop` on the tour is 2.933:

```
next_stop <- sample(c(proposal, current),
                    size = 1, prob = c(alpha, 1-alpha))
next_stop
[1] 2.933
```

This is merely one of countless possible outcomes for a single iteration of the Metropolis-Hastings algorithm for our Normal posterior. To streamline this process, we'll write our own **R function**, `one_mh_iteration()`, which implements a single Metropolis-Hastings iteration starting from any given `current` tour stop and utilizing a Uniform proposal model with any given half-width `w`. If you are new to writing functions, we encourage you to focus on the structure over the details of this code. Some things to pick up on:

- We first specify that `one_mh_iteration` is a `function()` of two arguments: the Uniform half-width `w` and the `current` chain value.
- We open and close the definition of the function with { }.
- Inside the function, we carry out the same steps as above and `return()` 3 pieces of information: the `proposal`, acceptance probability `alpha`, and `next_stop`.

```
one_mh_iteration <- function(w, current){
 # STEP 1: Propose the next chain location
 proposal <- runif(1, min = current - w, max = current + w)

 # STEP 2: Decide whether or not to go there
 proposal_plaus <- dnorm(proposal, 0, 1) * dnorm(6.25, proposal, 0.75)
 current_plaus  <- dnorm(current, 0, 1) * dnorm(6.25, current, 0.75)
 alpha <- min(1, proposal_plaus / current_plaus)
```

```
next_stop <- sample(c(proposal, current),
                    size = 1, prob = c(alpha, 1-alpha))

# Return the results
return(data.frame(proposal, alpha, next_stop))
}
```

Let's try it out. Running `one_mh_iteration()` from our current tour stop of 3 under a seed of 8 reproduces the results from above:

```
set.seed(8)
one_mh_iteration(w = 1, current = 3)
  proposal alpha next_stop
1    2.933 0.824     2.933
```

If we use a seed of 83, the proposed next tour stop is 2.018, which has a low corresponding acceptance probability of 0.017:

```
set.seed(83)
one_mh_iteration(w = 1, current = 3)
  proposal   alpha next_stop
1    2.018 0.01709         3
```

This makes sense. We see from Figure 7.3 that the posterior plausibility of 2.018 is much lower than that of our current location of 3. Though we do want to explore such extreme values, we don't want to do so often. In fact, we see that upon the flip of our coin, the proposal was rejected and the tour again visits location 3 on its `next_stop`. As a final example, we can confirm that when the posterior plausibility of the proposed next stop (here 3.978) is greater than that of our current location, the acceptance probability is 1 and the proposal is automatically accepted:

```
set.seed(7)
one_mh_iteration(w = 1, current = 3)
  proposal alpha next_stop
1    3.978     1     3.978
```

7.3   Implementing the Metropolis-Hastings

We've spent much energy understanding how to implement *one iteration* of the Metropolis-Hastings algorithm. That was the hard part! To construct an entire Metropolis-Hastings tour of our $N(4, 0.6^2)$ posterior, we now just have to repeat this process over and over and over. And by "we," we mean the computer. To this end, the `mh_tour()` function below constructs a Metropolis-Hastings tour of any given length `N` utilizing a Uniform proposal model with any given half-width `w`:

```
mh_tour <- function(N, w){
  # 1. Start the chain at location 3
  current <- 3

  # 2. Initialize the simulation
  mu <- rep(0, N)

  # 3. Simulate N Markov chain stops
  for(i in 1:N){
    # Simulate one iteration
    sim <- one_mh_iteration(w = w, current = current)

    # Record next location
    mu[i] <- sim$next_stop

    # Reset the current location
    current <- sim$next_stop
  }

  # 4. Return the chain locations
  return(data.frame(iteration = c(1:N), mu))
}
```

Again, this code is a big leap if you're new to *for loops* and functions. Let's focus on the general structure indicated by the # comment blocks. In one call to this function:

1. Start the tour at location 3, a somewhat arbitrary choice based on our prior understanding of μ.
2. Initialize the tour simulation by setting up an "empty" vector in which we'll eventually store the N tour stops (mu).
3. Utilizing a *for loop*, at each tour stop i from 1 to N, run one_mh_iteration() and store the resulting next_stop in the ith element of the mu vector. Before closing the *for loop*, update the current tour stop to serve as a starting point for the next iteration.
4. Return a data frame with the iteration numbers and corresponding tour stops mu.

To see this function in action, use mh_tour() to simulate a Markov chain tour of length N = 5000 utilizing a Uniform proposal model with half-width $w = 1$:

```
set.seed(84735)
mh_simulation_1 <- mh_tour(N = 5000, w = 1)
```

A trace plot and histogram of the tour results are shown below. Notably, this tour produces a remarkably accurate approximation of the $N(4, 0.6^2)$ posterior. You might need to step out of the weeds we've waded into in order to reflect upon the mathemagic here. Through a rigorous and formal process, we utilized *dependent* draws from a *Uniform* model to approximate a *Normal* model. And it worked!

```r
ggplot(mh_simulation_1, aes(x = iteration, y = mu)) +
  geom_line()

ggplot(mh_simulation_1, aes(x = mu)) +
  geom_histogram(aes(y = ..density..), color = "white", bins = 20) +
  stat_function(fun = dnorm, args = list(4,0.6), color = "blue")
```

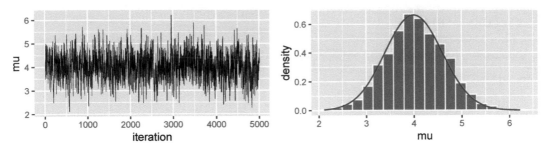

FIGURE 7.7: A trace plot (left) and histogram (right) of 5000 Metropolis chain values of μ. The target posterior pdf is superimposed in blue.

7.4 Tuning the Metropolis-Hastings algorithm

Let's wade into one final weed patch. In implementing the Metropolis-Hastings algorithm above, we utilized a Uniform proposal model $\mu'|\mu \sim \text{Unif}(\mu - w, \mu + w)$ with half-width $w = 1$. Our selection of $w = 1$ defined the *neighborhood* or range of potential proposals (Figure 7.5). Naturally, the choice of w impacts the performance of our Markov chain tour. Check your intuition about w with this quiz.

> ❓ **Quiz Yourself!**
>
> Figure 7.8 presents trace plots and histograms of three separate Metropolis-Hastings tours of the $N(4, 0.6^2)$ posterior. Each tour utilizes a Uniform proposal model, but with different half-widths: $w = 0.01$, $w = 1$, or $w = 100$. Match each tour to the w with which it was generated.

The answers are below.[2] The main punchline is that the Metropolis-Hastings algorithm can work – we've seen as much – but we have to **tune** it. In our example, this means that we have to pick an appropriate half-width w for the Uniform proposal model. The tours in the quiz above illustrate the **Goldilocks challenge** this presents: we don't want w to be too small or too large, but *just right*.[3] Tour 2 illustrates what can go wrong when w is too small (here $w = 0.01$). You can reproduce these results with the following code:

---

[2]Answers: Tour 1 uses $w = 100$, Tour 2 uses $w = 0.01$, Tour 3 uses $w = 1$.

[3]This technical term was inspired by the "Goldilocks and the three bears" fairy tale in which, for some reason, a child (Goldilocks) taste tests three different bears' porridge while trespassing in the bears' house. The first bowl of porridge is too hot, the second is too cold, and the third is just right.

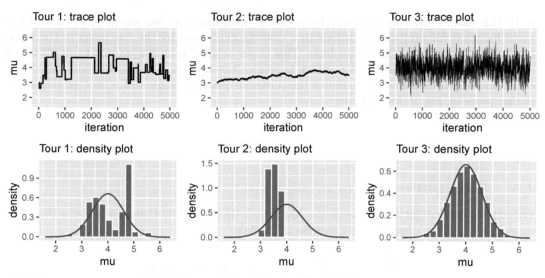

FIGURE 7.8: Trace plots (top row) and histograms (bottom row) for three different Metropolis-Hastings tours, where each tour utilizes a different proposal model. The shared target posterior pdf is superimposed in blue.

```
set.seed(84735)
mh_simulation_2 <- mh_tour(N = 5000, w = 0.01)
ggplot(mh_simulation_2, aes(x = iteration, y = mu)) +
  geom_line() +
  lims(y = c(1.6, 6.4))
```

In this case, the Uniform proposal model places a tight neighborhood around the tour's current location – the chain can only move within 0.01 units at a time. Proposals will therefore tend to be very close to the current location, and thus have similar posterior plausibility and a high probability of being accepted. Specifically, when a proposal $\mu' \approx \mu$, it's typically the case that $f(\mu')L(\mu'|y) \approx f(\mu)L(\mu|y)$ hence

$$\alpha = \min\left\{1, \frac{f(\mu')L(\mu'|y)}{f(\mu)L(\mu|y)}\right\} \approx \min\{1, 1\} = 1.$$

The result is a Markov chain that's almost always moving, but takes such excruciatingly small steps that it will take an excruciatingly long time to explore the entire posterior plausible region of μ values. Tour 1 illustrates the other extreme in which w is too large (here $w = 100$):

```
set.seed(7)
mh_simulation_3 <- mh_tour(N = 5000, w = 100)
ggplot(mh_simulation_3, aes(x = iteration, y = mu)) +
  geom_line() +
  lims(y = c(1.6,6.4))
```

By utilizing a Uniform proposal model with such a large neighborhood around the tour's current location, proposals can be far flung – far outside the region of posterior plausible

μ. In this case, proposals will often be rejected, resulting in a tour which gets stuck at the same location for multiple iterations in a row (as evidenced by the flat parts of the trace plot). Tour 3 presents a happy medium. This is our original tour (`mh_simulation_1`) which utilized $w = 1$, a neighborhood size which is neither too small nor too big but just right. The corresponding tour efficiently explores the posterior plausible region of μ, not getting stuck in one place or region for too long.

7.5   A Beta-Binomial example

For extra practice, let's implement the Metropolis-Hastings algorithm for a Beta-Binomial model in which we observe $Y = 1$ success in 2 trials:[4]

$$\begin{matrix} Y|\pi \sim \text{Bin}(2,\pi) \\ \pi \sim \text{Beta}(2,3) \end{matrix} \quad \Rightarrow \quad \pi|(Y = 1) \sim \text{Beta}(3,4).$$

Again, *pretend* we were only able to define the posterior pdf up to some missing normalizing constant,

$$f(\pi|y = 1) \propto f(\pi)L(\pi|y = 1),$$

where $f(\pi)$ is the Beta(2,3) prior pdf and $L(\pi|y = 1)$ is the Bin$(2,\pi)$ likelihood function. Our goal then is to construct a Metropolis-Hastings tour of the posterior, $\{\pi^{(1)}, \pi^{(2)}, \ldots, \pi^{(N)}\}$, utilizing a two-step iterative process: (1) at the current tour stop π, take a random draw π' from a proposal model; and then (2) decide whether or not to move there. In step 1, the proposed tour stops would ideally be restricted to be between 0 and 1, just like π itself. Since the Uniform proposal model we used above might propose values outside this range, we'll instead tune and utilize a Beta(a,b) proposal model. Further, we'll utilize the *same* Beta(a,b) proposal model at each step in the chain. As such, our proposal strategy does *not* depend on the current tour stop (though whether or not we *accept* a proposal still will). This special case of the Metropolis-Hastings is referred to as the **independence sampling algorithm**.

> **Independence sampling algorithm**
>
> The independence sampling algorithm is a special case of the Metropolis-Hastings in which the same proposal model is utilized at each iteration, *independent* of the chain's current location. That is, from a current location π, a proposal π' is drawn from a proposal model with pdf $q(\pi')$ (as opposed to $q(\pi'|\pi)$). Thus, the acceptance probability (7.3) simplifies to
>
> $$\alpha = \min\left\{1, \; \frac{f(\pi')L(\pi'|y)}{f(\pi)L(\pi|y)} \frac{q(\pi)}{q(\pi')}\right\}. \tag{7.6}$$

We can rewrite α for the independence sampler to emphasize its dependence on the relative posterior plausibility of proposal π' versus current location π:

---

[4]By Chapter 3, the Beta posterior has parameters 3 ($\alpha + y = 2 + 1$) and 4 ($\beta + n - y = 3 + 2 - 1$).

$$\alpha = \min\left\{1, \ \frac{f(\pi')L(\pi'|y)/f(y)}{f(\pi)L(\pi|y)/f(y)}\frac{q(\pi)}{q(\pi')}\right\} = \min\left\{1, \ \frac{f(\pi'|y)}{f(\pi|y)}\frac{q(\pi)}{q(\pi')}\right\}. \tag{7.7}$$

Like the acceptance probability for the Metropolis algorithm (7.5), (7.7) includes the posterior pdf ratio $f(\pi'|y)/f(\pi|y)$. This ensures that the independence sampler weighs the relative posterior plausibility of π' versus π in making its moves. Yet unlike the Metropolis acceptance probability, $q(\pi)$ and $q(\pi')$ aren't typically equal, and thus do not cancel out of the formula. Rather, the inclusion of their ratio serves as a corrective measure: the probability α of accepting a proposal π' decreases as $q(\pi')$ increases. Conceptually, we place a penalty on common proposal values, ensuring that our tour doesn't float toward these values simply because we keep proposing them.

The `one_iteration()` function below implements a single iteration of this independence sampling algorithm, starting from any `current` value π and utilizing a Beta(a, b) proposal model for any given `a` and `b`. In the calculation of acceptance probability α (`alpha`), notice that we utilize `dbeta()` to evaluate the prior and proposal pdfs as well as `dbinom()` to evaluate the Binomial likelihood function with data $Y = 1$, $n = 2$, and unknown probability π:

```
one_iteration <- function(a, b, current){
  # STEP 1: Propose the next chain location
  proposal <- rbeta(1, a, b)

  # STEP 2: Decide whether or not to go there
  proposal_plaus <- dbeta(proposal, 2, 3) * dbinom(1, 2, proposal)
  proposal_q     <- dbeta(proposal, a, b)
  current_plaus  <- dbeta(current, 2, 3) * dbinom(1, 2, current)
  current_q      <- dbeta(current, a, b)
  alpha <- min(1, proposal_plaus / current_plaus * current_q / proposal_q)
  next_stop <- sample(c(proposal, current),
                      size = 1, prob = c(alpha, 1-alpha))

  return(data.frame(proposal, alpha, next_stop))
}
```

Subsequently, we write a `betabin_tour()` function which constructs an N-length Markov chain tour for any Beta(a, b) proposal model, utilizing `one_iteration()` to determine each stop:

```
betabin_tour <- function(N, a, b){
  # 1. Start the chain at location 0.5
  current <- 0.5

  # 2. Initialize the simulation
  pi <- rep(0, N)

  # 3. Simulate N Markov chain stops
  for(i in 1:N){
    # Simulate one iteration
    sim <- one_iteration(a = a, b = b, current = current)
```

```
    # Record next location
    pi[i] <- sim$next_stop

    # Reset the current location
    current <- sim$next_stop
  }

  # 4. Return the chain locations
  return(data.frame(iteration = c(1:N), pi))
}
```

We encourage you to try different tunings of the Beta(a, b) proposal model. To keep it simple here, we run a 5,000 step tour of the Beta-Binomial posterior using a Beta($1, 1$), i.e., Unif($0, 1$), proposal model. As such, each proposal is equally likely to be anywhere between 0 and 1, no matter the chain's current location. The results are summarized in Figure 7.9. You're likely not surprised by now that this worked. The tour appears to be stable, random, and provides an excellent approximation of the Beta(3,4) posterior model (which in practice we wouldn't have access to for comparison). That is, the Metropolis-Hastings algorithm allowed us to utilize draws from the Beta(1,1) proposal model to approximate our Beta(3,4) posterior. Cool.

```
set.seed(84735)
betabin_sim <- betabin_tour(N = 5000, a = 1, b = 1)

# Plot the results
ggplot(betabin_sim, aes(x = iteration, y = pi)) +
  geom_line()
ggplot(betabin_sim, aes(x = pi)) +
  geom_histogram(aes(y = ..density..), color = "white") +
  stat_function(fun = dbeta, args = list(3, 4), color = "blue")
```

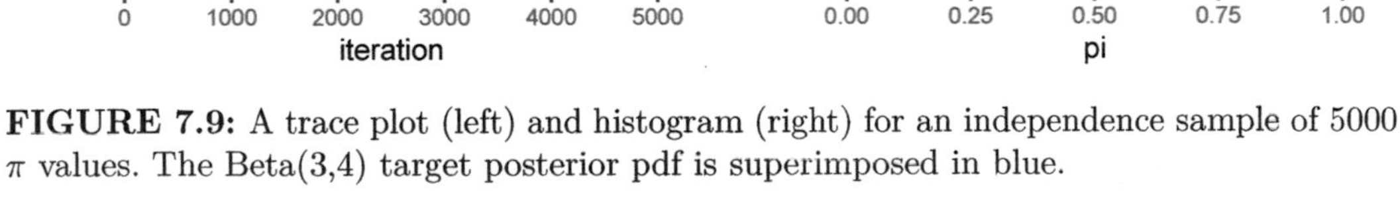

FIGURE 7.9: A trace plot (left) and histogram (right) for an independence sample of 5000 π values. The Beta(3,4) target posterior pdf is superimposed in blue.

7.6 Why the algorithm works

We've now seen *how* the Metropolis-Hastings algorithm works. But we haven't yet discussed *why* it works. Though we'll skip a formal proof, we'll examine the underlying spirit. Throughout, consider constructing a Metropolis-Hastings tour of some *generic* posterior pdf $f(\mu|y)$ utilizing a proposal pdf $q(\mu'|\mu)$. For simplicity, let's also assume μ is a discrete variable. In general, the Metropolis-Hastings tour moves between different pairs of values, μ and μ'. Specifically, the chain can move from μ to μ' with probability

$$P(\mu \to \mu') = P\left(\mu^{(i+1)} = \mu' \mid \mu^{(i)} = \mu\right)$$

or from μ' to μ with probability

$$P(\mu' \to \mu) = P\left(\mu^{(i+1)} = \mu \mid \mu^{(i)} = \mu'\right).$$

In order for the Metropolis-Hastings algorithm to *work* (i.e., produce a good posterior approximation), its tour must preserve the relative posterior plausibility of any μ' and μ pair. That is, the relative probabilities of moving between these two values must satisfy

$$\frac{P(\mu \to \mu')}{P(\mu' \to \mu)} = \frac{f(\mu'|y)}{f(\mu|y)}.$$

This is indeed the case for any Metropolis-Hastings algorithm. And we can prove it. First, we need to calculate the movement probabilities $P(\mu \to \mu')$ and $P(\mu' \to \mu)$. Consider the former. For the tour to move from μ to μ', two things must happen: we need to propose μ' and then accept the proposal. The associated probabilities of these two events are described by $q(\mu'|\mu)$ and α (7.3), respectively. It follows that the chain moves from μ to μ' with probability

$$P(\mu \to \mu') = q(\mu'|\mu) \cdot \min\left\{1, \ \frac{f(\mu'|y)}{f(\mu|y)} \frac{q(\mu|\mu')}{q(\mu'|\mu)}\right\}.$$

Similarly, the chain moves from μ' to μ with probability

$$P(\mu' \to \mu) = q(\mu|\mu') \cdot \min\left\{1, \ \frac{f(\mu|y)}{f(\mu'|y)} \frac{q(\mu'|\mu)}{q(\mu|\mu')}\right\}.$$

To simplify the ratio of movement probabilities $P(\mu \to \mu')/P(\mu' \to \mu)$, consider the two potential scenarios:

1. Scenario 1: $f(\mu'|y) \geq f(\mu|y)$
   When μ' is at least as plausible as μ, $P(\mu \to \mu')$ simplifies to $q(\mu'|\mu)$ and
   $$\frac{P(\mu \to \mu')}{P(\mu' \to \mu)} = \frac{q(\mu'|\mu)}{q(\mu|\mu')\frac{f(\mu|y)}{f(\mu'|y)}\frac{q(\mu'|\mu)}{q(\mu|\mu')}} = \frac{f(\mu'|y)}{f(\mu|y)}.$$

2. Scenario 2: $f(\mu'|y) < f(\mu|y)$
   When μ' is less plausible than μ, $P(\mu' \to \mu)$ simplifies to $q(\mu|\mu')$ and
   $$\frac{P(\mu \to \mu')}{P(\mu' \to \mu)} = \frac{q(\mu'|\mu)\frac{f(\mu'|y)}{f(\mu|y)}\frac{q(\mu|\mu')}{q(\mu'|\mu)}}{q(\mu|\mu')} = \frac{f(\mu'|y)}{f(\mu|y)}.$$

Thus, no matter the scenario, the Metropolis-Hastings algorithm preserves the relative likelihood of any pair of values μ and μ'!

7.7   Variations on the theme

We've merely considered one MCMC algorithm here, the Metropolis-Hastings. Though powerful, this algorithm has its limits. In Chapter 7, we used the Metropolis-Hastings to simulate Normal-Normal and Beta-Binomial posteriors with **single parameters** μ and π, respectively. In future chapters, our Bayesian models will grow to include *lots* of parameters. Tuning a Metropolis-Hastings algorithm to adequately explore *each* of these parameters gets unwieldy. Yet even when it reaches its own limits of utility, the Metropolis-Hastings serves as the foundation for a more flexible set of MCMC tools, including the adaptive Metropolis-Hastings, Gibbs, and Hamiltonian Monte Carlo (HMC) algorithms. Among these, HMC is the algorithm utilized by the `rstan` and `rstanarm` packages.

As we noted at the top of Chapter 7, studying these alternative algorithms would require a book itself. From here on out, we'll rely on **rstan** with a fresh confidence in what's going on under the hood. If you're curious to learn a little more, McElreath (2019) provides an excellent video introduction to the HMC algorithm and how it compares to the Metropolis-Hastings. For a deeper dive, Brooks et al. (2011) provides a comprehensive overview of the broader MCMC landscape.

7.8   Chapter summary

In Chapter 7, you built a strong **conceptual understanding** of the foundational **Metropolis-Hastings** MCMC algorithm. You also implemented this algorithm to study the familiar Normal-Normal and Beta-Binomial models. Whether in these relatively simple one-parameter model settings, or in more complicated model settings, the Metropolis-Hastings algorithm produces an approximate sample from the posterior by iterating between two steps:

1.   **Propose a new chain location** by drawing from a proposal pdf which is, perhaps, dependent upon the current location.
2.   **Determine whether to accept the proposal.** Simply put, whether or not we accept a proposal depends on how favorable its posterior plausibility is relative to the posterior plausibility of the current location.

7.9   Exercises

Chapter 7 continued to shift focus towards more computational methods. Change can be uncomfortable, but it also spurs growth. As you work through these exercises, know that if you are struggling a bit, that just means you are engaging with some new concepts and you are putting in the work to grow as a Bayesian. Our advice: verbalize what you do and don't understand. Don't rush yourself. Take a break and come back to exercises that you feel stuck on. Work with a buddy. Ask for help, and help others when you can.

7.9.1 Conceptual exercises

Exercise 7.1 (Getting to know Monte Carlo).

a. What are the steps to the Monte Carlo Algorithm?
b. Name one pro and one con for the Monte Carlo Algorithm.

Exercise 7.2 (Getting to know Metropolis-Hastings).

a) What are the steps to the Metropolis-Hastings Algorithm?
b) What's the difference between the Monte Carlo and Metropolis-Hastings algorithms?
c) What's the difference between the Metropolis and Metropolis-Hastings algorithms?

Exercise 7.3 (Metropolis vs Metropolis Hastings). Select all of the correct endings to this sentence: *The difference between the Metropolis and Metropolis-Hastings algorithms is that...*

a) Metropolis accepts all proposals, whereas Metropolis-Hastings accepts only some proposals.
b) Metropolis uses a symmetric proposal model, whereas a Metropolis-Hastings proposal model is not necessarily symmetric.
c) The acceptance probability is simpler to calculate for the Metropolis than for the Metropolis-Hastings.

Exercise 7.4 (Tuning the Metropolis-Hastings). In this exercise you will consider how to tune a Uniform proposal model with half-width w for a Metropolis-Hastings algorithm.

a) Draw a trace plot for a tour where the Uniform proposal model uses a very small w.
b) Why is it problematic if w is too small, and hence defines the neighborhood around the current chain value too narrowly?
c) Draw a trace plot for a tour where the Uniform proposal model uses a very large w.
d) Why is it problematic if w is too large, and hence defines the neighborhood too widely?
e) Draw a trace plot for a tour where the Uniform proposal model uses a w that is neither too small or too large.
f) Describe how you would go about finding an appropriate half-width w for a Uniform proposal model.

Exercise 7.5 (Independence sampling: True or False). Identify whether the below statements about independence sampling are True or False. If False, explain.

a) The proposal model depends on the current value of the chain.
b) The proposal model is the same every time.

  c)   It is a special case of the Metropolis-Hastings algorithm.
  d)   It is a special case of Metropolis algorithm.

Exercise 7.6 (Proposing a new location). In each situation below, complete Step 1 of the Metropolis-Hastings Algorithm. That is, starting from the given current chain value $\lambda^{(i)} = \lambda$ and with `set.seed(84735)`, use the given proposal model to draw a λ' proposal value for the next chain value $\lambda^{(i+1)}$.

  a) $\lambda = 4.6$, $\lambda'|\lambda \sim N(\lambda, 2^2)$
  b) $\lambda = 2.1$, $\lambda'|\lambda \sim N(\lambda, 7^2)$
  c) $\lambda = 8.9$, $\lambda'|\lambda \sim \text{Unif}(\lambda - 2, \lambda + 2)$
  d) $\lambda = 1.2$, $\lambda'|\lambda \sim \text{Unif}(\lambda - 0.5, \lambda + 0.5)$
  e) $\lambda = 7.7$, $\lambda'|\lambda \sim \text{Unif}(\lambda - 3, \lambda + 3)$

Exercise 7.7 (Calculate the acceptance probability: Part I). Suppose that a Markov chain is currently at $\lambda^{(i)} = 2$ and that the proposal for $\lambda^{(i+1)}$ is $\lambda' = 2.1$. For each pair of unnormalized posterior pdf $f(\lambda)L(\lambda|y)$ and proposal pdf $q(\lambda'|\lambda)$, calculate the acceptance probability α used in Step 2 of the Metropolis-Hastings algorithm (7.3).

  a) $f(\lambda)L(\lambda|y) = \lambda^{-2}$, $\lambda'|\lambda \sim N(\lambda, 1^2)$ with pdf $q(\lambda'|\lambda)$
  b) $f(\lambda)L(\lambda|y) = e^{\lambda}$, $\lambda'|\lambda \sim N(\lambda, 0.2^2)$ with pdf $q(\lambda'|\lambda)$
  c) $f(\lambda)L(\lambda|y) = e^{-10\lambda}$, $\lambda'|\lambda \sim \text{Unif}(\lambda - 0.5, \lambda + 0.5)$ with pdf $q(\lambda'|\lambda)$
  d) $f(\lambda)L(\lambda|y) = e^{-\lambda^4}$, $\lambda'|\lambda \sim \text{Exp}(\lambda)$ with pdf $q(\lambda'|\lambda)$
  e)   For which of these scenarios is there a 100% acceptance probability? Explain why we'd certainly want to accept λ' in these scenarios.

Exercise 7.8 (Calculate the acceptance probability: Part II). Suppose that a Markov chain is currently at $\lambda^{(i)} = 1.8$ and that the proposal for $\lambda^{(i+1)}$ is $\lambda' = 1.6$. For each pair of unnormalized posterior pdf $f(\lambda)L(\lambda|y)$ and proposal pdf $q(\lambda'|\lambda)$, calculate the acceptance probability α used in Step 2 of the Metropolis-Hastings algorithm (7.3).

  a) $f(\lambda)L(\lambda|y) = \lambda^{-1}$, $\lambda'|\lambda \sim N(\lambda, 2^2)$ with pdf $q(\lambda'|\lambda)$
  b) $f(\lambda)L(\lambda|y) = e^{3\lambda}$, $\lambda'|\lambda \sim N(\lambda, 0.5^2)$ with pdf $q(\lambda'|\lambda)$
  c) $f(\lambda)L(\lambda|y) = e^{-1.9\lambda}$, $\lambda'|\lambda \sim \text{Unif}(\lambda - 0.3, \lambda + 0.3)$ with pdf $q(\lambda'|\lambda)$
  d) $f(\lambda)L(\lambda|y) = e^{-\lambda^4}$, $\lambda'|\lambda \sim \text{Exp}(\lambda)$ with pdf $q(\lambda'|\lambda)$
  e)   For which of these scenarios is there a 100% acceptance probability? Explain why we'd certainly want to accept λ' in these scenarios.

7.9.2   Practice: Normal-Normal simulation

In the next set of exercises, return to the Bayesian model setting of (7.1),

$$Y|\mu \sim N(\mu, 0.75^2)$$
$$\mu \sim N(0, 1^2).$$

Assume we observe $Y = 6.25$ and wish to construct a Metropolis-Hastings simulation of the corresponding posterior model for λ.

Exercise 7.9 (One iteration with a Uniform proposal model). The function `one_mh_iteration()` from the text utilizes a Uniform proposal model, $\mu'|\mu \sim \text{Unif}(\mu - w, \mu + w)$ with half-width $w = 1$. Starting from a current value of $\mu = 3$ and using `set.seed(1)`, run the code below and comment on the returned `proposal`, `alpha`, and `next_stop` values.

a)   `one_mh_iteration(w = 0.01, current = 3)`
b)   `one_mh_iteration(w = 0.5, current = 3)`
c)   `one_mh_iteration(w = 1, current = 3)`
d)   `one_mh_iteration(w = 3, current = 3)`

Exercise 7.10 (An entire tour with a Uniform proposal model). Implement the Metropolis-Hastings function `mh_tour()` defined in Section 7.3 to construct tours of μ under each of the following scenarios. Construct trace plots and histograms for each tour.

a)   50 iterations, $w = 50$
b)   50 iterations, $w = 0.01$
c)   1000 iterations, $w = 50$
d)   1000 iterations, $w = 0.01$
e)   Contrast the trace plots in parts a and b. Explain why changing w has this effect.
f)   Consider the results in parts c and d. Is the w value as important when the number of iterations is much larger? Explain.

Exercise 7.11 (Changing the proposal model). For this exercise, modify `one_mh_iteration()` to create a new function, `one_mh_iteration_normal()`, which utilizes a symmetric Normal proposal model, centered at the current chain value μ with standard deviation s:

$$\mu'|\mu \sim N(\mu, s^2).$$

Subsequently, starting from a current value of $\mu = 3$ and `set.seed(1)`, run this function under each setting below. Comment on the returned `proposal`, `alpha`, and `next_stop` values.

a)   `one_mh_iteration_normal(s = 0.01, current = 3)`
b)   `one_mh_iteration_normal(s = 0.5, current = 3)`
c)   `one_mh_iteration_normal(s = 1, current = 3)`
d)   `one_mh_iteration_normal(s = 3, current = 3)`

Exercise 7.12 (Metropolis-Hastings tour with Normal proposals). Upon completing the previous exercise, modify `mh_tour()` to create a new function, `mh_tour_normal()`, which constructs a chain of μ values using a Normal proposal model with standard deviation s. Subsequently, using `set.seed(84735)`, run this function under each setting below and construct a trace plot of the chain.

a)   20 iterations, $s = 0.01$
b)   20 iterations, $s = 10$
c)   1000 iterations, $s = 0.01$
d)   1000 iterations, $s = 10$
e)   Contrast the trace plots in a and b. Explain in simple terms why changing the standard deviation of the Normal proposal model causes these differences.

   f)   Reflecting on the above results, *tune* your Metropolis-Hastings algorithm. That is, identify a reasonable value for standard deviation s and provide a trace plot as proof.

Exercise 7.13 (Change the Normal prior). The `one_mh_iteration(w, current)` function from the text is tailored to the Normal-Normal model (7.1) with a $N(0,1^2)$ prior for μ. For this exercise, create a new function `new_mh_iteration(w, current, m, s)` which can utilize *any* $N(m,s^2)$ prior. Subsequently, starting from a current value of $\mu = 3$ and using `set.seed(84735)`, run this function under each setting below. Comment on the resulting proposal, `alpha`, and `next_stop` values.

   a)   `new_mh_iteration(w = 1, current = 3, m = 0, s = 10)`
   b)   `new_mh_iteration(w = 1, current = 3, m = 20, s = 1)`
   c)   `new_mh_iteration(w = 0.1, current = 3, m = 20, s = 1)`
   d)   `new_mh_iteration(w = 0.1, current = 3, m = -15, s = 10)`

Exercise 7.14 (A Gamma-Poisson model). Consider a Gamma-Poisson model in which rate λ has a Gamma(1,0.1) prior and you observe one Poisson data point, $Y = 4$. In this exercise, you will simulate the posterior model of λ using an **independence sampler**.

   a)   Which of the following would make the most reasonable proposal model to use in your independence sampler for λ: Normal, Beta, or Exponential? Hint: Recall that $\lambda > 0$.
   b)   Using the proposal model that you identified in part a, simulate a tour of 1000 λ values. Tune your algorithm until you are satisfied and produce a trace plot of your final results.
   c)   This is a situation where we *can* derive the exact posterior model of λ. What is it?
   d)   Plot a histogram of your tour and overlay the exact model. How good is your Markov chain approximation?

7.9.3   Practice: Simulating more Bayesian models

In this section you will apply your new simulation skills to answer your *own* research question using your own data! In each case, identify a question for which data collection takes fewer than 5 minutes of effort. (That part is supposed to be fun, not overwhelming.)

Exercise 7.15 (Using your own data: Beta-Binomial Metropolis-Hastings).

   a)   Identify a question that could be answered with a Beta-Binomial model for π, some probability or proportion of interest. *Consider scenarios for which data collection would be easy, such as the proportion of emails that you reply to within 24 hours, the proportion of phone calls that you answer, etc.*
   b)   Tune and describe a Beta prior model for π.
   c)   Collect and record your data here.
   d)   Below you will simulate the posterior model of π using a Metropolis-Hastings algorithm. What is a reasonable proposal model to use in this algorithm? Explain.
   e)   Simulate a tour of 2000 π values. Tune your proposal model until you are satisfied and construct the resulting trace plot.
   f)   Plot a histogram of your tour, and comment on the posterior model approximation within the context of your question.

Exercise 7.16 (Using your own data: Normal-Normal MCMC).

a)   Identify a question that could be answered with a Normal-Normal model for μ, some average of interest. *Consider scenarios for which data collection would be easy, such as the average high temperature this time of year, your average hourly screen time, etc.*

b)   Tune and describe a Normal prior model for μ.

c)   Collect and record your data here.

d)   Below you will simulate the posterior model of μ using a Metropolis-Hastings algorithm. What is a reasonable proposal model to use in this algorithm? Explain.

e)   Simulate a tour of 2000 μ values. Tune your proposal model until you are satisfied and construct the resulting trace plot.

f)   Plot a histogram of your tour, and comment on the posterior model approximation within the context of your question.

Exercise 7.17 (Using your own data: Normal-Normal Independence Sampler). Repeat the previous exercise, this time using an independence sampler.

8

Posterior Inference & Prediction

Imagine you find yourself standing at the Museum of Modern Art (MoMA) in New York City, captivated by the artwork in front of you. While understanding that "modern" art doesn't necessarily mean "new" art, a question still bubbles up: *what are the chances that this modern artist is Gen X or even younger, i.e., born in 1965 or later?* In this chapter, we'll perform a Bayesian analysis with the goal of answering this question. To this end, let π denote the proportion of artists represented in major U.S. modern art museums that are Gen X or younger. The Beta(4,6) prior model for π (Figure 8.1) reflects our own very vague prior assumption that major modern art museums disproportionately display artists born before 1965, i.e., π most likely falls below 0.5. After all, "modern art" dates back to the 1880s and it can take a while to attain such high recognition in the art world.

To learn more about π, we'll examine $n = 100$ artists sampled from the MoMA collection. This `moma_sample` dataset in the **bayesrules** package is a subset of data made available by MoMA itself (MuseumofModernArt, 2020).

```
# Load packages
library(bayesrules)
library(tidyverse)
library(rstan)
library(bayesplot)
library(broom.mixed)
library(janitor)

# Load data
data("moma_sample")
```

Among the sampled artists, $Y = 14$ are Gen X or younger:

```
moma_sample %>%
  group_by(genx) %>%
  tally()
# A tibble: 2 x 2
  genx      n
  <lgl>  <int>
1 FALSE     86
2 TRUE      14
```

Recognizing that the dependence of Y on π follows a Binomial model, our analysis follows the Beta-Binomial framework. Thus, our updated posterior model of π in light of the observed art data follows from (3.10):

DOI: 10.1201/9780429288340-8

$$Y|\pi \sim \text{Bin}(100, \pi)$$
$$\pi \sim \text{Beta}(4, 6) \quad \Rightarrow \quad \pi|(Y = 14) \sim \text{Beta}(18, 92)$$

with corresponding posterior pdf

$$f(\pi|y = 14) = \frac{\Gamma(18 + 92)}{\Gamma(18)\Gamma(92)} \pi^{18-1}(1 - \pi)^{92-1} \quad \text{for } \pi \in [0, 1]. \tag{8.1}$$

The evolution in our understanding of π is exhibited in Figure 8.1. Whereas we started out with a vague understanding that under half of displayed artists are Gen X, the data has swayed us toward some certainty that this figure likely falls below 25%.

```
plot_beta_binomial(alpha = 4, beta = 6, y = 14, n = 100)
```

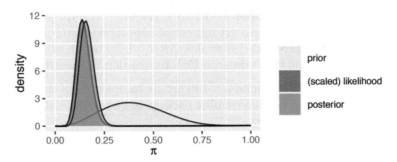

FIGURE 8.1: Our Bayesian model of π, the proportion of modern art museum artists that are Gen X or younger.

After celebrating our success in *constructing* the posterior, please recognize that there's a lot of work ahead. We must be able to utilize this posterior to perform a rigorous **posterior analysis** of π. There are three common tasks in posterior analysis: **estimation, hypothesis testing**, and **prediction**. For example, what's our *estimate* of π? Does our model support the claim that fewer than 20% of museum artists are Gen X or younger? If we sample 20 more museum artists, how many do we predict will be Gen X or younger?

◎ **Goals**

- Establish the *theoretical* foundations for the three posterior analysis tasks: estimation, hypothesis testing, and prediction.
- Explore how Markov chain simulations can be used to approximate posterior features, and hence be utilized in posterior analysis.

8.1   Posterior estimation

Reexamine the Beta(18, 92) posterior model for π, the proportion of modern art museum artists that are Gen X or younger (Figure 8.1). In a Bayesian analysis, we can think of this *entire* posterior model as an estimate of π. After all, this model of posterior plausible values provides a complete picture of the central tendency and uncertainty in π. Yet in specifying and communicating our posterior understanding, it's also useful to compute simple posterior summaries of π. Check in with your gut on how we might approach this task.

❓ Quiz Yourself!

What best describes your posterior *estimate* of π?

a) Roughly 16% of museum artists are Gen X or younger.
b) It's most likely the case that roughly 16% of museum artists are Gen X or younger, but that figure could plausibly be anywhere between 9% and 26%.

If you responded with answer b, your thinking is Bayesian in spirit. To see why, consider Figure 8.2, which illustrates our Beta(18, 92) posterior for π (left) alongside a different analyst's Beta(4, 16) posterior (right). This analyst started with the same Beta(4, 6) prior but only observed 10 artists, 0 of which were Gen X or younger. Though their different data resulted in a different posterior, the central tendency is similar to ours. Thus, the other analyst's *best guess* of π agrees with ours: roughly 16-17% of represented artists are Gen X or younger. However, reporting only this shared "best guess" would make our two posteriors seem misleadingly similar. In fact, whereas we're quite confident that the representation of younger artists is between 10% and 24%, the other analyst is only willing to put that figure somewhere in the much wider range from 6% to 40%. Their relative uncertainty makes sense – they only collected 10 artworks whereas we collected 100.

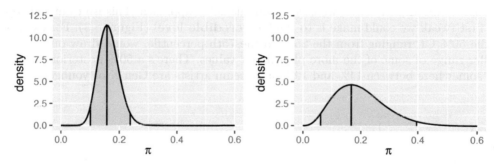

FIGURE 8.2: Our Beta(18, 92) posterior model for π (left) is shown alongside an alternative Beta(4, 16) posterior model (right). The shaded regions represent the corresponding 95% posterior credible intervals for π.

The punchline here is that posterior estimates should reflect both the **central tendency and variability** in π. The **posterior mean and mode** of π provide quick summaries of the central tendency alone. These features for our Beta(18, 92) posterior follow from the general Beta properties (3.2) and match our above observation that Gen X representation is most likely around 16%:

$$E(\pi|Y = 14) = \frac{18}{18 + 92} \approx 0.164$$
$$\text{Mode}(\pi|Y = 14) = \frac{18 - 1}{18 + 92 - 2} \approx 0.157.$$

Better yet, to capture *both* the central tendency and variability in π, we can report a *range* of posterior plausible π values. This range is called a **posterior credible interval (CI)** for

π. For example, we noticed earlier that the proportion of museum artists that are Gen X or younger is most likely between 10% and 24%. This range captures the more plausible values of π while eliminating the more extreme and unlikely scenarios (Figure 8.2). In fact, 0.1 and 0.24 are the 2.5th and 97.5th **posterior percentiles** (i.e., 0.025th and 0.975th **posterior quantiles**), and thus mark the middle 95% of posterior plausible π values. We can confirm these Beta(18,92) posterior quantile calculations using qbeta():

```
# 0.025th & 0.975th quantiles of the Beta(18,92) posterior
qbeta(c(0.025, 0.975), 18, 92)
[1] 0.1009 0.2379
```

The resulting **95% credible interval** for π, (0.1, 0.24), is represented by the shaded region in Figure 8.2 (left). Whereas the area under the entire posterior pdf is 1, the area of this shaded region, and hence the fraction of π values that fall into this region, is 0.95. This reveals an intuitive **interpretation** of the CI. There's a 95% posterior probability that somewhere between 10% and 24% of museum artists are Gen X or younger:

$$P(\pi \in (0.1, 0.24)|Y = 14) = \int_{0.1}^{0.24} f(\pi|y = 14)d\pi = 0.95.$$

Please stop for a moment. Does this interpretation feel natural and intuitive? Thus, a bit anticlimactic? If so, we're happy you feel that way – it means you're thinking like a Bayesian. In Section 8.5 we'll come back to just how special this result is.

In constructing the CI above, we used a "middle 95%" approach. This isn't our *only* option. The first tweak we could make is to the 95% **credible level** (Figure 8.3). For example, a middle 50% CI, ranging from the 25th to the 75th percentile, would draw our focus to a smaller range of some of the more plausible π values. There's a 50% posterior probability that somewhere between 14% and 19% of museum artists are Gen X or younger:

```
# 0.25th & 0.75th quantiles of the Beta(18,92) posterior
qbeta(c(0.25, 0.75), 18, 92)
[1] 0.1388 0.1862
```

In the other direction, a wider middle 99% CI would range from the 0.5th to the 99.5th percentile, and thus kick out only the extreme 1%. As such, a 99% CI would provide us with a fuller picture of *plausible*, though in some cases *very unlikely*, π values:

```
# 0.005th & 0.995th quantiles of the Beta(18,92) posterior
qbeta(c(0.005, 0.995), 18, 92)
[1] 0.0853 0.2647
```

Though a 95% level is a common choice among practitioners, it is somewhat arbitrary and simply ingrained through decades of tradition. **There's no one "right" credible level.** Throughout this book, we'll sometimes use 50% or 80% or 95% levels, depending upon the context of the analysis. Each provide a different snapshot of our posterior understanding.

Consider a second possible tweak to our construction of the CI: it's not necessary to report the *middle* 95% of posterior plausible values. In fact, the middle 95% approach can eliminate some of the more plausible values from the CI. A close inspection of the 50% and 95% credible intervals in Figure 8.3 reveals evidence of this possibility in the ever-so-slightly

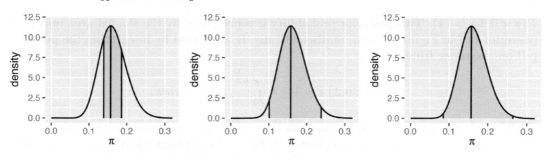

FIGURE 8.3: 50%, 95%, and 99% posterior credible intervals for π.

lopsided nature of the shaded region in our ever-so-slightly non-symmetric posterior. In the 95% CI, values included in the upper end of the CI are less plausible than lower values of π below 0.1 that were left out of the CI. If this lopsidedness were more extreme, we should consider forming a 95% CI for π using not the *middle*, but the 95% of posterior values with the **highest posterior density**. You can explore this idea in the exercises, though we won't lose sleep over it here. Mainly, this method will only produce meaningfully different results than the middle 95% approach in extreme cases, when the posterior is extremely skewed.

Posterior Credible Intervals

Let parameter π have posterior pdf $f(\pi|y)$. A posterior credible interval (CI) provides a range of posterior plausible values of π, and thus a summary of both posterior central tendency and variability. For example, a middle 95% CI is constructed by the 2.5th and 97.5th posterior percentiles,

$$(\pi_{0.025}, \pi_{0.975}).$$

Thus, there's a 95% posterior probability that π is in this range:

$$P(\pi \in (\pi_{0.025}, \pi_{0.975})|Y = y) = \int_{\pi_{0.025}}^{\pi_{0.975}} f(\pi|y)d\pi = 0.95.$$

8.2 Posterior hypothesis testing

8.2.1 One-sided tests

Hypothesis testing is another common task in posterior analysis. For example, suppose we read an article claiming that fewer than 20% of museum artists are Gen X or younger. Two clues we've observed from our posterior model of π point to this claim being at least partially *plausible*:

1.   The majority of the posterior pdf in Figure 8.3 falls below 0.2.
2.   The 95% credible interval for π, (0.1, 0.24), is *mostly* below 0.2.

These observations are a great start. Yet we can be even more precise. To evaluate exactly *how* plausible it is that $\pi < 0.2$, we can calculate the **posterior probability** of this scenario,

$P(\pi < 0.2|Y = 14)$. This posterior probability is represented by the shaded area under the posterior pdf in Figure 8.4 and, mathematically, is calculated by integrating the posterior pdf on the range from 0 to 0.2:

$$P(\pi < 0.2 \mid Y = 14) = \int_0^{0.2} f(\pi|y = 14)d\pi.$$

We'll bypass the integration and obtain this Beta(18,92) posterior probability using `pbeta()` below. The result reveals strong evidence in favor of our claim: there's a roughly 84.9% posterior chance that Gen Xers account for fewer than 20% of modern art museum artists.

```
# Posterior probability that pi < 0.20
post_prob <- pbeta(0.20, 18, 92)
post_prob
[1] 0.849
```

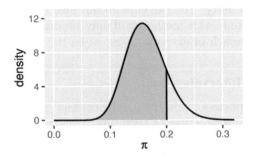

FIGURE 8.4: The Beta(18,92) posterior probability that π is below 0.20 is represented by the shaded region under the posterior pdf.

This analysis of our claim is refreshingly straightforward. We simply calculated the posterior probability of the scenario of interest. Though not always necessary, practitioners often *formalize* this procedure into a **hypothesis testing** framework. For example, we can frame our analysis as two competing hypotheses: the **null hypothesis** H_0 contends that *at least* 20% of museum artists are Gen X or younger (the status quo here) whereas the **alternative hypothesis** H_a (our claim) contends that this figure is *below* 20%. In mathematical notation:

$$H_0 : \pi \geq 0.2$$
$$H_a : \pi < 0.2$$

Note that H_a claims that π lies on *one* side of 0.2 ($\pi < 0.2$) as opposed to just being *different* than 0.2 ($\pi \neq 0.2$). Thus, we call this a **one-sided hypothesis test**.

We've already calculated the posterior probability of the alternative hypothesis to be $P(H_a \mid Y = 14) = 0.849$. Thus, the posterior probability of the null hypothesis is $P(H_0 \mid Y = 14) = 0.151$. Putting these together, the **posterior odds** that $\pi < 0.2$ are roughly 5.62. That is, our posterior assessment is that π is nearly 6 times more likely to be *below* 0.2 than to be *above* 0.2:

$$\text{posterior odds} = \frac{P(H_a \mid Y = 14)}{P(H_0 \mid Y = 14)} \approx 5.62.$$

```
# Posterior odds
post_odds <- post_prob / (1 - post_prob)
post_odds
[1] 5.622
```

Of course, these posterior odds represent our *updated* understanding of π upon observing the survey data, $Y = 14$ of $n = 100$ sampled artists were Gen X or younger. *Prior* to sampling these artists, we had a much higher assessment of Gen X representation at major art museums (Figure 8.5).

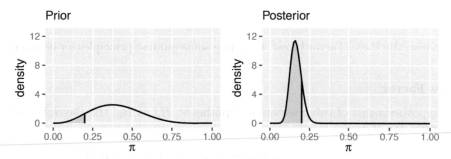

FIGURE 8.5: The posterior probability that π is below 0.2 (right) is contrasted against the prior probability of this scenario (left).

Specifically, the **prior probability** that $\pi < 0.2$, calculated by the area under the Beta(4,6) *prior* pdf $f(\pi)$ that falls below 0.2, was only 0.0856:

$$P(H_a) = \int_0^{0.2} f(\pi)d\pi \approx 0.0856.$$

```
# Prior probability that pi < 0.2
prior_prob <- pbeta(0.20, 4, 6)
prior_prob
[1] 0.08564
```

Thus, the prior probability of the null hypothesis is $P(H_0) = 0.914$. It follows that the prior odds of Gen X representation being below 0.2 were roughly only 1 in 10:

$$\text{Prior odds } = \frac{P(H_a)}{P(H_0)} \approx 0.093.$$

```
# Prior odds
prior_odds <- prior_prob / (1 - prior_prob)
prior_odds
[1] 0.09366
```

The **Bayes Factor (BF)** compares the posterior odds to the prior odds, and hence provides insight into just how much our understanding about Gen X representation *evolved* upon observing our sample data:

$$\text{Bayes Factor} = \frac{\text{posterior odds}}{\text{prior odds}}.$$

In our example, the Bayes Factor is roughly 60. Thus, upon observing the artwork data, the posterior odds of our hypothesis about Gen Xers are roughly 60 times higher than the prior odds. Or, our confidence in this hypothesis jumped quite a bit.

```
# Bayes factor
BF <- post_odds / prior_odds
BF
[1] 60.02
```

We summarize the Bayes Factor below, including some guiding principles for its interpretation.

Bayes Factor

In a hypothesis test of two competing hypotheses, H_a vs H_0, the Bayes Factor is an odds ratio for H_a:

$$\text{Bayes Factor} = \frac{\text{posterior odds}}{\text{prior odds}} = \frac{P(H_a|Y)/P(H_0|Y)}{P(H_a)/P(H_0)}.$$

As a ratio, it's meaningful to compare the Bayes Factor (BF) to 1. To this end, consider three possible scenarios:

1. BF = 1: The plausibility of H_a *didn't change* in light of the observed data.
2. BF > 1: The plausibility of H_a *increased* in light of the observed data. Thus, the greater the Bayes Factor, the more convincing the evidence for H_a.
3. BF < 1: The plausibility of H_a *decreased* in light of the observed data.

Bringing it all together, the posterior probability (0.85) and Bayes Factor (60) establish *fairly convincing* evidence in favor of the claim that fewer than 20% of artists at major modern art museums are Gen X or younger. Did you wince in reading that sentence? The term "fairly convincing" might seem a little wishy-washy. In the past, you might have learned specific cut-offs that distinguish between "statistically significant" and "not statistically significant" results, or allow you to "reject" or "fail to reject" a hypothesis. However, this practice provides false comfort. Reality is not so clear-cut. For this reason, across the frequentist and Bayesian spectrum, the broader statistics community advocates *against* making rigid conclusions using universal rules and *for* a more nuanced practice which takes into account the context and potential implications of each individual hypothesis test. Thus, there is no magic, one-size-fits-all cut-off for what Bayes Factor or posterior probability evidence is big enough to filter claims into "true" or "false" categories. In fact, what we have is more powerful than a binary decision – we have a holistic measure of our level of uncertainty about the claim. This level of uncertainty can inform our next steps. In our art example, do we have ample evidence for our claim? *We're* convinced.

8.2.2 Two-sided tests

Especially when working in new settings, it's not always the case that we wish to test a one-directional claim about some parameter π. For example, consider a new art researcher that simply wishes to test *whether or not* 30% of major museum artists are Gen X or younger. This hypothesis is **two-sided**:

$$H_0 : \pi = 0.3$$
$$H_a : \pi \neq 0.3$$

When we try to hit this two-sided hypothesis test with the same hammer we used for the one-sided hypothesis test, we quickly run into a problem. Since π is *continuous*, the prior and posterior probabilities that π is *exactly* 0.3 (i.e., that H_0 is true) are both *zero*. For example, the posterior probability that $\pi = 0.3$ is calculated by the area of the *line* under the posterior pdf at 0.3. As is true for any line, this area is 0:

$$P(\pi = 0.3 | Y = 14) = \int_{0.3}^{0.3} f(\pi|y = 14)d\pi = 0.$$

Thus, the posterior odds, prior odds, and consequently the Bayes factor are all undefined:

$$\text{Posterior odds} = \frac{P(H_a \,|\, Y = 14)}{P(H_0 \,|\, Y = 14)} = \frac{1}{0} = \text{nooooo!}$$

No problem. There's not one recipe for success. To that end, try the following quiz.[1]

> ❷ **Quiz Yourself!**
>
> Recall that the 95% posterior credible interval for π is (0.1, 0.24). Does this CI provide ample evidence that π differs from 0.3?

If you answered "yes," then you intuited a reasonable approach to two-sided hypothesis testing. The hypothesized value of π (here 0.3) is "substantially" outside the posterior credible interval, thus we have ample evidence in favor of H_a. The fact that 0.3 is so far above the range of plausible π values makes us pretty confident that the proportion of museum artists that are Gen X or younger is *not* 0.3. Yet what's "substantial" or clear in one context might be different than what's "substantial" in another. With that in mind, it is best practice to define "substantial" *ahead of time*, *before* seeing any data. For example, in the context of artist representation, we might consider any proportion outside the 0.05 window around 0.3 to be meaningfully different from 0.3. This essentially adds a little buffer into our hypotheses, π is either *around* 0.3 (between 0.25 and 0.35) or it's not:

$$H_0 : \pi \in (0.25, 0.35)$$
$$H_a : \pi \notin (0.25, 0.35)$$

With this defined buffer in place, we can more rigorously claim belief in H_a since the *entire* hypothesized range for π, (0.25, 0.35), lies above its 95% credible interval. Note also that since H_0 no longer includes a singular hypothesized value of π, its corresponding posterior and prior probabilities are no longer 0. Thus, just as we did in the *one-sided* hypothesis testing setting, we could (but won't here) supplement our above posterior credible interval analysis with posterior probability and Bayes Factor calculations.

---

[1]Answer: yes

8.3   Posterior prediction

Beyond posterior estimation and hypothesis testing, a third common task in a posterior analysis is to **predict** the outcome of new data Y'.

❓ Quiz Yourself!

Suppose we get our hands on data for 20 more artworks displayed at the museum. Based on the posterior understanding of π that we've developed throughout this chapter, what number would you *predict* are done by artists that are Gen X or younger?

Your knee-jerk reaction to this quiz might be: "I *got* this one. It's 3!" This is a very reasonable place to start. After all, our best posterior guess was that roughly 16% of museum artists are Gen X or younger and 16% of 20 new artists is roughly 3. However, this calculation ignores two sources of potential variability in our prediction:

- **Sampling variability** in the data
  When we randomly sample 20 artists, we don't expect *exactly* 3 (16%) of these to be Gen X or younger. Rather, the number will fluctuate depending upon which random sample we happen to get.

- **Posterior variability** in π
  0.16 isn't the *only* posterior plausible value of π, the underlying proportion of museum artists that are Gen X. Rather, our 95% posterior credible interval indicated that π might be anywhere between roughly 0.1 and 0.24. Thus, when making predictions about the 20 new artworks, we need to consider what outcomes we'd expect to see under *each* possible π while accounting for the fact that some π are more plausible than others.

Let's specify these concepts with some math. First, let $Y' = y'$ be the (yet unknown) number of the 20 new artworks that are done by Gen X or younger artists, where y' can be any number of artists in $\{0, 1, ..., 20\}$. Conditioned on π, the randomness or **sampling variability** in Y' can be modeled by $Y'|\pi \sim \text{Bin}(20, \pi)$ with pdf

$$f(y'|\pi) = P(Y' = y'|\pi) = \binom{20}{y'}\pi^{y'}(1 - \pi)^{20-y'}. \tag{8.2}$$

Thus, the random outcome of Y' depends upon π, which too can vary – π might be any value between 0 and 1. To this end, the Beta(18, 92) posterior model of π given the *original* data ($Y = 14$) describes the potential **posterior variability** in π, i.e., which values of π are more plausible than others.

For an overall understanding of how many of the next 20 artists will be Gen X, we must *combine* the sampling variability in Y' with the posterior variability in π. To this end, weighting $f(y'|\pi)$ (8.2) by posterior pdf $f(\pi|y = 14)$ (8.1) captures the chance of observing $Y' = y'$ Gen Xers for a given π *while taking into account the posterior plausibility of that π value*:

$$f(y'|\pi)f(\pi|y = 14). \tag{8.3}$$

Figure 8.6 illustrates this idea, plotting the weighted behavior of Y' (8.3) for just three possible values of π: the 2.5th posterior percentile (0.1), posterior mode (0.16), and 97.5th posterior percentile (0.24). Naturally, we see that the greater π is, the greater Y' tends to be: when $\pi = 0.1$ the most likely value of Y' is 2, whereas when $\pi = 0.24$, the most likely value of Y' is 5. Also notice that since π values as low as 0.1 or as high as 0.24 are not very plausible, the values of Y' that might be generated under these scenarios are given less weight (i.e., the sticks are much shorter) than those that are generated under $\pi = 0.16$, the *most* plausible π value.

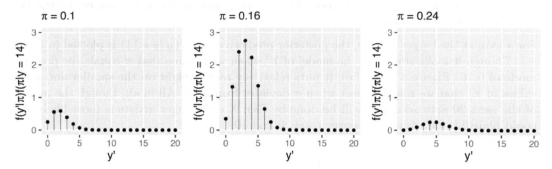

FIGURE 8.6: Possible Y' outcomes are plotted for $\pi \in \{0.10, 0.16, 0.24\}$ and weighted by the corresponding posterior plausibility of π.

Putting this all together, the **posterior predictive model** of Y', the number of the 20 new artists that are Gen X, takes into account both the sampling variability in Y' and posterior variability in π. Specifically, the posterior predictive pmf calculates the overall chance of observing $Y' = y'$ across *all* possible π from 0 to 1 by averaging across (8.3), the chance of observing $Y' = y'$ for any *given* π:

$$f(y'|y = 14) = P(Y' = y' \mid Y = y) = \int_0^1 f(y'|\pi)f(\pi|y = 14)d\pi.$$

Posterior predictive model

Let Y' denote a new outcome of variable Y. Further, let pdf $f(y'|\pi)$ denote the dependence of Y' on π and posterior pdf $f(\pi|y)$ denote the posterior plausibility of π given the original data $Y = y$. Then the posterior predictive model for Y' has pdf

$$f(y'|y) = \int f(y'|\pi)f(\pi|y)d\pi. \tag{8.4}$$

In words, the overall chance of observing $Y' = y'$ weights the chance of observing this outcome under *any* possible π ($f(y'|\pi)$) by the posterior plausibility of π ($f(\pi|y)$).

An exact formula for the pmf of Y' follows from some calculus (which we don't show here but is fun and we encourage you to try if you have calculus experience):

$$f(y'|y = 14) = \binom{20}{y'} \frac{\Gamma(110)}{\Gamma(18)\Gamma(92)} \frac{\Gamma(18 + y')\Gamma(112 - y')}{\Gamma(130)} \quad \text{for } y' \in \{0, 1, \ldots, 20\}. \tag{8.5}$$

Though this formula is unlike any we've ever seen (e.g., it's not Binomial or Poisson or anything else we've learned), it still specifies *what* values of Y' we might observe and the *probability* of each. For example, plugging $y' = 3$ into this formula, there's a 0.2217 posterior predictive probability that 3 of the 20 new artists will be Gen X:

$$f(y' = 3|y = 14) = \binom{20}{3} \frac{\Gamma(110)}{\Gamma(18)\Gamma(92)} \frac{\Gamma(18+3)\Gamma(112-3)}{\Gamma(130)} = 0.2217.$$

The pmf formula also reflects the influence of our Beta(18,92) posterior model for π (through parameters 18 and 92), and hence the original prior and data, on our posterior understanding of Y'. That is, like any posterior operation, our posterior predictions balance information from both the prior and data.

For a look at the bigger picture, the posterior predictive pmf $f(y'|y = 14)$ is plotted in Figure 8.7. Though it looks similar to the model of Y' when we assume that π equals the posterior mode of 0.16 (Figure 8.6 middle), it puts relatively more weight on the smaller and larger values of Y' that we might expect when π deviates from 0.16. All in all, though the number of the next 20 artworks that will be done by Gen X or younger artists is most likely 3, it could plausibly be anywhere between, say, 0 and 10.

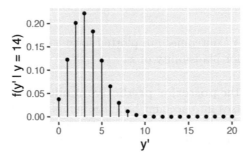

FIGURE 8.7: The posterior predictive model of Y', the number of the next 20 artworks that are done by Gen X or younger artists.

Finally, after building and examining the posterior predictive model of Y', the number of the next 20 artists that will be Gen X, we might have some follow-up questions. For example, what's the posterior probability that at least 5 of the 20 artists are Gen X, $P(Y' \geq 5|Y = 14)$? How many of the next 20 artists do we *expect* to be Gen X, $E(Y'|Y = 14)$? We *can* answer these questions, it's just a bit tedious. Since the posterior predictive model for Y' isn't familiar, we can't calculate posterior features using pre-built formulas or R functions like we did for the Beta posterior model of π. Instead, we have to calculate these features from scratch. For example, we can calculate the posterior probability that at least 5 of the 20 artists are Gen X by adding up the pmf (8.5) evaluated at each of the 16 y' values in this range. The result of this large sum, the details of which would fill a whole page, is 0.233:

$$\begin{aligned} P(Y' \geq 5|y = 14) &= \sum_{y'=5}^{20} f(y'|y = 14) \\ &= f(y' = 5|y = 14) + f(y' = 6|y = 14) + \cdots + f(y' = 20|y = 14) \\ &= 0.233. \end{aligned}$$

Similarly, though this isn't a calculation we've had to do yet (and won't do again), the expected number of the next 20 artists that will be Gen X can be obtained by the posterior weighted average of possible Y' values. That is, we can add up each Y' value from 0 to 20 weighted by their posterior probabilities $f(y'|y = 14)$. The result of this large sum indicates that we should *expect* roughly 3 of the 20 artists to be Gen X:

$$E(Y'|y = 14) = \sum_{y'=0}^{20} y' f(y'|y = 14)$$
$$= 0 \cdot f(y' = 0|y = 14) + 1 \cdot f(y' = 1|y = 14) + \cdots + 20 \cdot f(y' = 20|y = 14)$$
$$= 3.273.$$

But we don't want to get too distracted by these types of calculations. In this book, we'll never need to do something like this again. Starting in Chapter 9, our models will be complicated enough so that even tedious formulas like these will be unattainable and we'll need to rely on simulation to *approximate* posterior features.

8.4 Posterior analysis with MCMC

It's great to know that there's some theory behind Bayesian posterior analysis. And when we're working with models that are as straightforward as the Beta-Binomial, we can directly implement this theory — that is, we can calculate exact posterior credible intervals, probabilities, and predictive models. Yet in Chapter 9 we'll leave this nice territory and enter scenarios in which we cannot specify posterior models, let alone calculate exact summaries of their features. Recall from Chapter 6, that in these scenarios we can *approximate* posteriors using MCMC methods. In this section, we'll explore how the resulting Markov chain sample values can also be used to approximate specific posterior *features*, and hence be used to conduct posterior analysis.

8.4.1 Posterior simulation

Below we run four parallel Markov chains of π for 10,000 iterations each. After tossing out the first 5,000 iterations of each chain, we end up with four separate Markov chain samples of size 5,000, $\left\{\pi^{(1)}, \pi^{(2)}, \ldots, \pi^{(5000)}\right\}$, or a combined Markov chain sample size of 20,000.

```
# STEP 1: DEFINE the model
art_model <- "
  data {
    int<lower = 0, upper = 100> Y;
  }
  parameters {
    real<lower = 0, upper = 1> pi;
  }
  model {
    Y ~ binomial(100, pi);
    pi ~ beta(4, 6);
  }
"

# STEP 2: SIMULATE the posterior
```

```
art_sim <- stan(model_code = art_model, data = list(Y = 14),
                chains = 4, iter = 5000*2, seed = 84735)
```

Check out the numerical and visual diagnostics in Figure 8.8. First, the randomness in the trace plots (left), the agreement in the density plots of the four parallel chains (middle), and an Rhat value of effectively 1 suggest that our simulation is extremely stable. Further, our dependent chains are behaving "enough" like an independent sample. The autocorrelation, shown at right for just one chain, drops off quickly and the effective sample size ratio is satisfyingly high – our 20,000 Markov chain values are as effective as 7600 independent samples $(0.38 \cdot 20000)$.

```
# Parallel trace plots & density plots
mcmc_trace(art_sim, pars = "pi", size = 0.5) +
  xlab("iteration")
mcmc_dens_overlay(art_sim, pars = "pi")

# Autocorrelation plot
mcmc_acf(art_sim, pars = "pi")
```

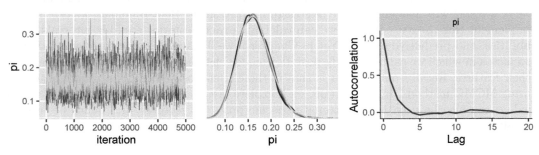

FIGURE 8.8: MCMC simulation results for the posterior model of π, the proportion of museum artists that are Gen X or younger, are exhibited by trace and density plots for the four parallel chains (left and middle) and an autocorrelation plot for a single chain (right).

```
# Markov chain diagnostics
rhat(art_sim, pars = "pi")
[1] 1
neff_ratio(art_sim, pars = "pi")
[1] 0.378
```

8.4.2   Posterior estimation & hypothesis testing

We can now use the *combined* 20,000 Markov chain values, with confidence, to **approximate** the Beta(18, 92) posterior model of π. Indeed, Figure 8.9 confirms that the complete MCMC approximation (right) closely mimics the actual posterior (left).

```
# The actual Beta(18, 92) posterior
plot_beta(alpha = 18, beta = 92) +
  lims(x = c(0, 0.35))
```

```
# MCMC posterior approximation
mcmc_dens(art_sim, pars = "pi") +
  lims(x = c(0,0.35))
```

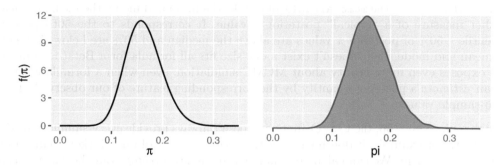

FIGURE 8.9: The actual Beta(18, 92) posterior pdf of π (left) alongside an MCMC approximation (right).

As such, we can approximate any feature of the Beta(18, 92) posterior model by the corresponding feature of the Markov chain. For example, we can approximate the posterior mean by the mean of the MCMC sample values, or approximate the 2.5th posterior percentile by the 2.5th percentile of the MCMC sample values. To this end, the `tidy()` function in the **broom.mixed** package (Bolker and Robinson, 2021) provides some handy statistics for the combined 20,000 Markov chain values stored in `art_sim`:

```
tidy(art_sim, conf.int = TRUE, conf.level = 0.95)
# A tibble: 1 x 5
  term  estimate std.error conf.low conf.high
  <chr>    <dbl>     <dbl>    <dbl>     <dbl>
1 pi       0.162    0.0352    0.101     0.239
```

And the `mcmc_areas()` function in the **bayesplot** package provides a visual complement (Figure 8.10):

```
# Shade in the middle 95% interval
mcmc_areas(art_sim, pars = "pi", prob = 0.95)
```

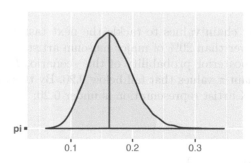

FIGURE 8.10: The median and middle 95% of the approximate posterior model of π.

In the `tidy()` summary, `conf.low` and `conf.high` report the 2.5th and 97.5th **percentiles** of the Markov chain values, 0.101 and 0.239, respectively. These form an *approximate* middle 95% credible interval for π which is represented by the shaded region in the `mcmc_areas()` plot. Further, the `estimate` reports that the **median** of our 20,000 Markov chain values, and thus our *approximation* of the actual posterior median, is 0.162. This median is represented by the vertical line in the `mcmc_areas()` plot. Like the mean and mode, the median provides another measure of a "typical" posterior π value. It corresponds to the 50th posterior percentile – 50% of posterior π values are above the median and 50% are below. Yet *unlike* the mean and mode, there doesn't exist a one-size-fits-all formula for a Beta(α, β) median. This exposes even more beauty about MCMC simulation: even when a formula is elusive, we can *estimate* a posterior quantity by the corresponding feature of our observed Markov chain sample values.

Though a nice first stop, the `tidy()` function doesn't always provide every summary statistic of interest. For example, it doesn't report the *mean* or *mode* of our Markov chain sample values. No problem. We can calculate summary statistics *directly* from the Markov chain values. The first step is to convert an *array* of the four parallel chains into a single *data frame* of the combined chains:

```
# Store the 4 chains in 1 data frame
art_chains_df <- as.data.frame(art_sim, pars = "lp__", include = FALSE)
dim(art_chains_df)
[1] 20000      1
```

With the chains in data frame form, we can proceed as usual, using our `dplyr` tools to transform and summarize. For example, we can directly calculate the sample mean, median, mode, and quantiles of the combined Markov chain values. The median and quantile values are precisely those reported by `tidy()` above, and thus eliminate any mystery about that function!

```
# Calculate posterior summaries of pi
art_chains_df %>%
  summarize(post_mean = mean(pi),
            post_median = median(pi),
            post_mode = sample_mode(pi),
            lower_95 = quantile(pi, 0.025),
            upper_95 = quantile(pi, 0.975))
  post_mean post_median post_mode lower_95 upper_95
1    0.1642      0.1624    0.1598   0.1011   0.2388
```

We can also use the raw chain values to tackle the next task in our posterior analysis – testing the claim that fewer than 20% of major museum artists are Gen X. To this end, we can **approximate** the posterior probability of this scenario, $P(\pi < 0.20|Y = 14)$, by the proportion of Markov chain π values that fall below 0.20. By this approximation, there's an 84.6% chance that Gen X artist representation is under 0.20:

```
# Tabulate pi values that are below 0.20
art_chains_df %>%
  mutate(exceeds = pi < 0.20) %>%
  tabyl(exceeds)
```

```
exceeds      n percent
  FALSE   3080   0.154
   TRUE  16920   0.846
```

Soak it in and remember the point. We've used our MCMC simulation to **approximate** the posterior model of π along with its features of interest. For comparison, Table 8.1 presents the Beta(18,92) posterior features we calculated in Section 8.1 alongside their corresponding MCMC approximations. The punchline is this: MCMC worked. The approximations are *quite* accurate. Let this bring you peace of mind as you move through the next chapters – though the models therein will be too complicated to specify, we can be confident in our MCMC approximations of these models (so long as the diagnostics check out!).

TABLE 8.1: Beta(18,92) posterior model features alongside their corresponding MCMC approximations.

	mean	mode	2.5th percentile	97.5 percentile
posterior value	0.16	0.16	0.1	0.24
MCMC approximation	0.1642	0.1598	0.1011	0.2388

8.4.3 Posterior prediction

Finally, we can utilize our Markov chain values to approximate the posterior predictive model of Y', the number of the next 20 sampled artists that will be Gen X or younger. Bonus: simulating this model also helps us build intuition for the theory underlying posterior prediction. Recall that the posterior predictive model reflects two sources of variability:

- **Sampling variability in the data**
  Y' might be any number of artists in $\{0, 1, \ldots, 20\}$ and depends upon the underlying proportion of artists that are Gen X, π: $Y'|\pi \sim \text{Bin}(20, \pi)$.

- **Posterior variability in π**
  The collection of 20,000 Markov chain π values provides an *approximate* sense for the variability and range in plausible π values.

To capture both sources of variability in posterior predictions Y', we can use `rbinom()` to simulate one $\text{Bin}(20, \pi)$ outcome Y' from *each* of the 20,000 π chain values. The first three results reflect a general trend: smaller values of π will *tend* to produce smaller values of Y'. This makes sense. The lower the underlying representation of Gen X artists in the museum, the fewer Gen X artists we should expect to see in our next sample of 20 artworks.

```
# Set the seed
set.seed(1)

# Predict a value of Y' for each pi value in the chain
art_chains_df <- art_chains_df %>%
  mutate(y_predict = rbinom(length(pi), size = 20, prob = pi))

# Check it out
art_chains_df %>%
```

```
head(3)
      pi y_predict
1 0.1301          2
2 0.1755          3
3 0.2214          5
```

The resulting collection of 20,000 predictions closely approximates the true posterior predictive distribution (Figure 8.7). It's *most likely* that 3 of the 20 artists will be Gen X or younger, though this figure might reasonably range between 0 and, say, 10:

```
# Plot the 20,000 predictions
ggplot(art_chains_df, aes(x = y_predict)) +
  stat_count()
```

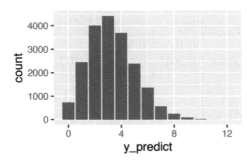

FIGURE 8.11: A histogram of 20,000 simulated posterior predictions of the number among the next 20 artists that will be Gen X or younger.

We can also utilize the posterior predictive sample to *approximate* features of the actual posterior predictive model that were burdensome to specify mathematically. For example, we can approximate the posterior mean prediction, $E(Y'|Y = 14)$, and the more comprehensive **posterior prediction interval** for Y'. To this end, we expect roughly 3 of the next 20 artists to be Gen X or younger, but there's an 80% chance that this figure is somewhere between 1 and 6 artists:

```
art_chains_df %>%
  summarize(mean = mean(y_predict),
            lower_80 = quantile(y_predict, 0.1),
            upper_80 = quantile(y_predict, 0.9))
    mean lower_80 upper_80
1 3.283        1        6
```

8.5   Bayesian benefits

In Chapter 1, we highlighted how Bayesian analyses compare to frequentist analyses. Now that we've worked through some concrete examples, let's revisit some of those ideas. As you've likely experienced, often the toughest part of a Bayesian analysis is building or simulating the posterior model. Once we have that piece in place, it's fairly straightforward to utilize this posterior for estimation, hypothesis testing, and prediction. In contrast, building up the formulas to perform the analogous frequentist calculations is often less intuitive.

We can also bask in the ease with which Bayesian results can be interpreted. In general, a **Bayesian analysis** assesses the uncertainty regarding an *unknown* parameter π in light of observed data Y. For example, consider the artist study. In light of observing that $Y = 14$ of 100 sampled artists were Gen X or younger, we determined that there was an 84.9% posterior chance that Gen X representation at the entire museum, π, falls below 0.20:

$$P(\pi < 0.20 \mid Y = 14) = 0.849.$$

This calculation doesn't make sense in a frequentist analysis. Flipping the script, a **frequentist analysis** assesses the uncertainty of the observed data Y in light of *assumed* values of π. For example, the frequentist counterpart to the Bayesian posterior probability above is the **p-value**, the formula for which we won't dive into here:

$$P(Y \leq 14 \mid \pi = 0.20) = 0.08.$$

The opposite order of the conditioning in this probability, Y given π instead of π given Y, leads to a different calculation and interpretation than the Bayesian probability: **if** π were only 0.20, then there's only an 8% chance we'd have observed a sample in which at most $Y = 14$ of 100 artists were Gen X. It's not our writing here that's awkward, it's the p-value. Though it does provide us with some interesting information, the question it answers is a little less natural for the human brain: since we actually *observed* the data but *don't* know π, it can be a mind bender to interpret a calculation that assumes the opposite. Mainly, when testing hypotheses, it's more natural to ask "how probable is my hypothesis?" (what the Bayesian probability answers) than "how probable is my data if my hypothesis weren't true?" (what the frequentist probability answers). Given how frequently p-values are misinterpreted, and hence misused, they're increasingly being de-emphasized across the entire frequentist and Bayesian spectrum.

8.6 Chapter summary

In Chapter 8, you learned how to turn a posterior model into answers. That is, you utilized posterior models, exact or approximate, to perform three posterior analysis tasks for an unknown parameter π:

1. **Posterior estimation**
   A posterior **credible interval** provides a range of posterior plausible values of π, and thus a sense of both the posterior typical values and uncertainty in π.
2. **Posterior hypothesis testing**
   Posterior probabilities provide insight into corresponding hypotheses regarding π.
3. **Posterior prediction**
   The posterior predictive model for a new data point Y takes into account both the *sampling variability* in Y and the *posterior variability* in π.

8.7 Exercises

8.7.1 Conceptual exercises

Exercise 8.1 (Posterior analysis). What are the three common tasks in a posterior analysis?

Exercise 8.2 (Warming up).

a)   In estimating some parameter λ, what are some drawbacks to only reporting the central tendency of the λ posterior model?
b)   The 95% credible interval for λ is (1,3.4). How would you interpret this?

Exercise 8.3 (Hypothesis testing?). In each situation below, indicate whether the issue at hand could be addressed using a *hypothesis test*.

a)   Your friend Trichelle claims that more than 40% of dogs at the dog park do not have a dog license.
b)   Your professor is interested in learning about the proportion of students at a large university who have heard of Bayesian statistics.
c)   An environmental justice advocate wants to know if more than 60% of voters in their state support a new regulation.
d)   Sarah is studying Ptolemy's *Syntaxis Mathematica* text and wants to investigate the number of times that Ptolemy uses a certain mode of argument per page of text. Based on Ptolemy's other writings she thinks it will be about 3 times per page. Rather than reading all 13 volumes of *Syntaxis Mathematica*, Sarah takes a random sample of 90 pages.

Exercise 8.4 (Bayes Factor). Answer questions about Bayes Factors from your friend Enrique who has a lot of frequentist statistics experience, but is new to Bayes.

a)   What are *posterior odds*?
b)   What are *prior odds*?
c)   What's a *Bayes Factor* and why we might want to calculate it?

Exercise 8.5 (Posterior prediction: concepts).

a)   What two types of variability do posterior predictive models incorporate? Define each type such that your non-Bayesian statistics friends could understand.
b)   Describe a real-life situation in which it would be helpful to carry out posterior prediction.
c)   Is a posterior predictive model conditional on just the data, just the parameter, or on both the data and the parameter?

8.7.2 Practice exercises

Exercise 8.6 (Credible intervals: Part I). For each situation, find the appropriate credible interval using the "middle" approach.

a)   A 95% credible interval for π with $\pi|y \sim \text{Beta}(4,5)$

  b)   A 60% credible interval for π with $\pi|y \sim \text{Beta}(4,5)$
  c)   A 95% credible interval for λ with $\lambda|y \sim \text{Gamma}(1,8)$

Exercise 8.7 (Credible intervals: Part II). For each situation, find the appropriate credible interval using the "middle" approach.

  a)   A 99% credible interval of λ with $\lambda|y \sim \text{Gamma}(1,5)$
  b)   A 95% credible interval of μ with $\mu|y \sim N(10, 2^2)$
  c)   An 80% credible interval of μ with $\mu|y \sim N(-3, 1^2)$

Exercise 8.8 (Credible intervals: highest posterior density). There's more than one approach to constructing a 95% credible interval. The "middle 95%" approach reports the range of the middle 95% of the posterior density, from the 2.5th to the 97.5th percentile. The "highest posterior density" approach reports the 95% of posterior values with the highest posterior densities.

  a)   Let $\lambda|y \sim \text{Gamma}(1,5)$. Construct the 95% highest posterior density credible interval for λ. Represent this interval on a sketch of the posterior pdf. Hint: The sketch itself will help you identify the appropriate CI.
  b)   Repeat part a using the middle 95% approach.
  c)   Compare the two intervals from parts a and b. Are they the same? If not, how do they differ and which is more appropriate here?
  d)   Let $\mu|y \sim N(-13, 2^2)$. Construct the 95% highest posterior density credible interval for μ.
  e)   Repeat part d using the middle 95% approach.
  f)   Compare the two intervals from parts d and e. Are they the same? If not, why not?

Exercise 8.9 (Hypothesis tests: Part I). For parameter π, suppose you have a $\text{Beta}(1,0.8)$ prior model and a $\text{Beta}(4,3)$ posterior. You wish to test the null hypothesis that $\pi \leq 0.4$ versus the alternative that $\pi > 0.4$.

  a)   What is the posterior probability for the alternative hypothesis?
  b)   Calculate and interpret the posterior odds.
  c)   Calculate and interpret the prior odds.
  d)   Calculate and interpret the Bayes Factor.
  e)   Putting this together, explain your conclusion about these hypotheses to someone who is unfamiliar with Bayesian statistics.

Exercise 8.10 (Hypothesis tests: Part II). Repeat Exercise 8.9 for the following scenario. For parameter μ, suppose you have a $N(10, 10^2)$ prior model, a $N(5, 3^2)$ posterior, and you wish to test $H_0 : \mu \geq 5.2$ versus $H_a : \mu < 5.2$.

Exercise 8.11 (Posterior predictive Beta-Binomial: with calculus). As discussed in Section 8.3, it is sometimes possible to derive an exact posterior predictive model. Such is the case with the conjugate models we have studied thus far. To begin, suppose we observe $Y = y$ successes in n trials where $Y|\pi \sim \text{Bin}(n, \pi)$ and π has a $\text{Beta}(\alpha, \beta)$ prior.

  a)   Identify the posterior pdf of π given the observed data $Y = y$, $f(\pi|y)$. NOTE: This will depend upon $(y, n, \alpha, \beta, \pi)$.

b)   Suppose we conduct n' *new* trials (where n' might differ from our original number of trials n) and let $Y' = y'$ be the observed number of successes in these new trials. Identify the conditional pmf of Y' given π, $f(y'|\pi)$. NOTE: This will depend upon (y', n', π).

c)   Identify the posterior predictive pmf of Y', $f(y'|y)$. NOTE: This pmf, found using (8.4), will depend upon $(y, n, y', n', \alpha, \beta)$.

d)   As with the example in Section 8.3, suppose your posterior model of π is based on a prior model with $\alpha = 4$ and $\beta = 6$ and an observed $y = 14$ successes in $n = 100$ original trials. We plan to conduct $n' = 20$ *new* trials. Specify the posterior predictive pmf of Y', the number of successes we might observe in these 20 trials. NOTE: This should match (8.5).

e)   Continuing part d, suppose instead we plan to conduct $n' = 4$ *new* trials. Specify *and sketch* the posterior predictive pmf of Y', the number of successes we might observe in these 4 trials.

Exercise 8.12 (Posterior predictive Gamma-Poisson: with calculus). Suppose we have observed count data, $Y = y$, where $Y|\lambda \sim \text{Pois}(\lambda)$ and λ has a $\text{Gamma}(s, r)$ prior.

a)   Identify the posterior pdf of λ given the observed data $Y = y$, $f(\lambda|y)$.

b)   Let $Y' = y'$ be the number of events that will occur in a *new* observation period. Identify the conditional pmf of Y' given λ, $f(y'|\lambda)$.

c)   Identify the posterior predictive pmf of Y', $f(y'|y)$. NOTE: This will depend upon (y, y', s, r).

d)   Suppose your posterior model of λ is based on a prior model with $s = 50$ and $r = 50$ and $y = 7$ events in your original observation period. Specify the posterior predictive pmf of Y', the number of events we might observe in the next observation period.

e)   Sketch and discuss the posterior predictive pmf from part d.

Exercise 8.13 (Posterior predictive Normal-Normal: with calculus). Let $Y = y$ be an observed data point from a $N(\mu, \sigma^2)$ model. Further, suppose σ is known and that μ is unknown with a $N(\theta, \tau^2)$ prior.

a)   Identify the posterior pdf of μ given the observed data, $f(\mu|y)$.

b)   Let $Y' = y'$ be the value of a *new* data point. Identify the conditional pdf of Y' given μ, $f(y'|\mu)$.

c)   Identify the posterior predictive pdf of Y', $f(y'|y)$.

d)   Suppose $y = -10$, $\sigma = 3$, $\theta = 0$, and $\tau = 1$. Specify and sketch the posterior predictive pdf of Y'.

8.7.3   Applied exercises

Exercise 8.14 (Climate change: estimation). Let π denote the proportion of U.S. adults that do not believe in climate change. To learn about π, we'll use survey data on n adults and count up the number of these that don't believe in climate change, Y.

a)   Explain which Bayesian model is appropriate for this analysis: Beta-Binomial, Gamma-Poisson, or Normal-Normal.

b)   Specify and discuss your own prior model for π.

c)   For the remainder of the exercise, we'll utilize the authors' Beta(1,2) prior for π. How does your prior understanding differ from that of the authors?

d) Using the `pulse_of_the_nation` data from the **bayesrules** package, report the sample proportion of surveyed adults with the opinion that `climate_change` is `Not Real At All`.

e) In light of the Beta(1,2) prior and data, calculate and interpret a (middle) 95% posterior credible interval for π. NOTE: You'll first need to specify your posterior model of π.

Exercise 8.15 (Climate change: hypothesis testing). Continuing the analysis from Exercise 8.14, suppose you wish to test a researcher's claim that more than 10% of people believe in climate change: $H_0 : \pi \le 0.1$ versus $H_a : \pi > 0.1$.

a) What decision might you make about these hypotheses utilizing the credible interval from the previous exercise?

b) Calculate and interpret the posterior probability of H_a.

c) Calculate and interpret the Bayes Factor for your hypothesis test.

d) Putting this together, explain your conclusion about π.

Exercise 8.16 (Climate change with MCMC: simulation). In the next exercises, you'll repeat and build upon your climate change analysis using MCMC simulation.

a) Simulate the posterior model of π, the proportion of U.S. adults that do not believe in climate change, with **rstan** using 4 chains and 10000 iterations per chain.

b) Produce and discuss trace plots, overlaid density plots, and autocorrelation plots for the four chains.

c) Report the effective sample size ratio and R-hat values for your simulation, explaining what these values mean in context.

Exercise 8.17 (Climate change with MCMC: estimation and hypothesis testing).

a) Utilize your MCMC simulation to *approximate* a (middle) 95% posterior credible interval for π. Do so using the `tidy()` shortcut function as well as a direct calculation from your chain values.

b) Utilize your MCMC simulation to *approximate* the posterior probability that $\pi > 0.1$.

c) How close are the approximations in parts a and b to the actual corresponding posterior values you calculated in Exercises 8.14 and 8.15?

Exercise 8.18 (Climate change with MCMC: prediction).

a) Suppose you were to survey 100 more adults. Use your MCMC simulation to approximate the posterior predictive model of Y', the number that don't believe in climate change. Construct a histogram visualization of this model.

b) Summarize your observations of the posterior predictive model of Y'.

c) Approximate the probability that at least 20 of the 100 people don't believe in climate change.

Exercise 8.19 (Penguins: estimation). Let μ denote the typical flipper length (in mm) among the *Adelie* penguin species. To learn about μ, we'll utilize flipper measurements $(Y_1, Y_2, \ldots, Y_n)$ on a sample of Adelie penguins.

a)   Explain which Bayesian model is appropriate for this analysis: Beta-Binomial, Gamma-Poisson, or Normal-Normal.

b)   Your prior understanding is that the average flipper length for all Adelie penguins is about 200mm, but you aren't very sure. It's plausible that the average could be as low a 140mm or as high as 260mm. Specify an appropriate prior model for μ.

c)   The penguins_bayes data in the **bayesrules** package contains data on the flipper lengths for a sample of three different penguin species. For the *Adelie* species, how many data points are there and what's the sample mean flipper_length_mm?

d)   In light of your prior and data, calculate and interpret a (middle) 95% posterior credible interval for μ. NOTE: You'll first need to specify your posterior model of μ.

Exercise 8.20 (Penguins: hypothesis testing). Let's continue our analysis of μ, the typical flipper length (in mm) among the *Adelie* penguin species.

a)   You hypothesize that the average Adelie flipper length is somewhere between 200mm and 220mm. State this as a formal hypothesis test (using H_0, H_a, and μ notation). **NOTE:** This is a two-sided hypothesis test!

b)   What decision might you make about these hypotheses utilizing the credible interval from the previous exercise?

c)   Calculate and interpret the posterior probability that your hypothesis is true.

d)   Putting this together, explain your conclusion about μ.

Exercise 8.21 (Loons: estimation). The *loon* is a species of bird common to the Ontario region of Canada. Let λ denote the typical number of loons observed by a birdwatcher across a 100-hour observation period. To learn about λ, we'll utilize bird counts $(Y_1, Y_2, \ldots, Y_n)$ collected in n different outings.

a)   Explain which Bayesian model is appropriate for this analysis: Beta-Binomial, Gamma-Poisson, or Normal-Normal.

b)   Your prior understanding is that the typical rate of loon sightings is 2 per 100 hours with a standard deviation of 1 per 100-hours. Specify an appropriate prior model for λ and explain your reasoning.

c)   The loons data in the **bayesrules** package contains loon counts in different 100-hour observation periods. How many data points do we have and what's the average loon count per 100 hours?

d)   In light of your prior and data, calculate and interpret a (middle) 95% posterior credible interval for λ. NOTE: You'll first need to specify your posterior model of λ.

Exercise 8.22 (Loons: hypothesis testing). Let's continue our analysis of λ, the typical rate of loon sightings in a 100-hour observation period.

a)   You hypothesize that birdwatchers should anticipate a rate of less than 1 loon per observation period. State this as a formal hypothesis test (using H_0, H_a, and λ notation).

b)   What decision might you make about these hypotheses utilizing the credible interval from the previous exercise?

c)   Calculate and interpret the posterior probability that your hypothesis is true.

d)   Putting this together, explain your conclusion about λ.

Exercise 8.23 (Loons with MCMC: simulation). In the next exercises, you'll repeat your loon analysis using MCMC simulation.

a) Simulate the posterior model of λ, the typical rate of loon sightings per observation period, with **rstan** using 4 chains and 10000 iterations per chain.

b) Perform some MCMC diagnostics to confirm that your simulation has stabilized.

c) Utilize your MCMC simulation to *approximate* a (middle) 95% posterior credible interval for λ. Do so using the `tidy()` shortcut function as well as a direct calculation from your chain values.

d) Utilize your MCMC simulation to *approximate* the posterior probability that $\lambda < 1$.

e) How close are the approximations in parts c and d to the actual corresponding posterior values you calculated in Exercises 8.21 and 8.22?

Exercise 8.24 (Loons with MCMC: prediction).

a) Use your MCMC simulation to approximate the posterior predictive model of Y', the number of loons that a birdwatcher will spy in their next observation period. Construct a histogram visualization of this model.

b) Summarize your observations of the posterior predictive model of Y'.

c) Approximate the probability that the birdwatcher observes 0 loons in their next observation period.

Unit III

Bayesian Regression & Classification

9

Simple Normal Regression

In Chapter 9 we'll start with the foundational **Normal regression model** for a quantitative response variable Y. Consider the following data story. *Capital Bikeshare* is a bike sharing service in the Washington, D.C. area. To best serve its registered members, the company must understand the demand for its service. To help them out, we can analyze the number of rides taken on a random sample of n days, $(Y_1, Y_2, ..., Y_n)$. Since Y_i is a *count* variable, you might assume that ridership might be well modeled by a Poisson. However, past bike riding seasons have exhibited bell-shaped daily ridership with a variability in ridership that far exceeds the typical ridership, grossly violating the Poisson assumption of equal mean and variance (5.4). Thus, we'll assume instead that, independently from day to day, the number of rides varies **normally** around some **typical** ridership, μ, with **standard deviation** σ (Figure 9.1): $Y_i|\mu, \sigma \stackrel{ind}{\sim} N(\mu, \sigma^2)$.

DOI: 10.1201/9780429288340-9

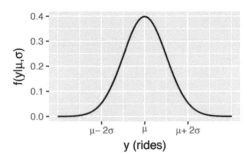

FIGURE 9.1: A Normal model of bike ridership.

Utilizing the Normal-Normal model from Chapter 5, we *could* conduct a posterior analysis of the typical ridership μ in light of the observed data by (1) tuning a Normal prior model for μ; and (2) assuming the variability in ridership σ is known:

$$
\begin{aligned}
Y_i | \mu &\stackrel{ind}{\sim} N(\mu, \sigma^2) \\
\mu &\sim N(\theta, \tau^2).
\end{aligned}
\tag{9.1}
$$

Yet we can greatly extend the power of this model by tweaking its assumptions. First, you might have scratched your head at the assumption that we *don't* know the typical ridership μ but *do* know the variability in ridership from day to day, σ. You'd be right. This assumption typically breaks down outside textbook examples. No problem. We can generalize the Normal-Normal model (9.1) to accommodate the reality that σ is a second **unknown parameter** by including a corresponding prior model:

$$
\begin{aligned}
Y_i | \mu, \sigma &\stackrel{ind}{\sim} N(\mu, \sigma^2) \\
\mu &\sim N(\theta, \tau^2) \\
\sigma &\sim \text{ some prior model.}
\end{aligned}
\tag{9.2}
$$

We can do even better. Though this two-parameter Normal-Normal model is more flexible than the one-parameter model (9.1), it ignores a lot of potentially helpful information. After all, ridership is likely linked to factors or **predictors** such as the weather, day of the week, time of the year, etc. In Chapter 9 we'll focus on incorporating just one predictor into our analysis – *temperature* – which we'll label as X. Specifically, our goal will be to model the *relationship* between ridership and temperature: Does ridership tend to increase on warmer days? If so, by how much? And how strong is this relationship? Figuring out *how* to conduct this analysis, i.e., *how* exactly to get information about the temperature predictor (X) into our model of ridership (Y) (9.2), is the driver behind Chapter 9.

⊚ **Goals**

Upon **building a Bayesian simple linear regression model** of response variable Y versus predictor X, you will:
- interpret appropriate **prior models** for the regression parameters;
- **simulate** the posterior model of the regression parameters; and
- utilize simulation results to build a **posterior understanding** of the relationship between Y and X and to build **posterior predictive models** of Y.

To get started, load the following packages, which we'll utilize throughout the chapter:

```
# Load packages
library(bayesrules)
library(tidyverse)
library(rstan)
library(rstanarm)
library(bayesplot)
library(tidybayes)
library(janitor)
library(broom.mixed)
```

9.1 Building the regression model

In this section, we build up the framework of the Normal Bayesian regression model. These concepts will be solidified when applied to the bike data in Section 9.3.

9.1.1 Specifying the data model

Our analysis of the relationship between bike ridership (Y) and temperature (X) requires a sample of n data *pairs*,

$$\{(Y_1, X_1), (Y_2, X_2), ..., (Y_n, X_n)\}$$

where Y_i is the number of rides and X_i is the high temperature (in degrees Fahrenheit) on day i. We can check this assumption once we have some data, but our experience suggests there's a **positive linear relationship** between ridership and temperature – the warmer it is, the more likely people are to hop on their bikes. For example, we might see data like that in the two scenarios in Figure 9.2 where each dot reflects the ridership and temperature on a unique day. Thus, instead of focusing on the **global mean** ridership across all days combined (μ), we can refine our analysis to the **local mean** ridership on day i, μ_i, specific to the temperature on that day. Assuming the relationship between ridership and temperature is linear, we can write μ_i as

$$\mu_i = \beta_0 + \beta_1 X_i$$

where we can interpret the **model coefficients** β_0 and β_1 as follows:

- **Intercept coefficient** β_0 technically indicates the typical ridership on days in which the temperature was 0 degrees Fahrenheit ($X_i = 0$). Since this frigid temperature is *far* outside the norm for D.C., we shouldn't put stock into this interpretation. Rather, we can think of β_0 as providing a baseline for where our model "lives" along the y-axis.
- Temperature **coefficient** β_1 indicates the typical change in ridership for every one unit (degree) increase in temperature. In this particular model with only one quantitative predictor X, this coefficient is equivalent to a *slope*.

For example, the model lines in Figure 9.2 both have intercept $\beta_0 = -2000$ and slope $\beta_1 = 100$. The intercept just tells us that if we extended the line all the way down to 0 degrees

Fahrenheit, it would cross the y-axis at -2000. The slope is more meaningful, indicating that for every degree increase in temperature, we'd expect 100 more riders.

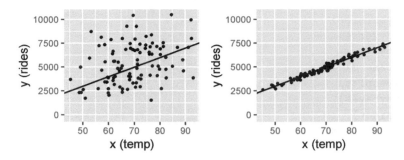

FIGURE 9.2: Two simulated scenarios for the relationship between ridership and temperature, utilizing $\sigma = 2000$ (left) and $\sigma = 200$ (right). In both cases, the model line is defined by $\beta_0 + \beta_1 x = -2000 + 100x$.

We can plunk this assumption of a linear relationship between Y_i and X_i right into our Bayesian model, by replacing the global mean μ in the **Normal data model**, $Y_i | \mu, \sigma \sim N(\mu, \sigma^2)$, with the temperature specific local mean μ_i:

$$Y_i | \beta_0, \beta_1, \sigma \stackrel{ind}{\sim} N\left(\mu_i, \sigma^2\right) \quad \text{with} \quad \mu_i = \beta_0 + \beta_1 X_i. \tag{9.3}$$

In this formulation of the data model, σ *also* takes on new meaning. It no longer measures the variability in ridership from the global mean μ across all days, but the **variability from the local mean** on days with similar temperature, $\mu_i = \beta_0 + \beta_1 X_i$. Pictures help. In Figure 9.2, the right plot reflects a scenario with a relatively small value of $\sigma = 20$. The observed data here deviates very little from the mean model line – we can expect the observed ridership on a given day to differ by only 20 rides from the mean ridership on days of the same temperature. This tightness around the mean model line indicates that temperature is a strong predictor of ridership. The opposite is true in the left plot which exhibits a larger $\sigma = 200$. There is quite a bit of variability in ridership among days of the same temperature, reflecting a weaker relationship between these variables. In summary, the formal assumptions encoded by data model (9.3) are included below.

Normal regression assumptions

The appropriateness of the Bayesian Normal regression model (9.3) depends upon the following assumptions.

- **Structure of the data**
  Accounting for predictor X, the observed data Y_i on case i is **independent** of the observed data on any other case j.
- **Structure of the relationship**
  The typical Y outcome can be written as a **linear function** of predictor X, $\mu = \beta_0 + \beta_1 X$.
- **Structure of the variability**
  At any value of predictor X, the observed values of Y will vary **normally** around their average μ with consistent standard deviation σ.

9.1.2 Specifying the priors

Considered alone, the modified data model (9.3) is precisely the frequentist "simple" linear regression model that you might have studied outside this book[1] – "simple" here meaning that our model has only one predictor variable X, not that you should find this model easy. To turn this into a *Bayesian* model, we must incorporate prior models for each of the unknown regression parameters. To practice distinguishing between model *parameters* and data *variables*, take the following quiz.[2]

> ❓ **Quiz Yourself!**
>
> Identify the regression *parameters* upon which the data model (9.3) depends.

In the data model (9.3), there are two data variables (Y and X) and *three* unknown regression parameters that encode the relationship between these variables: β_0, β_1, σ. We must specify prior models for each. There are countless approaches to this task. We won't and can't survey them all. Rather, throughout this book we'll utilize the default framework of the prior models used by the **rstanarm** package. Working within this framework will allow us to survey a broad range of modeling tools in Units 3 and 4. Once you're comfortable with the general modeling concepts therein and are ready to customize, you can take that leap. In doing so, we recommend Gabry and Goodrich (2020b), which provides an overview of all possible prior structures in **rstanarm**.

The first assumption we'll make is that our prior models of β_0, β_1, and σ are **independent**. That is, we'll assume that our prior understanding of where the model "lives" (β_0) has nothing to do with our prior understanding of the rate at which ridership increases with temperature (β_1). Similarly, we'll assume that our prior understanding of σ, and hence the strength of the relationship, is unrelated to both β_0 and β_1. Though in practice we might have some prior notion about the *combination* of these parameters, the assumption of independence greatly simplifies the model. It's also consistent with the **rstanarm** framework.

In specifying the structure of the independent priors, we must consider (as usual) the values that these parameters might take. To this end, the intercept and slope regression parameters, β_0 and β_1, can technically take any values in the real line. That is, a model line can cross anywhere along the y-axis and the slope of a line can be any positive or negative value (or even 0). Thus, it's reasonable to utilize Normal prior models for β_0 and β_1, which also live on the entire real line. Specifically,

$$\begin{aligned} \beta_0 &\sim N\left(m_0, s_0^2\right) \\ \beta_1 &\sim N\left(m_1, s_1^2\right) \end{aligned} \tag{9.4}$$

where we can tune the m_0, s_0, m_1, s_1 **hyperparameters** to match our prior understanding of β_0 and β_1. Similarly, since the standard deviation parameter σ must be positive, it's reasonable to utilize an **Exponential model** which is also restricted to positive values:

$$\sigma \sim \text{Exp}(l). \tag{9.5}$$

Thus, by (5.10), the prior mean and standard deviation of σ are:

---

[1] Equivalently, and perhaps more familiarly in frequentist analyses, we can write this model as $Y_i = \beta_0 + \beta_1 X_i + \varepsilon_i$ where residual errors $\varepsilon_i \overset{ind}{\sim} N(0, \sigma^2)$.

[2] Answer: β_0, β_1, σ

$$E(\sigma) = \frac{1}{l} \quad \text{and} \quad SD(\sigma) = \frac{1}{l}.$$

Though this Exponential prior is currently the default in **rstanarm**, popular alternatives include the half-Cauchy and inverted Gamma models.

9.1.3   Putting it all together

Combining the Normal data model (9.3) with our priors on the model parameters, (9.4) and (9.5), completes a common formulation of the **Bayesian simple linear regression model**:

$$
\begin{aligned}
\text{data:} \quad & Y_i | \beta_0, \beta_1, \sigma \overset{ind}{\sim} N\left(\mu_i, \sigma^2\right) \quad \text{with} \quad \mu_i = \beta_0 + \beta_1 X_i \\
\text{priors:} \quad & \beta_0 \sim N\left(m_0, s_0^2\right) \\
& \beta_1 \sim N\left(m_1, s_1^2\right) \\
& \sigma \sim \text{Exp}(l).
\end{aligned}
\tag{9.6}
$$

Step back and reflect upon how we got here. We built a regression model of Y, not all at once, but by starting with and building upon the simple Normal-Normal model one step at a time. In contrast, our human instinct often draws us into starting with the *most* complicated model we can think of. When this instinct strikes, resist it and remember this: complicated is not necessarily sophisticated. Complicated models are often wrong and difficult to apply.

Model building: One step at a time

In building any Bayesian model, it's important to start with the basics and build up one step at a time. Let Y be a response variable and X be a predictor or set of predictors. Then we can build a model of Y by X through the following general principles:

- Take note of whether Y is discrete or continuous. Accordingly, identify an appropriate model structure of data Y (e.g., Normal, Poisson, Binomial).
- Rewrite the mean of Y as a function of predictors X (e.g., $\mu = \beta_0 + \beta_1 X$).
- Identify all unknown model parameters in your model (e.g., β_0, β_1, σ).
- Take note of the values that each of these parameters might take. Accordingly, identify appropriate prior models for these parameters.

9.2   Tuning prior models for regression parameters

Let's now *apply* the Bayesian simple linear regression model (9.6) to our study of the relationship between Capital Bikeshare ridership (Y) and temperature (X). We'll begin by tuning the prior models for intercept coefficient β_0, temperature coefficient β_1, and regression standard deviation σ. Based on past bikeshare analyses, suppose we have the following prior understanding of this relationship:

1.   On an *average* temperature day, say 65 or 70 degrees for D.C., there are typically around 5000 riders, though this average could be somewhere between 3000 and 7000.

2. For every one degree increase in temperature, ridership typically increases by 100 rides, though this average increase could be as low as 20 or as high as 180.

3. At any given temperature, daily ridership will tend to vary with a moderate standard deviation of 1250 rides.

Prior assumption 1 tells us something about the model baseline or intercept, β_0. There's a twist though. This prior information has been *centered*. Whereas β_0 reflects the typical ridership on a 0-degree day (which doesn't make sense in D.C.), the *centered* intercept, which we'll denote β_{0c}, reflects the typical ridership at the *typical* temperature X. The distinction between β_0 and β_{0c} is illustrated in Figure 9.3. Processing the prior information about the model baseline in this way is more intuitive. In fact, it's this *centered* information that we'll supply when using **rstanarm** to simulate our regression model. With this, we can capture prior assumption 1 with a Normal model for β_{0c} which is centered at 5000 rides with a standard deviation of 1000 rides, and thus largely falls between 3000 and 7000 rides: $\beta_{0c} \sim N(5000, 1000^2)$. This prior is drawn in Figure 9.4.

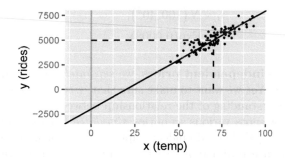

FIGURE 9.3: A simulated set of ridership data with intercept $\beta_0 = -2000$ and centered intercept $\beta_{0c} = 5000$ at an average temperature of 70 degrees.

Moving on, prior assumption 2 tells us about the rate of increase in ridership with temperature, and thus temperature coefficient β_1. This prior understanding is well represented by a Normal model centered at 100 with a standard deviation of 40, and thus largely falls between 20 and 180: $\beta_1 \sim N(100, 40^2)$. Finally, prior assumption 3 reflects our understanding of the regression standard deviation parameter σ. Accordingly, we can tune the Exponential model to have a mean that matches the expected standard deviation of $E(\sigma) = 1/l = 1250$ rides, and thus a rate of $l = 1/1250 = 0.0008$: $\sigma \sim \text{Exp}(0.0008)$. These independent β_1 and σ priors are drawn in Figure 9.4.

```
plot_normal(mean = 5000, sd = 1000) +
  labs(x = "beta_0c", y = "pdf")
plot_normal(mean = 100, sd = 40) +
  labs(x = "beta_1", y = "pdf")
plot_gamma(shape = 1, rate = 0.0008) +
  labs(x = "sigma", y = "pdf")
```

Plugging our tuned priors into (9.6), the Bayesian regression model of ridership (Y) by temperature (X) is specified as follows:

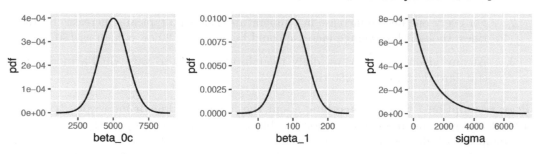

FIGURE 9.4: Prior models for the parameters in the regression analysis of bike ridership, $(\beta_{0c}, \beta_1, \sigma)$.

$$Y_i|\beta_0, \beta_1, \sigma \overset{ind}{\sim} N\left(\mu_i, \sigma^2\right) \quad \text{with} \quad \mu_i = \beta_0 + \beta_1 X_i$$
$$\beta_{0c} \sim N\left(5000, 1000^2\right)$$
$$\beta_1 \sim N\left(100, 40^2\right) \tag{9.7}$$
$$\sigma \sim \text{Exp}(0.0008).$$

Since our model utilizes **independent priors**, we *separately* processed our prior information on β_0, β_1, and σ above. Yet we want to make sure that, when *combined*, these priors actually reflect our current understanding of the relationship between ridership and temperature. To this end, Figure 9.5 presents various scenarios *simulated* from our prior models.[3] The 200 prior model lines, $\beta_0 + \beta_1 X$ (left), do indeed capture our prior understanding that ridership increases with temperature and tends to be around 5000 on average temperature days. The variability in these lines also adequately reflects our overall uncertainty about this association. Next, at right are four ridership datasets simulated from the Normal data model (9.3) using four prior plausible sets of $(\beta_0, \beta_1, \sigma)$ values. If our prior models are reasonable, then data simulated from these models should be consistent with ridership data we'd actually expect to see in practice. That's indeed the case here. The rate of increase in ridership with temperature, the baseline ridership, and the variability in ridership are consistent with our prior assumptions.

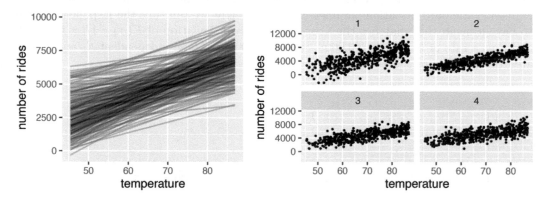

FIGURE 9.5: Simulated scenarios under the prior models of β_0, β_1, and σ. At left are 200 prior plausible model lines, $\beta_0 + \beta_1 X$. At right are 4 prior plausible datasets.

---

[3]You'll learn to construct these plots in Section 9.7. For now, we'll focus on the concepts.

9.3   Posterior simulation

In the next step of our Bayesian analysis, let's update our prior understanding of the relationship between ridership and temperature using **data**. The `bikes` data in the `bayesrules` package is a subset of the Bike Sharing dataset made available on the UCI Machine Learning Repository (2017) by Fanaee-T and Gama (2014).[4] For each of 500 days in the study, `bikes` contains the number of `rides` taken and a measure of what the temperature *felt* like when incorporating factors such as humidity (`temp_feel`). A scatterplot of `rides` by `temp_feel` supports our prior assumption of a positive *linear* relationship between the two, i.e., $\mu = \beta_0 + \beta_1 X$ with $\beta_1 > 0$. Further, the strength of this relationship appears *moderate* – σ is neither small nor big.

```
# Load and plot data
data(bikes)
ggplot(bikes, aes(x = temp_feel, y = rides)) +
  geom_point(size = 0.5) +
  geom_smooth(method = "lm", se = FALSE)
```

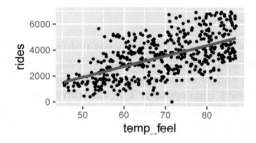

FIGURE 9.6: A scatterplot of ridership vs temperature on 500 different days.

We can now combine the information from this data with that from the prior to build a posterior model for parameters $(\beta_0, \beta_1, \sigma)$. Our inclination here is to jump straight to an MCMC **simulation** of this posterior, without even attempting to specify this model mathematically. However, we'll pause long enough to convince you that simulation is the reasonable choice here. Consider the general Bayesian simple regression model (9.6) and let $\vec{y} = (y_1, y_2, ..., y_n)$ denote a collection of n observed Y values. Assuming that the regression parameters are independent, their **joint prior pdf** is the product of their marginal pdfs which, in turn, are defined by their respective Normal and Exponential prior models:

$$f(\beta_0, \beta_1, \sigma) = f(\beta_0)f(\beta_1)f(\sigma).$$

Further, the **likelihood function** of the parameters given the *independent* data $\vec{y}$ is defined by the joint pdf of $\vec{y}$ which, in turn, is the product of the marginal pdfs defined by the Normal data structure (9.3):

$$L(\beta_0, \beta_1, \sigma|\vec{y}) = f(\vec{y}|\beta_0, \beta_1, \sigma) = \prod_{i=1}^{n} f(y_i|\beta_0, \beta_1, \sigma).$$

---

[4]The `bikes` data was extracted from a larger dataset to match our pedagogical goals, and thus should be used for illustration purposes only. Type `?bikes` in the console for a detailed codebook.

Thus, building the posterior model of $(\beta_0, \beta_1, \sigma)$ would require us to specify the **joint posterior pdf**

$$f(\beta_0, \beta_1, \sigma \mid \vec{y}) \propto \text{prior} \cdot \text{likelihood} = f(\beta_0)f(\beta_1)f(\sigma) \cdot \left[\prod_{i=1}^{n} f(y_i|\beta_0, \beta_1, \sigma) \right].$$

If you went through the tedious work of plugging in the formulas for the $3 + n$ pdfs in the product above, you wouldn't discover a familiar structure. Thus, if you *really* wanted to specify the posterior pdf, you'd need to calculate the normalizing constant. But you might not get far – the constant which guarantees that $f(\beta_0, \beta_1, \sigma|\vec{y})$ integrates to 1 across all possible sets of $(\beta_0, \beta_1, \sigma)$ is a triple integral of our complicated product:

$$
\begin{aligned}
f(\beta_0, \beta_1, \sigma \mid \vec{y}) &= \frac{\text{prior} \cdot \text{likelihood}}{\int \text{prior} \cdot \text{likelihood}} \\
&= \frac{f(\beta_0)f(\beta_1)f(\sigma) \cdot \left[\prod_{i=1}^{n} f(y_i|\beta_0, \beta_1, \sigma) \right]}{\int \int \int f(\beta_0)f(\beta_1)f(\sigma) \cdot \left[\prod_{i=1}^{n} f(y_i|\beta_0, \beta_1, \sigma) \right] d\beta_0 d\beta_1 d\sigma}
\end{aligned}
$$

Let's not. Instead, we can utilize Markov chain Monte Carlo simulation techniques to **approximate** the posterior.

9.3.1   Simulation via rstanarm

In Chapter 6, our MCMC approximation process unfolded in two steps using the **rstan** package: first *define* and then *simulate* the model using **rstan** syntax. Not only can this two-step process get more and more complicated as our Bayesian models get more and more sophisticated, it's not always unnecessary. The **rstanarm** package (Gabry and Goodrich, 2020c) tailors the power of **rstan** to applied regression models (**arm**). As such, it contains one-stop functions for simulating a broad set of *pre-defined* Bayesian regression models, i.e., no need for step 1! For example, to simulate our Normal Bayesian linear regression model of `rides` by `temp_feel`, we can utilize the `stan_glm()` function, which applies to the wider family of **generalized linear regression models (glm)**:

```
bike_model <- stan_glm(rides ~ temp_feel, data = bikes,
                       family = gaussian,
                       prior_intercept = normal(5000, 1000),
                       prior = normal(100, 40),
                       prior_aux = exponential(0.0008),
                       chains = 4, iter = 5000*2, seed = 84735)
```

The syntax above is common to the other **rstanarm** models we'll see in this book and looks more intimidating than it is. In general, `stan_glm()` requires three types of information:

- **Data information**
  The first three `stan_glm()` arguments specify the structure of our data: we want to model ridership by temperature (`rides ~ temp_feel`) using `data = bikes` and assuming a Normal data model, aka `family = gaussian`.

- **Prior information**
  The `prior_intercept`, `prior`, and `prior_aux` arguments specify the priors of β_{0c}, β_1, and σ, respectively. These match the priors defined by (9.7).

- **Markov chain information**
  The remaining arguments specify the structure of our MCMC simulation: the number of Markov `chains` to run, the length or number of `iterations` of each chain, and the random number `seed` to use.

After tossing out the first half of Markov chain values from the learning or *burn-in* phase, the `stan_glm()` simulation produces four parallel chains of length 5000 for *each* model parameter: $\left\{\beta_0^{(1)}, \beta_0^{(2)}, \ldots, \beta_0^{(5000)}\right\}$, $\left\{\beta_1^{(1)}, \beta_1^{(2)}, \ldots, \beta_1^{(5000)}\right\}$, and $\left\{\sigma_0^{(1)}, \sigma_0^{(2)}, \ldots, \sigma_0^{(5000)}\right\}$. These are stored as `(Intercept)`, `temp_feel`, and `sigma`, respectively. Some quick diagnostics indicate that these chains are trustworthy. The effective sample size ratios are slightly *above* 1 and the R-hat values are very close to 1, indicating that the chains are stable, mixing quickly, and behaving much like an independent sample.

```
# Effective sample size ratio and Rhat
neff_ratio(bike_model)
(Intercept)    temp_feel         sigma
      1.042        1.037         1.004
rhat(bike_model)
(Intercept)    temp_feel         sigma
     0.9999       0.9999        1.0000
```

We come to similar conclusions from the trace and density plots (Figure 9.7).

```
# Trace plots of parallel chains
mcmc_trace(bike_model, size = 0.1)

# Density plots of parallel chains
mcmc_dens_overlay(bike_model)
```

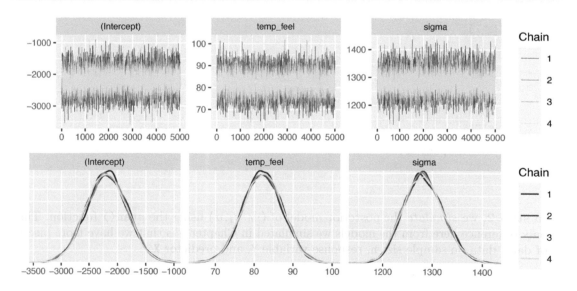

FIGURE 9.7: Trace and density plots for the bike model posterior simulation.

9.3.2 Optional: Simulation via rstan

Though the **rstanarm** package includes shortcut functions for the types of models we'll be building in this book, you might be curious how the **rstan** techniques from Chapter 6 scale up to Normal regression, and hence other models beyond the scope of Unit 1. If you *are* curious, please read on. Otherwise, please skip to the next section.

In **rstanarm**, the use of `stan_glm()` with `family = gaussian` both defined the exact structure of our Bayesian model *and* simulated the corresponding posterior. In **rstan** we have to separately define and simulate our model, steps 1 and 2. Consider the important pieces of information we must communicate in step 1:

- `data`: The data on variables `Y` and `X`, rides and temperature, will be vectors of length `n`.
- `parameters`: Our two regression coefficients `beta0` and `beta1` (β_0 and β_1) can both be any real number whereas the standard deviation parameter standard deviation parameter `sigma` (σ) must be non-negative.
- `model`: The data model of `Y` is `normal` with mean `beta0 + beta1 * X` and standard deviation `sigma`. Further, with the *exception* of `beta0`, the priors are similar to those in our `stan_glm()` syntax. Using **rstan**, we must directly express our prior understanding of the intercept β_0, not the *centered* intercept β_{0c}. In this case, we can extend our prior understanding that there are typically 5000 riders on a 70-degree day, to there being -2000 *hypothetical* riders on a 0-degree day (Figure 9.3).

```
# STEP 1: DEFINE the model
stan_bike_model <- "
  data {
    int<lower = 0> n;
    vector[n] Y;
    vector[n] X;
  }
  parameters {
    real beta0;
    real beta1;
    real<lower = 0> sigma;
  }
  model {
    Y ~ normal(beta0 + beta1 * X, sigma);
    beta0 ~ normal(-2000, 1000);
    beta1 ~ normal(100, 40);
    sigma ~ exponential(0.0008);
  }
"
```

In **step 2**, we *simulate* the posterior model of $(\beta_0, \beta_1, \sigma)$ using the `stan()` function. The only difference here from the models we simulated in Chapter 6 is that we have more pieces of data: data on sample size `n`, response variable `Y`, and predictor `X`:

```
# STEP 2: SIMULATE the posterior
stan_bike_sim <-
  stan(model_code = stan_bike_model,
```

```
        data = list(n = nrow(bikes), Y = bikes$rides, X = bikes$temp_feel),
        chains = 4, iter = 5000*2, seed = 84735)
```

Though they ultimately contain similar information about the approximate posterior model of $(\beta_0, \beta_1, \sigma)$, the structures of the **rstan** stan_bike_sim object and **rstanarm** bike_model object slightly differ. And now that we've made the connection between **rstan** and **rstanarm** here, moving forward we'll focus on the **rstanarm** shortcuts and their output. Should you wish to learn more about **rstan**, the Stan development team provides an excellent resource (Stan development team, 2019).

9.4 Interpreting the posterior

We've now successfully simulated the posterior Normal regression model of bikeshare ridership by temperature. So what does it all mean?! The mcmc_dens_overlay() plot in Figure 9.7 allows us to *visualize* and examine the approximate posterior models for each of the regression parameters β_0, β_1, and σ. Posterior summary statistics, calculated via the tidy() function in the **broom.mixed** package, provide complementary *numerical* summaries. Since our Normal regression model has more than one parameter, we must specify the "effects" or parameters of interest. Here, we summarize both the fixed regression coefficients, β_0 and β_1, and the aux (or auxiliary) parameter σ:

```
# Posterior summary statistics
tidy(bike_model, effects = c("fixed", "aux"),
      conf.int = TRUE, conf.level = 0.80)
# A tibble: 4 x 5
  term          estimate std.error conf.low conf.high
  <chr>            <dbl>     <dbl>    <dbl>     <dbl>
1 (Intercept)    -2194.      362.   -2656.    -1732.
2 temp_feel         82.2       5.15     75.6      88.8
3 sigma           1281.       40.7    1231.     1336.
4 mean_PPD        3487.       80.4    3385.     3591.
```

Let's first focus on the typical relationship between ridership and temperature, $\mu = \beta_0 + \beta_1 X$. Referring to the tidy() summary, the **posterior median relationship** is

$$-2194.24 + 82.16X. \qquad (9.8)$$

That is, for every one degree increase in temperature, we expect ridership to increase by roughly 82 rides. There is, of course, posterior *uncertainty* in this relationship. For example, the 80% posterior credible interval for β_1, (75.6, 88.8), indicates that this slope could range anywhere between 76 and 89. To combine this uncertainty in β_1 with that in β_0 for a better overall picture of our model, notice that the Markov chain simulations provide 20,000 posterior plausible pairs of β_0 and β_1 values:

```
# Store the 4 chains for each parameter in 1 data frame
bike_model_df <- as.data.frame(bike_model)

# Check it out
nrow(bike_model_df)
[1] 20000
head(bike_model_df, 3)
  (Intercept) temp_feel sigma
1       -2657     88.16  1323
2       -2188     83.01  1323
3       -1984     81.54  1363
```

These pairs provide 20,000 alternative scenarios for the typical relationship between ridership and temperature, $\beta_0 + \beta_1 X$, and thus capture our overall uncertainty about this relationship. For example, the first pair indicates the plausibility that $\beta_0 + \beta_1 X$ = -2657 + 88.2 X. The second pair has a higher intercept and a smaller slope. Below we plot just *50* of these 20,000 posterior plausible mean models, $\beta_0^{(i)} + \beta_1^{(i)} X$. This is a multi-step process:

- The `add_fitted_draws()` function from the **tidybayes** package (Kay, 2021) evaluates 50 posterior plausible relationships, $\beta_0^{(i)} + \beta_1^{(i)} X$, along the observed range of temperatures X. We encourage you to increase this number – plotting more than 50 lines here just doesn't print nicely.
- We then plot these 50 model lines, labeled by .draw, on top of the observed data points using `ggplot()`.

```
# 50 simulated model lines
bikes %>%
  add_fitted_draws(bike_model, n = 50) %>%
  ggplot(aes(x = temp_feel, y = rides)) +
    geom_line(aes(y = .value, group = .draw), alpha = 0.15) +
    geom_point(data = bikes, size = 0.05)
```

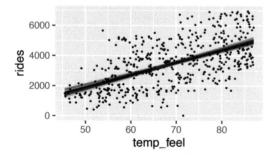

FIGURE 9.8: Model lines constructed from 50 posterior plausible sets of β_0 and β_1.

Comparing the posterior plausible models in Figure 9.8 to the *prior* plausible models in Figure 9.5 reveals the evolution in our understanding of ridership. First, the increase in ridership with temperature appears to be less steep than we had anticipated. Further, the posterior plausible models are far less variable, indicating that we're far more confident about the relationship between ridership and temperature upon observing some data. Once you've reflected on the results above, quiz yourself.

---

❷ Quiz Yourself!

Do we have ample posterior evidence that there's a positive association between ridership and temperature, i.e., that $\beta_1 > 0$? Explain.

The answer to the quiz is *yes*. We can support this answer with three types of evidence.

- **Visual evidence**
  In our visual examination of 50 posterior plausible scenarios for the relationship between ridership and temperature (Figure 9.8), *all* exhibited positive associations. A line exhibiting *no* relationship ($\beta_1 = 0$) would stick out like a sore thumb.

- **Numerical evidence from the posterior credible interval**
  More rigorously, the 80% credible interval for β_1 in the above `tidy()` summary, (75.6, 88.8), lies entirely and well above 0.

- **Numerical evidence from a posterior probability**
  To add one more unnecessary piece of evidence to the pile, a quick tabulation approximates that there's *almost certainly* a positive association, $P(\beta_1 > 0 \mid \vec{y}) \approx 1$. Of our 20,000 Markov chain values of β_1, *zero* are positive.

```
# Tabulate the beta_1 values that exceed 0
bike_model_df %>%
  mutate(exceeds_0 = temp_feel > 0) %>%
  tabyl(exceeds_0)
 exceeds_0     n percent
      TRUE 20000       1
```

Finally, let's examine the posterior results for σ, the degree to which ridership varies on days of the same temperature. Above we estimated that σ has a posterior median of 1281 and an 80% credible interval (1231, 1336). Thus, on average, we can expect the observed ridership on a given day to fall 1281 rides from the average ridership on days of the same temperature. Figure 9.8 adds some context, presenting four simulated sets of ridership data under four posterior plausible values of σ. At least visually, these plots exhibit similarly moderate relationships, indicating relative posterior certainty about the strength in the relationship between ridership and temperature. The syntax here is quite similar to that used for plotting the plausible regression lines $\beta_0 + \beta_1 X$ in Figure 9.8. The main difference is that we've replaced `add_fitted_draws` with `add_predicted_draws`.

```
# Simulate four sets of data
bikes %>%
  add_predicted_draws(bike_model, n = 4) %>%
  ggplot(aes(x = temp_feel, y = rides)) +
    geom_point(aes(y = .prediction, group = .draw), size = 0.2) +
    facet_wrap(~ .draw)
```

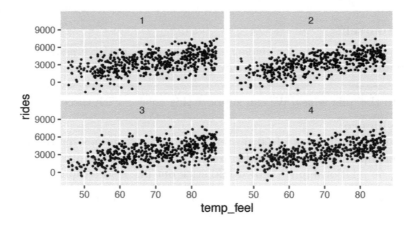

FIGURE 9.9: Four datasets simulated from the posterior models of β_0, β_1, and σ.

9.5 Posterior prediction

Our above examination of the regression parameters illuminates the relationship between ridership and temperature. Beyond building such insight, a common goal of regression analysis is to use our model to make **predictions**.

> **❓ Quiz Yourself!**
>
> Suppose a weather report indicates that tomorrow will be a 75-degree day in D.C. What's your posterior guess of the number of riders that Capital Bikeshare should anticipate?

Your natural first crack at this question might be to plug the 75-degree temperature into the posterior median model (9.8). Thus, we expect that there will be 3968 riders tomorrow:

$$-2194.24 + 82.16 \star 75 = 3967.76.$$

BUT, recall from Section 8.4.3 that this singular prediction ignores two potential sources of variability:

- **Sampling variability** in the data
  The observed ridership outcomes, Y, typically *deviate* from the model line. That is, we don't expect every 75-degree day to have the same exact number of rides.

- **Posterior variability** in parameters $(\beta_0, \beta_1, \sigma)$
  The posterior median model is merely the center in a *range* of plausible model lines $\beta_0 + \beta_1 X$ (Figure 9.8). We should consider this entire range as well as that in σ, the degree to which observations might deviate from the model lines.

The **posterior predictive model** of a new data point Y_{new} accounts for both sources of variability. Specifically, the posterior predictive pdf captures the *overall* chance of observing $Y_{\text{new}} = y_{\text{new}}$ by weighting the chance of this outcome under *any* set of possible parameters $(f(y_{new}|\beta_0, \beta_1, \sigma))$ by the posterior plausibility of these parameters $(f(\beta_0, \beta_1, \sigma|\vec{y}))$. Mathematically speaking:

$$f(y_{\text{new}}|\vec{y}) = \int \int \int f(y_{new}|\beta_0, \beta_1, \sigma) f(\beta_0, \beta_1, \sigma|\vec{y}) d\beta_0 d\beta_1 d\sigma.$$

Now, we don't actually have a nice, tidy formula for the posterior pdf of our regression parameters, $f(\beta_0, \beta_1, \sigma|\vec{y})$, and thus can't get a nice tidy formula for the posterior predictive pdf $f(y_{\text{new}}|\vec{y})$. What we *do* have is 20,000 sets of parameters in the Markov chain $\left(\beta_0^{(i)}, \beta_1^{(i)}, \sigma^{(i)}\right)$. We can then *approximate* the posterior predictive model for Y_{new} at $X = 75$ by simulating a ridership prediction from the Normal model evaluated each parameter set:

$$Y_{\text{new}}^{(i)}|\beta_0, \beta_1, \sigma \sim N\left(\mu^{(i)}, \left(\sigma^{(i)}\right)^2\right) \quad \text{with} \quad \mu^{(i)} = \beta_0^{(i)} + \beta_1^{(i)} \cdot 75.$$

Thus, each of the 20,000 parameter sets in our Markov chain (left) produces a unique prediction (right):

$$\begin{bmatrix} \beta_0^{(1)} & \beta_1^{(1)} & \sigma^{(1)} \\ \beta_0^{(2)} & \beta_1^{(2)} & \sigma^{(2)} \\ \vdots & \vdots & \vdots \\ \beta_0^{(20000)} & \beta_1^{(20000)} & \sigma^{(20000)} \end{bmatrix} \longrightarrow \begin{bmatrix} Y_{\text{new}}^{(1)} \\ Y_{\text{new}}^{(2)} \\ \vdots \\ Y_{\text{new}}^{(20000)} \end{bmatrix}$$

The resulting collection of 20,000 predictions, $\left\{Y_{\text{new}}^{(1)}, Y_{\text{new}}^{(2)}, \ldots, Y_{\text{new}}^{(20000)}\right\}$, *approximates* the posterior predictive model of ridership Y on 75-degree days. We will obtain this approximation both "by hand," which helps us build some powerful intuition, and using shortcut R functions.

9.5.1 Building a posterior predictive model

To really connect with the concepts, let's start by approximating posterior predictive models without the use of a shortcut function. To do so, we'll simulate 20,000 predictions of ridership on a 75-degree day, $\left\{Y_{\text{new}}^{(1)}, Y_{\text{new}}^{(2)}, \ldots, Y_{\text{new}}^{(20000)}\right\}$, one from each parameter set in `bike_model_df`. Let's start small with just the first posterior plausible parameter set:

```
first_set <- head(bike_model_df, 1)
first_set
  (Intercept) temp_feel sigma
1       -2657     88.16  1323
```

Under this particular scenario, $\left(\beta_0^{(1)}, \beta_1^{(1)}, \sigma^{(1)}\right) = (-2657, 88.16, 1323)$, the average ridership at a given temperature is defined by

$$\mu = \beta_0^{(1)} + \beta_1^{(1)} X = -2657 + 88.16X.$$

As such, we'd expect an **average** of $\mu = 3955$ riders on a 75-degree day:

```
mu <- first_set$`(Intercept)` + first_set$temp_feel * 75
mu
[1] 3955
```

To capture the **sampling variability** around this average, i.e., the fact that not all 75-degree days have the same ridership, we can simulate our first official prediction $Y_{\text{new}}^{(1)}$ by taking a random draw from the Normal model specified by this first parameter set:

$$Y_{\text{new}}^{(1)} | \beta_0, \beta_1, \sigma \sim N\left(3955, 1323^2\right).$$

Taking a draw from this model using `rnorm()`, we happen to observe an above average 4838 rides on the 75-degree day:

```
set.seed(84735)
y_new <- rnorm(1, mean = mu, sd = first_set$sigma)
y_new
[1] 4838
```

Now let's do this 19,999 more times. That is, let's follow the same two-step process to simulate a prediction of ridership from each of the 20,000 sets of regression parameters i in `bike_model_df`: (1) calculate the *average* ridership on 75-degree days, $\mu^{(i)} = \beta_0^{(i)} + \beta_1^{(i)} \cdot 75$; then (2) sample from the Normal model centered at this average with standard deviation $\sigma^{(i)}$:

```
# Predict rides for each parameter set in the chain
set.seed(84735)
predict_75 <- bike_model_df %>%
  mutate(mu = `(Intercept)` + temp_feel*75,
         y_new = rnorm(20000, mean = mu, sd = sigma))
```

The first 3 sets of average ridership (`mu`) and predicted ridership on a specific day (`y_new`) are shown here along with the first 3 posterior plausible parameter sets from which they were generated (`(Intercept)`, `temp_feel`, `sigma`):

```
head(predict_75, 3)
  (Intercept) temp_feel sigma   mu y_new
1       -2657     88.16  1323 3955  4838
2       -2188     83.01  1323 4038  3874
3       -1984     81.54  1363 4132  5196
```

Whereas the collection of 20,000 `mu` values approximates the posterior model for the *typical* ridership on 75-degree days, $\mu = \beta_0 + \beta_1 * 75$, the 20,000 `y_new` values approximate the **posterior predictive model** of ridership for tomorrow, an *individual* 75-degree day,

$$Y_{\text{new}} | \beta_0, \beta_1, \sigma \sim N\left(\mu, \sigma^2\right) \quad \text{with} \quad \mu = \beta_0 + \beta_1 \cdot 75.$$

In the plots of these two posterior models (Figure 9.10), you'll immediately pick up the fact that, though they're centered at roughly the same value, the posterior predictive model for `mu` is much narrower than that of `y_new`. Specifically, the 95% credible interval for the **typical** number of rides on a 75-degree day, μ, ranges from 3843 to 4095. In contrast, the 95% **posterior prediction interval** for the number of rides *tomorrow* has a much *wider* range from 1500 to 6482.

```
# Construct 80% posterior credible intervals
predict_75 %>%
  summarize(lower_mu = quantile(mu, 0.025),
            upper_mu = quantile(mu, 0.975),
            lower_new = quantile(y_new, 0.025),
            upper_new = quantile(y_new, 0.975))
  lower_mu upper_mu lower_new upper_new
1     3843     4095      1500      6482
```

```
# Plot the posterior model of the typical ridership on 75 degree days
ggplot(predict_75, aes(x = mu)) +
  geom_density()
```

```
# Plot the posterior predictive model of tomorrow's ridership
ggplot(predict_75, aes(x = y_new)) +
  geom_density()
```

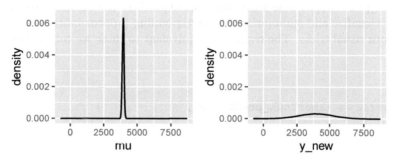

FIGURE 9.10: The posterior model of μ, the typical ridership on a 75-degree day (left), and the posterior predictive model of the ridership tomorrow, a specific 75-degree day (right).

These two 95% intervals are represented on a scatterplot of the observed data (Figure 9.11), clarifying that the posterior model for μ merely captures the uncertainty in the *average* ridership on *all* $X = 75$-degree days. Since there is so little uncertainty about this average, this interval visually appears like a wee dot! In contrast, the posterior predictive model for the number of rides tomorrow (a *specific* day) accounts for not only the *average* ridership on a 75-degree day, but the *individual variability* from this average. The punchline? There's more accuracy in anticipating the *average* behavior across multiple data points than the *unique* behavior of a single data point.

9.5.2 Posterior prediction with rstanarm

Simulating the posterior predictive model from scratch allowed you to really connect with the concept, but moving forward we can utilize the `posterior_predict()` function in the **rstanarm** package:

```
# Simulate a set of predictions
set.seed(84735)
```

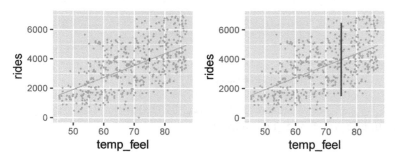

FIGURE 9.11: 95% posterior credible intervals (blue) for the **average** ridership on 75-degree days (left) and the **predicted** ridership for tomorrow, an individual 75-degree day (right).

```
shortcut_prediction <-
  posterior_predict(bike_model, newdata = data.frame(temp_feel = 75))
```

The `shortcut_prediction` object contains 20,000 predictions of ridership on 75-degree days. We can both visualize and summarize the corresponding (approximate) posterior predictive model using our usual tricks. The results are equivalent to those we constructed from scratch above:

```
# Construct a 95% posterior credible interval
posterior_interval(shortcut_prediction, prob = 0.95)
    2.5% 97.5%
1 1500  6482

# Plot the approximate predictive model
mcmc_dens(shortcut_prediction) +
  xlab("predicted ridership on a 75 degree day")
```

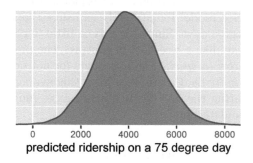

FIGURE 9.12: Posterior predictive model of ridership on a 75-degree day.

9.6 Sequential regression modeling

Our analysis above spotlighted the newest concept of the book: modeling the *relationship* between two variables Y and X. Yet our Bayesian thinking, second nature by now, provided the foundation for this analysis. Here, let's consider another attractive feature of Bayesian thinking, *sequentiality*, in the context of regression modeling (Chapter 4). Above we analyzed the complete `bikes` data that spanned 500 different days, arranged by `date`:

```
bikes %>%
  select(date, temp_feel, rides) %>%
  head(3)
        date temp_feel rides
1 2011-01-01     64.73   654
2 2011-01-03     49.05  1229
3 2011-01-04     51.09  1454
```

Suppose instead that we were given access to this data in bits and pieces, as it became available: after the first 30 days of data collection, then again after the first 60 days, and finally after all 500 days.

```
phase_1 <- bikes[1:30, ]
phase_2 <- bikes[1:60, ]
phase_3 <- bikes
```

After each data collection phase, we can re-simulate the posterior model by plugging in the accumulated data (phase_1, phase_2, or phase_3):

```
my_model <- stan_glm(rides ~ temp_feel, data = ___, family = gaussian,
                     prior_intercept = normal(5000, 1000),
                     prior = normal(100, 40),
                     prior_aux = exponential(0.0008),
                     chains = 4, iter = 5000*2, seed = 84735)
```

Figure 9.13 displays the posterior models for the temperature coefficient β_1 after each phase of data collection, and thus the evolution in our understanding of the relationship between ridership and temperature. What started in Phase 1 as a vague understanding that there might be *no* relationship between ridership and temperature (β_1 values near 0 are plausible), evolved into clear understanding by Phase 3 that ridership tends to increase by roughly 80 rides per one degree increase in temperature.

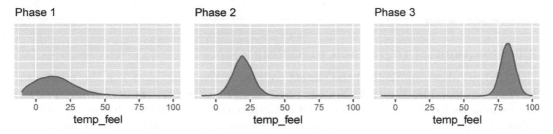

FIGURE 9.13: Approximate posterior models for the temperature coefficient β_1 after three phases of data collection.

Figure 9.14 provides more insight into this evolution, displaying the accumulated data and 100 posterior plausible models at each phase in our analysis. Having observed only 30 data points, the Phase 1 posterior models are all over the map. Further, since these 30 data points happened to land on cold days in the winter, our Phase 1 information did not yet reveal that ridership tends to increase on warmer days. Over Phases 2 and 3, we not only gathered *more* data, but data which allows us to examine the ridership across the full spectrum of temperatures in Washington, D.C. By the end of Phase 3, we have great posterior certainty of the positive association between these two variables. Again, this kind of evolution in our understanding is how learning, science, progress happen. Knowledge is built up, piece by piece, over time.

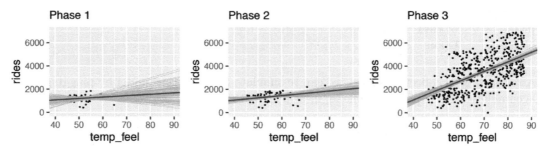

FIGURE 9.14: 100 simulated posterior median models of ridership vs temperature after three phases of data collection.

9.7 Using default rstanarm priors

It's not always the case that we'll have such strong prior information as we did in the bike analysis. As such, we'll want to tune our prior models to reflect this great uncertainty. To this end, we recommend utilizing the default priors in the **rstanarm** package. The default tuning for our Bayesian regression model of Capital Bikeshare ridership (Y) by temperature (X) is shown below (9.9). The only holdover from the original priors (9.7) is our assumption that the average ridership tends to be around 5000 on average temperature days. Even when we don't have strong prior information, we often have a sense of the baseline.

$$Y_i | \beta_0, \beta_1, \sigma \overset{ind}{\sim} N\left(\mu_i, \sigma^2\right) \quad \text{with} \quad \mu_i = \beta_0 + \beta_1 X_i$$
$$\beta_{0c} \sim N\left(5000, 3937^2\right)$$
$$\beta_1 \sim N\left(0, 351^2\right)$$
$$\sigma \sim \text{Exp}(0.00064). \tag{9.9}$$

The prior tunings here might seem bizarrely specific. We'll explain. When we don't have specific prior information, we can utilize **rstanarm**'s defaults through the following syntax. (We could also omit all prior information whatsoever, in which case `stan_glm()` would automatically assign priors. However, given different versions of the software, the results may not be reproducible.)

```
bike_model_default <- stan_glm(
  rides ~ temp_feel, data = bikes,
  family = gaussian,
  prior_intercept = normal(5000, 2.5, autoscale = TRUE),
  prior = normal(0, 2.5, autoscale = TRUE),
  prior_aux = exponential(1, autoscale = TRUE),
  chains = 4, iter = 5000*2, seed = 84735)
```

This syntax specifies the following priors: $\beta_{0c} \sim N(5000, 2.5^2)$, $\beta_1 \sim N(0, 2.5^2)$, and $\sigma \sim$ Exp(1). With a twist. Consider the priors for β_{0c} and β_1. Assuming we have a weak prior understanding of these parameters, and hence their scales, we're not really sure whether a standard deviation of 2.5 is relatively small or relatively large. Thus, we're not really sure if these priors are more specific than we want them to be. This is why we also set `autoscale = TRUE`. By doing so, `stan_glm()` adjusts or *scales* our default priors to optimize the study of parameters which have different scales.[5] These `Adjusted` priors are specified by the `prior_summary()` function and match our reported model formulation (9.9):

```
prior_summary(bike_model_default)
Priors for model 'bike_model_default'
------
Intercept (after predictors centered)
  Specified prior:
    ~ normal(location = 5000, scale = 2.5)
  Adjusted prior:
    ~ normal(location = 5000, scale = 3937)

Coefficients
  Specified prior:
    ~ normal(location = 0, scale = 2.5)
  Adjusted prior:
    ~ normal(location = 0, scale = 351)

Auxiliary (sigma)
```

---

[5]If you have some experience with Bayesian modeling, you might be wondering about whether or not we should be *standardizing* predictor X. The **rstanarm** manual recommends against this, noting that the same ends are achieved through the default scaling of the prior models (Gabry and Goodrich, 2020c).

```
Specified prior:
  ~ exponential(rate = 1)
Adjusted prior:
  ~ exponential(rate = 0.00064)
-------
See help('prior_summary.stanreg') for more details
```

In this scaling process, **rstanarm** seeks to identify **weakly informative priors** (Gabry and Goodrich, 2020b). The idea is similar to vague priors: weakly informative priors reflect general prior uncertainty about the model parameters. However, whereas a vague prior might be so vague that it puts weight on *non-sensible* parameter values (e.g., β_1 values that assume ridership could increase by 1 billion rides for every one degree increase in temperature), weakly informative priors are a bit more focused. They reflect general prior uncertainty across a range of *sensible* parameter values. As such, weakly informative priors foster computationally efficient posterior simulation since the chains don't have to waste time exploring non-sensible parameter values.

There's also a catch. Weakly informative priors are tuned to identify "sensible" parameter values by considering the scales of our data, here ridership and temperature. Though it seems odd to tune priors using data, the process merely takes into account the scales of the variables (e.g., what are temperatures like in D.C.? what's the variability in ridership from day to day?). It does not consider the *relationship* among these variables.

Let's see how these weakly informative priors compare to our original informed priors. Instead of starting from scratch, to simulate our new priors, we update() the bike_model_default with prior_PD = TRUE, thereby indicating we're interested in the *prior* not *posterior* models of $(\beta_0, \beta_1, \sigma)$.

```
# Perform a prior simulation
bike_default_priors <- update(bike_model_default, prior_PD = TRUE)
```

We then plot 200 plausible model lines $(\beta_0 + \beta_1 X)$ and 4 datasets simulated under the weakly informative priors using the add_fitted_draws() and add_predicted_draws() functions (Figure 9.15). In contrast to the original priors (Figure 9.5), the weakly informative priors in Figure 9.15 reflect much greater uncertainty about the relationship between ridership and temperature. As the Normal prior for β_1 is centered at 0, the model lines indicate that the association between ridership and temperature might be *positive* $(\beta_1 > 0)$, *non-existent* $(\beta_1 = 0)$, or *negative* $(\beta_1 < 0)$. Some of the simulated data points even include negative ridership values! Further, the simulated datasets reflect our uncertainty about whether the relationship is *strong* (with σ near 0) or *weak* (with large σ). Yet, by utilizing weakly informative priors instead of totally vague priors, our prior uncertainty is still in the right ballpark. Our priors focus on ridership being in the thousands (reasonable), not in the millions or billions (unreasonable for a city of Washington D.C.'s size).

```
# 200 prior model lines
bikes %>%
  add_fitted_draws(bike_default_priors, n = 200) %>%
  ggplot(aes(x = temp_feel, y = rides)) +
    geom_line(aes(y = .value, group = .draw), alpha = 0.15)
```

```
# 4 prior simulated datasets
set.seed(3)
bikes %>%
  add_predicted_draws(bike_default_priors, n = 4) %>%
  ggplot(aes(x = temp_feel, y = rides)) +
    geom_point(aes(y = .prediction, group = .draw)) +
    facet_wrap(~ .draw)
```

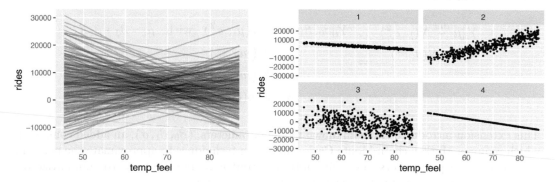

FIGURE 9.15: Simulated scenarios under the default prior models of β_0, β_1, and σ. At left are 200 prior plausible model lines, $\beta_0 + \beta_1 x$. At right are 4 prior plausible datasets.

Default vs non-default priors

There are pros and cons to utilizing the default priors in **rstanarm**. In terms of drawbacks, weakly informative priors are tuned with information from the data (through a fairly minor consideration of scale). But on the plus side:

- Unless we have strong prior information, utilizing the defaults will typically lead to more stable simulation results than if we tried tuning our own vague priors.
- The defaults can help us get up and running with Bayesian modeling. In future chapters, we'll often utilize the defaults in order to focus on the new modeling concepts.

9.8 You're not done yet!

Taken alone, Chapter 9 is quite dangerous! Thus, far you've explored how to build, simulate, and interpret a simple Bayesian Normal regression model. The next crucial step in any analysis is to ask: but is this a *good* model? This question is deserving of its own chapter. Thus, we strongly encourage you to review Chapter 10 before applying what you've learned here.

9.9   Chapter summary

In Chapter 9 you explored how to build a **Bayesian simple Normal regression** model of a **quantitative response variable** Y by a **quantitative predictor variable** X. We can view this model as an extension of the Normal-Normal model from Chapter 5, where we replace the **global mean** μ by the **local mean** $\beta_0 + \beta_1 X$ which incorporates the linear dependence of Y on X:

$$
\begin{array}{ccc}
Y_i|\mu \overset{ind}{\sim} N(\mu,\sigma^2) & & Y_i|\beta_0,\beta_1,\sigma \overset{ind}{\sim} N(\mu_i,\ \sigma^2) \quad \text{with} \quad \mu_i = \beta_0 + \beta_1 X_i \\
\mu \sim N(\theta,\tau^2) & \Rightarrow & \beta_0 \sim N\left(m_0,s_0^2\right) \\
& & \beta_1 \sim N\left(m_1,s_1^2\right) \\
& & \sigma \sim \mathrm{Exp}(l)
\end{array}
$$

This extension introduces three unknown model parameters, (β_0,β_1,σ), an added complexity which requires Markov chain Monte Carlo simulation techniques to approximate the posterior model of these parameters. This simulation output can then be used to summarize our posterior understanding of the relationship between Y and X.

9.10   Exercises

9.10.1   Conceptual exercises

Exercise 9.1 (Normal regression priors). For the Normal regression model (9.6) with $Y_i|\beta_0,\beta_1,\sigma \sim N(\mu_i,\sigma)$ where $\mu_i = \beta_0 + \beta_1 X_i$, we utilized Normal priors for β_0,β_1 and an Exponential prior on σ.

  a)   Why is a Normal prior a reasonable choice for β_0 and β_1?
  b)   Why *isn't* a Normal prior a reasonable choice for σ?
  c)   What's the difference between weakly informative and vague priors?

Exercise 9.2 (Identify the variable). Identify the response variable (Y) and predictor variable (X) in each given relationship of interest.

  a)   We want to use a person's arm length to understand their height.
  b)   We want to predict a person's carbon footprint (in annual CO_2 emissions) with the distance between their home and work.
  c)   We want to understand how a child's vocabulary level might increase with age.
  d)   We want to use information about a person's sleep habits to predict their reaction time.

Exercise 9.3 (Interpreting coefficients). In each situation below, suppose that the typical relationship between the given response variable Y and predictor X can be described by $\beta_0 + \beta_1 X$. Interpret the meaning of β_0 and β_1 and indicate whether your prior understanding suggests that β_1 is negative or positive.

a) Y = height in cm of a baby kangaroo, X = its age in months
b) Y = a data scientist's number of GitHub followers, X = their number of GitHub commits in the past week
c) Y = number of visitors to a local park on a given day, X = rainfall in inches on that day
d) Y = the daily hours of Netflix that a person watches, X = the typical number of hours that they sleep

Exercise 9.4 (Deviation from the average). Consider the Normal regression model (9.6). Explain in one or two sentences, in a way that one of your non-stats friends could understand, how σ is related to the strength of the relationship between a response variable Y and predictor X.

Exercise 9.5 (Bayesian model building: Part I). A researcher wants to use a person's age (in years) to predict their annual orange juice consumption (in gallons). Here you'll build up a relevant Bayesian regression model, step by step.

a) Identify the Y and X variables in this study.
b) Use mathematical notation to specify an appropriate structure for the model of data Y (ignoring X for now).
c) Rewrite the structure of the data model to incorporate information about predictor X. In doing so, assume there's a linear relationship between Y and X.
d) Identify all unknown parameters in your model. For each, indicate the values the parameter can take.
e) Identify and tune suitable prior models for your parameters. Explain your rationale.

Exercise 9.6 (Bayesian model building: Part II). Repeat the above exercise for the following scenario. A researcher wishes to predict tomorrow's high temperature by today's high temperature.

Exercise 9.7 (Posterior simulation T/F). Mark each statement about posterior regression simulation as *True* or *False*.

a) MCMC provides the exact posterior model of our regression parameters $(\beta_0, \beta_1, \sigma)$.
b) MCMC allows us to avoid complicated mathematical derivations.

Exercise 9.8 (Posterior simulation). For each situation, specify the appropriate `stan_glm()` syntax for simulating the Normal regression model using 4 chains, each of length 10000. (You won't actually run any code.)

a) X = `age`; Y = `height`; dataset name: `bunnies`
b) $\text{Clicks}_i | \beta_0, \beta_1, \sigma \sim N\left(\mu_i, \sigma^2\right)$ with $\mu_i = \beta_0 + \beta_1 \text{Snaps}_i$; dataset name: `songs`.
c) $\text{Happiness}_i | \beta_0, \beta_1, \sigma \sim N\left(\mu_i, \sigma^2\right)$ with $\mu_i = \beta_0 + \beta_1 \text{Age}_i$; dataset name: `dogs`.

9.10.2 Applied exercises

Exercise 9.9 (How humid is too humid: model building). Throughout this chapter, we explored how bike ridership fluctuates with temperature. But what about humidity? In the next exercises, you will explore the Normal regression model of `rides` (Y) by `humidity` (X) using the `bikes` dataset. Based on past bikeshare analyses, suppose we have the following prior understanding of this relationship:

- On an *average* humidity day, there are typically around 5000 riders, though this average could be somewhere between 1000 and 9000.

- Ridership tends to decrease as humidity increases. Specifically, for every one percentage point increase in humidity level, ridership tends to decrease by 10 rides, though this average decrease could be anywhere between 0 and 20.

- Ridership is only weakly related to humidity. At any given humidity, ridership will tend to vary with a large standard deviation of 2000 rides.

a) Tune the Normal regression model (9.6) to match our prior understanding. Use careful notation to write out the complete Bayesian structure of this model.

b) To explore our combined **prior** understanding of the model parameters, simulate the Normal regression prior model with 5 chains run for 8000 iterations each. HINT: You can use the same `stan_glm()` syntax that you would use to simulate the posterior, but include `prior_PD = TRUE`.

c) Plot 100 prior plausible model lines $(\beta_0 + \beta_1 X)$ and 4 datasets simulated under the priors.

d) Describe our overall prior understanding of the relationship between ridership and humidity.

Exercise 9.10 (How humid is too humid: data). With the priors in place, let's examine the data.

a) Plot and discuss the observed relationship between ridership and humidity in the `bikes` data.

b) Does simple Normal regression seem to be a reasonable approach to modeling this relationship? Explain.

Exercise 9.11 (How humid is too humid: posterior simulation). We can now simulate our posterior model of the relationship between ridership and humidity, a balance between our prior understanding and the data.

a) Use `stan_glm()` to simulate the Normal regression posterior model. Do so with 5 chains run for 8000 iterations each. HINT: You can either do this from scratch or `update()` your prior simulation from Exercise 9.9 using `prior_PD = FALSE`.

b) Perform and discuss some MCMC diagnostics to determine whether or not we can "trust" these simulation results.

c) Plot 100 posterior model lines for the relationship between ridership and humidity. Compare and contrast these to the prior model lines from Exercise 9.9.

Exercise 9.12 (How humid is too humid: posterior interpretation). Finally, let's dig deeper into our posterior understanding of the relationship between ridership and humidity.

a) Provide a `tidy()` summary of your posterior model, including 95% credible intervals.

b) Interpret the posterior median value of the σ parameter.

c) Interpret the 95% posterior credible interval for the `humidity` coefficient, β_1.

d) Do we have ample posterior evidence that there's a negative association between ridership and humidity? Explain.

Exercise 9.13 (How humid is too humid: prediction). Tomorrow is supposed to be 90% humidity in Washington, D.C. What levels of ridership should we expect?

a) *Without* using the `posterior_predict()` shortcut function, simulate two posterior models:
    - the posterior model for the *typical* number of riders on 90% humidity days; and
    - the posterior predictive model for the number of riders *tomorrow*.
b) Construct, discuss, and compare density plot visualizations for the two separate posterior models in part a.
c) Calculate and interpret an 80% posterior prediction interval for the number of riders tomorrow.
d) Use `posterior_predict()` to confirm the results from your posterior predictive model of tomorrow's ridership.

Exercise 9.14 (On your own: Part I). Temperature and humidity aren't the only possible weather factors in ridership. Let's explore the relationship between ridership (Y) and `windspeed` (X).

a) Describe your own prior understanding of the relationship between ridership and wind speed.
b) Tune the Normal regression model (9.6) to match your prior understanding. Use careful notation to write out the complete Bayesian structure of this model.
c) Plot and discuss 100 prior plausible model lines ($\beta_0 + \beta_1 X$) and 4 datasets simulated under the priors.
d) Construct and discuss a plot of `rides` versus `windspeed` using the `bikes` data. How consistent are the observed patterns with your prior understanding of this relationship?

Exercise 9.15 (On your own: Part II). In this open-ended exercise, conduct a posterior analysis of the relationship between ridership (Y) and `windspeed` (X). This should include a discussion of your posterior understanding of this relationship along with supporting evidence.

Exercise 9.16 (Penguins: model building and simulation). The `penguins_bayes` dataset in the **bayesrules** package includes data on 344 penguins. In the next exercises, you will use this data to model the length of penguin flippers in mm (Y) by the length of their bills in mm (X). We have a general sense that the average penguin has flippers that are somewhere between 150mm and 250mm long. Beyond that, we don't have a strong understanding of the relationship between flipper and bill length, and thus will otherwise utilize weakly informative priors.

a) Simulate the Normal regression **prior** model of `flipper_length_mm` by `bill_length_mm` using 4 chains for 10000 iterations each. HINT: You can use the same `stan_glm()` syntax that you would use to simulate the posterior, but include `prior_PD = TRUE`.
b) Check the `prior_summary()` and use this to write out the complete structure of your Normal regression model (9.6).
c) Plot 100 prior plausible model lines ($\beta_0 + \beta_1 X$) and 4 datasets simulated under the priors.

d)   Summarize your weakly informative prior understanding of the relationship be-
     tween flipper and bill length.

Exercise 9.17 (Penguins: data). With the priors in place, let's examine the data.

a)   Plot and discuss the observed relationship between `flipper_length_mm` and
     `bill_length_mm` among the 344 sampled penguins.
b)   Does simple Normal regression seem to be a reasonable approach to modeling this
     relationship? Explain.

Exercise 9.18 (Penguins: posterior analysis). Finally, let's examine our posterior understanding of the relationship between flipper and bill length, a balance between our prior
understanding and the data.

a)   Use `stan_glm()` to simulate the Normal regression posterior model. HINT: You
     can either do this from scratch or `update()` your prior simulation from Exercise
     9.16 using `prior_PD = FALSE`.
b)   Plot 100 posterior model lines for the relationship between flipper and bill length.
c)   Provide a `tidy()` summary of your posterior model, including 90% credible
     intervals.
d)   Interpret the 90% posterior credible interval for the `bill_length_mm` coefficient,
     β_1.
e)   Do we have ample posterior evidence that penguins with longer bills tend to have
     longer flippers? Explain.

Exercise 9.19 (Penguins: prediction). A researcher comes across Pablo the penguin. They're
able to ascertain that Pablo's bill is 51mm long but the penguin waddles off before they get
to measure the flipper.

a)   *Without* using the `posterior_predict()` shortcut function, simulate the posterior
     model for the *typical* flipper length among penguins with 51mm bills as well as
     the posterior predictive model for *Pablo's* flipper length.
b)   Construct, discuss, and compare density plot visualizations for the two separate
     posterior models in part a.
c)   Calculate and interpret an 80% posterior prediction interval for Pablo's flipper
     length.
d)   Would the 80% credible interval for the typical flipper length among all penguins
     with 51mm bills be wider or narrower? Explain.
e)   Use `posterior_predict()` to confirm your results to parts b and c.

Exercise 9.20 (More penguins). Instead of bill length, consider the Normal regression
model of penguin flipper lengths (Y) by body mass in grams (X).

a)   Based on their study of penguins in other regions, suppose that researchers are
     quite certain about the relationship between flipper length and body mass, prior
     to seeing any data: $\beta_1 \sim N(0.01, 0.002^2)$. Describe their prior understanding.
b)   Plot and discuss the observed relationship between `flipper_length_mm` and
     `body_mass_g` among the 344 sampled penguins.
c)   In a simple Normal regression model of flipper length Y by one predictor X, do
     you think that the σ parameter is bigger when X = bill length or when X = body

mass? Explain your reasoning and provide some evidence from the information
you already have.

d)   Use `stan_glm()` to simulate the Normal regression posterior model of flipper
     length by body mass *using the researchers' prior* for β_1 and weakly informative
     priors for β_{0c} and σ. Do so with 4 chains run for 10000 iterations each.

e)   Plot the posterior model of the β_1 `body_mass_g` coefficient. Use this to describe
     the researchers' posterior understanding of the relationship between flippers and
     mass and how, if at all, this evolved from their prior understanding.

10

Evaluating Regression Models

Imagine that we, the authors, invite you over for dinner. It took us *hours* to forage mushrooms and cook them up into something delicious. Putting niceties aside, you might have some questions for us: Do you know what you're doing? Are these mushrooms safe to eat? After dinner, we offer to drive you home in a new car that we just built. Before obliging, it would be wise to check: Is this car *safe*? How did it perform in crash tests? Just as one should never eat a foraged mushroom or get in a new car without questioning their safety, one should *never* apply a model without first evaluating its quality. No matter whether we're talking about frequentist or Bayesian models, "simple" or "big" models, there are three critical questions to ask. Examining these questions is the goal of Chapter 10.

◎ **Goals**

1. **How fair is the model?**
   How was the data collected? By whom and for what purpose? How might the results of the analysis, or the data collection itself, impact individuals and society? What biases or power structures might be baked into this analysis?
2. **How wrong is the model?**
   George Box famously said: "All models are wrong, but some are useful." What's important to know then is, *how* wrong is our model? Are our model assumptions reasonable?
3. **How accurate are the posterior predictive models?**
   How far are the posterior predictive models from reality?

```
# Load packages
library(bayesrules)
library(tidyverse)
library(bayesplot)
library(rstanarm)
```

10.1 Is the model fair?

We'll explore the model evaluation process in the context of the Normal regression analysis from Chapter 9. Letting Y_i denote the number of Capital Bikeshare rides and X_i denote the temperature on day i, we built and simulated the following model:

DOI: 10.1201/9780429288340-10

$$Y_i|\beta_0, \beta_1, \sigma \overset{ind}{\sim} N\left(\mu_i, \sigma^2\right) \quad \text{with} \quad \mu_i = \beta_0 + \beta_1 X_i$$
$$\beta_{0c} \sim N\left(5000, 1000^2\right)$$
$$\beta_1 \sim N\left(100, 40^2\right)$$
$$\sigma \sim \text{Exp}(0.0008).$$

The first question in evaluating this or any other Bayesian model is context specific and gets at the underlying ethical implications: *Is the model fair?* We must *always* ask this question, even when the consideration is uneventful, as it is in our bike example. Let's break it down into a series of smaller questions:

- **How was the data collected?**
  Presumably, data is electronically recorded for each bike ride, without any harm to the riders. By the time the data gets into our hands, it has been anonymized. Seems fair.

- **By whom and for what purpose was the data collected?**
  A company like Capital Bikeshare tracks their own data for the purpose of better serving their customers and informing their business decisions.

- **How might the results of the analysis, or the data collection itself, impact individuals and society?**
  We can't imagine how the data collection process or analysis could *negatively* impact individuals or society. In fact, helping improve Capital Bikeshare's service might help get some cars off the road. *However*, just because we, the authors, cannot imagine negative societal impacts does not mean that none exist. In any analysis, it's critical to recognize your own perspective, and that no perspective is natural or neutral. It's then easier to admit that you might not perceive any negative consequences simply because they do not affect you. To truly determine the impact of our analysis, we'd need to engage and partner with D.C. residents who *are* impacted by Capital Bikeshare's decisions, especially those that have not been part of the decision-making processes.

- **What biases might be baked into this analysis?**
  We only have data on rides taken by people that use the Capital Bikeshare service. Thus, our analysis might better inform how to serve *this* population at the exclusion of serving others.

Upon these reflections, we don't find anything ethically dubious or biased about the data collection or analysis for bike ridership in D.C. We think this analysis could be *fair*.

Models aren't always fair

Unfair models are unfortunately abundant. Consider just three recent examples. These complex examples are overly simplified here, thus we encourage you to read more outside this book.

1. As recently as 2015, a major corporation reportedly utilized a *model* to evaluate applicants' résumés for technical posts. They scrapped this model upon discovering that, by building this model using résumé data from its current technical employees (mostly men), it reinforced a preference for male job applicants (Dastin, 2018).

2. Facial recognition *models*, increasingly used in police surveillance, are often built using image data of researchers that do not represent the whole of society. Thus, when applied in practice, misidentification is more common among people that are underrepresented in the research process. Given the severe consequences of misidentification, including false arrest, citizens are pushing back on the use of this technology in their communities (Harmon, 2019).

3. In 2020, the New York Civil Liberties Union filed a lawsuit against the U.S. Immigration and Customs Enforcement's (ICE) use of a "risk classification assessment" *model* that evaluates whether a subject should be detained or released (Hadavas, 2020). This model is notably unfair, recommending detention in *nearly all cases*.

10.2 How wrong is the model?

All models are wrong. Mainly, statistical models are idealistic representations of more complex realities. Even so, good statistical models can still be useful and inform our understanding of the world's complexities. Thus, the next question to ask in evaluating our model is not, *is* the model wrong, but *how* wrong is the model? Specifically, to what extent do the assumptions behind our Bayesian linear regression model (9.6) match reality?

10.2.1 Checking the model assumptions

Recall from Chapter 9 the three assumptions baked into the data model,

$$Y_i|\beta_0, \beta_1, \sigma \overset{ind}{\sim} N(\mu_i, \sigma^2) \quad \text{with} \quad \mu_i = \beta_0 + \beta_1 X_i.$$

1. Conditioned on X, the observed **data** Y_i on case i is *independent* of the observed data on any other case j.
2. The typical Y outcome can be written as a *linear function* of X, $\mu = \beta_0 + \beta_1 X$.
3. At any X value, Y **varies normally** around μ with consistent variability σ.

Though formal hypothesis tests for **assumption 1** do exist, we can typically evaluate its appropriateness by the data collection context. In our `bikes` analysis, we have 500 daily observations within the two-year period from 2011 to 2012. *When taken alone*, ridership Y is likely correlated over time – today's ridership likely tells us something about tomorrow's ridership. Yet much of this correlation, or dependence, can be explained by the time of year and features associated with the time of year – like temperature X. Thus, knowing the *temperature* on two subsequent days may very well "cancel out" the time correlation in their ridership data. For example, if we know that today was 70 degrees and tomorrow will be 75 degrees, then information about today's ridership might *not* inform us of tomorrow's ridership. In short, it's *reasonable* to assume that, in light of the temperature X, ridership data Y is independent from day to day. ("Reasonable" does not mean "perfect," but rather "good enough to proceed.")

Revisiting a scatterplot of the raw data provides insight into **assumptions 2 and 3**. The relationship between ridership and temperature *does* appear to be linear. Further, with the slight exception of colder days on which ridership is uniformly small, the variability in ridership *does* appear to be roughly consistent across the range of temperatures X:

```
ggplot(bikes, aes(y = rides, x = temp_feel)) +
  geom_point(size = 0.2) +
  geom_smooth(method = "lm", se = FALSE)
```

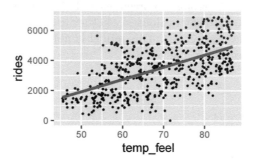

FIGURE 10.1: A scatterplot of daily bikeshare ridership vs temperature.

Now, not only are humans known to misinterpret visualizations, relying upon such visualizations of Y vs X to evaluate assumptions ceases to be possible for more complicated models that have more than one predictor X. To rigorously evaluate assumptions 2 and 3 in a way that scales up to other model settings, we can conduct a **posterior predictive check**. We'll provide a more general definition below, but the basic idea is this. If the *combined* model assumptions are reasonable, then our posterior model should be able to simulate ridership data that's similar to the original 500 `rides` observations. To assess whether this is the case, for each of the 20,000 posterior plausible parameter sets $(\beta_0, \beta_1, \sigma)$ in our Markov chain simulation from Chapter 9, we can predict 500 days of ridership data from the 500 days of observed temperature data. The end result is 20,000 unique sets of predicted ridership data, each of size 500, here represented by the rows of the right matrix:

Markov chain parameter sets   Simulated samples

$$
\begin{bmatrix}
\beta_0^{(1)} & \beta_1^{(1)} & \sigma^{(1)} \\
\beta_0^{(2)} & \beta_1^{(2)} & \sigma^{(2)} \\
\vdots & \vdots & \vdots \\
\beta_0^{(20000)} & \beta_1^{(20000)} & \sigma^{(20000)}
\end{bmatrix}
\longrightarrow
\begin{bmatrix}
Y_1^{(1)} & Y_2^{(1)} & \cdots & Y_{500}^{(1)} \\
Y_1^{(2)} & Y_2^{(2)} & \cdots & Y_{500}^{(2)} \\
\vdots & \vdots & & \vdots \\
Y_1^{(20000)} & Y_2^{(20000)} & \cdots & Y_{500}^{(20000)}
\end{bmatrix}
$$

Specifically, for each parameter set $j \in \{1, 2, ..., 20000\}$, we predict the ridership on day $i \in \{1, 2, ..., 500\}$ by drawing from the Normal data model evaluated at the observed temperature X_i on day i:

$$
Y_i^{(j)} | \beta_0, \beta_1, \sigma \sim N\left(\mu^{(j)}, \left(\sigma^{(j)}\right)^2\right) \quad \text{with} \quad \mu^{(j)} = \beta_0^{(j)} + \beta_1^{(j)} X_i.
$$

Consider this process for the first parameter set in `bike_model_df`, a data frame with the Markov chain values we simulated in Chapter 9:

```
first_set <- head(bike_model_df, 1)
first_set
  (Intercept) temp_feel sigma
1       -2657     88.16  1323
```

From the observed `temp_feel` on each of the 500 days in the `bikes` data, we simulate a ridership outcome from the Normal data model tuned to this first parameter set:

```
beta_0 <- first_set$`(Intercept)`
beta_1 <- first_set$temp_feel
sigma  <- first_set$sigma
set.seed(84735)
one_simulation <- bikes %>%
  mutate(mu = beta_0 + beta_1 * temp_feel,
         simulated_rides = rnorm(500, mean = mu, sd = sigma)) %>%
  select(temp_feel, rides, simulated_rides)
```

Check out the original `rides` alongside the `simulated_rides` for the first two data points:

```
head(one_simulation, 2)
  temp_feel rides simulated_rides
1     64.73   654            3932
2     49.05  1229            1503
```

Of course, on any given day, the simulated ridership is off (very off in the case of the first day in the dataset). The question is whether, *on the whole*, the simulated data is similar to the observed data. To this end, Figure 10.2 compares the density plots of the 500 days of simulated ridership (light blue) and observed ridership (dark blue):

```
ggplot(one_simulation, aes(x = simulated_rides)) +
  geom_density(color = "lightblue") +
  geom_density(aes(x = rides), color = "darkblue")
```

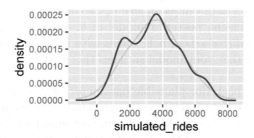

FIGURE 10.2: One posterior simulated dataset of ridership (light blue) along with the actual observed ridership data (dark blue).

Though the simulated data doesn't *exactly* replicate all original ridership features, it *does* capture the big things such as the general center and spread. And before further picking apart this plot, recall that we generated the simulated data here from merely one of 20,000 posterior plausible parameter sets. The **rstanarm** and **bayesplot** packages make it easy to repeat the data simulation process for each parameter set. The `pp_check()` function plots 50 of these 20,000 simulated datasets (labeled y_{rep}) against a plot of the original ridership data (labeled y). (It would be computationally and logically excessive to examine all 20,000 sets.)

```
# Examine 50 of the 20000 simulated samples
pp_check(bike_model, nreps = 50) +
  xlab("rides")
```

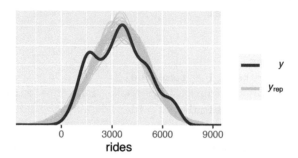

FIGURE 10.3: 50 datasets of ridership simulated from the posterior (light blue) alongside the actual observed ridership data (dark blue).

This plot highlights both things to cheer and lament. What's there to cheer about? Well, the 50 sets of predictions well capture the *typical* ridership as well as the observed *range* in ridership. However, most sets don't pick up the apparent **bimodality** in the original bike data. This doesn't necessarily mean that our model of bike ridership is *bad* – we certainly know more about bikeshare demand than we did before. It just means that it could be *better*, or *less wrong*. We conclude this section with a general summary of the **posterior predictive check** process.

Posterior predictive check

Consider a regression model with response variable Y, predictor X, and a set of regression parameters θ. For example, in the model above $\theta = (\beta_0, \beta_1, \sigma)$. Further, suppose that based on a data sample of size n, we simulate an N-length Markov chain $\{\theta^{(1)}, \theta^{(2)}, \ldots, \theta^{(N)}\}$ for the posterior model of θ. Then a "good" Bayesian model produces *posterior predicted sets* of n Y values with features similar to the *original* Y data. To evaluate whether your model satisfies this goal, do the following:

1. At each set of posterior plausible parameters $\theta^{(i)}$, simulate a sample of Y values from the data model, one corresponding to each X in the original sample of size n. This produces N separate samples of size n.
2. Compare the features of the N simulated Y samples, or a subset of these samples, to those of the original Y data.

10.2.2   Dealing with wrong models

Our discussion has thus far focused on an ideal setting in which our model assumptions are reasonable. Stopping here would be like an auto mechanic only learning about mint condition cars. Not a great idea. Though it's impossible to enumerate everything that might go wrong in modeling, we can address some basic principles. Let's begin with **assumption 1**, that of independence. There are a few common scenarios in which this assumption is unreasonable. In Unit 4 we address scenarios where Y values are correlated across multiple data points observed on the same "group" or subject. For example, to explore the association between memory and age, researchers might give each of n adults three tests in which they're asked to recall a sequence of 10 words:

subject	age (X)	number of words remembered (Y)
1	51	9
1	51	7
1	51	10
2	38	6
2	38	8
2	38	8

Naturally, how well any particular adult performs in one trial tells us something about how well they might perform in another. Thus, within any given subject, the observed Y values are dependent. You will learn how to incorporate and address such dependent, grouped data using **hierarchical Bayesian models** in Unit 4.

Correlated data can also pop up when modeling changes in Y over time, space, or time *and* space. For example, we might study historical changes in temperatures Y in different parts of the world. In doing so, it would be unreasonable to assume that the temperatures in one location are independent of those in neighboring locations, or that the temperatures in one month don't tell us about the next. Though applying our Normal Bayesian regression model to study these spatiotemporal dynamics might produce misleading results, there do exist Bayesian models that are tailor-made for this task: times series models, spatial regression models, and their combination, spatiotemporal models. Though beyond the scope of this book, we encourage the interested reader to check out Blangiardo and Cameletti (2015).

Next, let's consider violations of **assumptions 2 and 3**, which often go hand in hand. Figure 10.4 provides an example. The relationship between Y and X is nonlinear (violating assumption 2) *and* the variability in Y increases with X (violating assumption 3).

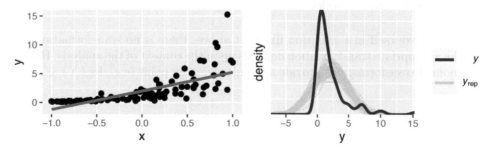

FIGURE 10.4: Simulated data with corresponding Bayesian posterior predictive model checks.

Even if we hadn't seen this raw data, the `pp_check()` (right) would confirm that a Normal regression model of this relationship is wrong – the posterior simulations of Y exhibit higher central tendency and variability than the observed Y values. There are a few common approaches to addressing such violations of assumptions 2 and 3:

- **Assume a different data structure.** Not all data and relationships are Normal. In Chapters 12 and 13 we will explore models in which the data structure of a regression model is better described by a Poisson, Negative Binomial, or Binomial than by a Normal.

- **Make a transformation.** When the data model structure isn't the issue we might do the following:

  - **Transform Y.** For some function $g(\cdot)$, assume $g(Y_i)|\beta_0,\beta_1,\sigma \overset{ind}{\sim} N(\mu_i,\sigma^2)$ with $\mu_i = \beta_0 + \beta_1 X_i$.

  - **Transform X.** For some function of $h(\cdot)$, assume $Y_i|\beta_0,\beta_1,\sigma \overset{ind}{\sim} N(\mu_i,\sigma^2)$ with $\mu_i = \beta_0 + \beta_1 h(X_i)$.

  - **Transform Y and X.** Assume $g(Y_i)|\beta_0,\beta_1,\sigma \overset{ind}{\sim} N(\mu_i,\sigma^2)$ with $\mu_i = \beta_0 + \beta_1 h(X_i)$.

In our simulated example, the second approach proves to be the winner. Based on the skewed pattern in the raw data, we take a **log transform** of Y:

$$\log(Y_i)|\beta_0,\beta_1,\sigma \overset{ind}{\sim} N(\mu_i,\sigma^2) \quad \text{with} \quad \mu_i = \beta_0 + \beta_1 X_i.$$

Figure 10.5 confirms that this transformation addresses the violations of both assumptions 2 and 3: the relationship between $\log(Y)$ and X *is* linear *and* the variability in Y is *consistent* across the range of X values. The ideal `pp_check()` at right further confirms that this transformation turns our model from wrong to good. Better yet, when transformed off the log scale, we can *still* use this model to learn about the relationship between Y and X.

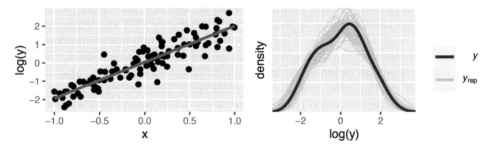

FIGURE 10.5: Simulated data, after a log transform of Y, with corresponding Bayesian posterior predictive model checks.

When you find yourself in a situation like that above, there is no transformation recipe to follow. The appropriate transformation depends upon the context of the analysis. However, log and polynomial transformations provide a popular starting point (e.g., $\log(X), \sqrt{X}, X^2, X^3$).

10.3   How accurate are the posterior predictive models?

In an ideal world, not only is our Bayesian model fair and not *too* wrong, it can be used to accurately *predict* the outcomes of new data Y. That is, a good model will generalize beyond the data we used to build it. We'll explore three approaches to evaluating predictive quality, starting from the most straightforward to the most technical.

10.3.1 Posterior predictive summaries

Our ultimate goal is to determine how well our Bayesian model predicts ridership on days in the *future*. Yet we can first assess how well it predicts the sample data points that we used to *build* this model. For starters, consider the sample data point from October 22, 2012:

```
bikes %>%
  filter(date == "2012-10-22") %>%
  select(temp_feel, rides)
  temp_feel rides
1     75.46  6228
```

On this 75-degree day, Capital Bikeshare happened to see 6228 rides. To see how this observation compares to the posterior *prediction* of ridership on this day, we can simulate and plot the posterior predictive model using the same code from Section 9.5.

```
# Simulate the posterior predictive model
set.seed(84735)
predict_75 <- bike_model_df %>%
  mutate(mu = `(Intercept)` + temp_feel*75,
         y_new = rnorm(20000, mean = mu, sd = sigma))

# Plot the posterior predictive model
ggplot(predict_75, aes(x = y_new)) +
  geom_density()
```

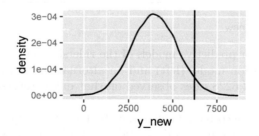

FIGURE 10.6: The posterior predictive model of ridership on October 22, 2012, a 75-degree day. The actual 6228 riders observed that day are marked by the vertical line.

So how good was this predictive model at anticipating what we actually observed? There are multiple ways to answer this question. First, in examining Figure 10.6, notice how far the *observed* $Y = 6228$ rides fall from the *center* of the predictive model, i.e., the *expected* outcome of Y. We can measure this expected outcome by the **posterior predictive mean**, denoted Y'. Then the distance between the observed Y and its posterior predictive mean provides one measurement of posterior **prediction error**:

$$Y - Y'.$$

For example, our model *under*-predicted ridership. The $Y = 6228$ rides we *observed* were 2261 rides greater than the $Y' = 3967$ rides we *expected*:

```
predict_75 %>%
  summarize(mean = mean(y_new), error = 6228 - mean(y_new))
  mean error
1 3967   2261
```

It's also important to consider the *relative*, not just *absolute*, distance between the observed value and its posterior predictive mean.

---

❷ Quiz Yourself!

Figure 10.7 compares the posterior predictive model of Y, the ridership on October 22, 2012, from *our* Bayesian model to that of an alternative Bayesian model. In *both* cases, the posterior predictive mean is 3967. Which model produces the better posterior predictive model of the 6228 rides we actually observed on October 22?

    a.   Our Bayesian model
    b.   The alternative Bayesian model
    c.   The quality of these models' predictions are equivalent

---

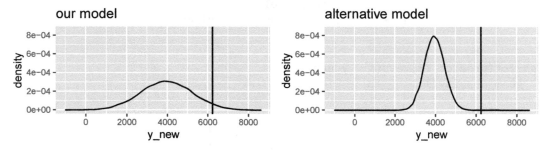

FIGURE 10.7: Posterior predictive models of the ridership on October 22, 2012 are shown for our Bayesian regression model (left) and an alternative model (right), both having a mean of 3967. The *actual* 6228 rides observed on that day is represented by the black line.

If you answered **a**, you were correct. Though the observed ridership ($Y = 6228$) is equidistant from the posterior predictive mean ($Y' = 3967$) in both scenarios, the *scale* of this error is much larger for the alternative model – the observed ridership falls entirely outside the plausible range of its posterior predictive model. We can formalize this observation by *standardizing* our posterior prediction errors. To this end, we can calculate how many posterior *standard deviations* our observed value falls from the posterior predictive mean:

$$\frac{Y - Y'}{\text{sd}}.$$

When a posterior predictive model is roughly symmetric, absolute values beyond 2 or 3 on this standardized scale indicate that an observation falls quite far, more than 2 or 3 standard deviations, from its posterior mean prediction. In our October 22 example, the posterior predictive model of ridership has a standard deviation of 1280 rides. Thus, the 6228 rides we observed were 2261 *rides*, or 1.767 *standard deviations*, above the mean prediction:

```
predict_75 %>%
  summarize(sd = sd(y_new), error = 6228 - mean(y_new),
            error_scaled = error / sd(y_new))
    sd error error_scaled
1 1280  2261         1.767
```

In comparison, the alternative model in Figure 10.7 has a standard deviation of 500. Thus, the observed 6228 rides fell a whole 4.5 standard deviations (far!) above its mean prediction. This clarifies the comparison of our model and the alternative model. Though the *absolute* prediction error is the same for both models, the *scaled* prediction errors indicate that *our* model provided the more accurate posterior prediction of ridership on October 22.

Though we're getting somewhere now, these absolute and scaled error calculations don't capture the full picture. They give us a mere sense of how far an observation falls from the *center* of its posterior predictive model. In Bayesian statistics, the *entire* posterior predictive model serves as a prediction of Y. Thus, to complement the error calculations, we can track whether an observed Y value falls into its **posterior prediction interval**, and hence the general range of the posterior predictive model. For our October 22 case, our posterior predictive model places a 0.5 probability on ridership being between 3096 and 4831, and a 0.95 probability on ridership being between 1500 and 6482. The observed 6228 rides lie *outside* the 50% interval but *inside* the 95% interval. That is, though we didn't think this was a very likely ridership outcome, it was still within the realm of possibility.

```
predict_75 %>%
  summarize(lower_95 = quantile(y_new, 0.025),
            lower_50 = quantile(y_new, 0.25),
            upper_50 = quantile(y_new, 0.75),
            upper_95 = quantile(y_new, 0.975))
  lower_95 lower_50 upper_50 upper_95
1     1500     3096     4831     6482
```

Great – we now understand the posterior predictive accuracy for *one* case in our dataset. We can take these same approaches to evaluate the accuracy for *all* 500 cases in our `bikes` data. As discussed in Section 10.2, at each set of model parameters in the Markov chain, we can predict the 500 ridership values Y from the corresponding temperature data X:

Markov chain parameter sets

$$
\begin{bmatrix}
\beta_0^{(1)} & \beta_1^{(1)} & \sigma^{(1)} \\
\beta_0^{(2)} & \beta_1^{(2)} & \sigma^{(2)} \\
\vdots & \vdots & \vdots \\
\beta_0^{(20000)} & \beta_1^{(20000)} & \sigma^{(20000)}
\end{bmatrix}
\longrightarrow
$$

Simulated posterior predictive models

$$
\begin{bmatrix}
Y_1^{(1)} & Y_2^{(1)} & \cdots & Y_{500}^{(1)} \\
Y_1^{(2)} & Y_2^{(2)} & \cdots & Y_{500}^{(2)} \\
\vdots & \vdots & & \vdots \\
Y_1^{(20000)} & Y_2^{(20000)} & \cdots & Y_{500}^{(20000)}
\end{bmatrix}
$$

The result is represented by the 20000×500 matrix of posterior predictions (right matrix). Whereas the 20,000 *rows* of this matrix provide 20,000 simulated sets of ridership data which provide insight into the validity of our model assumptions (Section 10.2), each of the 500 *columns* provides 20,000 posterior predictions of ridership for a unique day in the `bikes` data. That is, each column provides an approximate posterior predictive model for the corresponding day. We can obtain these sets of 20,000 predictions per day by applying `posterior_predict()` to the full `bikes` dataset:

```
set.seed(84735)
predictions <- posterior_predict(bike_model, newdata = bikes)
dim(predictions)
[1] 20000   500
```

The `ppc_intervals()` function in the **bayesplot** package provides a quick visual summary of the 500 approximate posterior predictive models stored in `predictions`. For each data point in the `bikes` data, `ppc_intervals()` plots the bounds of the 95% prediction interval (narrow, long blue bars), 50% prediction interval (wider, short blue bars), and posterior predictive median (light blue dots). This information offers a glimpse into the center and spread of each posterior predictive model, as well as the compatibility of each model with the corresponding *observed* outcome (dark blue dots). For illustrative purposes, the posterior predictive models for just 25 days in the dataset (Figure 10.8 right) provide a cleaner picture. In general, notice that almost all of the observed ridership data points fall within the bounds of their 95% prediction interval, and fewer fall within the bounds of their 50% interval. However, the 95% prediction intervals are quite wide – the posterior predictions of ridership on a given day span a range of more than 4000 rides.

```
ppc_intervals(bikes$rides, yrep = predictions, x = bikes$temp_feel,
              prob = 0.5, prob_outer = 0.95)
```

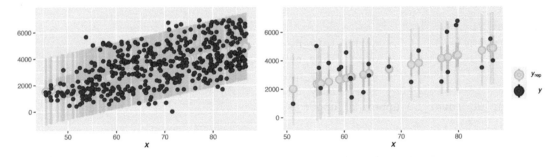

FIGURE 10.8: At left are the posterior predictive medians (light blue dots), 50% prediction intervals (wide, short blue bars), and 95% prediction intervals (narrow, long blue bars) for each day in the bikes dataset, along with the corresponding observed data points (dark blue dots). This same information is shown at right for just 25 days in the dataset.

The `prediction_summary()` function in the **bayesrules** package formalizes these visual observations with four numerical posterior predictive summaries.

Posterior predictive summaries

Let $Y_1, Y_2, \ldots, Y_n$ denote n *observed* outcomes. Then each Y_i has a corresponding posterior predictive model with *mean* Y_i' and *standard deviation* sd_i. We can evaluate the overall posterior predictive model quality by the following measures:

- mae
  The **median absolute error (MAE)** measures the *typical* difference between the observed Y_i and their posterior predictive means Y_i',

$$\text{MAE} = \text{median}|Y_i - Y_i'|.$$

- `mae_scaled`
  The **scaled median absolute error** measures the *typical* number of standard deviations that the observed Y_i fall from their posterior predictive means Y_i':

$$\text{MAE scaled} = \text{median}\frac{|Y_i - Y_i'|}{\text{sd}_i}.$$

- `within_50` and `within_95`
  The `within_50` statistic measures the proportion of observed values Y_i that fall within their 50% posterior prediction interval. The `within_95` statistic is similar, but for 95% posterior prediction intervals.

  NOTE: For stability across potentially skewed posteriors, we could measure the *center* of the posterior predictive model by the median (instead of mean) and the *scale* by the median absolute deviation (instead of standard deviation). This option is provided by setting `stable = TRUE` in the `prediction_summary()` function, though we don't utilize it here.

Let's examine the posterior predictive summaries for our data:

```
# Posterior predictive summaries
set.seed(84735)
prediction_summary(bike_model, data = bikes)
    mae mae_scaled within_50 within_95
1 989.7     0.7712     0.438     0.968
```

Among all 500 days in the dataset, we see that the observed ridership is *typically* 990 rides, or 0.77 standard deviations, from the respective posterior predictive mean. Further, only 43.8% of test observations fall within their respective 50% prediction interval whereas 96.8% fall within their 95% prediction interval. This is compatible with what we saw in the `ppc_intervals()` plot above: almost all dark blue dots are within the span of the corresponding 95% predictive bars and fewer are within the 50% bars (naturally). So what can we conclude in light of these observations: Does our Bayesian model produce accurate predictions? The answer to this question is context dependent and somewhat subjective. For example, knowing whether a typical prediction error of 990 rides is reasonable would require a conversation with Capital Bikeshare. As is a theme in this book, there's not a yes or no answer.

10.3.2 Cross-validation

The posterior prediction summaries in Section 10.3.1 can provide valuable insight into our Bayesian model's predictive accuracy. They can also be flawed. Consider an analogy. Suppose you want to open a new taco stand. You build all of your recipes around Reem, your friend who prefers that every meal include anchovies. You test your latest "anchov-ladas" dish on her and it's a hit. Does this imply that this dish will enjoy broad success among the general public? Probably not! Not everybody shares Reem's particular tastes.[1] Similarly, a *model* is optimized to capture the features in the *data* with which it's trained or built. Thus,

---

[1] We assume here that readers agree that not everybody likes anchovies.

evaluating a model's predictive accuracy on this *same* data, as we did above, can produce overly optimistic assessments. Luckily, we don't have to go out and collect new data with which to evaluate our model. Rather, only for the purposes of model evaluation, we can split our *existing* `bikes` data into different pieces that play distinct **"training"** and **"testing"** roles in our analysis. The basic idea is this:

- **Train the model.**
  Randomly select only, say, 90% (or 450) of our 500 `bikes` data points. Use these **training data** points to build the regression model of ridership.
- **Test the model.**
  Test or *evaluate* the training model using the other 10% of (or 50) data points. Specifically, use the training model to predict our **testing data** points, and record the corresponding measures of model quality (e.g., MAE).

Figure 10.9 provides an example. The scatterplot at left exhibits the relationship between ridership and temperature for 450 randomly selected training data points. Upon building a model from this training data, we then use it to predict the *other* 50 data points that we left out for testing. The `ppc_intervals()` plot at right provides a visual evaluation of the posterior predictive accuracy. In general, our training model did a decent job of predicting the outcomes of the testing data – all 50 testing data points fall within their 95% prediction intervals.

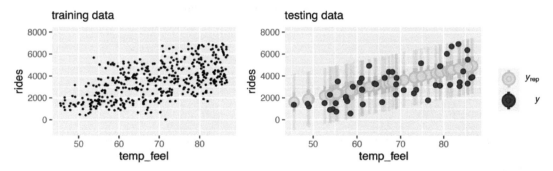

FIGURE 10.9: At left is a scatterplot of the relationship between ridership and temperature for 450 training data points. A model built using this training data is evaluated against 50 testing data points (right).

Since it trains and tests our model using different portions of the `bikes` data, this procedure would provide a more *honest* or *conservative estimate* of how well our model generalizes *beyond* our particular bike sample, i.e., how well it predicts *future* ridership. But there's a catch. Performing just one round of training and testing can produce an unstable estimate of posterior predictive accuracy – it's based on only one random split of our `bikes` data and uses only 50 data points for testing. A different random split might paint a different picture. The *k*-**fold cross-validation** algorithm, outlined below, provides a more *stable* approach by repeating the training / testing process *multiple* times and averaging the results.

The k-fold cross-validation algorithm

1. **Create folds.**
   Let k be some integer from 2 to our original sample size n. Split the data into k non-overlapping **folds**, or subsets, of roughly equal size.

2. **Train and test the model.**
   - *Train* the model using the combined data in the first $k - 1$ folds.
   - *Test* this model on the kth data fold.
   - Measure the prediction quality (e.g., by MAE).

3. **Repeat.**
   Repeat step 2 $k - 1$ more times, each time leaving out a different fold for testing.

4. **Calculate cross-validation estimates.**
   Steps 2 and 3 produce k different training models and k corresponding measures of prediction quality. *Average* these k measures to obtain a single cross-validation estimate of prediction quality.

Consider the cross-validation procedure using $k = 10$ folds. In this case, our 500 `bikes` data points will be split into 10 subsets, each of size 50. We will then build 10 separate training models. Each model will be *built* using 450 days of data (9 folds), and then *tested* on the other 50 days (1 fold). To be clear, $k = 10$ is *not* some magic number. However, it strikes a nice balance in the **Goldilocks challenge**. Using only $k = 2$ folds would split the data in half, into 2 subsets of only 250 days each. Using so little data to train our models would likely underestimate the posterior predictive accuracy we achieve when using all 500 data points. At the other extreme, suppose we used $k = 500$ folds where each fold consists of only one data point. This special case of the cross-validation procedure is known as **leave-one-out** (LOO) – each training model is built using 499 data points and tested on the *one* we left out. Further, since we have to use each of the 500 folds for testing, leave-one-out cross-validation would require us to train 500 different models! That would take a long time.

Now that we've settled on 10-fold cross-validation, we can implement this procedure using the `prediction_summary_cv()` function in the **bayesrules** package. Note: Since this procedure trains 10 different models of ridership, you will have to be a little patient.

```
set.seed(84735)
cv_procedure <- prediction_summary_cv(
  model = bike_model, data = bikes, k = 10)
```

Below are the resulting posterior prediction metrics corresponding to each of the 10 testing folds in this cross-validation procedure. Since the splits are random, the training models perform better on some test sets than on others, essentially depending on how similar the testing data is to the training data. For example, the `mae` was as low as 786.8 rides for one fold and as high as 1270.9 for another:

```
cv_procedure$folds
    fold     mae mae_scaled within_50 within_95
1      1   990.2     0.7699      0.46      0.98
2      2   963.8     0.7423      0.40      1.00
3      3   951.3     0.7300      0.42      0.98
4      4  1018.6     0.7910      0.46      0.98
5      5  1161.5     0.9091      0.36      0.96
6      6   937.6     0.7327      0.46      0.94
```

7	7	1270.9	1.0061	0.32	0.96
8	8	1111.9	0.8605	0.36	1.00
9	9	1098.7	0.8679	0.40	0.92
10	10	786.8	0.6060	0.56	0.96

Averaging across each set of 10 `mae`, `mae_scaled`, `within_50`, and `within_95` values produces the ultimate cross-validation estimates of posterior predictive accuracy:

```
cv_procedure$cv
   mae mae_scaled within_50 within_95
1 1029     0.8015     0.42     0.968
```

Having split our data into distinct training and testing roles, these cross-validated summaries provide a fairer assessment of how well our Bayesian model will predict the outcomes of new cases, not just those on which it's trained. For a point of comparison, recall our posterior predictive assessment based on using the full `bikes` dataset for *both* training and testing:

```
# Posterior predictive summaries for original data
set.seed(84735)
prediction_summary(bike_model, data = bikes)
    mae mae_scaled within_50 within_95
1 989.7     0.7712    0.438     0.968
```

In light of the original and cross-validated posterior predictive summaries above, take the following quiz.[2]

---

❷ Quiz Yourself!

If we were to apply our model to predict ridership *tomorrow*, we should expect that our prediction will be off by:

   a.   1029 rides
   b.   990 rides

Remember Reem and the anchovies? Remember how we thought she'd like your anchov-lada recipe better than a new customer would? The same is true here. Our original posterior model was optimized to our full `bikes` dataset. Thus, evaluating its posterior predictive accuracy on this same dataset seems to have produced an overly rosy picture – the stated typical prediction error (990 rides) is smaller than when we apply our model to predict the outcomes of *new* data (1029 rides). In the spirit of "better safe than sorry," it's thus wise to supplement any measures of model quality with their cross-validated counterparts.

---

[2]Answer: a

10.3.3 Expected log-predictive density

Our running goal in Section 10.3 has been to evaluate the posterior predictive accuracy of a Bayesian regression model. Of interest is the compatibility of Y_{new}, the yet unobserved ridership on a future day, with its corresponding posterior predictive model. This model is specified by pdf

$$f(y'|\bar{y}) = \int \int \int f(y'|\beta_0,\beta_1,\sigma)\, f(\beta_0,\beta_1,\sigma|y)d\beta_0 d\beta_1 d\sigma$$

where y' represents a *possible* Y_{new} value and $\bar{y} = (y_1, y_2, ..., y_n)$ represents our original $n = 500$ ridership observations. Two hypothetical posterior predictive pdfs of Y_{new} are plotted in Figure 10.10. Though the eventual observed value of Y_{new}, y_{new}, is squarely within the range of both pdfs, it's closer to the posterior predictive mean in Scenario 1 (left) than in Scenario 2 (right). That is, Y_{new} is more compatible with Scenario 1.

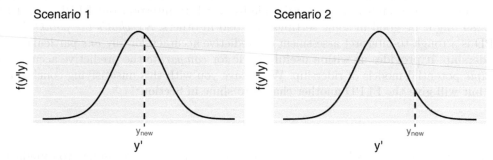

FIGURE 10.10: Two hypothetical posterior predictive pdfs for Y_{new}, the yet unobserved ridership on a new day. The eventual observed value of Y_{new}, y_{new}, is represented by a dashed vertical line.

Also notice that the posterior predictive pdf is relatively *higher* at y_{new} in Scenario 1 than in Scenario 2, providing more evidence in favor of Scenario 1. In general, the *greater* the posterior predictive pdf evaluated at y_{new}, $f(y_{new}|\bar{y})$, the more *accurate* the posterior prediction of Y_{new}. Similarly, the greater the *logged* pdf at y_{new}, $\log(f(y_{new}|\bar{y}))$, the more accurate the posterior prediction. With this, we present you with a final numerical assessment of posterior predictive accuracy.

Expected log-predictive density (ELPD)

ELPD measures the average log posterior predictive pdf, $\log(f(y_{new}|\bar{y}))$, across all possible new data points y_{new}. The higher the ELPD, the better. Higher ELPDs indicate greater posterior predictive accuracy when using our model to predict *new* data points.

The `loo()` function in the **rstanarm** package utilizes leave-one-out cross-validation to estimate the ELPD of a given model:

```
model_elpd <- loo(bike_model)
model_elpd$estimates
          Estimate       SE
elpd_loo   -4289.0  13.1186
p_loo          2.5   0.1638
looic       8578.1  26.2371
```

From the `loo()` output, we learn that the ELPD for our Bayesian Normal regression model of bike ridership is -4289. *But what does this mean?!?* Per our earlier warning, this final approach to evaluating posterior predictions is the most technical. It isn't easy to interpret. As is true of any pdf, the ELPD scale can be different in each analysis. Thus, though *the higher the better*, there's not a general scale by which to interpret our ELPD – an ELPD of -4289 might be good in some settings and bad in others. So *what's the point?* Though ELPD is a tough-to-interpret assessment of predictive accuracy for any *one* particular model of ridership, it provides us with a useful metric for *comparing* the predictive accuracies of *multiple* potential models of ridership. We'll leave you with this unsatisfying conclusion for now but will give the ELPD another chance to shine in Section 11.5.

10.3.4 Improving posterior predictive accuracy

With all of this talk of *evaluating* a model's posterior predictive accuracy, you might have questioned how to *improve* it. Again, we can provide some common principles, but no rule will apply to all situations:

1. **Collect more data.** Though gathering more data can't change an underlying weak relationship between Y and X, it *can* improve our *model* or understanding of this relationship, and hence the model's posterior predictions.

2. **Use different or more predictors.** If the goal of your Bayesian analysis is to accurately predict Y, you can play around with different predictors. For example, humidity level might be better than temperature at predicting ridership. Further, you will learn in Chapter 11 that we needn't restrict our models to include just *one* predictor. Rather, we might improve our model and its predictions of Y by including 2 or more predictors (e.g., we might model ridership by *both* temperature and humidity). Though throwing more predictors into a model is a common and enticing strategy, you'll also learn that there are limits to its effectiveness.

10.4 How good is the MCMC simulation vs how good is the model?

In Chapters 6 and 10, we've learned two big questions to ask ourselves when performing a Bayesian analysis: How good is the *MCMC simulation* and how good is the *model*? These are two different questions. The first question is concerned with how well our MCMC simulation *approximates* the model. It ponders whether our MCMC simulation is long enough and mixes quickly enough to produce a reliable picture of the model. The second question is concerned with the fitness of our *model*. It ponders whether the assumptions of our model are reasonable, whether the model is fair, and whether it produces good predictions. Ideally, the answer to both questions is *yes*. Yet this isn't always the case. We might have a good model framework, but an unstable chain that produces inaccurate approximations of this model. Or, we might do a good job of approximating a bad model. Please keep this in mind moving forward.

10.5 Chapter summary

In Chapter 10, you learned how to rigorously evaluate a Bayesian regression model:

1. The determination of whether a model is **fair** is context dependent and requires careful attention to the potential implications of your analysis, from the data collection to the conclusions.

2. To determine how **wrong** a model is, we can conduct a posterior predictive check using `pp_check()`. If Y values simulated from the Bayesian model are similar in feature to the original Y data, we have reason to believe that our model assumptions are reasonable.

3. To determine our model's posterior predictive **accuracy**, we can calculate **posterior predictive summaries** of (a) the typical distance between observed Y values and their posterior predictive means Y'; and (b) the frequency with which observed Y values fall within the scope of their posterior predictive models. For a more honest assessment of how well our model generalizes to *new* data beyond our original sample, we can estimate these properties using **cross-validation** techniques.

10.6 Exercises

10.6.1 Conceptual exercises

Exercise 10.1 (The Big Three). When evaluating a model, what are the big three questions to ask yourself?

Exercise 10.2 (Model fairness). Give an example of a model that will *not* be fair for each of the reasons below. Your examples don't have to be from real life, but try to keep them in the realm of plausibility.

   a) How the data was collected.
   b) The purpose of the data collection.
   c) Impact of analysis on society.
   d) Bias baked into the analysis.

Exercise 10.3 (Your perspective). Everyone has a standpoint or perspective that influences how they perceive the world. Rather than pretend that our standpoint is neutral, it is important to recognize it and how it might limit our ability to perceive the potential harms or benefits of our analyses.

a)   Identify two aspects of your life experience that inform your standpoint or perspective. (Example: Bill grew up in Maryland on the east coast of the United States and he is 45 years old.)
b)   Identify how those two aspects of your life experience might *limit* your evaluation of future analyses.
c)   Identify how those two aspects of your life experience might *benefit* your evaluation of future analyses.

Exercise 10.4 (Neutrality?). There are several instances in which data scientists might falsely consider themselves neutral. In this exercise, you will practice how to challenge the false idea of neutrality, an underappreciated skill for data scientists.

a)   How would you respond if your colleague were to tell you "I'm just a neutral observer, so there's no bias in my data analysis"?
b)   Your colleague now admits that they are not *personally* neutral, but they say "my model is neutral." How do you respond to your colleague now?
c)   Give an example of when your personal experience or perspective has informed a data analysis.

Exercise 10.5 (That famous quote). George Box famously said: "All models are wrong, but some are useful." Write an explanation of what this quote means so that one of your non-statistical friends can understand.

Exercise 10.6 (Assumptions). Provide 3 assumptions of the Normal Bayesian linear regression model with $Y_i | \beta_0, \beta_1, \sigma \stackrel{ind}{\sim} N(\mu_i, \sigma^2)$ where $\mu_i = \beta_0 + \beta_1 X_i$.

Exercise 10.7 (Mini posterior predictive check). Suppose we have a small dataset where predictor X has values $\vec{x} = (12, 10, 4, 8, 6)$, and response variable Y has values $\vec{y} = (20, 17, 4, 11, 9)$. Based on this data, we built a Bayesian linear regression model of Y vs X.

a)   In our first simulated parameter set, $\left(\beta_0^{(1)}, \beta_1^{(1)}, \sigma^{(1)} \right) = (-1.8, 2.1, 0.8)$. Explain how you would use these values, combined with the data, to generate a prediction for Y_1.
b)   Using the first parameter set, generate predictions for $(Y_1, Y_2, \ldots, Y_5)$. Comment on the difference between the predictions and the observed values $\vec{y}$.

Exercise 10.8 (Explain to a friend: posterior predictive check). Congratulations! You have just completed a posterior predictive check for your regression model. Your friend Shefali has a lot of questions about this. Explain the following to Shefali in plain language:

a)   The goal of a posterior predictive check.
b)   How to interpret the posterior predictive check results.

Exercise 10.9 (Explain to a friend: posterior predictive summary). Shefali really appreciated your explanations from the previous exercise. She now wants you to explain posterior predictive summaries. Explain the following in plain language:

a) What the median absolute error tells us about your model.

b) What the scaled median absolute error is, and why it might be an improvement over median absolute error.

c) What the within-50 statistic tells us about your model.

Exercise 10.10 (Posterior predictive checks).

a) In `pp_check()` plots, what does the darker density represent? What does a single lighter-colored density represent?

b) If our model fits well, describe how its `pp_check()` will appear. Explain why a good fitting model will produce a plot like this.

c) If our model fits poorly, describe how its `pp_check()` might appear.

Exercise 10.11 (Cross-validation and tacos). Recall this example from the chapter: *Suppose you want to open a new taco stand. You build all of your recipes around Reem, your friend who prefers that anchovies be a part of every meal. You test your latest "anchov-ladas" dish on her and it's a hit.*

a) What is the "data" in this analogy?

b) What is the "model" in this analogy?

c) How could you use cross-validation to evaluate a new taco recipe?

d) Why would cross-validation help you develop a successful recipe?

Exercise 10.12 (Cross-validation).

a) What are the four steps for the k-fold cross-validation algorithm?

b) What problems can occur when you use the same exact data to train *and* test a model?

c) What questions do you have about k-fold cross-validation?

10.6.2 Applied exercises

Some people take their coffee very seriously. And not every coffee bean is of equal quality. In the next set of exercises, you'll model a bean's rating (on a 0–100 scale) by grades on features such as its aroma and aftertaste using the `coffee_ratings` data in the **bayesrules** package. This data was originally processed by James LeDoux (@jmzledoux) and distributed through the R for Data Science (2020a) TidyTuesday project:

```
library(bayesrules)
data("coffee_ratings")
coffee_ratings <- coffee_ratings %>%
  select(farm_name, total_cup_points, aroma, aftertaste)
```

Exercise 10.13 (Getting started with coffee ratings). Before doing any modeling, let's get to know the `coffee_ratings` data.

a) The `coffee_ratings` data includes ratings and features of 1339 different batches of beans grown on 571 different farms. Explain why using this data to model ratings (`total_cup_points`) by `aroma` or `aftertaste` likely violates the *independence* assumption of the Bayesian linear regression model. NOTE: Check out the `head()` of the dataset.

b)   You'll learn how to utilize this type of data in Unit 4. But solely for the purpose of simplifying things here, take just one observation per farm. Use this `new_coffee` data for the remaining exercises.

```
set.seed(84735)
new_coffee <- coffee_ratings %>%
    group_by(farm_name) %>%
    sample_n(1) %>%
    ungroup()
dim(new_coffee)
[1] 572    4
```

Exercise 10.14 (Coffee ratings: model it). In this exercise you will build a Bayesian Normal regression model of a coffee bean's rating (Y) by its aroma grade (X) with $\mu = \beta_0 + \beta_1 X$. In doing so, assume that our only prior understanding is that the average cup of coffee has a 75-point rating, though this might be anywhere between 55 and 95. Beyond that, utilize weakly informative priors.

a)   Plot and discuss the relationship between a coffee's rating (`total_cup_points`) and its `aroma` grade (the higher the better).
b)   Use `stan_glm()` to simulate the Normal regression posterior model.
c)   Provide visual and numerical posterior summaries for the `aroma` coefficient β_1.
d)   Interpret the posterior median of β_1.
e)   Do you have significant posterior evidence that, the better a coffee bean's aroma, the higher its rating tends to be? Explain.

Exercise 10.15 (Coffee ratings: Is it wrong?). Before putting too much stock into your regression analysis, step back and consider whether it's *wrong*.

a)   Your posterior simulation contains multiple sets of posterior plausible parameter sets, $(\beta_0, \beta_1, \sigma)$. Use the *first* of these to simulate a sample of 572 new coffee ratings from the observed aroma grades.
b)   Construct a density plot of your simulated sample and superimpose this with a density plot of the actual observed `total_cup_points` data. Discuss.
c)   Think bigger. Use `pp_check()` to implement a more complete posterior predictive check.
d)   Putting this together, do you think that assumptions 2 and 3 of the Normal regression model are reasonable? Explain.

Exercise 10.16 (Coffee ratings: Are the posterior predictions accurate? (Part 1)). Next, let's explore how well our posterior model predicts coffee bean ratings.

a)   The first batch of coffee beans in `new_coffee` has an `aroma` grade of 7.67. Without using `posterior_predict()`, simulate and plot a posterior predictive model for the rating of this batch.

b) In reality, this batch of beans had a rating of 84. Without using `prediction_summary()`, calculate and interpret two measures of the posterior predictive error for this batch: both the raw and standardized error.

c) To get a sense of the posterior predictive accuracy for all batches in `new_coffee`, construct and discuss a `ppc_intervals()` plot.

d) How many batches have ratings that are within their 50% posterior prediction interval? (Answer this using R code; don't try to visually count it up!)

Exercise 10.17 (Coffee ratings: Are the posterior predictions accurate? (Part 2)).

a) Use `prediction_summary_cv()` to obtain 10-fold cross-validated measurements of our model's posterior predictive quality.

b) Interpret each of the four cross-validated metrics reported in part a.

c) Verify the reported cross-validated MAE using information from the 10 folds.

Exercise 10.18 (Coffee ratings: Is it fair?). Is our coffee bean analysis fair?

Exercise 10.19 (Coffee ratings now with aftertaste). Aroma isn't the only possible predictor of a coffee bean's rating. What if, instead, we were to predict rating by a bean's aftertaste? In exploring this relationship, continue to utilize the same prior models.

a) Use `stan_glm()` to simulate the Normal regression posterior model of `total_cup_points` by `aftertaste`.

b) Produce a quick plot to determine whether this model is wrong.

c) Obtain 10-fold cross-validated measurements of this model's posterior predictive quality.

d) Putting it all together, if you could only pick one predictor of coffee bean ratings, would it be `aroma` or `aftertaste`? Why?

10.6.3 Open-ended exercises

Exercise 10.20 (Open-ended: more weather). In this exercise you will use the `weather_perth` data in the `bayesrules` package to explore the Normal regression model of the maximum daily temperature (`maxtemp`) by the minimum daily temperature (`mintemp`) in Perth, Australia. You can either tune or utilize weakly informative priors.

a) Fit the model using `stan_glm()`.

b) Summarize your posterior understanding of the relationship between `maxtemp` and `mintemp`.

c) Evaluate your model and summarize your findings.

Exercise 10.21 (Open-ended: more bikes). In this exercise you will use the `bikes` data in the `bayesrules` package to explore the Normal regression model of `rides` by `humidity`. You can either tune or utilize weakly informative priors.

a) Fit the model using `stan_glm()`.

b) Summarize your posterior understanding of the relationship between `rides` and `humidity`.

c) Evaluate your model and summarize your findings.

11

Extending the Normal Regression Model

The Bayesian simple Normal linear regression model in Chapters 9 and 10 is like a bowl of plain rice – delicious on its own, while also providing a strong foundation to build upon. In Chapter 11, you'll consider extensions of this model that greatly expand its flexibility in broader modeling settings. You'll do so in an analysis of weather in Australia, the ultimate goal being to predict how hot it will be in the *afternoon* based on information we have in the *morning*. We have a vague prior understanding here that, on an average day in Australia, the typical afternoon temperature is somewhere between 15 and 35 degrees Celsius. Yet beyond this baseline assumption, we are unfamiliar with Australian weather patterns, thus we will utilize weakly informative priors throughout our analysis.

To build upon our weak prior understanding of Australian weather, we'll explore the `weather_WU` data in the **bayesrules** package. This wrangled subset of the `weatherAUS` data in the `rattle` package (Williams, 2011) contains 100 days of weather data for each of two Australian cities: Uluru and Wollongong. A call to `?weather_WU` from the console will pull up a detailed codebook.

```
# Load some packages
library(bayesrules)
library(rstanarm)
library(bayesplot)
library(tidyverse)
library(broom.mixed)
library(tidybayes)

# Load the data
data(weather_WU)
weather_WU %>%
  group_by(location) %>%
  tally()
# A tibble: 2 x 2
  location      n
  <fct>     <int>
1 Uluru       100
2 Wollongong  100
```

To simplify things, we'll retain only the variables on afternoon temperatures (`temp3pm`) and a subset of possible predictors that we'd have access to in the morning:

DOI: 10.1201/9780429288340-11

```
weather_WU <- weather_WU %>%
  select(location, windspeed9am, humidity9am, pressure9am, temp9am, temp3pm)
```

Let's begin our analysis with the familiar: a simple Normal regression model of `temp3pm` with one *quantitative* predictor, the morning temperature `temp9am`, both measured in degrees Celsius. As you might expect, there's a positive association between these two variables – the warmer it is in the morning, the warmer it tends to be in the afternoon:

```
ggplot(weather_WU, aes(x = temp9am, y = temp3pm)) +
  geom_point(size = 0.2)
```

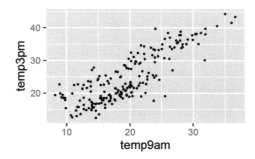

FIGURE 11.1: A scatterplot of 3 p.m. versus 9 a.m. temperatures, in degrees Celsius, collected in two Australian cities.

To *model* this relationship, let Y_i denote the 3 p.m. temperature and X_{i1} denote the 9 a.m. temperature on a given day i. Notice that we're representing our predictor by X_{i1} here, instead of simply X_i, in order to distinguish it from other predictors used later in the chapter. Then the Bayesian Normal regression model of Y by X_1 is represented by (11.1):

$$
\begin{aligned}
Y_i | \beta_0, \beta_1, \sigma &\stackrel{ind}{\sim} N\left(\mu_i, \sigma^2\right) \quad \text{with} \quad \mu_i = \beta_0 + \beta_1 X_{i1} \\
\beta_{0c} &\sim N\left(25, 5^2\right) \\
\beta_1 &\sim N\left(0, 3.1^2\right) \\
\sigma &\sim \text{Exp}(0.13).
\end{aligned}
\tag{11.1}
$$

Consider the independent priors utilized by this model:

- Since 0-degree mornings are rare in Australia, it's difficult to state our prior understanding of the typical afternoon temperature on such a rare day, β_0. Instead, the Normal prior model on the **centered intercept** β_{0c} reflects our prior understanding that the average afternoon temperature on a *typical* day is somewhere between 15 and 35 degrees ($25 \pm 2*5$).
- The **weakly informative priors** for β_1 and σ are auto-scaled by `stan_glm()` to reflect our lack of prior information about Australian weather, as well as *reasonable* ranges for these parameters based on the simple scales of our temperature data.
- The fact that the Normal prior for β_1 is centered around 0 reflects a default, conservative prior assumption that the relationship between 3 p.m. and 9 a.m. temperatures might be positive ($\beta_1 > 0$), negative ($\beta_1 < 0$), or even non-existent ($\beta_1 = 0$).

We simulate the model posterior below and encourage you to follow this up with a check of the prior model specifications and some MCMC diagnostics (which all look good!):

```
weather_model_1 <- stan_glm(
  temp3pm ~ temp9am,
  data = weather_WU, family = gaussian,
  prior_intercept = normal(25, 5),
  prior = normal(0, 2.5, autoscale = TRUE),
  prior_aux = exponential(1, autoscale = TRUE),
  chains = 4, iter = 5000*2, seed = 84735)

# Prior specification
prior_summary(weather_model_1)

# MCMC diagnostics
mcmc_trace(weather_model_1, size = 0.1)
mcmc_dens_overlay(weather_model_1)
mcmc_acf(weather_model_1)
neff_ratio(weather_model_1)
rhat(weather_model_1)
```

The simulation results provide finer insight into the association between afternoon and morning temperatures. Per the 80% credible interval for β_1, there's an 80% posterior probability that for every one degree increase in temp9am, the average increase in temp3pm is somewhere between 0.98 and 1.1 degrees. Further, per the 80% credible interval for standard deviation σ, this relationship is fairly strong – observed afternoon temperatures tend to fall somewhere between only 3.87 and 4.41 degrees from what we'd expect based on the corresponding morning temperature.

```
# Posterior credible intervals
posterior_interval(weather_model_1, prob = 0.80)
                10%    90%
(Intercept)  2.9498 5.449
temp9am      0.9803 1.102
sigma        3.8739 4.409
```

But is this a "good" model? We'll more carefully address this question in Section 11.5. For now, we'll leave you with a quick pp_check(), which illustrates that *we can do better* (Figure 11.2). Though the 50 sets of afternoon temperature data simulated from the weather_model_1 posterior (light blue) tend to capture the general center and spread of the afternoon temperatures we actually observed (dark blue), *none* capture the bimodality in these temperatures. That is, none reflect the fact that there's a batch of temperatures around 20 degrees and another batch around 35 degrees.

```
pp_check(weather_model_1)
```

This comparison of the observed and posterior simulated temperature data indicates that, though temp9am contains *some* useful information about temp3pm, it doesn't tell us *everything*. Throughout Chapter 11 we'll expand upon weather_model_1 in the hopes of improving our

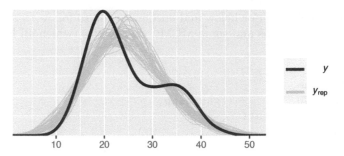

FIGURE 11.2: 50 posterior simulated sets of temperature data (light blue) alongside the actual observed temperature data (dark blue).

model of afternoon temperatures. The main idea is this. If we're in Australia at 9 a.m., the current temperature isn't the *only* factor that might help us predict 3 p.m. temperature. Our model might be improved by incorporating information about our location or humidity or air pressure and so on. It might be further improved by incorporating *multiple* predictors in the same model!

◎ **Goals**

- Extend the Normal linear regression model of a quantitative response variable Y to settings in which we have:
    - a **categorical** predictor X;
    - **multiple** predictors $(X_1, X_2, ..., X_p)$; or
    - predictors which **interact**.
- Compare and evaluate competing models of Y to answer the question, "Which model should I choose?"
- Consider the **bias-variance trade-off** in building a regression model of Y with multiple predictors.

11.1 Utilizing a categorical predictor

Again, suppose we find ourselves in Australia at 9 a.m. Our exact location can shed some light on what temperature we can expect at 3 p.m. Density plots of the temperatures in both locations indicate that it tends to be colder in the southeastern coastal city of Wollongong than in the desert climate of Uluru:

```
ggplot(weather_WU, aes(x = temp3pm, fill = location)) +
  geom_density(alpha = 0.5)
```

Given the useful information it provides, we're tempted to incorporate `location` into our temperature analysis. We can! But in doing so, it's important to recognize that `location` would be a **categorical predictor variable**. We are either in `Uluru` or `Wollongong`. Though this impacts the details, we can model the relationship between `temp3pm` and `location` much as we modeled the relationship between `temp3pm` and the quantitative `temp9am` predictor.

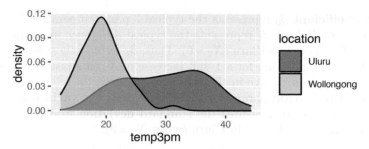

FIGURE 11.3: Density plots of afternoon temperatures in Uluru and Wollongong.

11.1.1 Building the model

For day i, let Y_i denote the 3 p.m. temperature and X_{i2} be an **indicator** for the location being Wollongong. That is, X_{i2} is 1 if we're in Wollongong and 0 otherwise:

$$X_{i2} = \begin{cases} 1 & \text{Wollongong} \\ 0 & \text{otherwise (i.e., Uluru).} \end{cases}$$

Since X_{i2} is 0 if we're in Uluru, we can think of Uluru as the **baseline** or **reference level** of the `location` variable. (Uluru is the default reference level since it's first alphabetically.) Then the **Bayesian Normal regression model** between 3 p.m. temperature and location has the same structure as (11.1), with a $N(25, 5^2)$ prior model for the centered intercept β_{0c} and weakly informative prior models for β_1 and σ tuned by `stan_glm()`:

$$\begin{aligned} \text{data:} \quad & Y_i | \beta_0, \beta_1, \sigma \overset{ind}{\sim} N(\mu_i, \sigma^2) \quad \text{with} \quad \mu_i = \beta_0 + \beta_1 X_{i2} \\ \text{priors:} \quad & \beta_{0c} \sim N\left(25, 5^2\right) \\ & \beta_1 \sim N\left(0, 38^2\right) \\ & \sigma \sim \text{Exp}(0.13). \end{aligned} \qquad (11.2)$$

Though the model *structure* is unchanged, the *interpretation* is impacted by the categorical nature of X_{i2}. The model mean $\mu_i = \beta_0 + \beta_1 X_{i2}$ no longer represents a *line* with intercept β_0 and slope β_1. In fact, $\mu_i = \beta_0 + \beta_1 X_{i2}$ more simply describes the typical 3 p.m. temperature under only two scenarios. **Scenario 1:** For Uluru, $X_{i2} = 0$ and the typical 3 p.m. temperature simplifies to

$$\beta_0 + \beta_1 \cdot 0 = \beta_0.$$

Scenario 2: For Wollongong, $X_{i2} = 1$ and the typical 3 p.m. temperature is

$$\beta_0 + \beta_1 \cdot 1 = \beta_0 + \beta_1.$$

That is, the typical 3 p.m. temperature in Wollongong is equal to that in Uluru (β_0) plus some *adjustment* or tweak β_1. Putting this together, we can reinterpret the model parameters as follows:

- **Intercept coefficient** β_0 represents the typical 3 p.m. temperature in Uluru ($X_2 = 0$).
- **Wollongong coefficient** β_1 represents the typical *difference* in 3 p.m. temperature in Wollongong ($X_2 = 1$) *versus* Uluru ($X_2 = 0$). Thus, β_1 *technically* still measures the typical change in 3 p.m. temperature for a 1-unit increase in X_2, it's just that this increase from 0 to 1 corresponds to a change in the location category.
- The **standard deviation** parameter σ still represents the **variability** in Y at a given X_2 value. Thus, here, σ measures the standard deviation in 3 p.m. temperatures in Wollongong (when $X_2 = 1$) and in Uluru (when $X_2 = 0$).

We can now interpret our prior models in (11.2) with this in mind. First, the Normal prior model on the **centered intercept** β_{0c} reflects a prior understanding that the average afternoon temperature in *Uluru* is somewhere between 15 and 35 degrees. Further, the **weakly informative** Normal prior for β_1 is centered around 0, reflecting a default, conservative prior assumption that the average 3 p.m. temperature in Wollongong might be greater ($\beta_1 > 0$), less ($\beta_1 < 0$), or even no different ($\beta_1 = 0$) from that in Uluru. Finally, the **weakly informative** prior for σ expresses our lack of understanding about the degree to which 3 p.m. temperatures vary at either location.

11.1.2   Simulating the posterior

To simulate the posterior model of (11.2) with weakly informative priors, we run four parallel Markov chains, each of length 10,000. Other than changing the formula argument from `temp3pm ~ temp9am` to `temp3pm ~ location`, the syntax is equivalent to that for `weather_model_1`:

```
weather_model_2 <- stan_glm(
  temp3pm ~ location,
  data = weather_WU, family = gaussian,
  prior_intercept = normal(25, 5),
  prior = normal(0, 2.5, autoscale = TRUE),
  prior_aux = exponential(1, autoscale = TRUE),
  chains = 4, iter = 5000*2, seed = 84735)
```

Trace plots and autocorrelation plots (omitted here for space), as well as density plots (Figure 11.4) suggest that our posterior simulation has sufficiently stabilized:

```
# MCMC diagnostics
mcmc_trace(weather_model_2, size = 0.1)
mcmc_dens_overlay(weather_model_2)
mcmc_acf(weather_model_2)
```

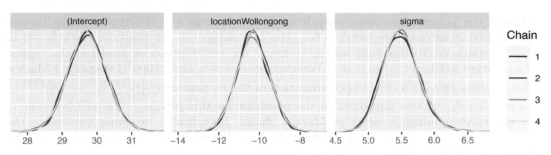

FIGURE 11.4: Four parallel Markov chain approximations of the `weather_model_2` posterior models.

These density plots and the below numerical posterior summaries for β_0 (`Intercept`), β_1 (`locationWollongong`), and σ (`sigma`) reflect our posterior understanding of 3 p.m. temperatures in Wollongong and Uluru. Consider the message of the posterior median values for β_0 and β_1: the typical 3 p.m. temperature is around 29.7 degrees in **Uluru** and, *comparatively*, around 10.3 degrees *lower* in **Wollongong**. Combined then, we can say that the typical 3 p.m. temperature in Wollongong is around 19.4 degrees (29.7 - 10.3). For context, Figure 11.5 frames these posterior median estimates of 3 p.m. temperatures in Uluru and Wollongong among the observed data.

```
# Posterior summary statistics
tidy(weather_model_2, effects = c("fixed", "aux"),
    conf.int = TRUE, conf.level = 0.80) %>%
  select(-std.error)
# A tibble: 4 x 4
  term               estimate conf.low conf.high
  <chr>                 <dbl>    <dbl>     <dbl>
1 (Intercept)            29.7     29.0      30.4
2 locationWollongong    -10.3    -11.3     -9.30
3 sigma                  5.48      5.14      5.86
4 mean_PPD               24.6     23.9      25.3
```

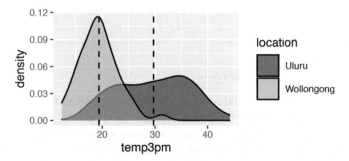

FIGURE 11.5: Density plots of afternoon temperatures in Uluru and Wollongong, with posterior median estimates of temperature (dashed lines).

Though the posterior medians provide a quick comparison of the typical temperatures in our two locations, they don't reflect the full picture or our posterior *uncertainty* in this comparison. To this end, we can compare the *entire* posterior model of the typical temperature in Uluru (β_0) to that in Wollongong ($\beta_0 + \beta_1$). The 20,000 (Intercept) Markov chain values provide a direct approximation of the β_0 posterior, $\left\{ \beta_0^{(1)}, \beta_0^{(2)}, \ldots, \beta_0^{(20000)} \right\}$. As for $\beta_0 + \beta_1$, we can approximate the posterior for this *function* of model parameters by the corresponding *function* of Markov chains. Specifically, we can approximate the $\beta_0 + \beta_1$ posterior using the chain produced by summing each pair of (Intercept) (β_0) and locationWollongong (β_1) chain values:

$$\left\{ \beta_0^{(1)} + \beta_1^{(1)}, \beta_0^{(2)} + \beta_1^{(2)}, \ldots, \beta_0^{(20000)} + \beta_1^{(20000)} \right\}.$$

The result below indicates that the typical Wollongong temperature is likely between 17 and 22 degrees, substantially below that in Uluru which is likely between 27 and 32 degrees.

```
as.data.frame(weather_model_2) %>%
  mutate(uluru = `(Intercept)`,
         wollongong = `(Intercept)` + locationWollongong) %>%
  mcmc_areas(pars = c("uluru", "wollongong"))
```

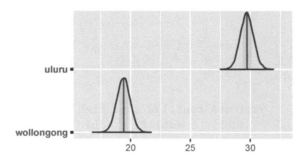

FIGURE 11.6: Posterior models of the typical 3 p.m. temperatures in Uluru and Wollongong.

11.2 Utilizing two predictors

We now know a couple of things about afternoon temperatures in Australia: (1) they're positively associated with morning temperatures and (2) they tend to be lower in Wollongong than in Uluru. This makes us think. If 9 a.m. temperature and location *both* tell us something about 3 p.m. temperature, then maybe *together* they would tell us even more! In fact, a plot of the sample data (Figure 11.7) reveals that, *no matter the location*, 3 p.m. temperature linearly increases with 9 a.m. temperature. Further, *at any given 9 a.m. temperature*, there is a consistent difference in 3 p.m. temperature in Wollongong and Uluru. Below, you will extend the **one-predictor** Normal regression models you've studied thus far to a **two-predictor** model of 3 p.m. temperature using both temp9am and location.

```
ggplot(weather_WU, aes(y = temp3pm, x = temp9am, color = location)) +
  geom_point() +
  geom_smooth(method = "lm", se = FALSE)
```

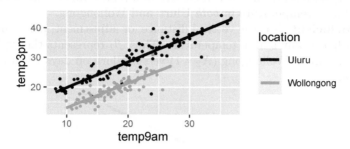

FIGURE 11.7: Scatterplot of 3 p.m. versus 9 a.m. temperatures in Uluru and Wollongong, superimposed with the observed linear relationships (solid lines).

11.2.1 Building the model

Letting X_{i1} denote the 9 a.m. temperature and X_{i2} the location on day i, the **multivariable Bayesian linear regression model** of 3 p.m. temperature versus these two predictors is a small extension of the one-predictor model (11.2). We state this model here and tune the weakly informative priors using `stan_glm()` below.

$$
\begin{aligned}
\text{data:}\quad & Y_i | \beta_0, \beta_1, \beta_2, \sigma \overset{ind}{\sim} N(\mu_i,\ \sigma^2) \quad \text{with} \quad \mu_i = \beta_0 + \beta_1 X_{i1} + \beta_2 X_{i2} \\
\text{priors:}\quad & \beta_{0c} \sim N\left(25, 5^2\right) \\
& \beta_1 \sim N\left(0, 3.11^2\right) \\
& \beta_2 \sim N\left(0, 37.52^2\right) \\
& \sigma \sim \text{Exp}(0.13).
\end{aligned}
\tag{11.3}
$$

Interpreting the 9 a.m. temperature and location coefficients in this new model, β_1 and β_2, requires some care. We can't simply interpret β_1 and β_2 as we did in our first two models, (11.1) and (11.2), when using either predictor alone. Rather, the meaning of our predictor coefficients changes depending upon the other predictors in the model. Let's again consider two scenarios. **Scenario 1:** In *Uluru*, $X_{i2} = 0$ and the relationship between 3 p.m. and 9 a.m. temperature simplifies to the following formula for a *line*:

$$
\beta_0 + \beta_1 X_{i1} + \beta_2 \cdot 0 = \beta_0 + \beta_1 X_{i1}.
$$

Scenario 2: In *Wollongong*, $X_{i2} = 1$ and the relationship between 3 p.m. and 9 a.m. temperature simplifies to the following formula for a *different* line:

$$
\beta_0 + \beta_1 X_{i1} + \beta_2 \cdot 1 = (\beta_0 + \beta_2) + \beta_1 X_{i1}.
$$

Figure 11.8 combines Scenarios 1 and 2 into one picture, thereby illuminating the meaning of our regression coefficients:

- **Intercept coefficient** β_0 is the intercept of the Uluru line. Thus, β_0 represents the typical 3 p.m. temperature in *Uluru* on a (theoretical) day in which it was *0 degrees* at 9 a.m., i.e., when $X_1 = X_2 = 0$.
- **9am temperature coefficient** β_1 is the common *slope* of the Uluru and Wollongong lines. Thus, β_1 represents the typical change in 3 p.m. temperature with every 1-degree increase in 9 a.m. temperature in *both* Uluru and Wollongong.
- **Location coefficient** β_2 is the vertical distance, or adjustment, between the Wollongong versus Uluru line at any 9 a.m. temperature. Thus, β_2 represents the typical *difference* in 3 p.m. temperature in Wollongong ($X_2 = 1$) versus Uluru on days when they have *equal 9 a.m. temperature* X_1.

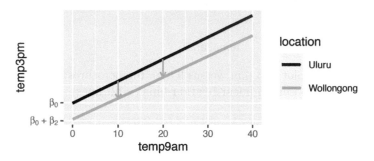

FIGURE 11.8: A graphical representation of the assumptions behind the multivariable model (11.3). The downward arrows reflect the meaning of the β_2 coefficient.

11.2.2   Understanding the priors

Now that we have a better sense for the overall structure and assumptions behind our multivariable model of 3 p.m. temperatures (11.3), let's go back and examine the meaning of our priors in this context. Beyond a prior for the centered intercept β_{0c} which reflected our basic understanding that the typical temperature on a typical day in Australia is likely between 15 and 35 degrees, our priors were weakly informative. We simulate these *priors* below using `stan_glm()` – the only difference between this and simulating the posteriors is the inclusion of `prior_PD = TRUE`. Also note that we represent the multivariable model structure by `temp3pm ~ temp9am + location`:

```
weather_model_3_prior <- stan_glm(
  temp3pm ~ temp9am + location,
  data = weather_WU, family = gaussian,
  prior_intercept = normal(25, 5),
  prior = normal(0, 2.5, autoscale = TRUE),
  prior_aux = exponential(1, autoscale = TRUE),
  chains = 4, iter = 5000*2, seed = 84735,
  prior_PD = TRUE)
```

From these priors, we then simulate and plot 100 different sets of 3 p.m. temperature data (Figure 11.9). First, notice from the left plot that our combined priors produce sets of 3 p.m. temperatures that are centered around 25 degrees (per the β_{0c} prior), yet tend to range widely, from roughly -75 to 125 degrees. This wide range is the result of our weakly informative priors and, though it spans unrealistic temperatures, it's at least in the right ballpark. After all, had we utilized *vague* instead of *weakly informative* priors, our prior simulated temperatures would span an even wider range, say on the order of *millions* of degrees Celsius.

Second, the plot at right displays our prior assumptions about the relationship between 3 p.m. and 9 a.m. temperature at each location. Per the prior model for slope β_1 being centered at 0, these model lines reflect a conservative assumption that 3 p.m. temperatures might be positively *or* negatively associated with 9 a.m. temperatures in both locations. Further, per the prior model for the Wollongong coefficient β_2 being centered at 0, the lack of a distinction among the model lines in the two locations reflects a conservative assumption that the typical 3 p.m. temperature in Wollongong might be hotter, cooler, or no different than in Uluru on days with similar 9 a.m. temperatures. In short, these prior assumptions reflect that when it comes to Australian weather, we're just not sure what's up.

```
set.seed(84735)
weather_WU %>%
  add_predicted_draws(weather_model_3_prior, n = 100) %>%
  ggplot(aes(x = .prediction, group = .draw)) +
    geom_density() +
    xlab("temp3pm")

weather_WU %>%
  add_fitted_draws(weather_model_3_prior, n = 100) %>%
  ggplot(aes(x = temp9am, y = temp3pm, color = location)) +
    geom_line(aes(y = .value, group = paste(location, .draw)))
```

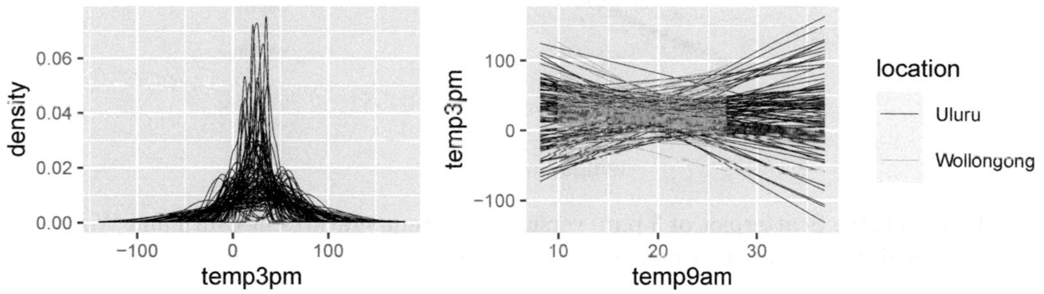

FIGURE 11.9: 100 datasets were simulated from the prior models. For each, we display a density plot of the 3 p.m. temperatures alone (left) and the relationship in 3 p.m. versus 9 a.m. temperatures by location (right).

11.2.3 Simulating the posterior

Finally, let's combine our prior assumptions with data to simulate the posterior of (11.3). Instead of starting the `stan_glm()` syntax from scratch, we can `update()` the `weather_model_3_prior` by setting `prior_PD = FALSE`:

```
weather_model_3 <- update(weather_model_3_prior, prior_PD = FALSE)
```

This simulation produces 20,000 posterior plausible sets of regression parameters β_0 (Intercept), 9 a.m. temperature coefficient β_1 (temp9am), location coefficient β_2 (locationWollongong), and regression standard deviation σ (sigma):

```
head(as.data.frame(weather_model_3), 3)
  (Intercept) temp9am locationWollongong sigma
1       13.05  0.8017            -7.663 2.392
2       12.73  0.8174            -7.839 2.445
3       11.81  0.8615            -7.648 2.414
```

Combined, these 20,000 parameter sets present us with 20,000 posterior plausible relationships between temperature and location. The 100 relationships plotted below provide a representative glimpse. Across *all* 100 scenarios, 3 p.m. temperature is positively associated with 9 a.m. temperature and tends to be higher in Uluru than in Wollongong. Further, relative to the prior simulated relationships in Figure 11.9, these posterior relationships are very consistent – we have a much clearer understanding of 3 p.m. temperatures in light of information from the `weather_WU` data.

```
weather_WU %>%
  add_fitted_draws(weather_model_3, n = 100) %>%
  ggplot(aes(x = temp9am, y = temp3pm, color = location)) +
    geom_line(aes(y = .value, group = paste(location, .draw)), alpha = .1) +
    geom_point(data = weather_WU, size = 0.1)
```

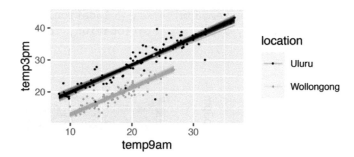

FIGURE 11.10: Scatterplot of 3 p.m. versus 9 a.m. temperatures in Uluru and Wollongong, superimposed with 100 posterior plausible models.

Inspecting the parameters themselves provides necessary details. In both Uluru and Wollongong, i.e., **when controlling for location**, there's an 80% chance that for every one degree increase in 9 a.m. temperature, the typical increase in 3 p.m. temperature is somewhere between 0.82 and 0.89 degrees. **When controlling for 9 a.m. temperature**, there's an 80% chance that the typical 3 p.m. temperature in Wollongong is somewhere between 6.6 and 7.51 degrees *lower* than that in Uluru.

```
# Posterior summaries
posterior_interval(weather_model_3, prob = 0.80,
                   pars = c("temp9am", "locationWollongong"))
                        10%      90%
temp9am               0.8196   0.8945
locationWollongong   -7.5068  -6.5999
```

11.2.4 Posterior prediction

Next, let's *use* this model to predict 3 p.m. temperature on specific days. For example, consider a day in which it's 10 degrees at 9 a.m. in both Uluru and Wollongong. To simulate and subsequently plot the posterior predictive models of 3 p.m. temperatures in these locations, we can call `posterior_predict()` and `mcmc_areas()`. Roughly speaking, we can anticipate 3 p.m. temperatures between 15 and 25 degrees in Uluru, and cooler temperatures between 8 and 18 in Wollongong:

```
# Simulate a set of predictions
set.seed(84735)
temp3pm_prediction <- posterior_predict(
  weather_model_3,
  newdata = data.frame(temp9am = c(10, 10),
                       location = c("Uluru", "Wollongong")))

# Plot the posterior predictive models
mcmc_areas(temp3pm_prediction) +
  ggplot2::scale_y_discrete(labels = c("Uluru", "Wollongong")) +
  xlab("temp3pm")
```

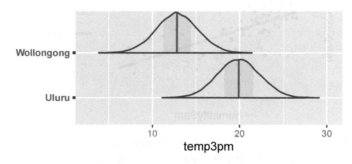

FIGURE 11.11: Posterior predictive models of the 3 p.m. temperatures in Uluru and Wollongong when the 9 a.m. temperature is 10 degrees.

11.3   Optional: Utilizing interaction terms

In the two-predictor model in Section 11.2, we assumed that the relationship between 3 p.m. temperature (Y) and 9 a.m. temperature (X_1) was the *same* in both locations (X_2). Or, visually, the relationship between Y and X_1 had *equal slopes* β_1 in both locations. This assumption isn't always appropriate – in some cases, our predictors **interact**. We'll explore interaction in this *optional* section, but you can skip right past to Section 11.4 if you choose.

Consider the relationship of 3 p.m. temperature (Y) with location (X_2) and 9 a.m. humidity (X_3). Visually, the relationships exhibit quite different slopes – afternoon temperatures and morning humidity are *negatively* associated in Uluru, yet weakly *positively* associated in Wollongong.

```
ggplot(weather_WU, aes(y = temp3pm, x = humidity9am, color = location)) +
  geom_point(size = 0.5) +
  geom_smooth(method = "lm", se = FALSE)
```

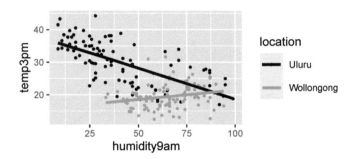

FIGURE 11.12: Scatterplot of 3 p.m. temperatures versus 9 a.m. humidity levels in Uluru and Wollongong, superimposed with the observed linear relationships (solid lines).

Think about this dynamic another way: the relationship between 3 p.m. temperature and 9 a.m. humidity *varies* by location. Equivalently, the relationship between 3 p.m. temperature and location *varies* by 9 a.m. humidity level. More technically, we say that *the location and humidity predictors* **interact**.

Interaction

Two predictors, X_1 and X_2, **interact** if the association between X_1 and Y varies depending upon the level of X_2.

11.3.1   Building the model

Consider modeling temperature (Y) by location (X_2) and humidity (X_3). If we were to assume that location and humidity do *not* interact, then we would describe their relationship by $\mu = \beta_0 + \beta_1 X_2 + \beta_2 X_3$. Yet to reflect the fact that the relationship between temperature and humidity is *modified* by location, we can incorporate a new predictor: the **interaction term**. This new predictor is simply the *product* of X_2 and X_3:

$$\mu = \beta_0 + \beta_1 X_2 + \beta_2 X_3 + \beta_3 X_2 X_3.$$

Thus, the complete structure for our **multivariable Bayesian linear regression model** *with* an **interaction term** is as follows, where the weakly informative priors on the non-intercept parameters are auto-scaled by `stan_glm()` below:

$$
\begin{aligned}
\text{data:} \quad & Y_i | \beta_0, \beta_1, \beta_2, \beta_3, \sigma \overset{ind}{\sim} N(\mu_i,\, \sigma^2) \quad \text{with} \quad \mu_i = \beta_0 + \beta_1 X_{i2} + \beta_2 X_{i3} + \beta_3 X_{i2} X_{i3} \\
\text{priors:} \quad & \beta_{0c} \sim N\left(25, 5^2\right) \\
& \beta_1 \sim N\left(0, 37.52^2\right) \\
& \beta_2 \sim N\left(0, 0.82^2\right) \\
& \beta_3 \sim N\left(0, 0.55^2\right) \\
& \sigma \sim \text{Exp}(0.13).
\end{aligned}
$$

(11.4)

To understand how our assumed structure for μ matches our observation of an interaction between 9 a.m. humidity and location, let's break this down into two scenarios, as usual. **Scenario 1:** In *Uluru*, $X_2 = 0$ and the relationship between temperature and humidity simplifies to

$$\mu = \beta_0 + \beta_2 X_3.$$

Scenario 2: In *Wollongong*, $X_2 = 1$ and the relationship between temperature and humidity simplifies to

$$\mu = \beta_0 + \beta_1 + \beta_2 X_3 + \beta_3 X_3 = (\beta_0 + \beta_1) + (\beta_2 + \beta_3) X_3.$$

Thus, we essentially have two types of regression coefficients:

- **Uluru coefficients**
  As we see in Scenario 1, intercept β_0 and humidity (slope) coefficient β_2 encode the relationship between temperature and humidity in *Uluru*.

- **Modifications to the Uluru coefficients**
  In comparing Scenario 2 to Scenario 1, the location and interaction coefficients, β_1 and β_3, encode how the relationship between temperature and humidity in Wollongong *compares* to that in Uluru. Location coefficient β_1 captures the *difference* in intercepts, and thus how the typical temperature in Wollongong compares to that in Uluru on days with 0% humidity ($X_3 = 0$). Interaction coefficient β_3 captures the *difference* in slopes, and thus how the change in temperature with humidity differs between the two cities.

11.3.2 Simulating the posterior

To simulate (11.4) using weakly informative priors, we can represent the multivariable model structure by `temp3pm ~ location + humidity9am + location:humidity9am`, the last piece of this formula representing the interaction term:

```
interaction_model <- stan_glm(
  temp3pm ~ location + humidity9am + location:humidity9am,
  data = weather_WU, family = gaussian,
  prior_intercept = normal(25, 5),
```

```
  prior = normal(0, 2.5, autoscale = TRUE),
  prior_aux = exponential(1, autoscale = TRUE),
  chains = 4, iter = 5000*2, seed = 84735)
```

Posterior summary statistics for our regression parameters $(\beta_0, \beta_1, \beta_2, \beta_3, \sigma)$ are printed below. The corresponding posterior median relationships between temperature and humidity by city reflect what we observed in the data – the association between temperature and humidity is *negative* in Uluru and slightly *positive* in Wollongong:

$$
\begin{aligned}
\text{Uluru:} \quad & \mu = 37.586 - 0.19 \text{ humidity9am} \\
\text{Wollongong:} \quad & \mu = (37.586 - 21.879) + (-0.19 + 0.246) \text{ humidity9am} \\
& = 15.707 + 0.056 \text{ humidity9am}
\end{aligned}
$$

```
# Posterior summary statistics
tidy(interaction_model, effects = c("fixed", "aux"))
# A tibble: 6 x 3
  term                              estimate std.error
  <chr>                                <dbl>     <dbl>
1 (Intercept)                           37.6     0.910
2 locationWollongong                   -21.9      2.33
3 humidity9am                         -0.190    0.0193
4 locationWollongong:humidity9am       0.246    0.0375
5 sigma                                 4.47     0.227
6 mean_PPD                              24.6     0.448
posterior_interval(interaction_model, prob = 0.80,
                pars = "locationWollongong:humidity9am")
                                      10%      90%
locationWollongong:humidity9am     0.1973   0.2941
```

The 200 posterior simulated models in Figure 11.13 capture our posterior uncertainty in these relationships and provide ample posterior evidence of a *significant* interaction between location and humidity – almost all Wollongong posterior models are *positive* and all Uluru posterior models are *negative*. We can back this up with numerical evidence. The 80% posterior credible interval for interaction coefficient β_3 is entirely and well above 0, ranging from 0.197 to 0.294, suggesting that the association between temperature and humidity is *significantly more positive* in Wollongong than in Uluru.

```
weather_WU %>%
  add_fitted_draws(interaction_model, n = 200) %>%
  ggplot(aes(x = humidity9am, y = temp3pm, color = location)) +
    geom_line(aes(y = .value, group = paste(location, .draw)), alpha = 0.1)
```

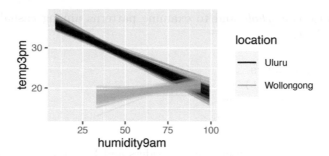

FIGURE 11.13: 200 posterior plausible relationships in 3 p.m. temperatures versus 9 a.m. humidity levels in Uluru and Wollongong.

11.3.3 Do you need an interaction term?

Now that you've seen that interaction terms allow us to capture even more nuance in a relationship, you might feel eager to include them in every model. Don't. There's nothing sophisticated about a model stuffed with unnecessary predictors. Though interaction terms *are* quite useful in some settings, they can overly complicate our models in others. Here are a few things to consider when weighing whether or not to go down the rabbit hole of interaction terms:

- **Context.** In the context of your analysis, does it make *sense* that the relationship between Y and one predictor X_1 *varies* depending upon the value of another predictor X_2?

- **Visualizations.** As with our example here, interactions might reveal themselves when visualizing the relationships between Y, X_1, and X_2.

- **Hypothesis tests.** Suppose we do include an interaction term in our model, $\mu = \beta_0 + \beta_1 X_1 + \beta_2 X_2 + \beta_3 X_1 X_2$. If there's significant posterior evidence that $\beta_3 \neq 0$, then it can make sense to include the interaction term. Otherwise, it's typically a good idea to get rid of it.

Let's practice the first two ideas with some informal examples. The `bike_users` data is from the same source as the `bikes` data in Chapter 9. Like the `bikes` data, it includes information about daily Capital Bikeshare ridership. Yet `bike_users` contains data on *both* registered, paying bikeshare members (who tend to ride more often and use the bikes for commuting) and casual riders (who tend to just ride the bikes every so often):

```
data(bike_users)
bike_users %>%
  group_by(user) %>%
  tally()
# A tibble: 2 x 2
  user          n
  <fct>     <int>
1 casual      267
2 registered  267
```

We'll use this data as a *whole* and to examine patterns among casual riders alone and registered riders alone:

```
bike_casual <- bike_users %>%
  filter(user == "casual")
bike_registered <- bike_users %>%
  filter(user == "registered")
```

To begin, let Y_c denote the number of casual riders and Y_r denote the number of registered riders on a given day. As with model (11.4), consider an analysis of ridership in which we have two predictors, one quantitative and one categorical: temperature (X_1) and weekend status (X_2). The observed relationships of Y_c and Y_r with X_1 and X_2 are shown below. Syntax is included for the former and is similar for the latter.

```
ggplot(bike_casual, aes(y = rides, x = temp_actual, color = weekend)) +
  geom_smooth(method = "lm", se = FALSE) +
  labs(title = "casual riders")
```

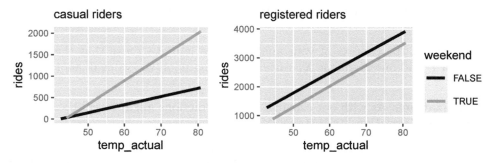

FIGURE 11.14: The observed relationships between ridership and temperature by weekend status for casual riders (left) and registered riders (right).

In their relationship with casual ridership Y_c (left), it seems that temperature and weekend status *do* interact. Ridership increases with temperature on both weekends and weekdays, yet the *rate* of this increase *differs*, being more rapid on weekends. In contrast, in their relationship with registered ridership Y_r (right), it seems that temperature and weekend status do *not* interact. Ridership increases with temperature at roughly the *same* rate on weekends and weekdays. Both observations *make sense*. Whereas registered riders tend to be more utility driven in their bikeshare use, casual riders tend to hop on for more occasional fun. Thus, casual ridership, more than registered ridership, might see disproportional increases in ridership on warm weekends than on warm weekdays.

Next, in a model of casual ridership Y_c, suppose that we have two quantitative predictors: temperature (X_1) and humidity (X_2). Figure 11.15 takes two approaches to visualizing this relationship. The left plot utilizes color, yet the fact that this relationship lives in three dimensions makes it more challenging to identify any trends. For visualization purposes then, one strategy is to "cut" one predictor (here humidity) into categories (here high and low) based on its quantitative scale (Figure 11.15 right). Though "cutting" or "discretizing" humidity greatly oversimplifies the relationship among our original variables, it does reveal a *potential* interaction between X_1 and X_2. Mainly, it appears that the increase in ridership with temperature is slightly quicker when the humidity is low. If you've ever biked in warm, humid weather, *this makes sense*. High humidity can make hot weather pretty miserable. Our plots provide evidence that other people might agree.

```
ggplot(bike_casual, aes(y = rides, x = temp_actual, color = humidity)) +
  geom_point()

ggplot(bike_casual,
       aes(y = rides, x = temp_actual,
           color = cut(humidity, 2, labels = c("low","high")))) +
  geom_smooth(method = "lm", se = FALSE) +
  labs(color = "humidity_level") +
  lims(y = c(0, 2500))
```

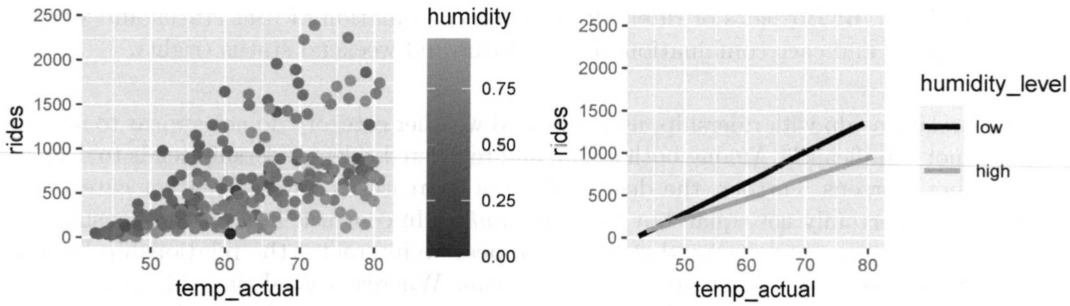

FIGURE 11.15: A scatterplot of ridership versus temperature, colored by humidity level (left). The relationships between ridership and temperature at two discretized humidity levels, low and high (right).

Finally, consider what it would mean for *categorical* predictors to interact. Let's model Y, the total ridership on any given day, by three potential predictors: user type X_1 (casual or registered), weather category X_2, and weekend status X_3. The three weather categories here range from "1," pleasant weather, to "3," more severe weather. Figure 11.16 plots Y versus X_1 and X_2 (left) and Y versus X_1 and X_3 (right). Upon examining the results, take the following quiz.

❓**Quiz Yourself!**

1. In their relationship with ridership, do user type and weather category appear to interact?
2. In their relationship with ridership, do user type and weekend status appear to interact?
3. In the context of our Capital Bikeshare analysis, explain why your answers to 1 and 2 make sense.

```
# Example syntax
ggplot(bike_users, aes(y = rides, x = user, fill = weather_cat)) +
  geom_boxplot()
```

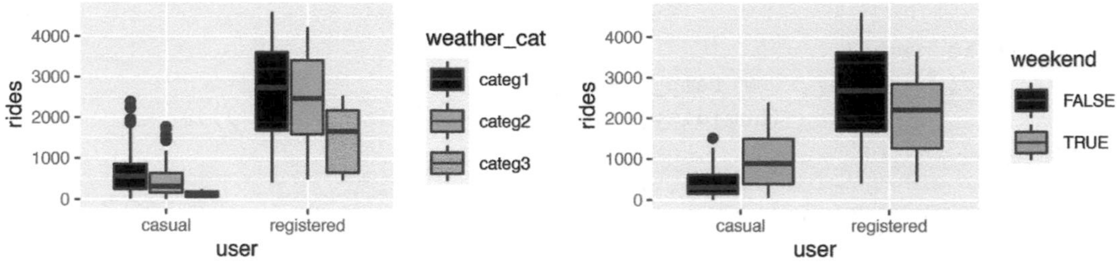

FIGURE 11.16: Boxplots of ridership for each combination of user status and weather category (left) and each combination of user status and weekend status (right).

In their relationship with ridership, user type and weather category do not appear to interact, at least not significantly. Among both casual and registered riders, ridership tends to decrease as weather worsens. Further, the degree of these decreases from one weather category to the next are certainly not equal, but they are *similar*. In contrast, in their relationship with ridership, user type and weekend status *do* appear to interact – the relationship between ridership and weekend status *varies* by user status. Whereas casual ridership is greater on weekends than on weekdays, registered ridership is greater on weekdays. Again, we might have *anticipated* this interaction given that casual and registered riders tend to use the bikeshare service to different ends.

The examples above have focused on examining interactions through visualizations and context. It remains to determine whether these interactions are actually *significant* or meaningful, thus whether we should include them in the corresponding models. To this end, we could simulate each model and formally test the significance of the interaction coefficient. In our opinion, after building up the necessary intuition, this formality is the easiest step.

11.4   Dreaming bigger: Utilizing more than 2 predictors!

Let's return to our Australian weather analysis. We can keep going! To improve our understanding and posterior predictive accuracy of afternoon temperatures, we can incorporate more and more predictors into our model. Now that you've explored models that include a quantitative predictor, a categorical predictor, and *both*, you are equipped to generalize regression models to include any number of predictors. Let's revisit the **weather_WU** data:

```
weather_WU %>%
  names()
[1] "location"     "windspeed9am" "humidity9am"
[4] "pressure9am"  "temp9am"      "temp3pm"
```

Beyond `temp9am` (X_1), `location` (X_2), and `humidity9am` (X_3), our sample data includes two more potential predictors of `temp3pm`: `windspeed9am` (km/hr) and atmospheric `pressure9am` (hpa) denoted X_4 and X_5, respectively. We'll put it all out there for our final model of Chapter 11 by including all five possible predictors of Y, without the complication of interaction terms. Simply to preserve space as our models grow, we do not write out the weakly informative priors for the regression coefficients ($\beta_1, \beta_2, \ldots, \beta_5$). These are all Normal and centered at 0, with prior standard deviations that can be obtained by the `prior_summary()` below:

$$
\begin{aligned}
\text{data:} \quad & Y_i | \beta_0, \beta_1, \ldots, \beta_5, \sigma \stackrel{ind}{\sim} N(\mu_i,\ \sigma^2) \quad \text{with} \quad \mu_i = \beta_0 + \beta_1 X_{i1} + \cdots + \beta_5 X_{i5} \\
\text{priors:} \quad & \beta_{0c} \sim N(25, 5^2) \\
& \beta_1, \ldots, \beta_5 \sim N(0, (\text{some weakly informative sd})^2) \\
& \sigma \sim \text{Exp}(0.13).
\end{aligned}
\tag{11.5}
$$

For our previous three models of `temp3pm`, we paused to define each and every model parameter. Here we have seven: ($\beta_0, \beta_1, \ldots, \beta_5, \sigma$). It's excessive and unnecessary to define each. Rather, we can turn to the *principles* that we developed throughout our first three model analyses.

Principles for interpreting multivariable model parameters

In a multivariable regression model of Y informed by p predictors ($X_1, X_2, \ldots, X_p$),

$$
\mu = \beta_0 + \beta_1 X_1 + \beta_2 X_2 + \cdots + \beta_p X_p,
$$

we can interpret the β_i regression coefficients as follows. First, the intercept coefficient β_0 represents the typical Y outcome when all predictors are 0, $X_1 = X_2 = \cdots = X_p = 0$. Further, *when controlling for* or *holding constant* the other X_j predictors in our model:

- the β_i coefficient corresponding to a **quantitative predictor** X_i can be interpreted as the typical change in Y per one unit increase in X_i;
- the β_i coefficient corresponding to a **categorical level** X_i can be interpreted as the typical *difference* in Y for level X_i versus the reference or baseline level.

As such, our interpretation of any predictor's coefficient depends in part on what other predictors are in our model.

Let's put these principles into practice in our analysis of (11.5). To simulate the model posterior, we define the model structure using the shortcut syntax `temp3pm ~ .` instead of `temp3pm ~ location + windspeed9am + (the rest of the predictors!)`. Without having to write them all out, this indicates that we'd like to use all `weather_WU` predictors in our model.[1]

---

[1]This shortcut should be used with caution! First make sure that you actually want to use every variable in the `weather_WU` data.

```
weather_model_4 <- stan_glm(
  temp3pm ~ .,
  data = weather_WU, family = gaussian,
  prior_intercept = normal(25, 5),
  prior = normal(0, 2.5, autoscale = TRUE),
  prior_aux = exponential(1, autoscale = TRUE),
  chains = 4, iter = 5000*2, seed = 84735)

# Confirm prior specification
prior_summary(weather_model_4)

# Check MCMC diagnostics
mcmc_trace(weather_model_4)
mcmc_dens_overlay(weather_model_4)
mcmc_acf(weather_model_4)
```

We can now pick through the posterior simulation results for our seven model parameters, here simplified to their corresponding 95% credible intervals:

```
# Posterior summaries
posterior_interval(weather_model_4, prob = 0.95)
                        2.5%      97.5%
(Intercept)        -23.92282   98.96185
locationWollongong  -7.19827   -5.66547
windspeed9am        -0.05491    0.02975
humidity9am         -0.05170   -0.01511
pressure9am         -0.08256    0.03670
temp9am              0.72893    0.87456
sigma                2.10497    2.57180
```

These intervals provide insight into some big picture questions. When controlling for the other model predictors, which predictors are **significantly associated** with temp3pm? Are these associations **positive or negative**? Try to answer these big picture questions for yourself in the quiz below. Though there's some gray area, one set of reasonable answers is in the footnotes.[2]

❓ Quiz Yourself!

When controlling for the other predictors included in weather_model_4, which predictors...

1. have a significant *positive* association with temp3pm?
2. have a significant *negative* association with temp3pm?
3. are *not* significantly associated with temp3pm?

---

[2]1. temp9am; 2. location, humidity9am; 3. windspeed9am, pressure9am

Let's see how you did. To begin, the 95% posterior credible interval for the `temp9am` coefficient β_1, (0.73, 0.87), is the only one that lies entirely *above* 0. This provides us with hearty evidence that, even when controlling for the four other predictors in the model, there's a significant **positive association** between morning and afternoon temperatures. In contrast, the 95% posterior credible intervals for the `locationWollongong` and `humidity9am` coefficients lie entirely *below* 0, suggesting that both factors are **negatively associated** with `temp3pm`. For example, when controlling for the other model factors, there's a 95% chance that the typical temperature in Wollongong is between 5.67 and 7.2 degrees lower than in Uluru. The `windspeed9am` and `pressure9am` coefficients are the only ones to have 95% credible intervals which straddle 0. Though both intervals lie mostly *below* 0, suggesting afternoon temperature is negatively associated with morning windspeed and atmospheric pressure when controlling for the other model predictors, the waffling evidence invites some skepticism and follow-up questions.

11.5   Model evaluation & comparison

Throughout Chapter 11, we've explored five different models of 3 p.m. temperatures in Australia. We'll consider four of these here:

Model	Formula
`weather_model_1`	`temp3pm ~ temp9am`
`weather_model_2`	`temp3pm ~ location`
`weather_model_3`	`temp3pm ~ temp9am + location`
`weather_model_4`	`temp3pm ~ .`

Naturally, you might have wondered which of these is the "best" model of `temp3pm`. To solve this mystery, we can explore our three model evaluation questions from Chapter 10:

1.   How *fair* is each model?
     Since the context for their analysis and data collection is the same, these four models are equally *fair*.

2.   How *wrong* is each model?
     Visual posterior predictive checks (Figure 11.17) suggest that the assumed structures underlying `weather_model_3` and `weather_model_4` better capture the bimodality in `temp3pm`. Thus, these models are *less wrong* than `weather_model_1` and `weather_model_2`.

3.   How *accurate* are each model's posterior predictions?
     We'll address this question below using the three approaches we learned in Chapter 10: visualization, cross-validation, and ELPD.

```
# Posterior predictive checks. For example:
pp_check(weather_model_1)
```

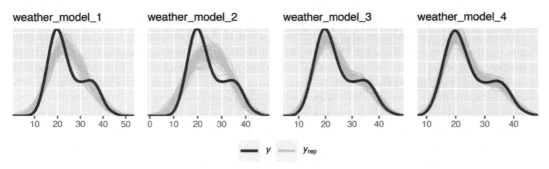

FIGURE 11.17: Posterior predictive checks for our four models of 3 p.m. temperature.

11.5.1   Evaluating predictive accuracy using visualizations

Visualizations provide a powerful first approach to assessing the quality of our models' posterior predictions. To begin, we'll use all four models to construct posterior predictive models for each case in the `weather_WU` dataset. Syntax is provided for `weather_model_1` and is similar for the other models.

```
set.seed(84735)
predictions_1 <- posterior_predict(weather_model_1, newdata = weather_WU)
```

Figure 11.18 illustrates the resulting posterior predictive models of `temp3pm` from `weather_model_1` (left) and `weather_model_2` (right). Both accurately capture the observed behavior in `temp3pm` – the majority of observed 3 p.m. temperatures fall within the bounds of their predictive models based on 9 a.m. temperatures alone (`weather_model_1`) or location alone (`weather_model_2`).

```
# Posterior predictive models for weather_model_1
ppc_intervals(weather_WU$temp3pm, yrep = predictions_1,
             x = weather_WU$temp9am, prob = 0.5, prob_outer = 0.95) +
  labs(x = "temp9am", y = "temp3pm")

# Posterior predictive models for weather_model_2
ppc_violin_grouped(weather_WU$temp3pm, yrep = predictions_2,
                   group = weather_WU$location, y_draw = "points") +
  labs(y = "temp3pm")
```

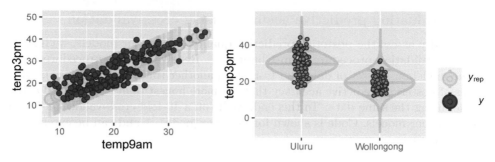

FIGURE 11.18: Posterior predictive models of 3 p.m. temperature corresponding to `weather_model_1` (left) and `weather_model_2` (right). In both plots, dark dots represent the observed 3 p.m. temperatures.

Further, not only do the `weather_model_3` posterior predictive models also well anticipate the observed 3 p.m. temperatures (Figure 11.19), they're much *narrower* than those from `weather_model_1` and `weather_model_2`. Essentially, using both `temp9am` and `location`, instead of either predictor alone, produces more precise posterior predictions of `temp3pm`.

```
# Posterior predictive models for weather_model_3
ppc_intervals_grouped(weather_WU$temp3pm, yrep = predictions_3,
                  x = weather_WU$temp9am, group = weather_WU$location,
                  prob = 0.5, prob_outer = 0.95,
                  facet_args = list(scales = "fixed")) +
  labs(x = "temp9am", y = "temp3pm")
```

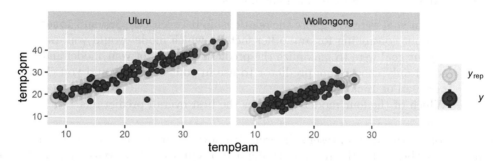

FIGURE 11.19: Posterior predictive models of 3 p.m. temperature corresponding to `weather_model_3`, alongside the observed 3 p.m. temperatures (dark dots).

What about `weather_model_4`?! Well, with five predictors, this high-dimensional model is simply too difficult to visualize. We thus have three motivations for not assessing posterior predictive accuracy using visual evidence alone: (1) we can't actually visualize the posterior predictive quality of every model; (2) even if we could, it can be difficult to ascertain the comparative predictive quality of our models from visuals alone; and (3) the visuals above illustrate how well our models predict the same data points we used to build them (not how well they'd predict temperatures on days we haven't yet observed), and thus might give overly optimistic assessments of predictive accuracy. This is where cross-validated and ELPD numerical summaries can really help.

11.5.2   Evaluating predictive accuracy using cross-validation

To numerically assess the posterior predictive quality of our four models, we'll apply the k-fold cross-validation procedure from Chapter 10. Recall that this procedure effectively trains the model on one set of data and tests it on another (multiple times over), producing estimates of posterior predictive accuracy that are more honest than if we trained and tested the model using the same data. To this end, we'll run a 10-fold cross-validation for each of our four weather models. Code is shown for `weather_model_1` and can be extended to the other three models:

```
set.seed(84735)
prediction_summary_cv(model = weather_model_1, data = weather_WU, k = 10)
```

Table 11.2 summarizes the resulting cross-validation statistics.

TABLE 11.2: Cross-validated posterior predictive summaries for four models of temp3pm.

model	mae	mae scaled	within 50	within 95
weather model 1	3.285	0.79	0.405	0.97
weather model 2	3.653	0.661	0.495	0.935
weather model 3	1.142	0.483	0.67	0.96
weather model 4	1.206	0.522	0.64	0.95

> ❓ **Quiz Yourself!**
>
> Utilize Table 11.2 to compare our four models.
>
> 1. If your goal were to explore the relationship between `temp3pm` and `location` without controlling for any other factors, which model would you use?
> 2. If your goal were to maximize the predictive quality of your model and you could only choose *one* predictor to use in your model, would you choose `temp9am` or `location`?
> 3. Which of the four models produces the best overall predictions?

Quiz questions 1 through 3 present different frameworks by which to define which of our four models is "best." By the framework defined in Question 1, `weather_model_2` is best – it's the only model which studies the exact relationship of interest (that of `temp3pm` versus `location`). By the framework defined in Question 2 in which we can only pick one predictor, `weather_model_1` seems best. That is, `temp9am` alone seems a better predictor of `temp3pm` than `location` alone. Across all test cases, the posterior mean predictions of `temp3pm` calculated from `weather_model_1` tend to be 3.3 degrees from the *observed* 3 p.m. temperature. The error is higher among the posterior mean predictions calculated from `weather_model_2` (3.7 degrees).[3]

Finally, by the framework defined by Question 3, we're assuming that we can utilize any number or combination of the available predictors. With this flexibility, `weather_model_3` appears to provide the "best" posterior predictions of `temp3pm` – it has the smallest `mae` and `mae_scaled` among the four models and among the highest `within_50` and `within_95`

---

[3]There's some nuance here. Though `weather_model_1` has a lower `mae` than `weather_model_2`, it's not best by *all* measures. However, the discrepancies aren't big enough to make us change our minds here.

coverage statistics. Let's put this into context. In utilizing both `temp9am` and `location` predictors, `weather_model_3` convincingly outperforms the models which use either predictor alone (`weather_model_1` and `weather_model_2`). Further, `weather_model_3` seems to *narrowly* outperform that which uses all five predictors (`weather_model_4`). In model building, *more isn't always better.* The fact that `weather_model_4` has higher prediction errors indicates that its additional three predictors (`windspeed9am`, `humidity9am`, `pressure9am`) don't substantially improve our understanding about `temp3pm` if we already have information about `temp9am` and `location`. Thus, in the name of simplicity and efficiency, we would be happy to pick the smaller `weather_model_3`.

11.5.3 Evaluating predictive accuracy using ELPD

In closing out this section, we'll compare the posterior predictive accuracy of our four models with respect to their **expected log-predictive densities (ELPD)**. Recall the basic idea from Section 10.3.3: the larger the expected logged posterior predictive pdf across a new set of data points y_{new}, $\log(f(y_{new}|y))$, the more accurate the posterior predictions of y_{new}. To begin, calculate the ELPD for each model:

```
# Calculate ELPD for the 4 models
set.seed(84735)
loo_1 <- loo(weather_model_1)
loo_2 <- loo(weather_model_2)
loo_3 <- loo(weather_model_3)
loo_4 <- loo(weather_model_4)

# Results
c(loo_1$estimates[1], loo_2$estimates[1],
  loo_3$estimates[1], loo_4$estimates[1])
[1] -568.4 -625.7 -461.1 -457.6
```

Though the ELPDs don't provide interpretable metrics for the posterior predictive accuracy of any *single* model, they *are* useful in *comparing* the posterior predictive accuracy of multiple models. We can do so using `loo_compare()`:

```
# Compare the ELPD for the 4 models
loo_compare(loo_1, loo_2, loo_3, loo_4)
                elpd_diff se_diff
weather_model_4    0.0      0.0
weather_model_3   -3.5      4.0
weather_model_1 -110.8     18.1
weather_model_2 -168.1     21.5
```

To begin, `loo_compare()` lists the models in order from best to worst, or from the highest ELPD to the lowest: `weather_model_4`, `weather_model_3`, `weather_model_1`, `weather_model_2`. The remaining output details the *extent* to which each model compares to the best model, `weather_model_4`. For example, the ELPD for `weather_model_3` is estimated to be 3.5 lower (worse) than that of `weather_model_4` (-461.1 - -457.6). Further, this *estimated* difference in ELPD has a corresponding **standard error** of 4.0. That is, we believe that the *true* difference in the ELPDs for `weather_model_3` and `weather_model_4` is within roughly 2 standard errors, or 8 units, of the estimated -3.5 difference.

The ELPD model comparisons are consistent with our conclusions based upon the 10-fold cross-validation comparisons. Mainly, `weather_model_1` outperforms `weather_model_2`, and `weather_model_3` outperforms both of them. Further, the distinction between `weather_model_3` and `weather_model_4` is cloudy. Though the ELPD for `weather_model_4` is slightly higher (better) than that of `weather_model_3`, it is not *significantly* higher. An ELPD difference of *zero* is within two standard errors of the estimated difference: -3.5 ± 2*4 = (-11.5, 4.5). Hence we don't have strong evidence that the posterior predictive accuracy of `weather_model_4` is significantly superior to that of `weather_model_3`, or vice versa. Again, since `weather_model_3` is *simpler* and achieves similar predictive accuracy, we'd personally choose to use it over the more complicated `weather_model_4`.

Comparing models via ELPD

Let $ELPD_1$ and $ELPD_2$ denote the estimated ELPDs for two different models, models 1 and 2. Further, let se_{diff} denote the standard error of the estimated difference in the ELPDs, $ELPD_2 - ELPD_1$. Then the posterior predictive accuracy of model 1 is *significantly* greater than that of model 2 if:

1. $ELPD_1 > ELPD_2$; and
2. $ELPD_2$ is at least two standard errors below $ELPD_1$. Equivalently, the difference in $ELPD_2$ and $ELPD_1$ is at least two standard errors below 0: $(ELPD_2 - ELPD_1) + 2se_{diff} < 0$.

11.5.4   The bias-variance trade-off

Our four weather models illustrate the fine balance we must strike in building statistical models. If our goal is to build a model that produces accurate predictions of our response variable Y, we want to include *enough* predictors so that we have ample information about Y, but not *so many* predictors that our model becomes **overfit** to our sample data. This riddle is known as the **bias-variance trade-off**. To examine this trade-off, we will take two *separate* 40-day samples of Wollongong weather:

```
# Take 2 separate samples
set.seed(84735)
weather_shuffle <- weather_australia %>%
  filter(temp3pm < 30, location == "Wollongong") %>%
  sample_n(nrow(.))
sample_1 <- weather_shuffle %>% head(40)
sample_2 <- weather_shuffle %>% tail(40)
```

We'll use both samples to separately explore the relationship between 3 p.m. temperature (Y) by day of year (X), from 1 to 365. The pattern that emerges in the `sample_1` data is what you might expect for a city in the southern hemisphere. Temperatures are higher at the beginning and end of the year, and lower in the middle of the year:

```
# Save the plot for later
g <- ggplot(sample_1, aes(y = temp3pm, x = day_of_year)) +
  geom_point()
g
```

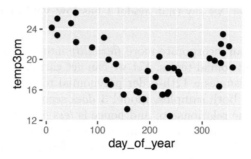

FIGURE 11.20: A scatterplot of 3 p.m. temperatures by the day of the year in Wollongong.

Consider three different **polynomial models** for the relationship between temperature and day of year:

$$
\begin{aligned}
\text{model 1:} \quad \mu &= \beta_0 + \beta_1 X \\
\text{model 2:} \quad \mu &= \beta_0 + \beta_1 X + \beta_2 X^2 \\
\text{model 3:} \quad \mu &= \beta_0 + \beta_1 X + \beta_2 X^2 + \beta_3 X^3 + \cdots + \beta_{12} X^{12}.
\end{aligned}
$$

Don't panic. We don't expect you to interpret the polynomial terms or their coefficients. Rather, we want you to take in the fact that each polynomial term, a transformation of X, is a separate predictor. Thus, model 1 has 1 predictor, model 2 has 2, and model 3 has 12. In obtaining posterior estimates of these three models, we know two things from past discussions. First, our two different data samples will produce different estimates (they're using different information!). Second, models 1, 2, and 3 vary in quality – one is better than the others. Before examining the nuances, take a quick quiz.[4]

> ❷ **Quiz Yourself!**
>
> 1. Which of models 1, 2, and 3 would be the most **variable** from sample to sample? That is, for which model will the sample 1 and 2 estimates of the relationship between temperature and day of year differ the most?
> 2. Which of models 1, 2, and 3 is the most **biased**? That is, which model will tend to be the furthest from the observed behavior in the data?

Figure 11.21 illustrates the three different models as estimated by both samples 1 and 2. For sample 1, these figures can be replicated as follows:

```
g + geom_smooth(method = "lm", se = FALSE)
g + stat_smooth(method = "lm", se = FALSE, formula = y ~ poly(x, 2))
g + stat_smooth(method = "lm", se = FALSE, formula = y ~ poly(x, 12))
```

---

[4]Answer: 1 = model 3; 2 = model 1

Let's connect what we see in this figure with some technical concepts and terminology. Starting at one extreme, model 1 assumes a linear relationship between temperature and day of year. Samples 1 and 2 produce similar estimates of this linear relationship. This stability is reassuring – no matter what data we happen to have, our posterior understanding of model 1 will be similar. However, model 1 turns out to be *overly simple* and *rigid*. It systematically underestimates temperatures on summer days and overestimates temperatures on winter days. Putting this together, we say that model 1 has **low variance** from sample to sample, but **high bias**.

To correct this bias, we can incorporate more *flexibility* into our model with the inclusion of more predictors, here polynomial terms. Yet we can get carried away. Consider the extreme case of model 3, which assumes a 12th-order polynomial relationship between temperature and day of year. Within both samples, model 3 *does* seem to be better than model 1 at following the trends in the relationship, and hence is *less biased*. However, this decrease in bias comes at a cost. Since model 3 is structured to pick up tiny, sample-specific details in the relationship, samples 1 and 2 produce quite different estimates of model 3, or two very distinct wiggly model lines. In this case, we say that model 3 has **low bias** but **high variance** from sample to sample. Utilizing this highly variable model would have two serious consequences. First, the results would be **unstable** – different data might produce very different model estimates, and hence conclusions about the relationship between temperature and day of year. Second, the results would be **overfit** to our sample data – the tiny, local trends in *our* sample likely don't extend to the general daily weather patterns in Wollongong. As a result, this model wouldn't do a good job of predicting temperatures for *future* days.

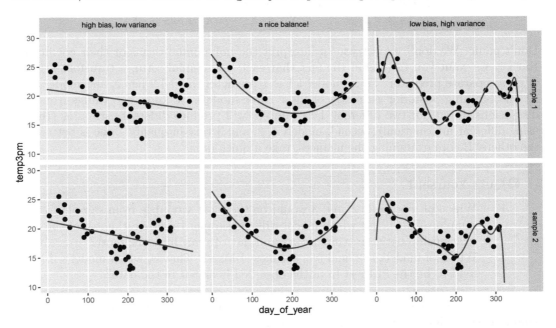

FIGURE 11.21: Sample 1 (top row) and sample 2 (bottom row) are used to model temperature by day of year under three assumptions: the relationship is linear (left), quadratic (center), or a 12th order polynomial (right).

Bringing this all together, in assuming a quadratic structure for the relationship between temperature and day of year, model 2 provides some good middle ground. It is neither too biased (simple) nor too variable (overfit). That is, it strikes a nice balance in the bias-variance trade-off.

Bias-variance trade-off

A model is said to have **high bias** if it tends to be "far" from the observed relationship in the data; and **high variance** if estimates of the model significantly differ depending upon what data is used. In model building, there are trade-offs between bias and variability:

- Overly simple models with few or no predictors tend to have *high bias* but *low variability* (high stability).
- Overly complicated models with lots of predictors tend to have *low bias* but *high variability* (low stability).

The goal is to build a model which strikes a good balance, enjoying relatively low bias *and* low variance.

Though the bias-variance trade-off speaks to the performance of a model across multiple *different* datasets, we don't actually have to go out and collect multiple datasets to hedge against the extremes of the bias-variance trade-off. Nor do we have to rely on intuition and guesswork. The raw and cross-validated posterior prediction summaries that we've utilized throughout Chapters 10 and 11 can help us identify overly simple models and overfit models. To explore this process with our three models of `temp3pm` by `day_of_year`, we first simulate the posteriors for each model. We provide complete syntax for `model_1`. The syntax for `model_2` and `model_3` is similar, but would use formulas `temp3pm ~ poly(day_of_year, 2)` and `temp3pm ~ poly(day_of_year, 12)`, respectively.

```
model_1 <- stan_glm(
  temp3pm ~ day_of_year,
  data = sample_1, family = gaussian,
  prior_intercept = normal(25, 5),
  prior = normal(0, 2.5, autoscale = TRUE),
  prior_aux = exponential(1, autoscale = TRUE),
  chains = 4, iter = 5000*2, seed = 84735)

# Ditto the syntax for models 2 and 3
model_2 <- stan_glm(temp3pm ~ poly(day_of_year, 2), ...)
model_3 <- stan_glm(temp3pm ~ poly(day_of_year, 12), ...)
```

Next, we calculate raw and cross-validated posterior prediction summaries for each model. We provide syntax for `model_1` – it's similar for `model_2` and `model_3`.

```
set.seed(84735)
prediction_summary(model = model_1, data = sample_1)
prediction_summary_cv(model = model_1, data = sample_1, k = 10)$cv
```

Check out the results in Table 11.3. By both measures, raw and cross-validated, `model_1` tends to have the greatest prediction errors. This suggests that `model_1` is **overly simple**, or possibly biased, in comparison to the other two models. At the other extreme, notice that the cross-validated prediction error for `model_3` (2.12 degrees) is roughly double the raw prediction error (1.06 degrees). Thus, `model_3` is doubly worse at predicting temperatures for days we haven't yet observed than temperatures for days in the `sample_1` data. This suggests that `model_3` is **overfit**, or overly optimized, to the `sample_1` data we used to build it. As such, the discrepancy in its raw and cross-validated prediction errors tips us off that `model_3` has low bias but **high variance** – a *different* sample of data might lead to very different posterior results. Between the extremes of `model_1` and `model_3`, `model_2` presents the best option with a *relatively* low raw prediction error and the *lowest* cross-validated prediction error.

TABLE 11.3: Raw and cross-validated posterior predictive summaries for three models of temp3pm.

model	raw mae	cross-validated mae
model 1	2.79	2.83
model 2	1.53	1.86
model 3	1.06	2.12

11.6   Chapter summary

In Chapter 11, we expanded our regression toolkit to include **multivariable** Bayesian Normal linear regression models. Letting Y denote a quantitative response variable and $(X_1, X_2, \ldots, X_p)$ a set of p predictor variables which can be quantitative, categorical, interaction terms, or any combination thereof:

$$
\begin{aligned}
\text{data:} \quad & Y_i | \beta_0, \beta_1, \ldots, \beta_p, \sigma \overset{ind}{\sim} N(\mu_i, \sigma^2) \quad \text{with} \quad \mu_i = \beta_0 + \beta_1 X_{i1} + \cdots + \beta_p X_{ip} \\
\text{priors:} \quad & \beta_{0c} \sim N(m_0, s_0^2) \\
& \beta_1 \sim N(m_1, s_1^2) \\
& \quad \vdots \\
& \beta_p \sim N(m_p, s_p^2) \\
& \sigma \sim \text{Exp}(l).
\end{aligned}
$$

By including more than one predictor, multivariable regression can improve our predictions and understanding of Y. Yet we can take it *too* far. This conundrum is known as the **bias-variance trade-off**. If our model is too simple or rigid (including, say, only one predictor), its estimate of the model mean μ can be **biased**, or far from the actual observed relationship in the data. If we greedily include more and more and more predictors into a regression model, its estimates of μ can become **overfit** to our sample data and highly **variable** from sample to sample. Thus, in model building, we aim to strike a balance – a model which enjoys relatively low bias and low variability.

11.7 Exercises

11.7.1 Conceptual exercises

Exercise 11.1 (Why multiple predictors?). Briefly explain why we might want to build a regression model with more than one predictor.

Exercise 11.2 (Categorical predictors: cars). Let's say that you want to model a car's miles per gallon in a city (Y) by the make of the car: Ford, Kia, Subaru, or Toyota. This relationship can be written as $\mu = \beta_0 + \beta_1 X_1 + \beta_2 X_2 + \beta_3 X_3$, where X_1, X_2, X_3 are indicators for whether or not the cars are Kias, Subarus, or Toyotas, respectively:

a) Explain why there is no indicator term for the Ford category.
b) Interpret the regression coefficient β_2.
c) Interpret the regression coefficient β_0.

Exercise 11.3 (Categorical and quantitative predictors: tomatoes). You have recently taken up the hobby of growing tomatoes and hope to learn more about the factors associated with bigger tomatoes. As such, you plan to model a tomato's weight in grams (Y) by the number of days it has been growing (X_1) and its type, Mr. Stripey or Roma. Suppose the expected weight of a tomato is a linear function of its age and type, $\mu = \beta_0 + \beta_1 X_1 + \beta_2 X_2$, where X_2 is an indicator for Roma tomatoes.

a) Interpret each regression coefficient, β_0, β_1, and β_2.
b) What would it mean if β_2 were equal to zero?

Exercise 11.4 (Interactions: tomatoes). Continuing your quest to understand tomato size, you incorporate an interaction term between the tomato grow time (X_1) and type (X_2) into your model: $\mu = \beta_0 + \beta_1 X_1 + \beta_2 X_2 + \beta_3 X_1 X_2$.

a) Explain, in context, what it means for X_1 and X_2 to interact.
b) Interpret β_3.

Exercise 11.5 (Interaction terms).

a) Sketch a model that *would* benefit from including an interaction term between a categorical and quantitative predictor.
b) Sketch a model that would *not* benefit from including an interaction term between a categorical and quantitative predictor.
c) Besides visualization, what are two other ways to determine if you should include interaction terms in your model?

Exercise 11.6 (Improving your model: shoe size). Let's say you model a child's shoe size (Y) by two predictors: the child's age in years (X_1) and an indicator of whether the child knows how to swim (X_2).

a) Generally speaking, why can it be beneficial to add predictors to models?
b) Generally speaking, why can it be beneficial to remove predictors from models?

c) What might you *add* to this model to improve your predictions of shoe size? Why?
d) What might you *remove* from this model to improve it? Why?

Exercise 11.7 (What makes a good model?). We don't expect our regression models to be perfect, but we do want to do our best. It can be helpful to think about what we want and expect from our models.

a) What are qualities of a good model?
b) What are qualities of a bad model?

Exercise 11.8 (Is our model good / better?). What techniques have you learned in this chapter to assess and compare your models? Give a brief explanation for each technique.

Exercise 11.9 (Bias-variance trade-off). In your own words, briefly explain what the bias-variance tradeoff is and why it is important.

11.7.2 Applied exercises

In the next exercises you will use the `penguins_bayes` data in the **bayesrules** package to build various models of penguin `body_mass_g`. Throughout, we'll utilize weakly informative priors and a basic understanding that the average penguin weighs somewhere between 3,500 and 4,500 grams. Further, one predictor of interest is penguin `species`: Adelie, Chinstrap, or Gentoo. We'll get our first experience with a 3-level predictor like this in Chapter 12. If you'd like to work with only 2 levels as you did in Chapter 11, you can utilize the `penguin_data` which includes only Adelie and Gentoo penguins:

```
# Alternative penguin data
penguin_data <- penguins_bayes %>%
  filter(species %in% c("Adelie", "Gentoo"))
```

Exercise 11.10 (Penguins! Main effects). Let's begin our analysis of penguin `body_mass_g` by exploring its relationship with `flipper_length_mm` and `species`.

a) Plot and summarize the observed relationships among these three variables.
b) Use `stan_glm()` to simulate a posterior Normal regression model of `body_mass_g` by `flipper_length_mm` and `species`, without an interaction term.
c) Create and interpret both visual and numerical diagnostics of your MCMC simulation.
d) Produce a `tidy()` summary of this model. Interpret the non-intercept coefficients' posterior median values in context.
e) Simulate, plot, and describe the posterior predictive model for the body mass of an Adelie penguin that has a flipper length of 197.

Exercise 11.11 (Penguins! Interaction). Building from the previous exercise, our next goal is to model `body_mass_g` by `flipper_length_mm` and `species` *with* an interaction term between these two predictors.

a) Use `stan_glm()` to simulate the posterior for this model, with four chains at 10,000 iterations each.

b)   Simulate and plot 50 posterior model lines. Briefly describe what you learn from this plot.

c)   Produce a `tidy()` summary for this model. Based on the summary, do you have evidence that the interaction terms are necessary for this model? Explain your reasoning.

Exercise 11.12 (Penguins! 3 predictors). Next, let's explore a model of `body_mass_g` by **three** predictors: `flipper_length_mm`, `bill_length_mm`, and `bill_depth_mm`. Do not use any interactions in this model.

a)   Use `stan_glm()` to simulate the posterior for this model.

b)   Use `posterior_interval()` to produce 95% credible intervals for the model parameters.

c)   Based on these 95% credible intervals, when controlling for the other predictors in the model, which predictors have a significant positive association with body mass, which have significant negative association with body mass, and which have no significant association?

Exercise 11.13 (Penguins! Comparing models). Consider 4 separate models of `body_mass_g`:

model	formula
1	`body_mass_g ~ flipper_length_mm`
2	`body_mass_g ~ species`
3	`body_mass_g ~ flipper_length_mm + species`
4	`body_mass_g ~ flipper_length_mm + bill_length_mm + bill_depth_mm`

a)   Simulate these four models using the `stan_glm()` function.

b)   Produce and compare the `pp_check()` plots for the four models.

c)   Use 10-fold cross-validation to assess and compare the posterior predictive quality of the four models using the `prediction_summary_cv()`. NOTE: We can only predict body mass for penguins that have complete information on our model predictors. Yet two penguins have NA values for multiple of these predictors. To remove these two penguins, we `select()` our columns of interest *before* removing penguins with NA values. This way, we don't throw out penguins just because they're missing information on variables we don't care about:

```
penguins_complete <- penguins_bayes %>%
  select(flipper_length_mm, body_mass_g, species,
         bill_length_mm, bill_depth_mm) %>%
  na.omit()
```

d)   Evaluate and compare the ELPD posterior predictive accuracy of the four models.

e)   In summary, which of these four models is "best"? Explain.

11.7.3 Open-ended exercises

Exercise 11.14 (Penguins on your own). Build three different models of penguin `bill_length_mm`. *You* get to choose which predictors to use, and whether to include interaction terms. Evaluate and compare these models. Which do you prefer and why?

Exercise 11.15 (Weather on your own). Use the `weather_perth` data in the **bayesrules** package to develop a "successful" model of afternoon temperatures (`temp3pm`) in Perth, Australia. *You* get to choose which predictors to use, and whether to include interaction terms. Be sure to evaluate your model and explain your modeling process.

12

Poisson & Negative Binomial Regression

Step back from the details of the previous few chapters and recall the big goal: to build **regression** models of **quantitative** response variables Y. We've only shared one regression tool with you so far, the *Bayesian Normal regression model*. The name of this "Normal" regression tool reflects its broad applicability. But (luckily!), not *every* model is "Normal." We'll expand upon our regression tools in the context of the following data story.

As of this book's writing, the *Equality Act* sits in the United States Senate awaiting consideration. If passed, this act or bill would ensure basic LGBTQ+ rights at the *national* level by prohibiting discrimination in education, employment, housing, and more. As is, each of the 50 individual states has their own set of unique anti-discrimination laws, spanning issues from anti-bullying to health care coverage. Our goal is to better understand how the *number* of laws in a state relates to its unique demographic features and political climate. For the former, we'll narrow our focus to the percentage of a state's residents that reside in an urban area. For the latter, we'll utilize historical voting patterns in presidential elections, noting whether a state has consistently voted for the *Democratic* candidate, consistently voted for the "GOP" Republican candidate,[1] or is a *swing* state that has flip flopped back and forth. Throughout our analysis, please recognize that the *number* of laws is *not* a perfect proxy for the *quality* of a state's laws – it merely provides a starting point in understanding how laws vary from state to state.

For each state $i \in \{1, 2, \ldots, 50\}$, let Y_i denote the number of anti-discrimination laws and predictor X_{i1} denote the percentage of the state's residents that live in urban areas. Further, our historical political climate predictor variable is categorical with three levels: Democrat, GOP, or swing. This is our first time working with a three-level variable, so let's set this up right. Recall from Chapter 11 that one level of a categorical predictor, here Democrat, serves as a **baseline** or **reference level** for our model. The *other* levels, GOP and swing, enter our model as *indicators*. Thus, X_{i2} indicates whether state i leans GOP and X_{i3} indicates a swing state:

$$X_{i2} = \begin{cases} 1 & \text{GOP} \\ 0 & \text{otherwise} \end{cases} \quad \text{and} \quad X_{i3} = \begin{cases} 1 & \text{swing} \\ 0 & \text{otherwise.} \end{cases}$$

Since it's the only technique we've explored thus far, our first approach to *understanding* the relationship between our **quantitative** response variable Y and our predictors X might be to build a regression model with a Normal data structure:

$$Y_i | \beta_0, \beta_1, \beta_2, \beta_3, \sigma \overset{ind}{\sim} N\left(\mu_i, \ \sigma^2\right) \quad \text{with} \quad \mu_i = \beta_0 + \beta_1 X_{i1} + \beta_2 X_{i2} + \beta_3 X_{i3}.$$

Other than an understanding that a state that's "typical" with respect to its urban population and historical voting patterns has around 7 laws, we have very little prior knowledge about

---

[1]"GOP" stands for the *grand old party*, a nickname for the Republican Party.

DOI: 10.1201/9780429288340-12

this relationship. Thus, we'll set a $N(7, 1.5^2)$ prior for the centered intercept β_{0c}, but utilize weakly informative priors for the other three parameters.

Next, let's consider some data. Each year, the Human Rights Campaign releases a "State Equality Index" which monitors the *number* of LQBTQ+ rights laws in each state. Among other state features, the `equality_index` dataset in the `bayesrules` package includes data from the 2019 index compiled by Sarah Warbelow, Courtnay Avant, and Colin Kutney (2019). To obtain a detailed codebook, type `?equality` in the console.

```
# Load packages
library(bayesrules)
library(rstanarm)
library(bayesplot)
library(tidyverse)
library(tidybayes)
library(broom.mixed)

# Load data
data(equality_index)
equality <- equality_index
```

The histogram below indicates that the number of laws ranges from as low as 1 to as high as 155, yet the majority of states have fewer than ten laws:

```
ggplot(equality, aes(x = laws)) +
  geom_histogram(color = "white", breaks = seq(0, 160, by = 10))
```

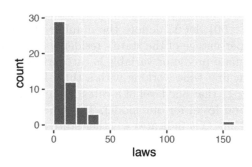

FIGURE 12.1: A histogram of the number of laws in each of the 50 states.

The state with 155 laws happens to be California. As a clear outlier, we'll remove this state from our analysis:

```
# Identify the outlier
equality %>%
  filter(laws == max(laws))
# A tibble: 1 x 6
  state   region gop_2016  laws historical percent_urban
  <fct>   <fct>     <dbl> <dbl> <fct>              <dbl>
1 calif~  west       31.6   155 dem                   95
```

```
# Remove the outlier
equality <- equality %>%
  filter(state != "california")
```

Next, in a scatterplot of the number of state `laws` versus its `percent_urban` population and `historical` voting patterns, notice that historically `dem` states and states with greater urban populations tend to have more LGBTQ+ anti-discrimination laws in place:

```
ggplot(equality, aes(y = laws, x = percent_urban, color = historical)) +
  geom_point()
```

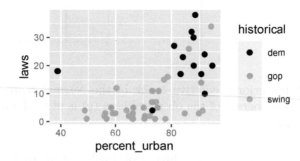

FIGURE 12.2: A scatterplot of the number of anti-discrimination laws in a state versus its urban population percentage and historical voting trends.

Using `stan_glm()`, we combine this data with our weak prior understanding to simulate the posterior Normal regression model of `laws` by `percent_urban` and `historical` voting trends. In a quick posterior predictive check of this `equality_normal_sim` model, we compare a histogram of the observed anti-discrimination laws to five posterior simulated datasets (Figure 12.3). (A histogram is more appropriate than a density plot here since our response variable is a non-negative count.) The results aren't good – the posterior predictions from this model do *not* match the features of the observed data. You might not be surprised. The observed number of anti-discrimination laws per state are right skewed (not Normal!). In contrast, the datasets simulated from the posterior Normal regression model are roughly symmetric. Adding insult to injury, these simulated datasets assume that it is quite common for states to have a *negative* number of laws (not possible!).

```
# Simulate the Normal model
equality_normal_sim <- stan_glm(laws ~ percent_urban + historical,
                    data = equality,
                    family = gaussian,
                    prior_intercept = normal(7, 1.5),
                    prior = normal(0, 2.5, autoscale = TRUE),
                    prior_aux = exponential(1, autoscale = TRUE),
                    chains = 4, iter = 5000*2, seed = 84735)


# Posterior predictive check
pp_check(equality_normal_sim, plotfun = "hist", nreps = 5) +
```

```
geom_vline(xintercept = 0) +
xlab("laws")
```

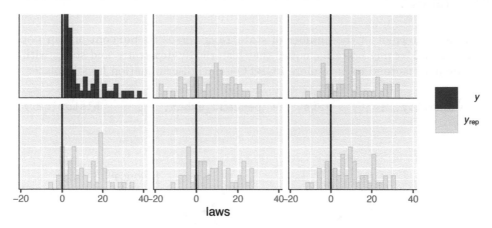

FIGURE 12.3: A posterior predictive check of the Normal regression model of anti-discrimination laws. A histogram of the observed laws (y) is plotted alongside five posterior simulated datasets (y_{rep}).

This sad result reveals the limits of our lone tool. Luckily, probability models come in all shapes, and we don't have to force something to be Normal when it's not.

> ◎ **Goals**
>
> You will extend the Normal regression model of a quantitative response variable Y to settings in which Y is a **count** variable whose dependence on predictors X is better represented by the **Poisson** or **Negative Binomial**, not Normal, models.

12.1 Building the Poisson regression model

12.1.1 Specifying the data model

Recall from Chapter 5 that the Poisson model is appropriate for modeling discrete counts of events (here anti-discrimination laws) that happen in a fixed interval of space or time (here states) and that, *theoretically*, have no upper bound. The Poisson is especially handy in cases like ours in which counts are right-skewed, and thus can't reasonably be approximated by a Normal model. Moving forward, let's assume a **Poisson data model** for the number of LGBTQ+ anti-discrimination laws in each state i (Y_i) where the *rate* of anti-discrimination laws λ_i depends upon demographic and voting trends (X_{i1}, X_{i2}, and X_{i3}):

$$Y_i|\lambda_i \overset{ind}{\sim} Pois\left(\lambda_i\right).$$

Under this Poisson structure, the *expected* or *average* number of laws Y_i in states with similar predictor values X is captured by λ_i:

$$E(Y_i|\lambda_i) = \lambda_i.$$

If we proceeded as in Normal regression, we might assume that the average number of laws is a linear combination of our predictors, $\lambda_i = \beta_0 + \beta_1 X_{i1} + \beta_2 X_{i2} + \beta_3 X_{i3}$. This assumption is illustrated by the lines in Figure 12.4, one per historical voting trend.

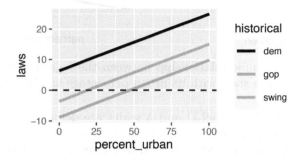

FIGURE 12.4: A graphical depiction of the (incorrect) assumption that, no matter the historical politics, the typical number of state laws is linearly related to urban population. The dashed horizontal line represents the x-axis.

❓ Quiz Yourself!

Figure 12.4 highlights a flaw in assuming that the expected number of laws in a state, λ_i, is a linear combination of `percent_urban` and `historical`. What is it?

When we assume that λ_i can be expressed by a linear combination of the X predictors, the model of λ_i spans both positive and negative values, and thus suggests that some states have a *negative* number of anti-discrimination laws. That doesn't make sense. Like the number of laws, a Poisson *rate* λ_i, must be *positive*. To avoid this violation, it is common to use a **log link function.**[2] That is, we'll assume that $log(\lambda_i)$, which *does* span both positive and negative values, is a linear combination of the X predictors:

$$Y_i|\beta_0, \beta_1, \beta_2, \beta_3 \overset{ind}{\sim} Pois(\lambda_i) \quad \text{with} \quad log(\lambda_i) = \beta_0 + \beta_1 X_{i1} + \beta_2 X_{i2} + \beta_3 X_{i3}.$$

At the risk of projecting, interpreting the *logged* number of laws isn't so easy. Instead, we can always transform the model relationship off the log scale by appealing to the fact that if $log(\lambda) = a$, then $\lambda = e^a$ for natural number e:

$$Y_i|\beta_0, \beta_1, \beta_2, \beta_3 \overset{ind}{\sim} Pois(\lambda_i) \quad \text{with} \quad \lambda_i = e^{\beta_0 + \beta_1 X_{i1} + \beta_2 X_{i2} + \beta_3 X_{i3}}.$$

Figure 12.5 presents a prior plausible outcome for this model on both the $log(\lambda)$ and λ scales. In both cases there are three curves, one per historical voting category. On the $log(\lambda)$ scale, these curves are *linear*. Yet when transformed to the λ or (unlogged) laws scale, these curves are *nonlinear* and restricted to be at or above 0. This was the point – we *want* our model to preserve the fact that a state can't have a negative number of laws.

---

[2]By convention, "log" refers to the *natural* log function "ln" throughout this book.

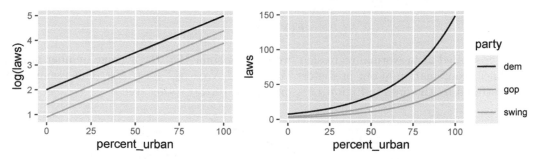

FIGURE 12.5: An example relationship between the logged number of laws (left) or number of laws (right) in a state with its urban percentage and historical voting trends.

The model curves on both the $\log(\lambda)$ and λ scales are defined by the $(\beta_0, \beta_1, \beta_2, \beta_3)$ parameters. When describing the *linear* model of the *logged* number of laws in a state, these parameters take on the usual meanings related to intercepts and slopes. Yet $(\beta_0, \beta_1, \beta_2, \beta_3)$ take on new meanings when describing the *nonlinear* model of the (unlogged) number of laws in a state.

Interpreting Poisson regression coefficients

Consider two equivalent formulations of the Poisson regression model: $Y|\beta_0, \beta_1, \ldots, \beta_p \sim \text{Pois}(\lambda)$ with

$$\log(\lambda) = \beta_0 + \beta_1 X_1 + \cdots \beta_p X_p$$
$$\lambda = e^{\beta_0 + \beta_1 X_1 + \cdots \beta_p X_p}.$$

$$(12.1)$$

Interpreting β_0

When $X_1, X_2, \ldots, X_p$ are all 0, β_0 is the *logged* average Y value and e^{β_0} is the average Y value.

Interpreting β_1 (and similarly $\beta_2, \ldots, \beta_p$)

Let λ_x represent the average Y value when $X_1 = x$ and λ_{x+1} represent the average Y value when $X_1 = x + 1$. When we control for all other predictors $X_2, \ldots, X_p$ and increase X_1 by 1, from x to $x + 1$: β_1 is the change in the *logged* average Y value and e^{β_1} is the *multiplicative change* in the (unlogged) average Y value. That is,

$$\beta_1 = \log(\lambda_{x+1}) - \log(\lambda_x) \quad \text{and} \quad e^{\beta_1} = \frac{\lambda_{x+1}}{\lambda_x}.$$

Let's apply these concepts to the hypothetical models of the logged and unlogged number of state laws in Figure 12.5. In this figure, the curves are defined by:

$$\log(\lambda) = 2 + 0.03 X_{i1} - 1.1 X_{i2} - 0.6 X_{i3}$$
$$\lambda = e^{2 + 0.03 X_{i1} - 1.1 X_{i2} - 0.6 X_{i3}}.$$

$$(12.2)$$

Consider the intercept parameter $\beta_0 = 2$. This intercept reflects the trends in anti-discrimination laws in historically Democratic states with 0 urban population, i.e., states with $X_{i1} = X_{i2} = X_{i3} = 0$. As such, we expect the *logged* number of laws in such states to be

β_0 = 2. More meaningfully, and equivalently, we expect historically Democratic states with 0 urban population to have roughly 7.4 anti-discrimination laws:

$$e^{\beta_0} = e^2 = 7.389.$$

Now, since there *are* no states in which the urban population is close to 0, it doesn't make sense to put too much emphasis on these interpretations of β_0. Rather, we can just understand β_0 as providing a baseline for our model, on both the $\log(\lambda)$ and λ scales.

Next, consider the urban percentage coefficient β_1 = 0.03. On the *linear* $\log(\lambda)$ scale, we can still interpret this value as the shared *slope* of the lines in Figure 12.5 (left). Specifically, no matter a state's `historical` voting trends, we expect the *logged* number of laws in states to increase by 0.03 for every extra percentage point in urban population. We can interpret the relationship between state laws and urban percentage more meaningfully on the *unlogged* scale of λ by examining

$$e^{\beta_1} = e^{0.03} = 1.03.$$

To this end, reexamine the *nonlinear* models of anti-discrimination laws in Figure 12.5 (right). Though the model curves don't increase by the same *absolute* amount for each incremental increase in `percent_urban`, they do increase by the same *percentage*. Thus, instead of representing a linear slope, e^{β_1} measures the nonlinear *percentage* or *multiplicative* increase in the average number of laws with urban percentage. In this case, if the urban population in one state is 1 percentage point greater than another state, we'd expect it to have 1.03 *times* the number of, or 3% more, anti-discrimination laws.

Finally, consider the GOP coefficient β_2 = −1.1. Recall from Chapter 11 that when interpreting a coefficient for a categorical indicator, we must do so relative to the reference category, here Democrat leaning states. On the *linear* $\log(\lambda)$ scale, β_2 serves as the vertical *difference* between the GOP versus Democrat model lines in Figure 12.5 (left). Thus, at any urban percentage, we'd expect the *logged* number of laws to be 1.1 lower in historically GOP states than Democrat states. Though we also see a difference between the GOP and Democrat curves on the *nonlinear* λ scale (Figure 12.5 right), the difference isn't constant – the gap in the number of laws widens as urban percentage increases. In this case, instead of representing a constant difference between two *lines*, e^{β_2} measures the *percentage* or *multiplicative* difference in the GOP versus Democrat curve. Thus, at any urban percentage level, we'd expect a historically GOP state to have 1/3 as many anti-discrimination laws as a historically Democrat state:

$$e^{\beta_2} = e^{-1.1} = 0.333.$$

We conclude this section with the formal assumptions encoded by the Poisson data model.

Poisson regression assumptions

Consider the Bayesian Poisson regression model with $Y_i|\beta_0, \beta_1, \ldots, \beta_p \overset{ind}{\sim} Pois(\lambda_i)$ where

$$\log(\lambda_i) = \beta_0 + \beta_1 X_{i1} + \cdots + \beta_p X_{ip}.$$

The appropriateness of this model depends upon the following assumptions.

- **Structure of the data**
  Conditioned on predictors X, the observed data Y_i on case i is **independent** of the observed data on any other case j.
- **Structure of variable Y**
  Response variable Y has a Poisson structure, i.e., is a discrete **count** of events that happen in a fixed interval of space or time.
- **Structure of the relationship**
  The *logged* average Y value can be written as a **linear combination** of the predictors, $\log(\lambda_i) = \beta_0 + \beta_1 X_{i1} + \beta_2 X_{i2} + \beta_3 X_{i3}$.
- **Structure of the variability in Y**
  A Poisson random variable Y with rate λ has equal mean and variance, $E(Y) = \text{Var}(Y) = \lambda$ (5.3). Thus, conditioned on predictors X, the *typical* value of Y should be roughly equivalent to the *variability* in Y. As such, the variability in Y increases as its mean increases. See Figure 12.6 for examples of when this assumption does and does not hold.

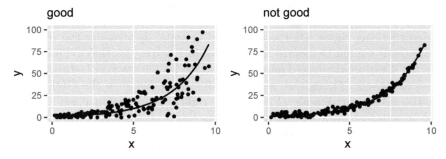

FIGURE 12.6: Two simulated datasets. The data on the left satisfies the Poisson regression assumption that, at any given X, the variability in Y is roughly on par with the average Y value. The data on the right does not, exhibiting consistently low variability in Y across all X values.

12.1.2   Specifying the priors

To complete our Bayesian model, we must express our prior understanding of regression coefficients $(\beta_0, \beta_1, \beta_2, \beta_3)$. Since these coefficients can each take on any value on the real line, it's again reasonable to utilize Normal priors. Further, as was the case for our Normal regression model, we'll assume these priors (i.e., our prior understanding of the model coefficients) are **independent**. The complete representation of our **Poisson regression model** of Y_i is as follows:

data: $\quad Y_i | \beta_0, \beta_1, \beta_2, \beta_3 \overset{ind}{\sim} Pois(\lambda_i) \quad \text{with} \quad \log(\lambda_i) = \beta_0 + \beta_1 X_{i1} + \beta_2 X_{i2} + \beta_3 X_{i3}$

priors: $\qquad\qquad\qquad \beta_{0c} \;\sim N\left(2, 0.5^2\right)$
$\qquad\qquad\qquad\quad\;\; \beta_1 \;\sim N(0, 0.17^2)$
$\qquad\qquad\qquad\quad\;\; \beta_2 \;\sim N(0, 4.97^2)$
$\qquad\qquad\qquad\quad\;\; \beta_3 \;\sim N(0, 5.60^2)$

$$(12.3)$$

First, consider the prior for the centered intercept β_{0c}. Recall our prior assumption that the average number of laws in "typical" states is around $\lambda = 7$. As such, we set the Normal prior mean for β_{0c} to 2 on the *log* scale:

$$\log(\lambda) = \log(7) \approx 1.95.$$

Further, the range of this Normal prior indicates our relative uncertainty about this baseline. Though the *logged* average number of laws is most likely around 2, we think it could range from roughly 1 to 3 ($2 \pm 2 * 0.5$). Or, more intuitively, we think that the average *number* of laws in typical states might be somewhere between 3 and 20:

$$(e^1, e^3) \approx (3, 20).$$

Beyond this baseline, we again used weakly informative default priors for $(\beta_1, \beta_2, \beta_3)$, tuned by `stan_glm()` below. Being centered at zero with relatively large standard deviation on the scale of our variables, these priors reflect a general uncertainty about whether and how the number of anti-discrimination laws is associated with a state's urban population and voting trends.

To examine whether these combined priors accurately reflect our uncertain understanding of state laws, we'll simulate 20,000 draws from the prior models of $(\beta_0, \beta_1, \beta_2, \beta_3)$. To this end, we can run the same `stan_glm()` function that we use to simulate the posterior with two new arguments: `prior_PD = TRUE` specifies that we wish to simulate the *prior*, and `family = poisson` indicates that we're using a Poisson data model (not Normal or `gaussian`).

```
equality_model_prior <- stan_glm(laws ~ percent_urban + historical,
                                 data = equality,
                                 family = poisson,
                                 prior_intercept = normal(2, 0.5),
                                 prior = normal(0, 2.5, autoscale = TRUE),
                                 chains = 4, iter = 5000*2, seed = 84735,
                                 prior_PD = TRUE)
```

A call to `prior_summary()` confirms the specification of our weakly informative priors in (12.3):

```
prior_summary(equality_model_prior)
```

Next, we plot 100 prior plausible models, $e^{\beta_0 + \beta_1 x_1 + \beta_2 x_2 + \beta_3 x_3}$. The mess of curves here certainly reflects our prior uncertainty! These are all over the map, indicating that the number of laws might increase *or* decrease with urban population and might or might not differ by historical voting trends. We don't really know.

```
equality %>%
  add_fitted_draws(equality_model_prior, n = 100) %>%
  ggplot(aes(x = percent_urban, y = laws, color = historical)) +
    geom_line(aes(y = .value, group = paste(historical, .draw))) +
    ylim(0, 100)
```

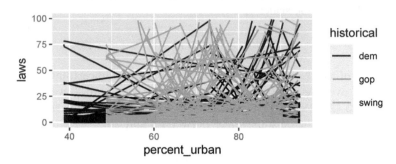

FIGURE 12.7: 100 prior plausible models of state anti-discrimination laws, $e^{\beta_0+\beta_1 x_1+\beta_2 x_2+\beta_3 x_3}$, under the prior models of $(\beta_0, \beta_1, \beta_2, \beta_3)$.

12.2   Simulating the posterior

With all the pieces in place, let's simulate the Poisson posterior regression model of anti-discrimination laws versus urban population and voting trends (12.3). Instead of starting from scratch with the `stan_glm()` function, we'll take a shortcut: we `update()` the `equality_model_prior` simulation, indicating `prior_PD = FALSE` (i.e., we wish to simulate the posterior, not the prior).

```
equality_model <- update(equality_model_prior, prior_PD = FALSE)
```

MCMC trace, density, and autocorrelation plots confirm that our simulation has stabilized:

```
mcmc_trace(equality_model)
mcmc_dens_overlay(equality_model)
mcmc_acf(equality_model)
```

And before we get too far into our analysis of these simulation results, a quick posterior predictive check confirms that we're now on the right track (Figure 12.8). First, histograms of just five posterior simulations of state law data exhibit similar skew, range, and trends as the observed law data. Second, though density plots aren't the best display of count data, they allow us to more *directly* compare a *broader* range of 50 posterior simulated datasets to the actual observed law data. These simulations aren't perfect, but they do reasonably capture the features of the observed law data.

```
set.seed(1)
pp_check(equality_model, plotfun = "hist", nreps = 5) +
  xlab("laws")
pp_check(equality_model) +
  xlab("laws")
```

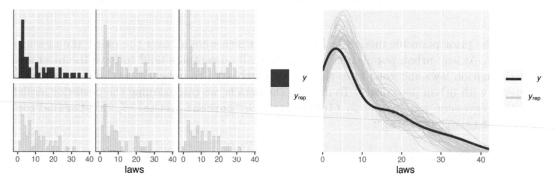

FIGURE 12.8: A posterior predictive check of the Poisson regression model of anti-discrimination laws compares the observed laws (y) to five posterior simulated datasets (y_{rep}) via histograms (left) and to 50 posterior simulated datasets via density plots (right).

12.3 Interpreting the posterior

With the reassurance that our model isn't too wrong, let's dig into the posterior results, beginning with the big picture. The 50 posterior plausible models in Figure 12.9, $\lambda = e^{\beta_0 + \beta_1 X_1 + \beta_2 X_2 + \beta_3 X_3}$, provide insight into our overall understanding of anti-discrimination laws.

```
equality %>%
  add_fitted_draws(equality_model, n = 50) %>%
  ggplot(aes(x = percent_urban, y = laws, color = historical)) +
    geom_line(aes(y = .value, group = paste(historical, .draw)),
              alpha = .1) +
    geom_point(data = equality, size = 0.1)
```

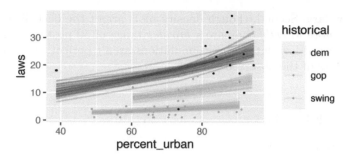

FIGURE 12.9: 50 posterior plausible models for the relationship of a state's number of anti-discrimination laws with its urban population rate and historical voting trends.

Unlike the prior plausible models in Figure 12.7, which were all over the place, the messages are clear. At any urban population level, historically **dem** states tend to have the most anti-discrimination laws and **gop** states the fewest. Further, the number of laws in a state tend to increase with urban percentage. To dig into the details, we can examine the posterior models for the regression parameters β_0 (Intercept), β_1 (percent_urban), β_2 (historicalgop), and β_3 (historicalswing):

```
tidy(equality_model, conf.int = TRUE, conf.level = 0.80)
# A tibble: 4 x 5
  term           estimate std.error conf.low conf.high
  <chr>             <dbl>     <dbl>    <dbl>     <dbl>
1 (Intercept)      1.71     0.303     1.31      2.09
2 percent_urban    0.0164   0.00353   0.0119    0.0210
3 historicalgop   -1.52     0.134    -1.69     -1.34
4 historicalswing -0.610    0.103    -0.745    -0.477
```

Thus, the posterior median relationship of a state's number of anti-discrimination laws with its urban population and historical voting trends can be described by:

$$\log(\lambda_i) = 1.71 + 0.0164X_{i1} - 1.52X_{i2} - 0.61X_{i3}$$
$$\lambda_i = e^{1.71 + 0.0164X_{i1} - 1.52X_{i2} - 0.61X_{i3}}.$$

(12.4)

Consider the percent_urban coefficient β_1, which has a posterior median of roughly 0.0164. Then, when controlling for historical voting trends, we expect the *logged* number of anti-discrimination laws in states to increase by 0.0164 for every extra percentage point in urban population. More meaningfully (on the *unlogged* scale), if the urban population in one state is 1 percentage point greater than another state, we'd expect it to have 1.0165 *times* the number of, or 1.65% more, anti-discrimination laws:

$$e^{0.0164} = 1.0165.$$

Or, if the urban population in one state is *25* percentage points greater than another state, we'd expect it to have roughly *one and a half* times the number of, or 51% more, anti-discrimination laws ($e^{25 \cdot 0.0164} = 1.5068$). Take a quick quiz to similarly interpret the β_3 coefficient for historically **swing** states.[3]

---

[3] Answer: b

---

❷ **Quiz Yourself!**

The posterior median of β_3 is roughly -0.61. Correspondingly, $e^{-0.61} = 0.54$. How can we interpret this value? When holding constant `percent_urban`...

    a.   The number of anti-discrimination laws tends to decrease by roughly 54 percent for every extra `swing` state.

    b.   `swing` states tend to have 54 percent as many anti-discrimination laws as `dem` leaning states.

    c.   `swing` states tend to have 0.54 fewer anti-discrimination laws than `dem` leaning states.

The key here is remembering that the categorical `swing` state indicator provides a *comparison* to the `dem` state reference level. Then, when controlling for a state's urban population, we'd expect the *logged* number of anti-discrimination laws to be 0.61 *lower* in a `swing` state than in a `dem` leaning state. Equivalently, on the *unlogged* scale, `swing` states tend to have 54 *percent* as many anti-discrimination laws as `dem` leaning states ($e^{-0.61} = 0.54$).

In closing out our posterior interpretation, notice that the 80% posterior credible intervals for $(\beta_1, \beta_2, \beta_3)$ in the above `tidy()` summary provide evidence that each coefficient significantly differs from 0. For example, there's an 80% posterior chance that the `percent_urban` coefficient β_1 is between 0.0119 and 0.021. Thus, when controlling for a state's `historical` political leanings, there's a *significant* positive association between the number of anti-discrimination laws in a state and its urban population. Further, when controlling for a state's `percent_urban` makeup, the number of anti-discrimination laws in `gop` leaning and `swing` states tend to be significantly *below* that of `dem` leaning states – the 80% credible intervals for β_2 and β_3 both fall below 0. These conclusions are consistent with the posterior plausible models in Figure 12.9.

12.4   Posterior prediction

To explore how the general Poisson model plays out in individual states, consider the state of Minnesota, a historically `dem` state with 73.3% of residents residing in urban areas and 4 anti-discrimination laws:

```
equality %>%
  filter(state == "minnesota")
# A tibble: 1 x 6
  state   region gop_2016  laws historical percent_urban
  <fct>   <fct>      <dbl> <dbl> <fct>              <dbl>
1 minne~  midwe~      44.9     4 dem                 73.3
```

Based on the state's demographics and political leanings, we can approximate a posterior predictive model for its number of anti-discrimination laws. Importantly, reflecting the *Poisson* data structure of our model, the 20,000 simulated posterior predictions are *counts*:

```
# Calculate posterior predictions
set.seed(84735)
mn_prediction <- posterior_predict(
  equality_model, newdata = data.frame(percent_urban = 73.3,
                                       historical = "dem"))
head(mn_prediction, 3)
        1
[1,]  20
[2,]  17
[3,]  21
```

The resulting posterior predictive model anticipates that Minnesota has between 10 and 30 anti-discrimination laws (roughly), a range that falls far above the actual number of laws in that state (4). This discrepancy reveals that a state's demographics and political leanings, and thus our Poisson model, don't tell the full story behind its anti-discrimination laws. It also reveals something about Minnesota itself. For a state with such a high urban population and historically Democratic voting trends, it has an unusually small number of anti-discrimination laws. (Again, the *number* of laws in a state isn't necessarily a proxy for the overall *quality* of laws.)

```
mcmc_hist(mn_prediction, binwidth = 1) +
  geom_vline(xintercept = 4) +
  xlab("Predicted number of laws in Minnesota")
```

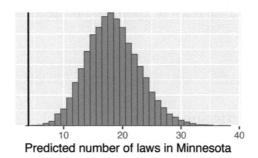

FIGURE 12.10: Posterior predictive model of the number of anti-discrimination laws in Minnesota. The vertical line marks the actual number of laws, 4.

Recall that we needn't use the `posterior_predict()` shortcut function to simulate the posterior predictive model for Minnesota. Rather, we could directly predict the number of laws in that state from each of the 20,000 posterior plausible parameter sets $\left(\beta_0^{(i)}, \beta_1^{(i)}, \beta_2^{(i)}, \beta_3^{(i)}\right)$. To this end, from each parameter set we:

1. calculate the logged average number of laws in states like Minnesota, with a 73.3% urban percentage and historically Democrat voting patterns:

$$\log(\lambda^{(i)}) = \beta_0^{(i)} + \beta_1^{(i)} \cdot 73.3 + \beta_2^{(i)} \cdot 0 + \beta_3^{(i)} \cdot 0;$$

2. transform $\log(\lambda^{(i)})$ to obtain the (*unlogged*) average number of laws in states like Minnesota, $\lambda^{(i)}$; and

3. simulate a $\text{Pois}(\lambda^{(i)})$ outcome for the number of laws in Minnesota, $Y_{\text{new}}^{(i)}$, using `rpois()`.

The result matches the shortcut!

```
# Predict number of laws for each parameter set in the chain
set.seed(84735)
as.data.frame(equality_model) %>%
  mutate(log_lambda = `(Intercept)` + percent_urban*73.3 +
           historicalgop*0 + historicalswing*0,
         lambda = exp(log_lambda),
         y_new = rpois(20000, lambda = lambda)) %>%
  ggplot(aes(x = y_new)) +
    stat_count()
```

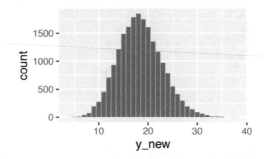

FIGURE 12.11: Posterior predictive model of the number of anti-discrimination laws in Minnesota, simulated from scratch.

12.5 Model evaluation

To close out our analysis, let's evaluate the quality of our Poisson regression model of anti-discrimination laws. The first two of our model evaluation questions are easy to answer. **How fair is our model?** Though we don't believe there to be bias in the data collection process, we certainly do warn against mistaking the *general state-level trends* revealed by our analysis as cause to make assumptions about *individuals* based on their voting behavior or where they live. Further, as we noted up front, an analysis of the *number* of state laws shouldn't be confused with an analysis of the *quality* of state laws. **How wrong is our model?** Our `pp_check()` in Figure 12.8 demonstrated that our Poisson regression assumptions are reasonable.

Consider the third model evaluation question: **How accurate are our model predictions?** As we did for the state of Minnesota, we can examine the posterior predictive models for each of the 49 states in our dataset. The plot below illustrates the posterior predictive credible intervals for the number of laws in each state alongside the actual observed data. Overall, our model does better at anticipating the number of laws in `gop` states than in `dem` or `swing` states – more of the observed `gop` data points fall within the bounds of their credible intervals. As exhibited by the narrower intervals, we also have more posterior certainty about the gop states where the number of laws tends to be consistently small.

```
# Simulate posterior predictive models for each state
set.seed(84735)
poisson_predictions <- posterior_predict(equality_model, newdata = equality)

# Plot the posterior predictive models for each state
ppc_intervals_grouped(equality$laws, yrep = poisson_predictions,
                      x = equality$percent_urban,
                      group = equality$historical,
                      prob = 0.5, prob_outer = 0.95,
                      facet_args = list(scales = "fixed"))
```

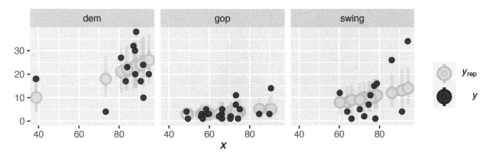

FIGURE 12.12: 50% and 95% posterior credible intervals (blue lines) for the number of anti-discrimination laws in a state. The actual number of laws are represented by the dark blue dots.

We can formalize our observations from Figure 12.12 with a `prediction_summary()`. Across the 49 states in our study, the observed numbers of anti-discrimination laws tend to fall only 3.4 laws, or 1.3 posterior standard deviations, from their posterior mean predictions. Given the range of the number of state laws, from 1 to 38, a typical prediction error of 3.4 is pretty good! Countering this positive with a negative, the observed number of laws for only roughly 78% of the states fall within their corresponding 95% posterior prediction interval. This means that our posterior predictive models "missed" or didn't anticipate the number of laws in 22%, or 11, of the 49 states.

```
prediction_summary(model = equality_model, data = equality)
    mae mae_scaled within_50 within_95
1 3.407      1.304    0.3265    0.7755
```

For due diligence, we also calculate the *cross-validated* posterior predictive accuracy in applying this model to a "new" set of 50 states, or the *same* set of 50 states under the recognition that our current data is just a random snapshot in time. In this case, the results are similar, suggesting that our model is *not* overfit to our sample data – we expect it to perform just as well at predicting "new" states.

```
# Cross-validation
set.seed(84735)
```

```
poisson_cv <- prediction_summary_cv(model = equality_model,
                                     data = equality, k = 10)
poisson_cv$cv
    mae mae_scaled within_50 within_95
1 3.788      1.264       0.3     0.725
```

12.6   Negative Binomial regression for overdispersed counts

In 2017, *Cards Against Humanity Saves America* launched a series of monthly surveys in order to get the "Pulse of the nation" (2017). We'll use their September 2017 poll results to model the number of books somebody has read in the past year (Y) by two predictors: their age (X_1) and their response to the question of whether they'd rather be *wise but unhappy* or *happy but unwise*:

$$X_2 = \begin{cases} 1 & \text{wise but unhappy} \\ 0 & \text{otherwise (i.e., happy but unwise).} \end{cases}$$

Because we really don't have any prior understanding of this relationship, we'll utilize weakly informative priors throughout our analysis. Moving on, let's load and process the **pulse_of_the_nation** data from the **bayesrules** package. In doing so, we'll remove some outliers, focusing on people that read fewer than 100 books:

```
# Load data
data(pulse_of_the_nation)
pulse <- pulse_of_the_nation %>%
  filter(books < 100)
```

Figure 12.13 reveals some basic patterns in readership. First, though most people read fewer than 11 books per year, there is a *lot* of variability in reading patterns from person to person. Further, though there appears to be a very weak relationship between book readership and one's age, readership appears to be slightly higher among people that would prefer wisdom over happiness (makes sense to us!).

```
ggplot(pulse, aes(x = books)) +
  geom_histogram(color = "white")
ggplot(pulse, aes(y = books, x = age)) +
  geom_point()
ggplot(pulse, aes(y = books, x = wise_unwise)) +
  geom_boxplot()
```

Given the skewed, count structure of our **books** variable Y, the Poisson regression tool is a reasonable *first approach* to modeling readership by a person's age and their prioritization of wisdom versus happiness:

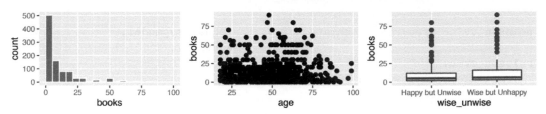

FIGURE 12.13: Survey responses regarding the number of books a person has read in the past year, and how this is associated with age and a person's prioritization of wisdom versus happiness.

```
books_poisson_sim <- stan_glm(
  books ~ age + wise_unwise,
  data = pulse, family = poisson,
  prior_intercept = normal(0, 2.5, autoscale = TRUE),
  prior = normal(0, 2.5, autoscale = TRUE),
  prior_aux = exponential(1, autoscale = TRUE),
  chains = 4, iter = 5000*2, seed = 84735)
```

BUT the results are definitely not great. Check out the `pp_check()` in Figure 12.14.

```
pp_check(books_poisson_sim) +
  xlab("books")
```

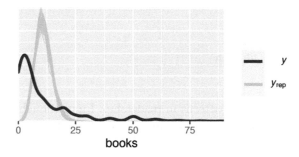

FIGURE 12.14: Posterior predictive summary for the Poisson regression model of readership.

Counter to the *observed* book readership, which is right skewed and tends to be below 11 books per year, the Poisson posterior simulations of readership are symmetric around 11 books year. Simply put, the Poisson regression model is wrong. *Why*? Well, recall that Poisson regression preserves the Poisson property of equal mean and variance. That is, it assumes that among subjects of similar age and perspectives on wisdom versus happiness (X_1 and X_2), the *typical* number of books read is roughly equivalent to the *variability* in books read. Yet the `pp_check()` highlights that, counter to this assumption, we actually observe *high variability* in book readership relative to a *low average* readership. We can confirm this observation with some numerical summaries. First, the discrepancy in the mean and variance in readership is true across *all* subjects in our survey. On average, people read 10.9 books per year, but the variance in book readership was a whopping 198 books[2]:

```
# Mean and variability in readership across all subjects
pulse %>%
  summarize(mean = mean(books), var = var(books))
# A tibble: 1 x 2
   mean   var
  <dbl> <dbl>
1  10.9  198.
```

When we `cut()` the age range into three groups, we see that this is also true among subjects in the same general age bracket and with the same take on wisdom versus happiness, i.e., among subjects with similar X_1 and X_2 values. For example, among respondents in the 45 to 72 year age bracket that prefer wisdom to happiness, the average readership was 12.5 books, a relatively small number in comparison to the variance of 270 books[2].

```
# Mean and variability in readership
# among subjects of similar age and wise_unwise response
pulse %>%
  group_by(cut(age,3), wise_unwise) %>%
  summarize(mean = mean(books), var = var(books))
# A tibble: 6 x 4
# Groups:    cut(age, 3) [3]
  `cut(age, 3)`  wise_unwise        mean   var
  <fct>          <fct>             <dbl> <dbl>
1 (17.9,45]      Happy but Unwise   9.23  138.
2 (17.9,45]      Wise but Unhappy  12.6   195.
3 (45,72]        Happy but Unwise   9.36  183.
4 (45,72]        Wise but Unhappy  12.5   270.
5 (72,99.1]      Happy but Unwise  12.6   236.
6 (72,99.1]      Wise but Unhappy  10.2    97.0
```

In this case, we say that book readership Y is **overdispersed**.

Overdispersion

A random variable Y is **overdispersed** if the *observed* variability in Y exceeds the variability *expected* by the assumed probability model of Y.

When our count response variable Y is too overdispersed to squeeze into the Poisson regression assumptions, we have some options. Two common options, which produce similar results, are to (1) include an *overdispersion parameter* in the Poisson data model or (2) utilize a non-Poisson data model. Since it fits nicely into the modeling framework we've established, we'll focus on the latter approach. To this end, the **Negative Binomial probability model** is a useful alternative to the Poisson when Y is overdispersed. *Like* the Poisson model, the Negative Binomial is suitable for count data $Y \in \{0, 1, 2, \ldots\}$. Yet *unlike* the Poisson, the Negative Binomial does *not* make the restrictive assumption that $E(Y) = \text{Var}(Y)$.

The Negative Binomial model

Let random variable Y be some count, $Y \in \{0,1,2,\ldots\}$, that can be modeled by the Negative Binomial with **mean parameter** μ and **reciprocal dispersion parameter** r:

$$Y|\mu,r \sim \text{NegBin}(\mu,r).$$

Then Y has conditional pmf

$$f(y|\mu,r) = \binom{y+r-1}{r}\left(\frac{r}{\mu+r}\right)^r\left(\frac{\mu}{\mu+r}\right)^y \quad \text{for } y \in \{0,1,2,\ldots\} \tag{12.5}$$

with *unequal* mean and variance:

$$E(Y|\mu,r) = \mu \quad \text{and} \quad \text{Var}(Y|\mu,r) = \mu + \frac{\mu^2}{r}.$$

Comparisons to the Poisson model

For *large* reciprocal dispersion parameters r, $\text{Var}(Y) \approx E(Y)$ and Y behaves much like a Poisson count variable. For *small* r, $\text{Var}(Y) > E(Y)$ and Y is **overdispersed** in comparison to a Poisson count variable with the same mean. Figure 12.15 illustrates these themes by example.

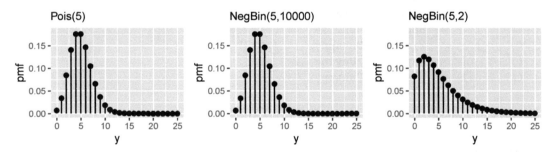

FIGURE 12.15: Pmfs for various Poisson and Negative Binomial random variables with a common mean of 5. The Negative Binomial model with a large dispersion parameter (r = 10000) is very similar to the Poisson, whereas that with a small dispersion parameter (r = 2) is relatively overdispersed.

To make the switch to the more flexible Negative Binomial regression model of readership, we can simply swap out a Poisson data model for a Negative Binomial data model. In doing so, we also pick up the extra reciprocal dispersion parameter, $r > 0$, for which the Exponential provides a reasonable prior structure. Our Negative Binomial regression model, along with the weakly informative priors scaled by `stan_glm()` and obtained by `prior_summary()`, follows:

data: $Y_i | \beta_0, \beta_1, \beta_2, r \overset{ind}{\sim} \text{NegBin}(\mu_i, r)$ with $\log(\mu_i) = \beta_0 + \beta_1 X_{i1} + \beta_2 X_{i2}$

priors: $\beta_{0c} \sim N(0, 2.5^2)$
$\beta_1 \sim N(0, 0.15^2)$
$\beta_2 \sim N(0, 5.01^2)$
$r \sim \text{Exp}(1)$

$$(12.6)$$

Similarly, to simulate the posterior of regression parameters $(\beta_0, \beta_1, \beta_2, r)$, we can swap out poisson for neg_binomial_2 in stan_glm():

```
books_negbin_sim <- stan_glm(
  books ~ age + wise_unwise,
  data = pulse, family = neg_binomial_2,
  prior_intercept = normal(0, 2.5, autoscale = TRUE),
  prior = normal(0, 2.5, autoscale = TRUE),
  prior_aux = exponential(1, autoscale = TRUE),
  chains = 4, iter = 5000*2, seed = 84735)

# Check out the priors
prior_summary(books_negbin_sim)
```

The results are fantastic. By incorporating more flexible assumptions about the variability in book readership, the posterior simulation of book readership very closely matches the observed behavior in our survey data (Figure 12.16). That is, the Negative Binomial regression model is not wrong (or, at least is much less wrong than the Poisson model).

```
pp_check(books_negbin_sim) +
  xlim(0, 75) +
  xlab("books")
```

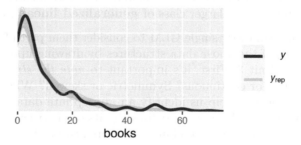

FIGURE 12.16: A posterior predictive check for the Negative Binomial regression model of readership.

With this peace of mind, we can continue just as we would with a Poisson analysis. Mainly, since it utilizes a log transform, the interpretation of the Negative Binomial regression coefficients follows the same framework as in the Poisson setting. Consider some posterior punchlines, supported by the `tidy()` summaries below:

- When controlling for a person's prioritization of wisdom versus happiness, there's no significant association between age and book readership – 0 is squarely in the 80% posterior credible interval for `age` coefficient β_1.
- When controlling for a person's age, people that prefer wisdom over happiness tend to read more than those that prefer happiness over wisdom – the 80% posterior credible interval for β_2 is comfortably above 0. Assuming they're the same age, we'd expect a person that prefers wisdom to read 1.3 *times* as many, or 30% more, books as somebody that prefers happiness ($e^{0.266} = 1.3$).

```
# Numerical summaries
tidy(books_negbin_sim, conf.int = TRUE, conf.level = 0.80)
# A tibble: 3 x 5
  term            estimate std.error conf.low conf.high
  <chr>              <dbl>     <dbl>    <dbl>     <dbl>
1 (Intercept)       2.23      0.131    2.07      2.41
2 age               0.000365  0.00239 -0.00270   0.00339
3 wise_unwiseWis~   0.266     0.0798   0.162     0.368
```

12.7   Generalized linear models: Building on the theme

Though the Normal, Poisson, and Negative Binomial data structures are common among Bayesian regression models, we're not limited to just these options. We can also use `stan_glm()` to fit models with Binomial, Gamma, inverse Normal, and other data structures. *All* of these options belong to a larger class of **generalized linear models** (GLMs).

We needn't march through every single GLM to consider them part of our toolbox. We can build a GLM with *any* of the above data structures by drawing upon the principles we've developed throughout Unit 3. First, it's important to *note the structure in our response variable Y*. Is Y discrete or continuous? Symmetric or skewed? What range of values can Y take? These questions can help us identify an appropriate data structure. Second, let $E(Y|\ldots)$ denote the average Y value as defined by its data structure. For all GLMs, the dependence of $E(Y|\ldots)$ on a linear combination of predictors $(X_1, X_2, \ldots, X_p)$ is expressed by

$$g\left(E(Y|\ldots)\right) = \beta_0 + \beta_1 X_1 + \beta_2 X_2 + \cdots + \beta_p X_p$$

where the appropriate **link function** $g(\cdot)$ depends upon the data structure. For example, in Normal regression, the data is modeled by

$$Y_i|\beta_0, \beta_1, \cdots, \beta_p, \sigma \sim N(\mu_i, \sigma^2)$$

and the dependence of $E(Y_i|\beta_0, \beta_1, \cdots, \beta_p, \sigma) = \mu_i$ on the predictors by

$$g(\mu_i) = \mu_i = \beta_0 + \beta_1 X_{i1} + \beta_2 X_{i2} + \cdots + \beta_p X_{ip}.$$

Thus, Normal regression utilizes an **identity link function** since $g(\mu_i)$ is equal to μ_i itself. In Poisson regression, the count data is modeled by

$$Y_i | \beta_0, \beta_1, \cdots, \beta_p \sim Pois(\lambda_i)$$

and the dependence of $E(Y_i | \beta_0, \beta_1, \cdots, \beta_p) = \lambda_i$ on the predictors by

$$g(\lambda_i) := \log(\lambda_i) = \beta_0 + \beta_1 X_{i1} + \beta_2 X_{i2} + \cdots + \beta_p X_{ip}.$$

Thus, Poisson regression utilizes a **log link function** since $g(\lambda_i) = \log(\lambda_i)$. The same is true for Negative Binomial regression. We'll dig into one more GLM, **logistic regression**, in the next chapter. We hope that our survey of these four specific tools (Normal, Poisson, Negative Binomial, and logistic regression) empowers you to implement other GLMs in your own Bayesian practice.

12.8 Chapter summary

Let response variable $Y \in \{0, 1, 2, \ldots\}$ be a discrete *count* of events that occur in a fixed interval of time or space. In this context, using Normal regression to model Y by predictors $(X_1, X_2, \ldots, X_p)$ is often inappropriate – it assumes that Y is symmetric and can be negative. The **Poisson regression model** offers a promising alternative:

$$
\begin{aligned}
\text{data:} \quad & Y_i | \beta_0, \beta_1, \ldots, \beta_p \stackrel{ind}{\sim} Pois(\lambda_i) \quad \text{with} \quad \log(\lambda_i) = \beta_0 + \beta_1 X_{i1} + \cdots + \beta_p X_{ip} \\
\text{priors:} \quad & \beta_{0c} \sim N(m_0, s_0^2) \\
& \beta_1 \sim N(m_1, s_1^2) \\
& \quad \vdots \\
& \beta_p \sim N(m_p, s_p^2)
\end{aligned}
$$

$$(12.7)$$

One major constraint of Poisson regression is its assumption that, at any set of predictor values, the typical value of Y and variability in Y are equivalent. Thus, when Y is **overdispersed**, i.e., its variability exceeds assumptions, we might instead utilize the more flexible **Negative Binomial regression model**:

$$
\begin{aligned}
\text{data:} \quad & Y_i | \beta_0, \beta_1, \ldots, \beta_p, r \stackrel{ind}{\sim} \text{NegBin}(\mu_i, r) \quad \text{with} \quad \log(\mu_i) = \beta_0 + \beta_1 X_{i1} + \cdots + \beta_p X_{ip} \\
\text{priors:} \quad & \beta_{0c} \sim N(m_0, s_0^2) \\
& \beta_1 \sim N(m_1, s_1^2) \\
& \quad \vdots \\
& \beta_p \sim N(m_p, s_p^2) \\
& r \sim \text{Exp}(\ldots)
\end{aligned}
$$

$$(12.8)$$

12.9 Exercises

12.9.1 Conceptual exercises

Exercise 12.1 (Warming up).

a)   Give a *new* example (i.e., not the same as from the chapter) in which we would want to use a Poisson, instead of Normal, regression model.
b)   The Poisson regression model uses a log link function, while the Normal regression model uses an identity link function. Explain in one or two sentences what a link function is.
c)   Explain why the log link function is used in Poisson regression.
d)   List the four assumptions for a Poisson regression model.

Exercise 12.2 (Poisson versus Negative Binomial). Specify whether Poisson regression, Negative Binomial regression, both, or neither fit with each situation described below.

a)   The response variable is a count.
b)   The link is a log.
c)   The link is the identity.
d)   We need to account for overdispersion.
e)   The response is a count variable, and as the expected response increases, the variability also increases.

Exercise 12.3 (Why use a Negative Binomial). You and your friend Nico are in a two-person *Bayes Rules!* book club. How lovely! Nico has read only part of this chapter, and now they know about Poisson regression, but not Negative Binomial regression. Be a good friend and answer their questions.

a)   What's the shortcoming of Poisson regression?
b)   How does Negative Binomial regression address the shortcoming of Poisson regression?
c)   Are there any situations in which Poisson regression would be a better choice than Negative Binomial?

Exercise 12.4 (Interpreting Poisson regression coefficients). As modelers, the ability to interpret regression coefficients is of utmost importance. Let Y be the number of "Likes" a tweet gets in an hour, X_1 be the number of followers the person who wrote the tweet has, and X_2 indicate whether the tweet includes an emoji ($X_2 = 1$ if there is an emoji, $X_2 = 0$ if there is no emoji). Further, suppose $Y|\beta_0, \beta_1, \beta_2 \sim \text{Pois}(\lambda)$ with

$$\log(\lambda) = \beta_0 + \beta_1 X_1 + \beta_2 X_2.$$

a)   Interpret e^{β_0} in context.
b)   Interpret e^{β_1} in context.
c)   Interpret e^{β_2} in context.
d)   Provide the model equation for the expected number of "Likes" for a tweet in one hour, when the person who wrote the tweet has 300 followers, and the tweet does not use an emoji.

12.9.2 Applied exercises

Exercise 12.5 (Eagles: get to know the data). In the next exercises, you will explore how the number of eagle sightings in Ontario, Canada has changed over time. Since this context is unfamiliar to us, we'll utilize weakly informative priors throughout. We'll balance this prior uncertainty by the `bald_eagles` data in the **bayesrules** package, which includes data on bald eagle sightings during 37 different one-week observation periods. First, get to know this data.

(a) Construct and discuss a univariate plot of `count`, the number of eagle sightings across the observation periods.

(b) Construct and discuss a plot of `count` versus `year`.

(c) In exploring the number of eagle sightings over time, it's important to consider the fact that the length of the observation periods vary from year to year, ranging from 134 to 248.75 hours. Update your plot from part b to also include information about the observation length in `hours` and comment on your findings.

Exercise 12.6 (Eagles: first model attempt). Our next goal is to model the relationship between bald eagle counts Y by year X_1 when controlling for the number of observation hours X_2. To begin, consider a *Normal* regression model of Y versus X_1 and X_2.

(a) Simulate the model posterior and check the `prior_summary()`.

(b) Use careful notation to write out the complete Bayesian structure of the Normal regression model of Y by X_1 and X_2.

(c) Complete a `pp_check()` for the Normal model. Use this to explain whether the model is "good" and, if not, what assumptions it makes that are inappropriate for the bald eagle analysis.

Exercise 12.7 (Eagles: second model attempt). Let's try to do better. Consider a *Poisson* regression model of Y versus X_1 and X_2.

(a) In the bald eagle analysis, why might a *Poisson* regression approach be more appropriate than a *Normal* regression approach?

(b) Simulate the posterior of the Poisson regression model of Y versus X_1 and X_2. Check the `prior_summary()`.

(c) Use careful notation to write out the complete Bayesian structure of the Poisson regression model of Y by X_1 and X_2.

(d) Complete a `pp_check()` for the Poisson model. Use this to explain whether the model is "good" and, if not, what assumptions it makes that are inappropriate for the bald eagle analysis.

Exercise 12.8 (Eagles: an even better model). The Poisson regression model of bald eagle counts (Y) by year (X_1) and observation hours (X_2), was pretty good. Let's see if a Negative Binomial approach is even better.

(a) Simulate the model posterior and use a `pp_check()` to confirm whether the Negative Binomial model is reasonable.

(b) Use careful notation to write out the complete Bayesian structure of the Negative Binomial regression model of Y by X_1 and X_2.

(c) Interpret the posterior median estimates of the regression coefficients on `year` and `hours`, β_1 and β_2. Do so on the unlogged scale.

(d)   Construct and interpret a 95% posterior credible interval for the `year` coefficient.

(e)   When controlling for the number of observation hours, do we have ample evidence that the rate of eagle sightings has increased over time?

Exercise 12.9 (Eagles: model evaluation). Finally, let's evaluate the quality of our Negative Binomial bald eagle model.

(a)   How fair is the model?

(b)   How wrong is the model?

(c)   How accurate are the model predictions?

Exercise 12.10 (Open exercise: AirBnB). The `airbnb_small` data in the **bayesrules** package contains information on AirBnB rentals in Chicago. This data was originally collated by Trinh and Ameri (2016) and distributed by Legler and Roback (2021). In this open-ended exercise, build, interpret, and evaluate a model of the number of `reviews` an AirBnB property has by its `rating`, `district`, `room_type`, and the number of guests it `accommodates`.

13

Logistic Regression

In Chapter 12 we learned that not every regression is *Normal*. In Chapter 13, we'll confront another fact: not every response variable Y is *quantitative*. Rather, we might wish to model Y, whether or not a singer wins a Grammy, by their album reviews. Or we might wish to model Y, whether or not a person votes, by their age and political leanings. Across these examples, and in **classification** settings in general, Y is *categorical*. In Chapters 13 and 14 we'll dig into two classification techniques: Bayesian logistic regression and naive Bayesian classification.

Consider the following data story. Suppose we again find ourselves in Australia, the city of Perth specifically. Located on the southwest coast, Perth experiences dry summers and wet winters. Our goal will be to predict whether or not it will rain tomorrow. That is, we want to model Y, a **binary categorical response variable**, converted to a 0-1 indicator for convenience:

$$Y = \begin{cases} 1 & \text{if rain tomorrow} \\ 0 & \text{otherwise} \end{cases}.$$

Though there are various potential predictors of rain, we'll consider just three:

$$X_1 = \text{today's humidity at 9 a.m. (percent)}$$
$$X_2 = \text{today's humidity at 3 p.m. (percent)}$$
$$X_3 = \text{whether or not it rained today.}$$

Our vague prior understanding is that on an average day, there's a roughly 20% chance of rain. We also have a vague sense that the chance of rain increases when preceded by a day with high humidity or rain itself, but we're foggy on the rate of this increase. Our eventual goal is to combine this prior understanding with data to model Y by one or more of the predictors above. Certainly, since Y is **categorical**, taking our Normal and Poisson regression hammers to this task would be the wrong thing. In Chapter 13, you will pick up a new tool: the **Bayesian logistic regression model** for binary response variables Y.

◎ **Goals**

- Build a Bayesian logistic regression model of a binary categorical variable Y by predictors $X = (X_1, X_2, ..., X_p)$.
- Utilize this model to *classify*, or *predict*, the outcome of Y for a given set of predictor values X.
- Evaluate the quality of this classification technique.

DOI: 10.1201/9780429288340-13

```
# Load packages
library(bayesrules)
library(rstanarm)
library(bayesplot)
library(tidyverse)
library(tidybayes)
library(broom.mixed)
```

13.1　Pause: Odds & probability

Before jumping into logistic regression, we'll pause to review the concept of **odds** and its relationship to **probability**. Throughout the book, we've used probability π to communicate the uncertainty of a given event of interest (e.g., rain tomorrow). Alternatively, we can cite the corresponding odds of this event, defined by the probability that the event happens relative to the probability that it doesn't happen:

$$\text{odds} = \frac{\pi}{1-\pi}. \tag{13.1}$$

Mathematically then, whereas a probability π is restricted to values between 0 and 1, odds can range from 0 on up to infinity. To *interpret* odds across this range, let π be the probability that it rains tomorrow and consider three different scenarios. If the probability of rain tomorrow is $\pi = 2/3$, then the probability it *doesn't rain* is $1 - \pi = 1/3$ and the *odds* of rain are 2:

$$\text{odds of rain } = \frac{2/3}{1-2/3} = 2.$$

That is, it's *twice* as likely to rain than to not rain. If the probability of rain tomorrow is $\pi = 1/3$, then the probability it *doesn't rain* is $2/3$ and the *odds* of rain are $1/2$:

$$\text{odds of rain } = \frac{1/3}{1-1/3} = \frac{1}{2}.$$

That is, it's *half* as likely to rain than to not rain tomorrow. Finally, if the chances of rain or no rain tomorrow are 50-50, then the *odds* of rain are 1:

$$\text{odds of rain } = \frac{1/2}{1-1/2} = 1.$$

That is, it's *equally likely* to rain or not rain tomorrow. These scenarios illuminate the general principles by which to interpret the odds of an event.

Interpreting odds

Let an event of interest have probability $\pi \in [0,1]$ and corresponding odds $\pi/(1-\pi) \in$

$[0, \infty)$. Across this spectrum, comparing the odds to 1 provides perspective on an event's uncertainty:

- The odds of an event are **less than 1** if and only if the event's chances are less than 50-50, i.e., $\pi < 0.5$.
- The odds of an event are **equal to 1** if and only if the event's chances are 50-50, i.e., $\pi = 0.5$.
- The odds of an event are **greater than 1** if and only if event's chances are greater than 50-50, i.e., $\pi > 0.5$.

Finally, just as we can convert probabilities to odds, we can convert odds to probability by rearranging (13.1):

$$\pi = \frac{\text{odds}}{1 + \text{odds}}. \tag{13.2}$$

Thus, if we learn that the odds of rain tomorrow are 4 to 1, then there's an 80% chance of rain:

$$\pi = \frac{4}{1 + 4} = 0.8.$$

13.2 Building the logistic regression model

13.2.1 Specifying the data model

Returning to our rain analysis, let Y_i be the 0/1 indicator of whether or not it rains tomorrow for any given day i. To begin, we'll model Y_i by a single predictor, today's 9 a.m. humidity level X_{i1}. As noted above, neither the Normal nor Poisson regression models are appropriate for this task. So what is?

> **? Quiz Yourself!**
>
> What's an appropriate model structure for data Y_i?
>
> a. Bernoulli (or, equivalently, Binomial with 1 trial)
> b. Gamma
> c. Beta

One simple question can help us narrow in on an appropriate model structure for any response variable Y_i: what values can Y_i take and what probability models assume this same set of values? Here, Y_i is a *discrete* variable which can only take two values, 0 or 1. Thus, the **Bernoulli probability model** is the best candidate for the data. Letting π_i denote the *probability* of rain on day i,

$$Y_i | \pi_i \sim \text{Bern}(\pi_i)$$

with expected value

$$E(Y_i|\pi_i) = \pi_i.$$

To complete the structure of this Bernoulli data model, we must specify how the expected value of rain π_i depends upon predictor X_{i1}. To this end, the **logistic regression** model is part of the broader class of generalized linear models highlighted in Section 12.7. Thus, we can identify an appropriate link function of π_i, $g(\cdot)$, that's linearly related to X_{i1}:

$$g(\pi_i) = \beta_0 + \beta_1 X_{i1}.$$

Keeping in mind the principles that went into building the Poisson regression model, take a quick quiz to reflect upon what $g(\pi_i)$ might be appropriate.[1]

---

❷ Quiz Yourself!

Let π_i and $\text{odds}_i = \pi_i/(1 - \pi_i)$ denote the probability and corresponding odds of rain tomorrow. What's a reasonable assumption for the dependence of tomorrow's rain probability π_i on today's 9 a.m. humidity X_{i1}?

   a. $\pi_i = \beta_0 + \beta_1 X_{i1}$
   b. $\text{odds}_i = \beta_0 + \beta_1 X_{i1}$
   c. $\log(\pi_i) = \beta_0 + \beta_1 X_{i1}$
   d. $\log(\text{odds}_i) = \beta_0 + \beta_1 X_{i1}$

Our goal here is to write the Bernoulli mean π_i, or a *function* of this mean $g(\pi_i)$, as a linear function of predictor X_{i1}, $\beta_0 + \beta_1 X_{i1}$. Among the options presented in the quiz above, the first three would be mistakes. Whereas the line defined by $\beta_0 + \beta_1 X_{i1}$ can span the entire real line, π_i (option a) is limited to values between 0 and 1 and odds_i (option b) are limited to positive values. Further, when evaluated at π_i values between 0 and 1, $\log(\pi_i)$ (option c) is limited to negative values. Among these options, $\log(\pi_i/(1 - \pi_i))$ (option d) is the only option that, like $\beta_0 + \beta_1 X_{i1}$, spans the entire real line. Thus, the most reasonable option is to assume that π_i depends upon predictor X_{i1} through the **logit link function** $g(\pi_i) = \log(\pi_i/(1 - \pi_i))$:

$$Y_i|\beta_0, \beta_1 \overset{ind}{\sim} \text{Bern}(\pi_i) \quad \text{with} \quad \log\left(\frac{\pi_i}{1 - \pi_i}\right) = \beta_0 + \beta_1 X_{i1}. \tag{13.3}$$

That is, we assume that the **log(odds of rain)** is linearly related to 9 a.m. humidity. To work on scales that are (much) easier to interpret, we can rewrite this relationship in terms of odds and probability, the former following from properties of the log function and the latter following from (13.2):

$$\frac{\pi_i}{1 - \pi_i} = e^{\beta_0 + \beta_1 X_{i1}} \quad \text{and} \quad \pi_i = \frac{e^{\beta_0 + \beta_1 X_{i1}}}{1 + e^{\beta_0 + \beta_1 X_{i1}}}. \tag{13.4}$$

There are two key features of these transformations, as illustrated by Figure 13.1. First, the relationships on the odds and probability scales are now represented by *nonlinear* functions. Further, and fortunately, these transformations preserve the properties of odds (which must be non-negative) and probability (which must be between 0 and 1).

---

[1]Answer: d

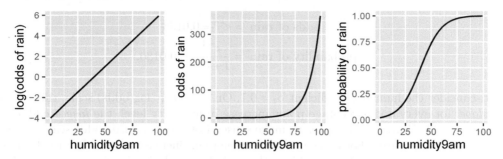

FIGURE 13.1: An example relationship between rain and humidity on the log(odds), odds, and probability scales.

In examining (13.3) and Figure 13.1, notice that parameters β_0 and β_1 take on the usual intercept and slope meanings when describing the *linear* relationship between 9 a.m. humidity (X_{i1}) and the *log*(odds of rain). Yet the parameter meanings change when describing the *nonlinear* relationship between 9 a.m. humidity and the *odds* of rain (13.4). We present a general framework for interpreting logistic regression coefficients here and apply it below.

Interpreting logistic regression coefficients

Let Y be a binary indicator of some event of interest which occurs with probability π. Consider the logistic regression model of Y with predictors $(X_1, X_2, \ldots, X_p)$:

$$\log(\text{odds}) = \log\left(\frac{\pi}{1-\pi}\right) = \beta_0 + \beta_1 X_1 + \cdots + \beta_p X_p.$$

Interpreting β_0

When $(X_1, X_2, \ldots, X_p)$ are all 0, β_0 is the **log odds** of the event of interest and e^{β_0} is the **odds**.

Interpreting β_1

When controlling for the other predictors $(X_2, \ldots, X_p)$, let odds_x represent the odds of the event of interest when $X_1 = x$ and odds_{x+1} the odds when $X_1 = x + 1$. Then when X_1 increases by 1, from x to $x + 1$, β_1 is the typical **change in log odds** and e^{β_1} is the typical **multiplicative change in odds**:

$$\beta_1 = \log(\text{odds}_{x+1}) - \log(\text{odds}_x) \quad \text{and} \quad e^{\beta_1} = \frac{\text{odds}_{x+1}}{\text{odds}_x}.$$

For example, the prior plausible relationship plotted in Figure 13.1 on the log(odds), odds, and probability scales assumes that

$$\log\left(\frac{\pi}{1-\pi}\right) = -4 + 0.1 X_{i1}.$$

Though humidity levels so close to 0 are merely hypothetical in this area of the world, we extended the plots to the y-axis for perspective on the intercept value $\beta_0 = -4$. This intercept indicates that on hypothetical days with 0 percent humidity, the log(odds of rain) would be -4. Or, on more meaningful scales, rain would be very unlikely if preceded by a day with 0 humidity:

$$\text{odds of rain } = e^{-4} = 0.0183$$

$$\text{probability of rain } = \frac{e^{-4}}{1 + e^{-4}} = 0.0180.$$

Next, consider the humidity coefficient $\beta_1 = 0.1$. On the *linear* log(odds) scale, this is simply the *slope*: for every one percentage point increase in humidity level, the *logged odds* of rain increase by 0.1. Huh? This is easier to make sense of on the *nonlinear* odds scale where the increase is *multiplicative*. For every one percentage point increase in humidity level, the *odds* of rain increase by 11%: $e^{0.1} = 1.11$. Since the probability relationship is a more complicated S-shaped curve, we cannot easily interpret β_1 on this scale. However, we can still gather some insights from the probability plot in Figure 13.1. The chance of rain is small when humidity levels hover below 20%, and rapidly increases as humidity levels inch up from 20% to 60%. Beyond 60% humidity, the probability model flattens out near 1, indicating a near certainty of rain.

13.2.2   Specifying the priors

To complete the **Bayesian logistic regression model** of Y, we must put prior models on our two regression parameters, β_0 and β_1. As usual, since these parameters can take any value in the real line, Normal priors are appropriate for both. We'll also assume independence among the priors and express our prior understanding of the model baseline β_0 through the centered intercept β_{0c}:

$$
\begin{aligned}
\text{data:} \quad & Y_i | \beta_0, \beta_1 \overset{ind}{\sim} \text{Bern}(\pi_i) \quad \text{with} \quad \log\left(\tfrac{\pi_i}{1-\pi_i}\right) = \beta_0 + \beta_1 X_{i1} \\
\text{priors:} \quad & \beta_{0c} \sim N\left(-1.4, 0.7^2\right) \\
& \beta_1 \sim N\left(0.07, 0.035^2\right).
\end{aligned}
\tag{13.5}
$$

Consider our prior tunings. Starting with the centered intercept β_{0c}, recall our prior understanding that on an average day, there's a roughly 20% chance of rain, i.e., $\pi \approx 0.2$. Thus, we set the prior mean for β_{0c} on the *log(odds)* scale to -1.4:

$$\log\left(\frac{\pi}{1-\pi}\right) = \log\left(\frac{0.2}{1-0.2}\right) \approx -1.4.$$

The range of this Normal prior indicates our vague understanding that the *log(odds of rain)* might also range from roughly -2.8 to 0 ($-1.4 \pm 2 * 0.7$). More meaningfully, we think that the *odds* of rain on an average day could be somewhere between 0.06 and 1:

$$\left(e^{-2.8}, e^0\right) \approx (0.06, 1)$$

and thus that the *probability* of rain on an average day could be somewhere between 0.057 and 0.50 (a pretty wide range in the context of rain):

$$\left(\frac{0.06}{1+0.06}, \frac{1}{1+1}\right) \approx (0.057, 0.50).$$

Next, our prior model on the humidity coefficient β_1 reflects our vague sense that the chance of rain increases when preceded by a day with high humidity, but we're foggy on the rate of this increase and are open to the possibility that it's nonexistent. Specifically, on the log(odds) scale, we assume that slope β_1 ranges somewhere between 0 and 0.14, and is most likely around 0.7. Or, on the odds scale, the odds of rain might increase anywhere from 0% to 15% for every extra percentage point in humidity level:

$$(e^0, e^{0.14}) = (1, 1.15).$$

In originally specifying these priors (before writing them down in the book), we combined a lot of trial and error with an understanding of how logistic regression coefficients work. Throughout this tuning process, we simulated data under a variety of prior models, each time asking ourselves if the results adequately reflected our understanding of rain. To this end, we followed the same framework used in Section 12.1.2 to tune Poisson regression priors.

First, we use the more and more familiar `stan_glm()` function with `prior_PD = TRUE` to simulate 20,000 sets of parameters (β_0, β_1) from the prior models. In doing so, we specify `family = binomial` to indicate that ours is a logistic regression model with a data structure specified by a Bernoulli / Binomial model (not `gaussian` or `poisson`). Even though our model (13.5) doesn't use default weakly informative priors, and thus no data is used in the prior simulation, `stan_glm()` still requires that we specify a dataset. Thus, we also load in the `weather_perth` data from the `bayesrules` package which contains 1000 days of Perth weather data.[2]

```
# Load and process the data
data(weather_perth)
weather <- weather_perth %>%
  select(day_of_year, raintomorrow, humidity9am, humidity3pm, raintoday)

# Run a prior simulation
rain_model_prior <- stan_glm(raintomorrow ~ humidity9am,
                  data = weather, family = binomial,
                  prior_intercept = normal(-1.4, 0.7),
                  prior = normal(0.07, 0.035),
                  chains = 4, iter = 5000*2, seed = 84735,
                  prior_PD = TRUE)
```

Each of the resulting 20,000 prior plausible pairs of β_0 and β_1 describe a prior plausible relationship between the probability of rain tomorrow and today's 9 a.m. humidity:

$$\pi = \frac{e^{\beta_0 + \beta_1 X_1}}{1 + e^{\beta_0 + \beta_1 X_1}}.$$

We plot just 100 of these prior plausible relationships below (Figure 13.2 left). These adequately reflect our prior understanding that the probability of rain increases with humidity, as well as our prior uncertainty around the rate of this increase. Beyond this relationship between rain and humidity, we also want to confirm that our priors reflect our understanding of the overall chance of rain in Perth. To this end, we simulate 100 datasets from the prior model. From each dataset (labeled by `.draw`) we utilize `group_by()` with `summarize()` to record the proportion of predicted outcomes Y (`.prediction`) that are 1, i.e., the proportion of days on which it rained. Figure 13.2 (right) displays a histogram of these 100 simulated rain rates from our 100 prior simulated datasets. The percent of days on which it rained ranged from as low as roughly 5% in one dataset to as high as roughly

---

[2] `weather_perth` is a wrangled subset of the `weatherAUS` dataset in the `rattle` package.

50% in another. This *does* adequately match our prior understanding and uncertainty about rain in Perth. In contrast, if our prior predictions tended to be centered around high values, we'd question our prior tuning since we don't believe that Perth is a rainy place.

```r
set.seed(84735)

# Plot 100 prior models with humidity
weather %>%
  add_fitted_draws(rain_model_prior, n = 100) %>%
  ggplot(aes(x = humidity9am, y = raintomorrow)) +
    geom_line(aes(y = .value, group = .draw), size = 0.1)

# Plot the observed proportion of rain in 100 prior datasets
weather %>%
  add_predicted_draws(rain_model_prior, n = 100) %>%
  group_by(.draw) %>%
  summarize(proportion_rain = mean(.prediction == 1)) %>%
  ggplot(aes(x = proportion_rain)) +
    geom_histogram(color = "white")
```

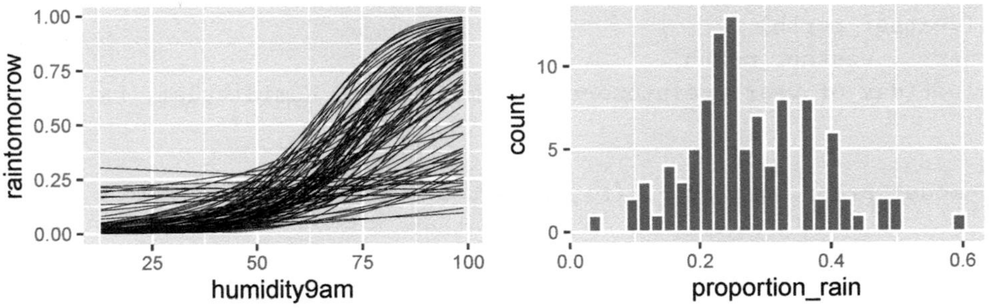

FIGURE 13.2: 100 datasets were simulated from the prior models. For each, we plot the relationship between the probability of rain and the previous day's humidity level (left) and the observed proportion of days on which it rained (right).

13.3   Simulating the posterior

With the hard work of prior tuning taken care of, consider our data. A visual examination illuminates some patterns in the relationship between rain and humidity. First, a quick jitter plot reveals that rainy days are much less common than non-rainy days. This is especially true at lower humidity levels. Second, though we don't have enough data to get a stable sense of the probability of rain at each possible humidity level, we *can* examine the probability of rain on days with *similar* humidity levels. To this end, we cut() humidity levels into brackets, and then calculate the proportion of days within each bracket that see rain. In general, notice that the chance of rain seems to hover around 10% when humidity levels are below 60%, and then increase from there. This observation is in sync with our prior understanding that rain and humidity are positively associated. However, whereas our prior understanding left open the possibility that the probability of rain might near 100% as humidity levels max out (Figure 13.2), we see in the data that the chance of rain at the highest humidity levels barely breaks 50%.

```
ggplot(weather, aes(x = humidity9am, y = raintomorrow)) +
  geom_jitter(size = 0.2)

# Calculate & plot the rain rate by humidity bracket
weather %>%
  mutate(humidity_bracket =
            cut(humidity9am, breaks = seq(10, 100, by = 10))) %>%
  group_by(humidity_bracket) %>%
  summarize(rain_rate = mean(raintomorrow == "Yes")) %>%
  ggplot(aes(x = humidity_bracket, y = rain_rate)) +
    geom_point() +
    theme(axis.text.x = element_text(angle = 45, vjust = 0.5))
```

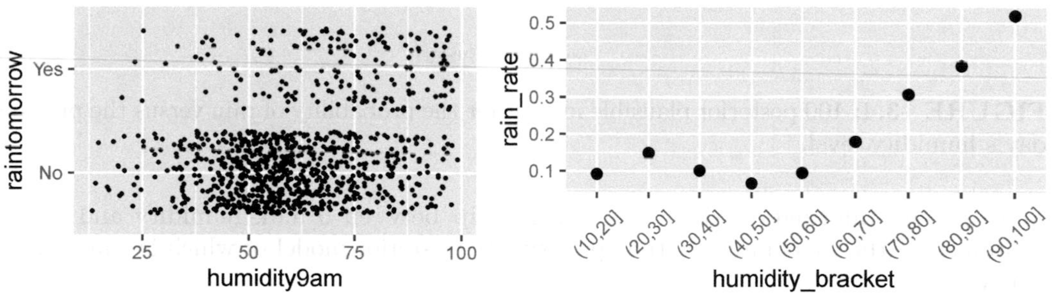

FIGURE 13.3: A jitter plot of rain outcomes versus humidity level (left). A scatterplot of the proportion of days that see rain by humidity bracket (right).

To simulate the posterior model of our logistic regression model parameters β_0 and β_1 (13.5), we can `update()` the `rain_model_prior` simulation in light of this new `weather` data:/indexfun%7Bupdate()}

```
# Simulate the model
rain_model_1 <- update(rain_model_prior, prior_PD = FALSE)
```

Not that you would forget (!), but we should check diagnostic plots for the stability of our simulation results before proceeding:

```
# MCMC trace, density, & autocorrelation plots
mcmc_trace(rain_model_1)
mcmc_dens_overlay(rain_model_1)
mcmc_acf(rain_model_1)
```

We plot 100 posterior plausible models in Figure 13.4. Naturally, upon incorporating the `weather` data, these models are much less variable than the prior counterparts in Figure 13.2 – i.e., we're much more certain about the relationship between rain and humidity. We now understand that the probability of rain steadily increases with humidity, yet it's not until a humidity of roughly 95% that we reach the tipping point when rain becomes more

likely than not. In contrast, when today's 9 a.m. humidity is below 25%, it's *very* unlikely
to rain tomorrow.

```
weather %>%
  add_fitted_draws(rain_model_1, n = 100) %>%
  ggplot(aes(x = humidity9am, y = raintomorrow)) +
    geom_line(aes(y = .value, group = .draw), alpha = 0.15) +
    labs(y = "probability of rain")
```

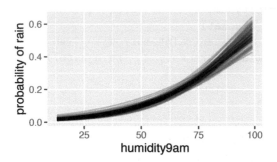

FIGURE 13.4: 100 posterior plausible models for the probability of rain versus the previous
day's humidity level.

More precise information about the relationship between 9 a.m. humidity and rain is
contained in the β_1 parameter, the approximate posterior model of which is summarized
below:

```
# Posterior summaries on the log(odds) scale
posterior_interval(rain_model_1, prob = 0.80)
                    10%       90%
(Intercept) -5.08785 -4.13450
humidity9am   0.04147   0.05487

# Posterior summaries on the odds scale
exp(posterior_interval(rain_model_1, prob = 0.80))
                    10%       90%
(Intercept) 0.006171 0.01601
humidity9am 1.042339 1.05641
```

For every one percentage point increase in today's 9 a.m. humidity, there's an 80% posterior
chance that the *log(odds of rain)* increases by somewhere between 0.0415 and 0.0549. This
rate of increase is *less* than our 0.07 prior mean for β_1 – the chance of rain *does* significantly
increase with humidity, just not to the degree we had anticipated. More meaningfully, for
every one percentage point increase in today's 9 a.m. humidity, the *odds* of rain increase by
somewhere between 4.2% and 5.6%: $(e^{0.0415}, e^{0.0549}) = (1.042, 1.056)$. Equivalently, for every
fifteen percentage point increase in today's 9 a.m. humidity, the odds of rain roughly *double*:
$(e^{15*0.0415}, e^{15*0.0549}) = (1.86, 2.28)$.

13.4   Prediction & classification

Beyond using our Bayesian logistic regression model to better understand the relationship between today's 9 a.m. humidity and tomorrow's rain, we also want to *predict* whether or not it will rain tomorrow. For example, suppose you're in Perth *today* and experienced 99% humidity at 9 a.m.. Yuk. To predict whether it will rain tomorrow, we can approximate the posterior predictive model for the *binary outcomes* of Y, whether or not it rains, where

$$Y|\beta_0, \beta_1 \sim \text{Bern}(\pi) \quad \text{with} \quad \log\left(\frac{\pi}{1-\pi}\right) = \beta_0 + \beta_1 * 99. \tag{13.6}$$

To this end, the `posterior_predict()` function simulates 20,000 rain outcomes Y, one per each of the 20,000 parameter sets in our Markov chain:

```
# Posterior predictions of binary outcome
set.seed(84735)
binary_prediction <- posterior_predict(
  rain_model_1, newdata = data.frame(humidity9am = 99))
```

To really connect with the prediction concept, we can also simulate the posterior predictive model from scratch. From each of the 20,000 pairs of posterior plausible pairs of β_0 (`Intercept`) and β_1 (`humidity9am`) in our Markov chain simulation, we calculate the log(odds of rain) (13.6). We then transform the log(odds) to obtain the odds and probability of rain. Finally, from each of the 20,000 probability values π, we simulate a Bernoulli outcome of rain, $Y \sim \text{Bern}(\pi)$, using `rbinom()` with `size = 1`:

```
# Posterior predictions of binary outcome - from scratch
set.seed(84735)
rain_model_1_df <- as.data.frame(rain_model_1) %>%
  mutate(log_odds = `(Intercept)` + humidity9am*99,
         odds = exp(log_odds),
         prob = odds / (1 + odds),
         Y = rbinom(20000, size = 1, prob = prob))

# Check it out
head(rain_model_1_df, 2)
  (Intercept) humidity9am log_odds    odds   prob Y
1      -4.244     0.04455  0.16577  1.1803 0.5413 0
2      -4.207     0.04210 -0.03934  0.9614 0.4902 1
```

For example, our first log(odds) and probability values are calculated by plugging $\beta_0 = -4.244$ and $\beta_1 = 0.0446$ into (13.6) and transforming it to the probability scale:

$$\log\left(\frac{\pi}{1-\pi}\right) = -4.244 + 0.0446 * 99 = 0.166$$

$$\pi = \frac{e^{-4.244+0.0446*99}}{1 + e^{-4.244+0.0446*99}} = 0.54.$$

Subsequently, we randomly draw a binary Y outcome from the Bern(0.54) model. Though the 0.54 probability of rain in this scenario exceeds 0.5, we ended up observing no rain, $Y = 0$. Putting all 20,000 predictions together, examine the posterior predictive models of rain obtained from our two methods below. On days with 99% humidity, our Bayesian logistic regression model suggests that rain ($Y = 1$) is slightly more likely than not ($Y = 0$).

```
mcmc_hist(binary_prediction) +
  labs(x = "Y")
ggplot(rain_model_1_df, aes(x = Y)) +
  stat_count()
```

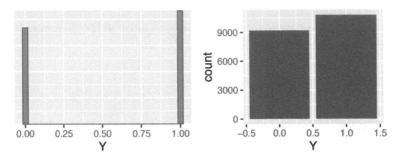

FIGURE 13.5: Posterior predictive model for the incidence of rain using a shortcut function (left) and from scratch (right).

As the name suggests, a common goal in a **classification** analysis is to turn our observations of the predicted probability of rain (π) or predicted outcome of rain (Y) into a binary, yes-or-no classification of Y. Take the following quiz to make this classification for yourself.

---

❷ Quiz Yourself!

Suppose it's 99% humidity at 9 a.m. today. Based on the posterior predictive model above, what binary classification would you make? Will it rain or not? Should we or shouldn't we carry an umbrella tomorrow?

Questions of classification are somewhat subjective, thus there's more than one reasonable answer to this quiz. *One* reasonable answer is this: yes, you should carry an umbrella. Among our 20,000 posterior predictions of Y, 10804 (or 54.02%) called for rain. Thus, since rain was more likely than no rain in the posterior predictive model of Y, it's reasonable to classify Y as "1" (rain).

```
# Summarize the posterior predictions of Y
table(binary_prediction)
binary_prediction
    0     1
 9196 10804
colMeans(binary_prediction)
     1
0.5402
```

In making this classification, we used the following **classification rule**:

- If more than half of our predictions predict $Y = 1$, classify Y as 1 (rain).
- Otherwise, classify Y as 0 (no rain).

Though it's a natural choice, we needn't use a 50% cut-off. We can utilize any cut-off between 0% and 100%.

Classification rule

Let Y be a binary response variable, 1 or 0. Further, let $\left(Y_{new}^{(1)}, Y_{new}^{(2)}, \ldots, Y_{new}^{(N)}\right)$ be N posterior predictions of Y calculated from an N-length Markov chain simulation and p be the proportion of these predictions for which $Y_{new}^{(i)} = 1$. Then for some user-chosen **classification cut-off** $c \in [0, 1]$, we can turn our posterior predictions into a binary classification of Y using the following rule:

- If $p \geq c$, then classify Y as 1.
- If $p < c$, then classify Y as 0.

An appropriate classification cut-off c should reflect the *context* of our analysis and the consequences of a misclassification: is it worse to underestimate the occurrence of $Y = 1$ or $Y = 0$? For example, suppose we dislike getting wet – we'd rather unnecessarily carry an umbrella than mistakenly leave it at home. To play it safe, we can then *lower* our classification cut-off to, say, 25%. That is, if even 25% of our posterior predictions call for rain, we'll classify Y as 1, and hence carry our umbrella. In contrast, suppose that we'd rather risk getting wet than to needlessly carry an umbrella. In this case, we can *raise* our classification cut-off to, say, 75%. Thus, we'll only dare classify Y as 1 if at least 75% of our posterior predictions call for rain. Though context can guide the process of selecting a classification cut-off c, examining the corresponding misclassification rates provides some guidance. We'll explore this process in the next section.

13.5   Model evaluation

Just as with regression models, it's critical to evaluate the quality of a classification model:

1.   How *fair* is the model?
2.   How *wrong* is the model?
3.   How *accurate* are the model's posterior **classifications**?

The first two questions have quick answers. We believe this weather analysis to be fair and innocuous in terms of its potential impact on society and individuals. To answer question (2), we can perform a posterior predictive check to confirm that data simulated from our posterior logistic regression model has features similar to the *original* data, and thus that the assumptions behind our Bayesian logistic regression model (13.5) are reasonable. Since the data outcomes we're simulating are binary, we must take a slightly different approach to this `pp_check()` than we did for Normal and Poisson regression. From each of 100 posterior simulated datasets, we record the proportion of outcomes Y that are 1, i.e., the proportion of days on which it rained, using the `proportion_rain()` function. A histogram of these simulated rain rates confirms that they are indeed consistent with the original data (Figure 13.6). Most of our posterior simulated datasets saw rain on roughly 18% of days, close to the observed rain incidence in the `weather` data, yet some saw rain on as few as 12% of the days or as many as 24% of the days.

```
proportion_rain <- function(x){mean(x == 1)}
pp_check(rain_model_1, nreps = 100,
         plotfun = "stat", stat = "proportion_rain") +
  xlab("probability of rain")
```

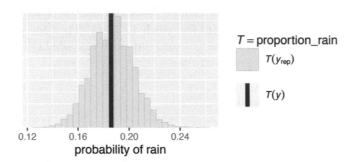

FIGURE 13.6: A posterior predictive check of the logistic regression model of rain. The histogram displays the proportion of days on which it rained in each of 100 posterior simulated datasets. The vertical line represents the observed proportion of days with rain in the weather data.

This brings us to question (3), how accurate are our posterior classifications? In the regression setting with quantitative Y, we answered this question by examining the typical difference between Y and its posterior predictions. Yet in the classification setting with categorical Y, our binary posterior classifications of Y are either *right* or *wrong*. What we're interested in is how *often* we're right. We will address this question in two ways: with and without cross-validation.

Let's start by evaluating the `rain_model_1` classifications of the same `weather` data we used to build this model. Though there's a shortcut function in the **bayesrules** package, it's important to pull back the curtain and try this by hand first. To begin, construct posterior predictive models of Y for each of the 1000 days in the `weather` dataset:

```
# Posterior predictive models for each day in dataset
set.seed(84735)
rain_pred_1 <- posterior_predict(rain_model_1, newdata = weather)
dim(rain_pred_1)
[1] 20000  1000
```

Each of the 1000 *columns* in `rain_pred_1` contain 20,000 1-or-0 predictions of whether or not it will rain on the corresponding day in the `weather` data. As we saw in Section 13.4, each column *mean* indicates the proportion of these predictions that are 1 – thus the 1,000 column means estimate the probability of rain for the corresponding 1,000 days in our data. We can then convert these proportions into binary classifications of rain by comparing them to a chosen classification cut-off. We'll start with a cut-off 0.5: if the probability of rain exceeds 0.5, then predict rain.

```
weather_classifications <- weather %>%
  mutate(rain_prob = colMeans(rain_pred_1),
         rain_class_1 = as.numeric(rain_prob >= 0.5)) %>%
  select(humidity9am, rain_prob, rain_class_1, raintomorrow)
```

The results for the first three days in our sample are shown below. Based on its 9 a.m. humidity level, only 12% of the 20,000 predictions called for rain on the first day (`rain_prob` = 0.122). Similarly, the simulated probabilities of rain for the second and third days are also amply below our 50% cut-off. As such, we predicted "no rain" for the first three sample days (as shown in `rain_class_1`). For the first two days, these classifications were correct – it didn't `raintomorrow`. For the third day, this classification was *incorrect* – it did `raintomorrow`.

```
head(weather_classifications, 3)
# A tibble: 3 x 4
  humidity9am rain_prob rain_class_1 raintomorrow
        <int>     <dbl>        <dbl> <fct>
1          55     0.122            0 No
2          43    0.0739            0 No
3          62     0.163            0 Yes
```

Finally, to estimate our model's overall posterior accuracy, we can compare the model classifications (`rain_class_1`) to the observed outcomes (`raintomorrow`) for each of our 1000 sample days. This information is summarized by the following table or **"confusion matrix"**:

```
# Confusion matrix
weather_classifications %>%
  tabyl(raintomorrow, rain_class_1) %>%
  adorn_totals(c("row", "col"))
 raintomorrow   0  1 Total
          No 803 11   814
         Yes 172 14   186
       Total 975 25  1000
```

Notice that our classification rule, in conjunction with our Bayesian model, correctly classified 817 of the 1000 total test cases (803 + 14). Thus, the **overall classification accuracy rate** is 81.7% (817 / 1000). At face value, this seems pretty good! But look closer. Our model is *much* better at anticipating when it *won't* rain than when it will. Among the 814 days on which it doesn't rain, we correctly classify 803, or 98.65%. This figure is referred to as the **true negative rate** or **specificity** of our Bayesian model. In stark contrast, among the 186 days on which it *does* rain, we correctly classify only 14, or 7.53%. This figure is referred to as the **true positive rate** or **sensitivity** of our Bayesian model. We can confirm these figures using the shortcut `classification_summary()` function in **bayesrules**:

```
set.seed(84735)
classification_summary(model = rain_model_1, data = weather, cutoff = 0.5)
$confusion_matrix
    y   0  1
   No 803 11
  Yes 172 14

$accuracy_rates

sensitivity      0.07527
specificity      0.98649
overall_accuracy 0.81700
```

Sensitivity, specificity, and overall accuracy

Let Y be a set of binary response values on n data points and $\hat{Y}$ represent the corresponding posterior classifications of Y. A confusion matrix summarizes the results of these classifications relative to the actual observations where $a + b + c + d = n$:

	$\hat{Y} = 0$	$\hat{Y} = 1$
$Y = 0$	a	b
$Y = 1$	c	d

The model's **overall accuracy** captures the proportion of all Y observations that are accurately classified:

$$\text{overall accuracy} = \frac{a + d}{a + b + c + d}.$$

Further, the model's **sensitivity** (true positive rate) captures the proportion of $Y = 1$ observations that are accurately classified and **specificity** (true negative rate) the proportion of $Y = 0$ observations that are accurately classified:

$$\text{sensitivity} = \frac{d}{c + d} \quad \text{and} \quad \text{specificity} = \frac{a}{a + b}.$$

Now, if you fall into the reasonable group of people that don't like walking around in wet clothes all day, the fact that our model is so bad at predicting rain is terrible news. BUT we can do better. Recall from Section 13.4 that we can adjust the classification cut-off to better suit the goals of our analysis. In our case, we can increase our model's sensitivity by *decreasing* the cut-off from 0.5 to, say, 0.2. That is, we'll classify a test case as rain if there's even a 20% chance of rain:

```
set.seed(84735)
classification_summary(model = rain_model_1, data = weather, cutoff = 0.2)
$confusion_matrix
    y   0  1
```

```
  No 580 234
  Yes  67 119
```

```
$accuracy_rates
```

```
sensitivity        0.6398
specificity        0.7125
overall_accuracy   0.6990
```

Success! By making it easier to classify rain, the sensitivity jumped from 7.53% to 63.98% (119 of 186). We're much less likely to be walking around with wet clothes. Yet this improvement is not without consequences. In lowering the cut-off, we make it more difficult to predict when it *won't* rain. As a result, the true negative rate dropped from 98.65% to 71.25% (580 of 814) and we'll carry around an umbrella more often than we need to.

Finally, to hedge against the possibility that the above model assessments are biased by training and testing `rain_model_1` using the same data, we can supplement these measures with cross-validated estimates of classification accuracy. The fact that these are so similar to our measures above suggests that the model is *not* overfit to our sample data – it does just as well at predicting rain on *new* days:[3]

```
set.seed(84735)
cv_accuracy_1 <- classification_summary_cv(
  model = rain_model_1, data = weather, cutoff = 0.2, k = 10)
cv_accuracy_1$cv
  sensitivity specificity overall_accuracy
1      0.6534      0.7185            0.705
```

Trade-offs in sensitivity and specificity

As analysts, we must utilize context to determine an appropriate classification rule with corresponding cut-off c. In doing so, there are some trade-offs to consider:

- As we lower c, sensitivity increases, but specificity decreases.
- As we increase c, specificity increases, but sensitivity decreases.

---

[3] A friendly reminder that this will be slow. We're training, testing, and combining the results of 10 models here!

In closing this section, was it worth it? That is, do the benefits of better classifying rain outweigh the consequences of mistakenly classifying no rain as rain? We don't know. This is a subjective question. As an analyst, you can continue to tweak the classification rule until the corresponding correct classification rates best match your goals.

13.6   Extending the model

As with Normal and Poisson regression models, logistic regression models can grow to accommodate more than one predictor. Recall from the introduction to this chapter that *morning* humidity levels at 9 a.m., X_1, aren't the only relevant weather observation we can make today that might help us anticipate whether or not it might rain tomorrow. We might also consider today's *afternoon* humidity at 3 p.m. (X_2) and whether or not it rains *today* (X_3), a *categorical* predictor with levels No and Yes. We can plunk these predictors right into our logistic regression model:

$$\log\left(\frac{\pi_i}{1 - \pi_i}\right) = \beta_0 + \beta_1 X_{i1} + \beta_2 X_{i2} + \beta_3 X_{i3}.$$

Our original prior understanding was that the chance of rain increases with each *individual* predictor in this model. Yet we have less prior certainty about how one predictor is related to rain *when controlling for* the other predictors – we're not meteorologists or anything. For example, if we know today's *3pm* humidity level, it could very well be the case that today's *9am* humidity doesn't add any additional information about whether or not it will rain tomorrow (i.e., $\beta_1 = 0$). With this, we'll maintain our original $N(-1.4, 0.7^2)$ prior for the centered intercept β_{0c}, but will utilize weakly informative priors for $(\beta_1, \beta_2, \beta_3)$:

$$Y_i | \beta_0, \beta_1, \beta_2, \beta_3 \sim Bern(\pi_i) \quad \text{with} \quad \log\left(\frac{\pi_i}{1 - \pi_i}\right) = \beta_0 + \beta_1 X_{i1} + \beta_2 X_{i2} + \beta_3 X_{i3}$$

$$\beta_{0c} \sim N(-1.4, 0.7^2)$$
$$\beta_1 \sim N(0, 0.14^2)$$
$$\beta_2 \sim N(0, 0.15^2)$$
$$\beta_3 \sim N(0, 6.45^2).$$

We encourage you to pause here to perform a prior simulation (e.g., Do prior simulations of rain rates match our prior understanding of rain in Perth?). Here we'll move on to simulating the corresponding posteriors for the humidity9am coefficient (β_1), humidity3pm coefficient (β_2), and raintodayYes coefficient (β_3) and confirm our prior specifications:

```
rain_model_2 <- stan_glm(
  raintomorrow ~ humidity9am + humidity3pm + raintoday,
  data = weather, family = binomial,
  prior_intercept = normal(-1.4, 0.7),
  prior = normal(0, 2.5, autoscale = TRUE),
  chains = 4, iter = 5000*2, seed = 84735)
```

```
# Obtain prior model specifications
prior_summary(rain_model_2)
```

A posterior `tidy()` summary is shown below. To begin, notice that the 80% credible intervals for the `humidity` and `raintodayYes` coefficients, β_2 and β_3, lie comfortably *above* 0. This suggests that, when controlling for the other predictors in the model, the chance of rain tomorrow increases with today's 3 p.m. humidity levels and if it rained today. Looking closer, the posterior median for the `raintodayYes` coefficient is 1.15, equivalently $e^{1.15} = 3.17$. Thus, when controlling for today's morning and afternoon humidity levels, we expect that the odds of rain tomorrow more than *triple* if it rains today.

```
# Numerical summaries
tidy(rain_model_2, effects = "fixed", conf.int = TRUE, conf.level = 0.80)
# A tibble: 4 x 5
   term          estimate std.error conf.low conf.high
   <chr>            <dbl>     <dbl>    <dbl>     <dbl>
1 (Intercept)     -5.46      0.483    -6.08     -4.85
2 humidity9am     -0.00693   0.00737  -0.0163    0.00251
3 humidity3pm      0.0796    0.00846   0.0689    0.0906
4 raintodayYes     1.15      0.220     0.874     1.44
```

In contrast, notice that the β_1 (`humidity9am`) posterior straddles 0 with an 80% posterior credible interval which ranges from -0.0163 to 0.0025. Let's start with what this observation *doesn't* mean. It does *not* mean that `humidity9am` isn't a significant predictor of tomorrow's rain. We saw in `rain_model_1` that it *is*. Rather, `humidity9am` isn't a significant predictor of tomorrow's rain *when controlling for afternoon humidity and whether or not it rains today*. Put another way, if we already know today's `humidity3pm` and rain status, then knowing `humidity9am` doesn't significantly improve our understanding of whether or not it rains tomorrow. This shift in understanding about `humidity9am` from `rain_model_1` to `rain_model_2` might not be much of a surprise – `humidity9am` is strongly associated with `humidity3pm` and `raintoday`, thus the information it holds about `raintomorrow` is somewhat redundant in `rain_model_2`.

Finally, which is the better model of tomorrow's rain, `rain_model_1` or `rain_model_2`? Using a classification cut-off of 0.2, let's compare the cross-validated estimates of classification accuracy for `rain_model_2` to those for `rain_model_1`:

```
set.seed(84735)
cv_accuracy_2 <- classification_summary_cv(
  model = rain_model_2, data = weather, cutoff = 0.2, k = 10)
```

```
# CV for the models
cv_accuracy_1$cv
  sensitivity specificity overall_accuracy
1      0.6534      0.7185            0.705
cv_accuracy_2$cv
  sensitivity specificity overall_accuracy
1      0.7522      0.8139            0.802
```

> ❷ **Quiz Yourself!**
>
> Which of `rain_model_1` and `rain_model_2` produces the most accurate posterior classifications of tomorrow's rain? Which model do you prefer?

The answer to this quiz is `rain_model_2`. The cross-validated estimates of sensitivity, specificity, and overall accuracy jump by roughly 10 percentage points from `rain_model_1` to `rain_model_2`. Thus, `rain_model_2` is superior to `rain_model_1` *both* in predicting when it will rain tomorrow and when it won't. If you want even more evidence, recall from Chapter 10 that the **expected log-predictive density (ELPD)** measures the overall compatibility of new data points with their posterior predictive models through an examination of the log posterior predictive pdfs – the higher the ELPD the better.

```
# Calculate ELPD for the models
loo_1 <- loo(rain_model_1)
loo_2 <- loo(rain_model_2)

# Compare the ELPD for the 2 models
loo_compare(loo_1, loo_2)
              elpd_diff se_diff
rain_model_2    0.0       0.0
rain_model_1  -80.2      13.5
```

Here, the estimated ELPD for `rain_model_1` is more than two standard errors below, and hence *worse than*, that of `rain_model_2`: $(-80.2 \pm 2 * 13.5)$. Thus, ELPD provides even more evidence in favor of `rain_model_2`.

13.7   Chapter summary

Let response variable $Y \in \{0, 1\}$ be a **binary categorical variable**. Thus, modeling the relationship between Y and a set of predictors $(X_1, X_2, \ldots, X_p)$ requires a *classification modeling approach*. To this end, we considered the **logistic regression model**:

$$
\begin{aligned}
\text{data:} \quad & Y_i | \beta_0, \beta_1, \ldots, \beta_p \overset{ind}{\sim} \text{Bern}(\pi_i) \quad \text{with} \quad \log\left(\frac{\pi_i}{1-\pi_i}\right) = \beta_0 + \beta_1 X_{i1} + \cdots + \beta_p X_{ip} \\
\text{priors:} \quad & \beta_{0c} \sim N(m_0, s_0^2) \\
& \beta_1 \sim N(m_1, s_1^2) \\
& \vdots \\
& \beta_p \sim N(m_p, s_p^2)
\end{aligned}
$$

(13.7)

where we can transform the model from the log(odds) scale to the more meaningful odds and probability scales:

$$
\frac{\pi_i}{1-\pi_i} = e^{\beta_0 + \beta_1 X_{i1} + \cdots + \beta_p X_{ip}} \quad \text{and} \quad \pi_i = \frac{e^{\beta_0 + \beta_1 X_{i1} + \cdots + \beta_p X_{ip}}}{1 + e^{\beta_0 + \beta_1 X_{i1} + \cdots + \beta_p X_{ip}}}.
$$

While providing insight into the relationship between the outcome of Y and predictors X, we can also utilize this logistic regression model to produce posterior classifications of Y. In evaluating the classification *quality*, we must consider the overall accuracy alongside **sensitivity** and **specificity**, i.e., our model's ability to anticipate when $Y = 1$ and $Y = 0$, respectively.

13.8 Exercises

13.8.1 Conceptual exercises

Exercise 13.1 (Normal vs logistic). For each scenario, identify whether *Normal* or *logistic* regression is the appropriate tool for modeling Y by X.

(a) Y = whether or not a person bikes to work, X = the distance from the person's home to work
(b) Y = the number of minutes it takes a person to commute to work, X = the distance from the person's home to work
(c) Y = the number of minutes it takes a person to commute to work, X = whether or not the person takes public transit to work

Exercise 13.2 (What are the odds?). Calculate and interpret the odds for each event of interest below.

(a) The probability of rain tomorrow is 0.8.
(b) The probability of flipping 2 Heads in a row is 0.25.
(c) The log(odds) that your bus will be on time are 0.
(d) The log(odds) that a person is left-handed are -1.386.

Exercise 13.3 (What's the probability?). Calculate and interpret the probability for each event of interest below.

(a) The odds that your team will win the basketball game are 20 to 1.
(b) The odds of rain tomorrow are 0.5.
(c) The log(odds) of a certain candidate winning the election are 1.
(d) The log(odds) that a person likes pineapple pizza are -2.

Exercise 13.4 (Logistic models). Let Y indicate whether or not somebody believes that climate change is real and caused by people (TRUE or FALSE), and X be their age. The simplified posterior median logistic regression model of Y by X provides insight into the relationship between the two. NOTE: This formula is loosely based on the `pulse_of_the_nation` survey results in the **bayesrules** package.

$$\log(\text{odds of belief in climate change}) = 1.43 - 0.02\text{age}$$

(a)   Express the posterior median model on the odds and probability scales.
(b)   Interpret the age coefficient on the *odds* scale.
(c)   Calculate the posterior median probability that a 60-year-old believes in climate change.
(d)   Repeat part c for a 20-year-old.

Exercise 13.5 (Sensitivity vs specificity). Continuing our climate change belief analysis, the confusion matrix below summarizes the performance of our logistic model in classifying the beliefs of 1000 survey respondents, using a probability cut-off of 0.5.

y	0	1
FALSE (0)	50	300
TRUE (1)	30	620

(a)   Calculate and interpret the model's overall accuracy.
(b)   Calculate and interpret the model's sensitivity.
(c)   Calculate and interpret the model's specificity.
(d)   Suppose that researchers want to improve their ability to identify people that do *not* believe in climate change. How might they adjust their probability cut-off: Increase it or decrease it? Why?

13.8.2   Applied exercises

Exercise 13.6 (Hotel bookings: getting started). Plans change. Hotel room bookings get canceled. In the next exercises, you'll explore whether hotel cancellations might be predicted based upon the circumstances of a reservation. Throughout, utilize weakly informative priors and the `hotel_bookings` data in the **bayesrules** package.[4] Your analysis will incorporate the following variables on hotel bookings:

variable	notation	meaning
`is_canceled`	Y	whether or not the booking was canceled
`lead_time`	X_1	number of days between the booking and scheduled arrival
`previous_cancellations`	X_2	number of previous times the guest has canceled a booking
`is_repeated_guest`	X_3	whether or not the booking guest is a repeat customer at the hotel
`average_daily_rate`	X_4	the average per day cost of the hotel

[4]This is a random sample of data collected by Antonio et al. (2019) and distributed by the *Hotels* TidyTuesday project (R for Data Science, 2020c).

a) What proportion of the sample bookings were canceled?
b) Construct and discuss plots of `is_canceled` vs each of the four potential predictors above.
c) Using formal mathematical notation, specify an appropriate Bayesian regression model of Y by predictors (X_1, X_2, X_3, X_4).
d) Explain your choice for the structure of the data model.

Exercise 13.7 (Hotel bookings: simulation).

a) Simulate the posterior model of your regression parameters $(\beta_0, \beta_1, \ldots, \beta_4)$. Construct trace plots, density plots, and a `pp_check()` of the chain output.
b) Report the posterior median model of hotel cancellations on each of the log(odds), odds, and probability scales.
c) Construct 80% posterior credible intervals for your model coefficients. Interpret those for β_2 and β_3 on the odds scale.
d) Among the four predictors, which are significantly associated with hotel cancellations, both statistically and meaningfully? Explain.

Exercise 13.8 (Hotel bookings: classification rules).

a) How good is your model at anticipating whether a hotel booking will be canceled? Evaluate the classification accuracy using both the in-sample and cross-validation approaches, along with a 0.5 probability cut-off.
b) Are the cross-validated and in-sample assessments of classification accuracy similar? Explain why this makes sense in the context of this analysis.
c) Interpret the cross-validated overall accuracy, sensitivity, and specificity measures in the context of this analysis.
d) Thinking like a hotel booking agent, you'd like to increase the sensitivity of your classifications to 0.75. Identify a probability cut-off that you could use to achieve this level while maintaining the highest possible specificity.

Exercise 13.9 (Hotel bookings: will they cancel?!).

a) A guest that is new to a hotel and has only canceled a booking 1 time before, has booked a $100 per day hotel room 30 days in advance. Simulate, plot, and discuss the posterior predictive model of Y, whether or not the guest will cancel this booking.
b) Come up with the features of another fictitious booking that's *more* likely to be canceled than the booking in part a. Support your claim by simulating, plotting, and comparing this booking's posterior predictive model of Y to that in part a.

Exercise 13.10 (Robots taking over: getting started). As engineers develop more and more sophisticated robots, there are opportunities for improved efficiency. But at what cost? Some people fear that their occupations will be replaced by robots. In the next exercises, you'll explore whether someone's view on robots might be associated with other survey factors. Throughout, utilize weakly informative priors and the `pulse_of_the_nation` survey data in the **bayesrules** package. Your analysis will incorporate the following variables:

variable	notation	meaning
robots	Y	$0 =$ likely, $1 =$ unlikely that their jobs will be replaced by robots within 10 years
transformers	X_1	number of *Transformers* films the respondent has seen
books	X_2	number of books read in past year
age	X_3	age in years
income	X_4	income in thousands of dollars

a) What proportion of people in the sample think their job is unlikely to be taken over by robots in the next ten years?

b) Construct and discuss plots of robots vs each of the four potential predictors above.

c) Using formal mathematical notation, specify an appropriate Bayesian regression model of Y by predictors (X_1, X_2, X_3, X_4).

d) Explain your choice for the structure of the data model.

Exercise 13.11 (Robots taking over: simulation).

a) Simulate the posterior model of your regression parameters $(\beta_0, \beta_1, \ldots, \beta_4)$. Construct trace plots, density plots, and a pp_check() of the chain output.

b) Report the posterior median model of a person's robot beliefs on each of the log(odds), odds, and probability scales.

c) Construct 80% posterior credible intervals for your model coefficients. Interpret those for β_3 and β_4 on the odds scale.

d) Among the four predictors, which are significantly associated with thinking robots are unlikely to take over their jobs in the next 10 years, both statistically and meaningfully? Explain.

Exercise 13.12 (Robots taking over: classification rules).

a) How good is your model at anticipating whether a person is unlikely to think that their job will be taken over by a robot in the next 10 years? Evaluate the classification accuracy using both the in-sample and cross-validation approaches, along with a 0.6 probability cut-off.

b) Are the cross-validated and in-sample assessments of classification accuracy similar? Explain why this makes sense in the context of this analysis.

c) Interpret the cross-validated overall accuracy, sensitivity, and specificity measures in the context of this analysis.

13.8.3   Open-ended exercises

Exercise 13.13 (Open exercise: fake news). The fake_news data in the **bayesrules** package contains information about 150 news articles, some real news and some fake news. In this open-ended exercise, complete a logistic regression analysis of article type, real or fake, using the following three predictors: whether or not the article's title includes an exclamation point (title_has_excl), the number of words in the title (title_words), and the **negative** sentiment rating.

Exercise 13.14 (Open exercise: ghosts are real?). We return to the `pulse_of_the_nation` survey data in the **bayesrules** package which includes a variable on whether or not a person believes in `ghosts`. In this open-ended exercise, complete a logistic regression analysis of whether or not someone believes in ghosts using at least three other variables from the dataset.

14

Naive Bayes Classification

There exist multiple penguin species throughout Antarctica, including the *Adelie*, *Chinstrap*, and *Gentoo*. When encountering one of these penguins on an Antarctic trip, we might *classify* its species

$$Y = \begin{cases} A & \text{Adelie} \\ C & \text{Chinstrap} \\ G & \text{Gentoo} \end{cases}$$

by examining various physical characteristics, such as whether the penguin weighs more than the average 4200g,

$$X_1 = \begin{cases} 1 & \text{above-average weight} \\ 0 & \text{below-average weight} \end{cases}$$

as well as measurements of the penguin's bill

$$X_2 = \text{bill length (in mm)}$$
$$X_3 = \text{flipper length (in mm)}$$

The `penguins_bayes` data, originally made available by Gorman et al. (2014) and distributed by Horst et al. (2020), contains the above species and feature information for a sample of 344 Antarctic penguins:

```
# Load packages
library(bayesrules)
library(tidyverse)
library(e1071)
library(janitor)

# Load data
data(penguins_bayes)
penguins <- penguins_bayes
```

Among these penguins, 152 are Adelies, 68 are Chinstraps, and 124 are Gentoos. We'll assume throughout that the proportional breakdown of these species in our dataset reflects the species breakdown in the wild. That is, our prior assumption about any new penguin is that it's most likely an Adelie and least likely a Chinstrap:

DOI: 10.1201/9780429288340-14

```
penguins %>%
  tabyl(species)
   species   n percent
    Adelie 152  0.4419
 Chinstrap  68  0.1977
    Gentoo 124  0.3605
```

Before proceeding with our analysis of this data, take a quick quiz.

> **❓ Quiz Yourself!**
>
> Explain why neither the Normal nor logistic regression model would be appropriate for classifying a penguin's species Y from its physical characteristics X.

Since species Y is *categorical*, Normal regression isn't an appropriate tool for understanding its relationship with physical characteristics X. And though logistic regression is (currently) our only tool for modeling and classifying categorical response variables, it too is inappropriate here. Recall that with a Bernoulli data structure, the logistic regression model is limited to classifying *binary* response variables. Yet our species variable Y has *three* categories. Luckily, there's a tool for everything. In Chapter 14 we'll explore an alternative to logistic regression: **naive Bayes classification**. Relative to Bayesian logistic regression, naive Bayes classification has a few advantages:

- it can classify categorical response variables Y with two *or more* categories;
- it doesn't require much theory beyond Bayes' Rule; and
- it's computationally efficient, i.e., doesn't require MCMC simulation.

Sounds fantastic. But as you might expect, these benefits don't come without a cost. We'll observe throughout this chapter that naive Bayes classification is called "naive" for a reason.

> **◎ Goals**
>
> - Explore the inner workings of naive Bayes classification.
> - Implement naive Bayes classification in R.
> - Develop strategies to evaluate the quality of naive Bayes classifications.

14.1   Classifying one penguin

We'll start our naive Bayes classification with just a single penguin. Suppose an Antarctic researcher comes across a penguin that weighs less than 4200g with a 195mm-long flipper and 50mm-long bill. Our goal is to help this researcher identify the species of this penguin, Adelie, Chinstrap, or Gentoo.

14.1.1   One categorical predictor

Let's begin by considering just the information we have on the *categorical* predictor X_1: the penguin weighs less than 4200g, i.e., is not `above_average_weight`. Take the following quiz to first build some intuition.[1]

> **❷ Quiz Yourself!**
>
> Based on the plot below, for which species is a below-average weight most likely?

```
ggplot(penguins %>% drop_na(above_average_weight),
       aes(fill = above_average_weight, x = species)) +
  geom_bar(position = "fill")
```

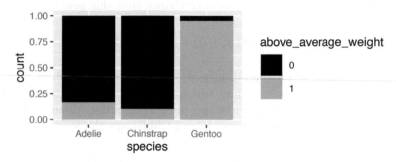

FIGURE 14.1: The proportion of each penguin species that's above average weight.

Notice in Figure 14.1 that Chinstrap penguins are the most likely to be below average weight. Yet before declaring this penguin to be a Chinstrap, we should consider that this is the *rarest* species to encounter in the wild. That is, we have to think like *Bayesians* by combining the information from our *data* with our *prior* information about species membership to construct a posterior model for our penguin's species. The naive Bayes classification approach to this task is nothing more than a direct appeal to the tried-and-true Bayes' Rule from Chapter 2. In general, to calculate the posterior probability that our penguin is species $Y = y$ in light of its weight status $X_1 = x_1$, we can plug into

$$f(y \mid x_1) = \frac{\text{prior} \cdot \text{likelihood}}{\text{normalizing constant}} = \frac{f(y)L(y \mid x_1)}{f(x_1)} \tag{14.1}$$

where, by the Law of Total Probability,

$$\begin{aligned} f(x_1) &= \sum_{\text{all } y'} f(y')L(y' \mid x_1) \\ &= f(y' = A)L(y' = A|x_1) + f(y' = C)L(y' = C|x_1) + f(y' = G)L(y' = G|x_1). \end{aligned} \tag{14.2}$$

A table which breaks down the `above_average_weight` weight status by `species` provides the necessary information to complete this Bayesian calculation:

---

[1] Answer: Chinstrap

```
penguins %>%
  select(species, above_average_weight) %>%
  na.omit() %>%
  tabyl(species, above_average_weight) %>%
  adorn_totals(c("row", "col"))
    species   0   1 Total
     Adelie 126  25   151
  Chinstrap  61   7    68
     Gentoo   6 117   123
      Total 193 149   342
```

In fact, we can directly calculate the posterior model of our penguin's species from this table. For example, notice that among the 193 penguins that are below average weight, 126 are Adelies. Thus, there's a roughly 65% posterior chance that this penguin is an Adelie:

$$f(y = A \mid x_1 = 0) = \frac{126}{193} \approx 0.6528.$$

Let's confirm this result by plugging information from our `tabyl` above into Bayes' Rule (14.1). This tedious step is not to annoy, but to practice for generalizations we'll have to make in more complicated settings. First, our prior information about species membership indicates that Adelies are the most common and Chinstraps the least:

$$f(y = A) = \frac{151}{342}, \quad f(y = C) = \frac{68}{342}, \quad f(y = G) = \frac{123}{342}. \tag{14.3}$$

Further, the likelihoods demonstrate that below-average weight is *most* common among Chinstrap penguins. For example, 89.71% of Chinstraps but only 4.88% of Gentoos are below average weight:

$$L(y = A \mid x_1 = 0) = \frac{126}{151} \approx 0.8344$$

$$L(y = C \mid x_1 = 0) = \frac{61}{68} \approx 0.8971$$

$$L(y = G \mid x_1 = 0) = \frac{6}{123} \approx 0.0488$$

Plugging these priors and likelihoods into (14.2), the *total* probability of observing a below-average weight penguin across all species is

$$f(x_1 = 0) = \frac{151}{342} \cdot \frac{126}{151} + \frac{68}{342} \cdot \frac{61}{68} + \frac{123}{342} \cdot \frac{6}{123} = \frac{193}{342}.$$

Finally, by Bayes' Rule we can confirm that there's a 65% posterior chance that this penguin is an Adelie:

$$f(y = A \mid x_1 = 0) = \frac{f(y = A)L(y = A \mid x_1 = 0)}{f(x_1 = 0)}$$

$$= \frac{(151/342) \cdot (126/151)}{193/342}$$

$$\approx 0.6528.$$

Similarly, we can show that

$$f(y = C \mid x_1 = 0) \approx 0.3161 \quad \text{and} \quad f(y = G \mid x_1 = 0) \approx 0.0311.$$

All in all, the posterior probability that this penguin is an Adelie is more than double that of the other two species. Thus, our naive Bayes classification, based on our prior information and the penguin's below-average weight alone, is that this penguin is an Adelie. Though below-average weight is relatively less common among Adelies than among Chinstraps, the final classification was pushed over the edge by the fact that Adelies are far more common.

14.1.2   One quantitative predictor

Let's ignore the penguin's weight for now and classify its species using only the fact that it has a 50mm-long bill. Take the following quiz to build some intuition.[2]

> **❓ Quiz Yourself!**
>
> Based on Figure 14.2, among which species is a 50mm-long bill the most common?

```
ggplot(penguins, aes(x = bill_length_mm, fill = species)) +
  geom_density(alpha = 0.7) +
  geom_vline(xintercept = 50, linetype = "dashed")
```

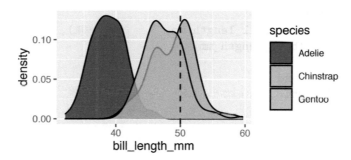

FIGURE 14.2: Density plots of the bill lengths (mm) observed among three penguin species.

Notice from the plot that a 50mm-long bill would be *extremely* long for an Adelie penguin. Though the distinction in bill lengths is less dramatic among the Chinstrap and Gentoo, Chinstrap bills tend to be a tad longer. In particular, a 50mm-long bill is fairly average for Chinstraps, though slightly long for Gentoos. Thus, again, our *data* points to our penguin being a Chinstrap. And again, we must weigh this data against the fact that Chinstraps are the rarest of these three species. To this end, we can appeal to Bayes' Rule to make a naive Bayes classification of Y from information that bill length $X_2 = x_2$:

$$f(y \mid x_2) = \frac{f(y)L(y \mid x_2)}{f(x_2)} = \frac{f(y)L(y \mid x_2)}{\sum_{\text{all } y'} f(y')L(y' \mid x_2)}. \tag{14.4}$$

Our question for you is this: What is the likelihood that we would observe a 50mm-long bill

---

[2]Answer: Chinstrap

if the penguin is an Adelie, $L(y = A|x_2 = 50)$? Unlike the penguin's *categorical* weight status (X_1), its bill length is *quantitative*. Thus, we can't simply calculate $L(y = A \mid x_2 = 50)$ from a table of `species` vs `bill_length_mm`. Further, we haven't assumed a model for data X_2 from which to define likelihood function $L(y|x_2)$. This is where one "naive" part of naive Bayes classification comes into play. The naive Bayes method typically assumes that any quantitative predictor, here X_2, is **continuous** and **conditionally Normal**. That is, within each species, bill lengths X_2 are Normally distributed with possibly different means μ and standard deviations σ:

$$X_2 \mid (Y = A) \sim N(\mu_A, \sigma_A^2)$$
$$X_2 \mid (Y = C) \sim N(\mu_C, \sigma_C^2).$$
$$X_2 \mid (Y = G) \sim N(\mu_G, \sigma_G^2)$$

Though this is a naive blanket assumption to make for all quantitative predictors, it turns out to be reasonable for bill length X_2. Reexamining Figure 14.2, notice that within each species, bill lengths are roughly bell shaped around different means and with slightly different standard deviations. Thus, we can tune the Normal model for each species by setting its μ and σ parameters to the observed sample means and standard deviations in bill lengths within that species. For example, we'll tune the Normal model of bill lengths for the Adelie species to have mean and standard deviation $\mu_A = 38.8$mm and $\sigma_A = 2.66$mm:

```
# Calculate sample mean and sd for each Y group
penguins %>%
  group_by(species) %>%
  summarize(mean = mean(bill_length_mm, na.rm = TRUE),
            sd = sd(bill_length_mm, na.rm = TRUE))
# A tibble: 3 x 3
  species     mean    sd
  <fct>      <dbl> <dbl>
1 Adelie      38.8  2.66
2 Chinstrap   48.8  3.34
3 Gentoo      47.5  3.08
```

Plotting the tuned Normal models for each species confirms that this naive Bayes assumption isn't *perfect* – it's a bit more idealistic than the density plots of the raw data in Figure 14.2. But it's fine enough to continue.

```
ggplot(penguins, aes(x = bill_length_mm, color = species)) +
  stat_function(fun = dnorm, args = list(mean = 38.8, sd = 2.66),
                aes(color = "Adelie")) +
  stat_function(fun = dnorm, args = list(mean = 48.8, sd = 3.34),
                aes(color = "Chinstrap")) +
  stat_function(fun = dnorm, args = list(mean = 47.5, sd = 3.08),
                aes(color = "Gentoo")) +
  geom_vline(xintercept = 50, linetype = "dashed")
```

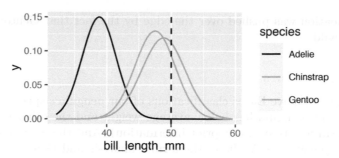

FIGURE 14.3: Normal pdfs tuned to match the observed mean and standard deviation in bill lengths (mm) among three penguin species.

Recall that this Normality assumption provides the mechanism we need to evaluate the likelihood of observing a 50mm-long bill among each of the three species, $L(y \mid x_2 = 50)$. Connecting back to Figure 14.3, these likelihoods correspond to the Normal density curve heights at a bill length of 50 mm. Thus, observing a 50mm-long bill is slightly more likely among Chinstrap than Gentoo, and *highly* unlikely among Adelie penguins. More specifically, we can calculate the likelihoods using `dnorm()`.

```
# L(y = A | x_2 = 50)
dnorm(50, mean = 38.8, sd = 2.66)
[1] 2.12e-05

# L(y = C | x_2 = 50)
dnorm(50, mean = 48.8, sd = 3.34)
[1] 0.112

# L(y = G | x_2 = 50)
dnorm(50, mean = 47.5, sd = 3.08)
[1] 0.09317
```

We now have everything we need to plug into Bayes' Rule (14.4). Weighting the likelihoods by the prior probabilities of each species (à la (14.2)), the marginal pdf of observing a penguin with a 50mm-long bill is

$$f(x_2 = 50) = \frac{151}{342} \cdot 0.0000212 + \frac{68}{342} \cdot 0.112 + \frac{123}{342} \cdot 0.09317 = 0.05579.$$

Thus, the posterior probability that our penguin is an Adelie is

$$f(y = A \mid x_2 = 50) = \frac{(151/342) \cdot 0.0000212}{0.05579} \approx 0.0002.$$

Similarly, we can calculate

$$f(y = C \mid x_2 = 50) \approx 0.3992 \quad \text{and} \quad f(y = G \mid x_2 = 50) \approx 0.6006.$$

It follows that our naive Bayes classification, based on our prior information and penguin's bill length alone, is that this penguin is a Gentoo – it has the highest posterior probability. Though a 50mm-long bill is relatively less common among Gentoo than among Chinstrap, the final classification was pushed over the edge by the fact that Gentoo are much more common in the wild.

14.1.3   Two predictors

We've now made two naive Bayes classifications of our penguin's species, one based solely on the fact that our penguin has below-average weight and the other based solely on its 50mm-long bill (in addition to our prior information). And these classifications disagree: we classified the penguin as Adelie in the former analysis and Gentoo in the latter. This discrepancy indicates that there's room for improvement in our naive Bayes classification method. In particular, instead of relying on any one predictor alone, we can incorporate *multiple* predictors into our classification process.

Consider the information that our penguin has a bill length of X_2 = 50mm *and* a flipper length of X_3 = 195mm. Either one of these measurements alone might lead to a misclassification. Just as it's tough to distinguish between the Chinstrap and Gentoo penguins based on their bill lengths alone, it's tough to distinguish between the Chinstrap and Adelie penguins based on their flipper lengths alone (Figure 14.4).

```
ggplot(penguins, aes(x = bill_length_mm, fill = species)) +
  geom_density(alpha = 0.6)
ggplot(penguins, aes(x = flipper_length_mm, fill = species)) +
  geom_density(alpha = 0.6)
```

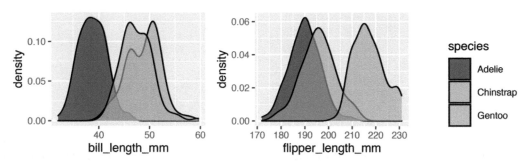

FIGURE 14.4: Density plots of the bill lengths (mm) and flipper lengths (mm) among our three penguin species.

BUT the species are fairly distinguishable when we *combine* the information about bill and flipper lengths. Our penguin with a 50mm-long bill and 195mm-long flipper, represented at the intersection of the dashed lines in Figure 14.5, now lies squarely among the Chinstrap observations:

```
ggplot(penguins,
       aes(x = flipper_length_mm, y = bill_length_mm, color = species)) +
  geom_point()
```

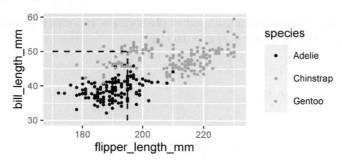

FIGURE 14.5: A scatterplot of the bill lengths (mm) versus flipper lengths (mm) among our three penguin species.

Let's use naive Bayes classification to balance this data with our prior information on species membership. To calculate the posterior probability that the penguin is species $Y = y$, we can adjust Bayes' Rule to accommodate our two predictors, $X_2 = x_2$ and $X_3 = x_3$:

$$f(y \mid x_2, x_3) = \frac{f(y)L(y \mid x_2, x_3)}{\sum_{\text{all } y'} f(y')L(y' \mid x_2, x_3)}. \tag{14.5}$$

This presents yet another new twist: How can we calculate the likelihood function that incorporates *two* variables, $L(y \mid x_2, x_3)$? This is where yet another "naive" assumption creeps in. Naive Bayes classification assumes that predictors are **conditionally independent**, thus

$$L(y \mid x_2, x_3) = f(x_2, x_3 \mid y) = f(x_2 \mid y)f(x_3 \mid y).$$

In words, within each species, we assume that the length of a penguin's bill has no relationship to the length of its flipper. Mathematically and computationally, this assumption makes the naive Bayes algorithm *efficient* and manageable. However, it might also make it *wrong*. Revisit Figure 14.5. Within each species, flipper length and bill length appear to be positively correlated, *not* independent. Yet, we'll naively roll with the imperfect independence assumption, and thus the possibility that our classification accuracy might be weakened.

Combined then, the multivariable naive Bayes model assumes that our two predictors are Normal and conditionally independent. We already tuned this Normal model for bill length X_2 in Section 14.1.2. Similarly, we can tune the species-specific Normal models for flipper length X_3 to match the corresponding sample means and standard deviations:

```
# Calculate sample mean and sd for each Y group
penguins %>%
  group_by(species) %>%
  summarize(mean = mean(flipper_length_mm, na.rm = TRUE),
            sd = sd(flipper_length_mm, na.rm = TRUE))
# A tibble: 3 x 3
  species    mean    sd
  <fct>     <dbl> <dbl>
1 Adelie     190.  6.54
2 Chinstrap  196.  7.13
3 Gentoo     217.  6.48
```

Accordingly, we can evaluate the likelihood of observing a 195-mm flipper length for each of the three species, $L(y \mid x_3 = 195) = f(x_3 = 195 \mid y)$:

```
# L(y = A | x_3 = 195)
dnorm(195, mean = 190, sd = 6.54)
[1] 0.04554

# L(y = C | x_3 = 195)
dnorm(195, mean = 196, sd = 7.13)
[1] 0.05541

# L(y = G | x_3 = 195)
dnorm(195, mean = 217, sd = 6.48)
[1] 0.0001934
```

For each species, we now have the likelihood of observing a bill length of $X_2 = 50$ mm (Section 14.1.2), the likelihood of observing a flipper length of $X_3 = 195$ mm (above), and the prior probability (14.3). Combined, the likelihoods of observing a 50mm-long bill *and* a 195mm-long flipper for each species $Y = y$, weighted by the prior probability of the species are:

$$f(y' = A)L(y' = A \mid x_2 = 50, x_3 = 195) = \frac{151}{342} \cdot 0.0000212 \cdot 0.04554$$

$$f(y' = C)L(y' = C \mid x_2 = 50, x_3 = 195) = \frac{68}{342} \cdot 0.112 \cdot 0.05541$$

$$f(y' = G)L(y' = G \mid x_2 = 50, x_3 = 195) = \frac{123}{342} \cdot 0.09317 \cdot 0.0001934$$

with a sum of

$$\sum_{\text{all } y'} f(y')L(y' \mid x_2 = 50, x_3 = 195) \approx 0.001241.$$

And plugging into Bayes' Rule (14.5), the posterior probability that the penguin is an Adelie is

$$f(y = A|x_2 = 50, x_3 = 195) = \frac{\frac{151}{342} \cdot 0.0000212 \cdot 0.04554}{0.001241} \approx 0.0003.$$

Similarly, we can calculate

$$f(y = C|x_2 = 50, x_3 = 195) \approx 0.9944$$
$$f(y = G|x_2 = 50, x_3 = 195) \approx 0.0052.$$

In conclusion, our penguin is *almost certainly* a Chinstrap. Though we didn't come to this conclusion using any physical characteristic alone, together they paint a pretty clear picture.

Naive Bayes classification

Let Y denote a *categorical* response variable with two or more categories and $(X_1, X_2, \ldots, X_p)$ be a set of p possible predictors of Y. Then the posterior probability that a new case with observed predictors $(X_1, X_2, \ldots, X_p) = (x_1, x_2, \ldots, x_p)$ belongs to class $Y = y$ is

$$f(y|x_1, x_2, \ldots, x_p) = \frac{f(y)L(y \mid x_1, x_2, \ldots, x_p)}{\sum_{\text{all } y'} f(y')L(y' \mid x_1, x_2, \ldots, x_p)}.$$

Naive Bayes classification makes the following naive assumptions about the likelihood function of Y:

- Predictors X_i are **conditionally independent**, thus

$$L(y \mid x_1, x_2, \ldots, x_p) = \prod_{i=1}^{p} f(x_i \mid y).$$

- For **categorical predictors** X_i, the conditional pmf

$$f(x_i \mid y) = P(X_i = x_i \mid Y = y)$$

is estimated by the sample proportion of observed $Y = y$ cases for which $X_i = x_i$.

- For **quantitative predictors** X_i, the conditional pmf / pdf $f(x_i \mid y)$ is defined by a Normal model:

$$X_i|(Y = y) \sim N(\mu_{iy}, \sigma_{iy}^2).$$

That is, within each category $Y = y$, we assume that X_i is Normally distributed around some mean μ_{iy} with standard deviation σ_{iy}. Further, we estimate μ_{iy} and σ_{iy} by the sample mean and standard deviation of the observed X_i values in category $Y = y$.

14.2 Implementing & evaluating naive Bayes classification

That was nice, but we needn't do all of this work by hand. To implement naive Bayes classification in R, we'll use the `naiveBayes()` function in the **e1071** package (Meyer et al., 2021). As with `stan_glm()`, we feed `naiveBayes()` the `data` and a formula indicating which data variables to use in the analysis. Yet, since naive Bayes calculates prior probabilities directly from the data and implementation doesn't require MCMC simulation, we don't have to worry about providing information regarding prior models or Markov chains. Below we build two naive Bayes classification algorithms, one using `bill_length_mm` alone and one that also incorporates `flipper_length_mm`:

```
naive_model_1 <- naiveBayes(species ~ bill_length_mm, data = penguins)
naive_model_2 <- naiveBayes(species ~ bill_length_mm + flipper_length_mm,
                            data = penguins)
```

Let's apply both of these to classify `our_penguin` that we studied throughout Section 14.1:

```
our_penguin <- data.frame(bill_length_mm = 50, flipper_length_mm = 195)
```

Beginning with `naive_model_1`, the `predict()` function returns the posterior probabilities of each species along with a final classification. This classification follows a simple rule: classify the penguin as whichever species has the highest posterior probability. The results of this process are similar to those that we got "by hand" in Section 14.1.2, just off a smidge due to some rounding error. Essentially, based on its bill length alone, our best guess is that this penguin is a Gentoo:

```
predict(naive_model_1, newdata = our_penguin, type = "raw")
        Adelie Chinstrap Gentoo
[1,] 0.000169    0.3978  0.602
predict(naive_model_1, newdata = our_penguin)
[1] Gentoo
Levels: Adelie Chinstrap Gentoo
```

And just as we concluded in Section 14.1.3, if we take into account *both* its bill length and flipper length, our best guess is that this penguin is a Chinstrap:

```
predict(naive_model_2, newdata = our_penguin, type = "raw")
         Adelie Chinstrap    Gentoo
[1,] 0.0003446    0.9949 0.004787
predict(naive_model_2, newdata = our_penguin)
[1] Chinstrap
Levels: Adelie Chinstrap Gentoo
```

We can similarly apply our naive Bayes models to classify any number of penguins. As with logistic regression, we'll take two common approaches to evaluating the *accuracy* of these classifications:

1. construct **confusion matrices** which compare the *observed* species of our sample penguins to their naive Bayes species *classifications*; and
2. for a better sense of how well our naive Bayes models classify *new* penguins, calculate **cross-validated** estimates of classification accuracy.

To begin with the first approach, we classify each of the `penguins` using both `naive_model_1` and `naive_model_2` and store these in `penguins` as `class_1` and `class_2`:

```
penguins <- penguins %>%
  mutate(class_1 = predict(naive_model_1, newdata = .),
         class_2 = predict(naive_model_2, newdata = .))
```

The classification results are shown below for four randomly sampled penguins, contrasted against the actual `species` of these penguins. For the last two penguins, the two models produce the *same* classifications (Adelie) and these classifications are correct. For the first two penguins, the two models lead to *different* classifications. In both cases, `naive_model_2` is correct.

```
set.seed(84735)
penguins %>%
  sample_n(4) %>%
  select(bill_length_mm, flipper_length_mm, species, class_1, class_2) %>%
  rename(bill = bill_length_mm, flipper = flipper_length_mm)
# A tibble: 4 x 5
  bill flipper species     class_1 class_2
  <dbl>   <int> <fct>       <fct>   <fct>
1 47.5      199 Chinstrap   Gentoo  Chinstrap
2 40.9      214 Gentoo      Adelie  Gentoo
3 41.3      194 Adelie      Adelie  Adelie
4 38.5      190 Adelie      Adelie  Adelie
```

Of course, it remains to be seen whether `naive_model_2` outperforms `naive_model_1` overall. To this end, the confusion matrices below summarize the models' classification accuracy across all penguins in our sample. Use these to complete the following quiz.[3]

> ❷ **Quiz Yourself!**
>
> 1. In `naive_model_1`, as what other species is the Chinstrap most commonly misclassified?
> 2. Which model is better at classifying Chinstrap penguins?
> 3. Which model has a higher overall accuracy rate?

```
# Confusion matrix for naive_model_1
penguins %>%
  tabyl(species, class_1) %>%
  adorn_percentages("row") %>%
  adorn_pct_formatting(digits = 2) %>%
  adorn_ns()
  species          Adelie Chinstrap         Gentoo
    Adelie 95.39% (145)  0.00% (0)    4.61%   (7)
 Chinstrap  5.88%   (4)  8.82% (6)   85.29%  (58)
    Gentoo  6.45%   (8)  4.84% (6)   88.71% (110)

# Confusion matrix for naive_model_2
penguins %>%
  tabyl(species, class_2) %>%
  adorn_percentages("row") %>%
  adorn_pct_formatting(digits = 2) %>%
  adorn_ns()
```

---

[3]Answers: 1. Gentoo; 2. `naive_model_2`; 3. `naive_model_2`

```
   species        Adelie    Chinstrap         Gentoo
    Adelie  96.05% (146)   2.63%   (4)   1.32%    (2)
 Chinstrap   7.35%   (5)  86.76%  (59)   5.88%    (4)
    Gentoo   0.81%   (1)   0.81%   (1)  98.39%  (122)
```

With these prompts in mind, let's examine the two confusion matrices. One quick observation is that `naive_model_2` does a better job across the board. Not only are its classification accuracy rates for *each* of the Adelie, Chinstrap, and Gentoo species higher than in `naive_model_1`, its *overall* accuracy rate is also higher. The `naive_model_2` correctly classifies 327 (146 + 59 + 122) of the 344 total penguins (95%), whereas `naive_model_1` correctly classifies only 261 (76%). Where `naive_model_2` enjoys the *greatest* improvement over `naive_model_1` is in the classification of Chinstrap penguins. In `naive_model_1`, only 9% of Chinstraps are correctly classified, with a whopping 85% being misclassified as Gentoo. At 87%, the classification accuracy rate for Chinstraps is *much* higher in `naive_model_2`.

Finally, for due diligence, we can utilize 10-fold **cross-validation** to evaluate and compare how well our naive Bayes classification models classify *new* penguins, not just those in our sample. We do so using the `naive_classification_summary_cv()` function in the **bayesrules** package:

```
set.seed(84735)
cv_model_2 <- naive_classification_summary_cv(
  model = naive_model_2, data = penguins, y = "species", k = 10)
```

The `cv_model_2$folds` object contains the classification accuracy rates for each of the 10 folds whereas `cv_model_2$cv` averages the results across all 10 folds:

```
cv_model_2$cv
   species        Adelie    Chinstrap         Gentoo
    Adelie  96.05% (146)   2.63%   (4)   1.32%    (2)
 Chinstrap   7.35%   (5)  86.76%  (59)   5.88%    (4)
    Gentoo   0.81%   (1)   0.81%   (1)  98.39%  (122)
```

The accuracy rates in this cross-validated confusion matrix are comparable to those in the non-cross-validated confusion matrix above. This implies that our naive Bayes model appears to perform nearly as well on *new* penguins as it does on the original penguin sample that we used to build this model.

14.3 Naive Bayes vs logistic regression

Given the *three* penguin species categories, our classification analysis above required naive Bayes classification – taking logistic regression to this task wasn't even an option. However, in scenarios with a *binary* categorical response variable Y, both logistic regression and naive Bayes are viable classification approaches. In fact, we encourage you to revisit and apply naive Bayes to the rain classification example from Chapter 13. In such binary settings, the question then becomes *which* classification tool to use. Both naive Bayes and logistic regression have their pros and cons. Though naive Bayes is certainly computationally efficient, it also makes some very rigid and often inappropriate assumptions about data structures. You might have also noted that we lose some nuance with naive Bayes. Unlike the logistic regression model with

$$\log\left(\frac{\pi}{1-\pi}\right) = \beta_0 + \beta_1 X_1 + \cdots + \beta_k X_p,$$

naive Bayes lacks regression coefficients β_i. Thus, though naive Bayes can turn information about predictors X into classifications of Y, it does so without much illumination of the relationships among these variables.

Whether naive Bayes or logistic regression is the right tool for a binary classification job depends upon the situation. In general, if the rigid naive Bayes assumptions are inappropriate or if you care about the specific connections between Y and X (i.e., you don't simply want a set of classifications), you should use logistic regression. Otherwise, naive Bayes might be just the thing. Better yet, don't choose! Try out and learn from both tools.

14.4 Chapter summary

Naive Bayes classification is a handy tool for classifying categorical response variables Y with *two or more* categories. Letting $(X_1, X_2, \ldots, X_p)$ be a set of p possible predictors of Y, naive Bayes calculates the posterior probability of each category membership via Bayes' Rule:

$$f(y|x_1, x_2, \ldots, x_p) = \frac{f(y)L(y \mid x_1, x_2, \ldots, x_p)}{\sum_{\text{all } y'} f(y')L(y' \mid x_1, x_2, \ldots, x_p)}.$$

In doing so, it makes some very naive assumptions about the data model from which we define the likelihood $L(y \mid x_1, x_2, \ldots, x_p)$: the predictors X_i are **conditionally independent** and the values of *quantitative* predictors X_i vary Normally within each category $Y = y$. When appropriate, these simplifying assumptions make the naive Bayes model computationally efficient and straightforward to apply. Yet when these simplifying assumptions are violated (which is common), the naive Bayes model can produce misleading classifications.

14.5   Exercises

14.5.1   Conceptual exercises

Exercise 14.1 (Naive). Why is naive Bayes classification called *naive*?

Exercise 14.2 (Which model?). For each scenario below, indicate whether you could classify Y by X using logistic regression, naive Bayes classification, or both.

a)   We want to classify a person's political affiliation Y (Democrat, Republican, or Other) by their age X.

b)   We want to classify whether a person owns a car Y (yes or no) by their age X.

c)   We want to classify whether a person owns a car Y (yes or no) by their location X (urban, suburban, or rural).

Exercise 14.3 (Pros and cons). Every modeling technique has some pros and cons.

a)   Name one pro of naive Bayes classification.

b)   Name one con of naive Bayes classification.

14.5.2   Applied exercises

Exercise 14.4 (Fake news: exclamation points). The `fake_news` data in the **bayesrules** package contains information about 150 news articles, some real news and some fake news. In the next exercises, our goal will be to develop a model that helps us classify an article's `type`, real or fake, using a variety of predictors. To begin, let's consider whether an article's title has an exclamation point (`title_has_excl`).

(a)   Construct and discuss a visualization of the relationship between article `type` and `title_has_excl`.

(b)   Suppose a new article is posted online and its title does *not* have an exclamation point. Utilize naive Bayes classification to calculate the posterior probability that the article is real. Do so from scratch, without using `naiveBayes()` with `predict()`.

(c)   Check your work to part b using `naiveBayes()` with `predict()`.

Exercise 14.5 (Fake news: title length). Consider another possible predictor of article type: the number of words in the title.

(a)   Construct and discuss a visualization of the relationship between article `type` and the number of `title_words`.

(b)   In using naive Bayes classification to classify an article's `type` based on its `title_words`, we assume that the number of `title_words` are conditionally Normal. Do you think this is a fair assumption in this analysis?

(c)   Suppose a new article is posted online and its title has 15 words. Utilize naive Bayes classification to calculate the posterior probability that the article is real. Do so from scratch, without using `naiveBayes()` with `predict()`.

(d)   Check your work to part c using `naiveBayes()` with `predict()`.

Exercise 14.6 (Fake news: title length and negative sentiment). Of course, we can use more than one feature to help us classify whether an article is real or fake. Here, let's consider both an article's title length (`title_words`) and the percent of words in the article that have a `negative` sentiment.

  (a)   Construct and discuss a visualization of the relationship between article `type` and `negative` sentiment.

  (b)   Construct a visualization of the relationship of article `type` with both `title_words` and `negative`.

  (c)   Suppose a new article is posted online – it has a 15-word title and 6% of its words have negative associations. Utilize naive Bayes classification to calculate the posterior probability that the article is real. Do so from scratch, without using `naiveBayes()` with `predict()`.

  (d)   Check your work to part c using `naiveBayes()` with `predict()`.

Exercise 14.7 (Fake news: three predictors). Suppose a new article is posted online – it has a 15-word title, 6% of its words have negative associations, and its title *doesn't* have an exclamation point. Based on these three features, utilize naive Bayes classification to calculate the posterior probability that the article is real. Do so using `naiveBayes()` with `predict()`.

Exercise 14.8 (Fake news: let's pick a model). We've now tried four different naive Bayes classification models of article `type`. In this exercise you'll evaluate and compare the performance of these four models.

model	formula
news_model_1	type ~ title_has_excl
news_model_2	type ~ title_words
news_model_3	type ~ title_words + negative
news_model_4	type ~ title_words + negative + title_has_excl

  a)   Construct a cross-validated confusion matrix for `news_model_1`.

  b)   Interpret each percentage in the confusion matrix.

  c)   Similarly, construct cross-validated confusion matrices for the other three models.

  d)   If our goal is to best detect when an article is *fake*, which of the four models should we use?

Exercise 14.9 (Logistic vs naive). Naive Bayes isn't the only approach we can take to classifying real vs fake news. Since article `type` is *binary*, we could also approach this task using **Bayesian logistic regression**.

  a)   Starting from weakly informative priors, use `stan_glm()` to simulate the posterior logistic regression model of news `type` by all three predictors: `title_words`, `negative`, and `title_has_excl`.

  b)   Perform and discuss some MCMC diagnostics.

  c)   Construct and discuss a `pp_check()`.

  d)   Using a probability cutoff of 0.5, obtain cross-validated estimates of the sensitivity and specificity for the logistic regression model.

  e)   Compare the cross-validated metrics of the logistic regression model to those of `naive_model_4`. Which model is better at detecting fake news?

14.5.3   Open-ended exercises

Exercise 14.10 (Open exercise: climate change). The `pulse_of_the_nation` data in the **bayesrules** package contains responses of 1000 adults to a wide-ranging survey. In one question, respondents were asked about their belief in `climate_change`, selecting from the following options: climate change is `Not Real at All`, `Real but not Caused by People`, or `Real and Caused by People`. In this open-ended exercise, complete a naive Bayesian analysis of a person's belief in `climate_change`, selecting a variety of possible predictors that are of interest to you.

Exercise 14.11 (Open exercise: vaccine safety). The `pulse_of_the_nation` survey also asked respondents about their agreement with the statement that `vaccines_are_safe`, selecting from the following options: `Neither Agree nor Disagree`, `Somewhat Agree`, `Somewhat Disagree`, `Strongly Agree` or `Strongly Disagree`. In this open-ended exercise, complete a naive Bayesian analysis of belief in the safety of vaccines, selecting a variety of possible predictors that are of interest to you.

Unit IV

Hierarchical Bayesian models

15

Hierarchical Models are Exciting

Welcome to Unit 4!

Unit 4 is all about hierarchies. Used in the sentence "my workplace is so hierarchical," this word might have negative connotations. In contrast, "my Bayesian model is so hierarchical" often connotes a good thing! Hierarchical models greatly expand the flexibility of our modeling toolbox by accommodating hierarchical, or *grouped* data. For example, our data might consist of:

- a sampled group of schools and data y on multiple individual students within each school; or
- a sampled group of labs and data y from multiple individual experiments within each lab; or
- a sampled group of people on whom we make multiple individual observations of information y over time.

Ignoring this type of underlying grouping structure violates the assumption of *independent* data behind our Unit 3 models and, in turn, can produce misleading conclusions. In Unit 4, we'll explore techniques that empower us to build this hierarchical structure into our models:

- Chapter 15: Why you should be excited about hierarchical modeling
- Chapter 16: (Normal) hierarchical models without predictors
- Chapter 17: (Normal) hierarchical models with predictors
- Chapter 18: Non-Normal hierarchical models
- Chapter 19: Adding more layers

Before we do, we want to point out that we're using "hierarchical" as a blanket term for group-structured models. Across the literature, these are referred to as **multilevel**, **mixed effects**, or **random effects** models. These terms can serve to confuse, especially since some people use them interchangeably and others use them to make minor distinctions in group-structured models. It's kind of like "pop" vs "soda" vs "cola." We'll avoid that confusion here but point it out so that you're able to make that connection outside this book.

Before pivoting to hierarchical models, it's crucial that we understand *why* we'd ever want to do so. In Chapter 15, our sole purpose is to address this question. We won't build or interpret or simulate any hierarchical models. We'll simply build up a case for the *importance* of these techniques in filling some gaps of our existing Bayesian methodology. To do so, we'll explore the `cherry_blossom_sample` data in the **bayesrules** package, shortened to `running` here:

```
# Load packages
library(bayesrules)
```

DOI: 10.1201/9780429288340-15

```
library(tidyverse)
library(rstanarm)
library(broom.mixed)

# Load data
data(cherry_blossom_sample)
running <- cherry_blossom_sample %>%
  select(runner, age, net)
nrow(running)
[1] 252
```

This data, a subset of the **Cherry** data in the **mdsr** package (Baumer et al., 2021), contains the **net** running times (in minutes) for 36 participants in the annual 10-mile Cherry Blossom race held in Washington, D.C.. Each runner is in their 50s or 60s and has entered the race in multiple years. The plot below illustrates the degree to which some runners are faster than others, as well as the variability in each runner's times from year to year.

```
ggplot(running, aes(x = runner, y = net)) +
  geom_boxplot()
```

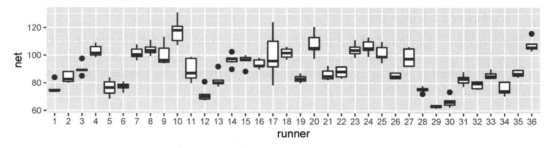

FIGURE 15.1: Boxplots of net running times (in minutes) for 36 runners that entered the Cherry Blossom race in multiple years.

For example, runner 17 tended to have slower times with a lot of variability from year to year. In contrast, runner 29 was the fastest in the bunch and had very consistent times. Looking more closely at the data, runner 1 completed the race in 84 minutes when they were 53 years old. In the subsequent year, they completed the race in 74.3 minutes:

```
head(running, 2)
# A tibble: 2 x 3
  runner   age    net
  <fct>  <int>  <dbl>
1 1         53   84.0
2 1         54   74.3
```

Our goal is to better understand the relationship between running time and age for runners in this age group. What we'll learn is that our current Bayesian modeling toolbox has some limitations.

> ◎ **Goals**
>
> - Explore the limitations of our current Bayesian modeling toolbox under two extremes, **complete pooling** and **no pooling**.
> - Examine the benefits of the **partial pooling** provided by **hierarchical Bayesian models**.
> - Focus on the big ideas and leave the details to subsequent chapters.

15.1 Complete pooling

We'll begin our analysis of the relationship between running time and age using a **complete pooling** technique: combine all 252 observations across our 36 runners into *one pool* of information. In doing so, notice that the relationship appears weak – there's quite a bit of variability in run times at each age with no clear trend as age increases:

```
ggplot(running, aes(y = net, x = age)) +
  geom_point()
```

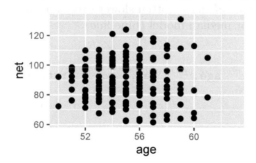

FIGURE 15.2: A scatterplot of net running time versus age for every race result.

To better understand this relationship, we can construct a familiar Normal regression model of running times by age. Letting Y_i and X_i denote the running time and age for the ith observation in our dataset, the data structure for this model is:

$$Y_i|\beta_0, \beta_1, \sigma \stackrel{ind}{\sim} N\left(\mu_i, \sigma^2\right) \quad \text{with} \quad \mu_i = \beta_0 + \beta_1 X_i. \tag{15.1}$$

And we can simulate this **complete pooled** model as usual, here using weakly informative priors:

```
complete_pooled_model <- stan_glm(
  net ~ age,
  data = running, family = gaussian,
  prior_intercept = normal(0, 2.5, autoscale = TRUE),
  prior = normal(0, 2.5, autoscale = TRUE),
  prior_aux = exponential(1, autoscale = TRUE),
  chains = 4, iter = 5000*2, seed = 84735)
```

The simulation results match our earlier observations of a weak relationship between running time and age. The posterior median model suggests that running times tend to increase by a mere 0.27 minutes for each year in age. And with an 80% posterior credible interval for β_1 which straddles 0 (-0.3, 0.84), this relationship is *not* significant:

```
# Posterior summary statistics
tidy(complete_pooled_model, conf.int = TRUE, conf.level = 0.80)
# A tibble: 2 x 5
  term        estimate std.error conf.low conf.high
  <chr>          <dbl>     <dbl>    <dbl>     <dbl>
1 (Intercept)    75.2      24.6     43.7     106.
2 age             0.268     0.446   -0.302     0.842
```

OK. We didn't want to be the first to say it, but this seems a bit strange. Our own experience does not support the idea that we'll continue to run at the same speed. In fact, check out Figure 15.3. If we examine the relationship between running time and age for each of our 36 runners (gray lines), *almost all* have gotten slower over time *and* at a more rapid rate than suggested by the posterior median model (blue line).

```
# Plot of the posterior median model
ggplot(running, aes(x = age, y = net, group = runner)) +
  geom_smooth(method = "lm", se = FALSE, color = "gray", size = 0.5) +
  geom_abline(aes(intercept = 75.2, slope = 0.268), color = "blue")
```

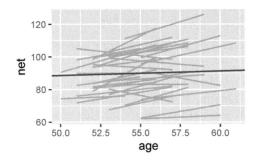

FIGURE 15.3: Observed trends in running time versus age for the 36 subjects (gray) along with the posterior median model (blue).

In Figure 15.4, we zoom in on the details for just three runners. Though the complete pooled model (blue line) does okay in some cases (e.g., runner 22), it's far from universal. The general speed, and changes in speed over time, *vary* quite a bit from runner to runner. For example, runner 20 tends to run slower than average *and*, with age, is slowing down at a more rapid rate. Runner 1 has consistently run faster than the average.

```
# Select an example subset
examples <- running %>%
  filter(runner %in% c("1", "20", "22"))

ggplot(examples, aes(x = age, y = net)) +
  geom_point() +
  facet_wrap(~ runner) +
  geom_abline(aes(intercept = 75.2242, slope = 0.2678),
              color = "blue")
```

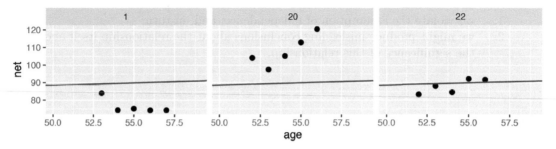

FIGURE 15.4: Scatterplots of running time versus age for 3 subjects, along with the posterior median model (blue).

With respect to these observations, our complete pooled model really misses the mark. As represented by the diagram in Figure 15.5, the complete pooled model lumps all observations of run time $(y_1, y_2, \ldots, y_n)$ into one population or one "pool." In doing so, it makes two assumptions:

1. each observation is *independent* of the others (a blanket assumption behind our Unit 3 models); and
2. information about the individual runners is irrelevant to our model of running time vs age.

We've now observed that these assumptions are inappropriate:

1. Though the observations on one runner might be independent of those on another, the observations *within* a runner are *correlated*. That is, how fast a runner ran in their previous race tells us something about how fast they'll run in the next.
2. With respect to the relationship between running time and age, people are inherently different.

These assumption violations had a significant consequence: our complete pooled model failed to pick up on the fact that people tend to get slower with age. Though we've explored the complete pooling drawbacks in the specific context of the Cherry Blossom race, they are true in general.

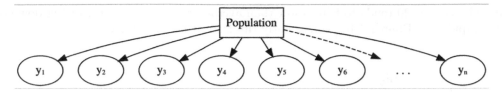

FIGURE 15.5: General framework for a complete pooled model.

Drawbacks of a complete pooling approach

There are two main drawbacks to taking a complete pooling approach to analyze group structured data:

1. we violate the assumption of independence; and, in turn,
2. we might produce misleading conclusions about the relationship itself and the significance of this relationship.

15.2   No pooling

Having failed with our complete pooled model, let's swing to the other extreme. Instead of lumping everybody together into one pool and ignoring any information about runners, the **no pooling** approach considers each of our $m = 36$ runners *separately*. This framework is represented by the diagram in Figure 15.6 where y_{ij} denotes the ith observation on runner j.

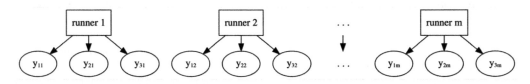

FIGURE 15.6: General framework for a no pooled model.

This framework also means that the no pooling approach builds a *separate* model for each runner. Specifically, let (Y_{ij}, X_{ij}) denote the observed run times and age for runner j in their ith race. Then the data structure for the Normal linear regression model of run time vs age for runner j is:

$$Y_{ij}|\beta_{0j},\beta_{1j},\sigma \sim N\left(\mu_{ij},\sigma^2\right) \quad \text{with} \quad \mu_{ij} = \beta_{0j} + \beta_{1j}X_{ij}. \tag{15.2}$$

This model allows for each runner j to have a unique intercept β_{0j} and age coefficient β_{1j}. Or, in the context of running, the no pooled models reflect the fact that some people tend to be faster than others (hence the different β_{0j}) and that *changes* in speed over time aren't the same for everyone (hence the different β_{1j}). Though we won't bother actually implementing these 36 no pooled models in R, if we utilized vague priors, the resulting *individual* posterior median models would be similar to those below:

```
ggplot(examples, aes(x = age, y = net)) +
  geom_point() +
  geom_smooth(method = "lm", se = FALSE, fullrange = TRUE) +
  facet_wrap(~ runner) +
  xlim(52, 62)
```

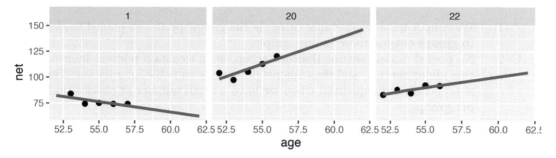

FIGURE 15.7: Scatterplots of running time versus age for 3 subjects, along with subject-specific trends (blue).

This *seems* great at first glance. The runner-specific models pick up on the runner-specific trends. However, there are two significant drawbacks to the no pooling approach. First, suppose that *you* planned to run in the Cherry Blossom race in each of the next few years. Based on the no pooling results for our three example cases, what do you anticipate your running times to be? Are you stumped? If not, you should be. The no pooling approach can't help you answer this question. Since they're tailored to the 36 individuals in our sample, the resulting 36 models don't reliably extend beyond these individuals. To consider a second wrinkle, take a quick quiz.[1]

> ❷ **Quiz Yourself!**
>
> Reexamine runner 1 in Figure 15.7. If they were to race a sixth time at age 62, 5 years after their most recent data point, what would you expect their **net** running time to be?
>
> a) Below 75 minutes
> b) Between 75 and 85 minutes
> c) Above 85 minutes

If you were utilizing the no pooled model to answer this question, your answer would be a. Runner 1's model indicates that they're getting faster with age and should have a running time under 75 minutes by the time they turn 62. Yet this no pooled conclusion exists in a vacuum, only taking into account data on runner 1. From the other *35* runners, we've observed that *most* people tend to get slower over time. It would be unfortunate to completely ignore this information, especially since we have a mere five race sample size for runner 1 (hence aren't in the position to disregard the extra data!). A more reasonable prediction might be option b: though they might not maintain such a steep downward trajectory, runner 1 will likely remain a fast runner with a race time between 75 and 85 minutes. Again, this would be the *reasonable* conclusion, not the conclusion we'd make if using our no pooled

---

[1] Answer: There's no definitive right answer here, but we think that b is the most reasonable.

models alone. Though we've explored the no pooling drawbacks in the specific context of the Cherry Blossom race, they are true in general.

Drawbacks of a no pooling approach

There are two main drawbacks to taking a no pooling approach to analyze group structured data:

1. We cannot reliably generalize or apply the group-specific no pooled models to groups outside those in our sample.
2. No pooled models assume that one group doesn't contain relevant information about another, and thus ignores potentially valuable information. This is especially consequential when we have a small number of observations per group.

15.3   Hierarchical data

The Cherry Blossom data has presented us with a new challenge: **hierarchical data.** The general structure for this hierarchy is represented in Figure 15.8. The top layer of this hierarchy is the population of *all* runners, not just those in our sample. In the next layer are our 36 runners, sampled from this population. In every analysis up to this chapter, this is where things have stopped. Yet in the `running` data, the final layer of the hierarchy holds the multiple observations per runner, often referred to as **repeated measures** or **longitudinal data.**

Hierarchical, or *grouped*, data structures are common in practice. These don't always arise from making multiple observations on a *person* as in the `running` data. In an analysis of new teaching technologies, our groups might be schools and our observations multiple students within each school:

```
  school_id tech score
1         1    A    90
2         1    A    94
3         1    B    82
4         2    A    85
5         2    B    89
```

In an analysis of a natural mosquito repellent, our groups might be a sample of labs and our observations multiple experiments within each lab:

```
  lab_id repellant score
1      1         A    90
2      1         A    94
3      1         B    82
4      2         A    85
5      2         B    89
```

There are good reasons to collect data in this hierarchical manner. To begin, it's often more sustainable with respect to time, money, logistics, and other resources. For example, though both data collection approaches produce 100 data points, it requires fewer resources to interview 20 students at each of 5 different schools than 1 student at each of 100 different schools. Hierarchical data also provides insights into two important features:

1. **Within-group variability**
   The degree of the variability among multiple observations *within* each group can be interesting on its own. For example, we can examine how *consistent* an *individual's* running times are from year to year.

2. **Between-group variability**
   Hierarchical data also allows us to examine the variability from group to group. For example, we can examine the degree to which running patterns *vary* from individual to individual.

15.4 Partial pooling with hierarchical models

Our existing Bayesian modeling toolbox presents two approaches to analyzing hierarchical data. We can ignore grouping structure entirely, lump all groups together, and assume that one model is appropriately universal through **complete pooling** (Figure 15.5). Or we can *separately* analyze each group and assume that one group's model doesn't tell us anything about another's through **no pooling** (Figure 15.6). In between these "no room for individuality" and "every person for themselves" attitudes, **hierarchical models** offer a welcome middle ground through **partial pooling** (Figure 15.8). In general, hierarchical models embrace the following idea: though each group is *unique*, having been sampled from the same population, all groups are connected and thus might contain valuable information about one another. We dig into that *idea* here, and will actually start *building* hierarchical models in Chapter 16.

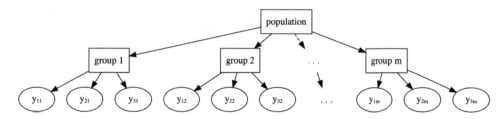

FIGURE 15.8: General framework for a hierarchical model.

Figure 15.9 compares hierarchical, no pooled, and complete pooled models of running time vs age for our three sample runners *and* you, a new runner that we haven't yet observed. Consider our three sample runners. Though they utilized the same data, our three techniques produced different posterior median models for these runners. Unlike the *complete* pooled model (black), the hierarchical model (dashed) allows each individual runner to enjoy a unique posterior trend. These trends detect the fact that runner 1 is the fastest among our example runners. Yet unlike the *no* pooled model (blue), the runner-specific hierarchical models do not ignore data on other runners. For example, examine the results for runner 1. Based on the data for that runner alone, the no pooled model assumes that their downward trajectory will continue. In contrast, in the face of only five data points on runner 1 yet strong evidence from our *other* 35 runners that people tend to slow down, the slope of the hierarchical trend has been pulled up. In general, in recognizing that one runner's data might contain valuable information about another, the hierarchical model trends are pulled *away* from the no pooled individual models and *toward* the complete pooled universal model.

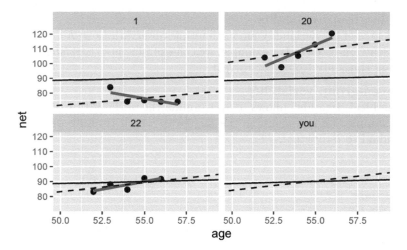

FIGURE 15.9: Posterior median models of the no pooled (blue), complete pooled (black), and hierarchical (dashed) models of running time vs age for three example runners and you.

Finally, consider the modeling results for you, a new runner. Notice that there's no blue line in your panel. Since you weren't in the original **running** sample, we can't use the no pooled model to predict your running trend. In contrast, since they don't ignore the broader population from which our runners were sampled, the complete pooled and hierarchical models can both be used to predict your trend. Yet the results are different! Since the complete pooled model wasn't able to detect the fact that runners in their 50s tend to get slower over time, your posterior median model (black) has a slope near zero. Though this is a nice thought, the hierarchical model likely offers a more realistic prediction. Based on the data across all 36 runners, you'll probably get a bit slower. Further, the anticipated rate at which you will slow down is comparable to the average rate across our 36 runners.

15.5   Chapter summary

In Chapter 15 we motivated the importance of incorporating another technique into our Bayesian toolbox: hierarchical models. Focusing on big ideas over details, we explored three approaches to studying group structured data:

- **Complete pooled models** lump all data points together, assuming they are independent and that a universal model is appropriate for all groups. This can produce misleading conclusions about the relationship of interest and the significance of this relationship. Simply put, the vibe of complete pooled models is "no room for individuality."

- **No pooled models** build a separate model for each group, assuming that one group's model doesn't tell us anything about another's. This approach underutilizes the data and cannot be generalized to groups outside our sample. Simply put, the vibe of no pooled models is "every group for themselves."

- **Partial pooled or hierarchical models** provide a middle ground. They acknowledge that, though each individual group might have its own model, one group can provide valuable information about another. That is, "let's learn from one another while celebrating our individuality."

15.6 Exercises

15.6.1 Conceptual exercises

Exercise 15.1 (Three pooling types: explain to your friend). In one to two sentences each, explain the following concepts to your friend Hakeem, who has taken an intro statistics course but otherwise is new to pooled data.

a) Complete pooling
b) No pooling
c) Partial pooling

Exercise 15.2 (Three pooling types: redux). Hakeem now understands the three pooling approaches thanks to your excellent explanations in the previous exercise. Now he has some follow-up questions! Answer these in your own words and using your own examples.

a) Can you give me an example when we might want to use the partial pooling approach?
b) Why wouldn't we always just use the complete pooling approach? It seems easier!
c) Does the no pooled approach have any drawbacks? It seems that there are fewer assumptions.
d) Can you remind me what the difference is between *within-group* variability and *between-group* variability? And how does that relate to partial pooling?

15.6.2 Applied exercises

Interested in the impact of sleep deprivation on reaction time, Belenky et al. (2003) enlisted 18 subjects in a study. The subjects got a regular night's sleep on "day 0" of the study, and were then restricted to 3 hours of sleep per night for the next 9 days. Each day, researchers recorded the subjects' reaction times (in ms) on a series of tests. The results are provided in the **sleepstudy** dataset in the **lme4** package.

Exercise 15.3 (Hierarchical data). The **sleepstudy** data is hierarchical. Draw a diagram in the spirit of Figure 15.8 that captures the hierarchical framework. Think: What are the "groups"?

Exercise 15.4 (Complete pooling: Part I). Suppose that we (incorrectly) took a complete pooling approach to modeling **Reaction** time (Y) by **Days** of sleep deprivation (X).

a. To explore the complete pooled behavior of the data points, construct and discuss a plot of **Reaction** by **Days**. In doing so, ignore the subjects and include the observed trend in the relationship.

b. Draw a diagram in the spirit of Figure 15.8 that captures the complete pooling framework.
c. Using careful notation, write out the structure for the complete pooled model of Y by X.

Exercise 15.5 (Complete pooling: Part II). In the context of the sleep study, what two incorrect assumptions does the complete pooled model make and why are these inappropriate in the sleep study analysis?

Exercise 15.6 (No pooling: Part I). Suppose instead that we (incorrectly) took a no pooling approach in our sleep study analysis.

a. To explore the no pooled behavior of the data points, construct and discuss separate scatterplots of `Reaction` by `Days` for each `Subject`, including subject-specific trend lines.
b. Draw a diagram in the spirit of Figure 15.6 that captures the no pooling framework.
c. Using careful notation, write out the structure for the no pooled model of Y by X.

Exercise 15.7 (No pooling: Part II). In the context of the sleep study, what are the two main drawbacks to analyzing the relationship between reaction time and sleep deprivation using a no pooling approach?

Exercise 15.8 (Complete vs no vs partial pooling). Suppose we only had data on the two subjects below. For both, provide a loose sketch of three separate trend lines corresponding to a complete pooled, no pooled, and partial pooled model of `Reaction` time by `Days`.

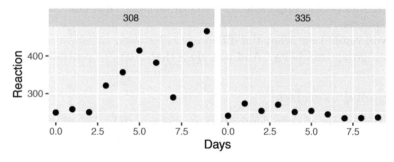

16

(Normal) Hierarchical Models without Predictors

In Chapter 16 we'll build our first hierarchical models upon the foundations established in Chapter 15. We'll start simply, by assuming that we have some response variable Y, but no predictors X. Consider the following data story. The Spotify music streaming platform provides listeners with access to more than 50 million songs. Of course, some songs are more popular than others. Let Y be a song's Spotify popularity rating on a 0-100 scale. In general, the more recent plays that a song has on the platform, the higher its popularity rating. Thus, the popularity rating *doesn't* necessarily measure a song's overall quality, long-term popularity, or popularity beyond the Spotify audience. And though it will be tough to resist asking which song features X can help us *predict* ratings, we'll focus on understanding Y alone. Specifically, we'd like to better understand the following:

- What's the typical popularity of a Spotify song?
- To what extent does popularity vary from artist to artist?
- For any single artist, how much might popularity vary from song to song?

Other than a *vague* sense that the average popularity rating is around 50, we don't have any strong prior understanding about these dynamics, and thus will be utilizing weakly informative priors throughout our Spotify analysis. With that, let's dig into the data. The **spotify** data in the **bayesrules** package is a subset of data analyzed by Kaylin Pavlik (2019) and distributed through the **#TidyTuesday** project (R for Data Science, 2020d). It includes information on 350 songs on Spotify.[1]

```
# Load packages
library(bayesrules)
library(tidyverse)
library(rstanarm)
library(bayesplot)
library(tidybayes)
library(broom.mixed)
library(forcats)

# Load data
data(spotify)
```

Here we select only the few variables that are important to our analysis, and reorder the **artist** levels according to their mean song popularity using `fct_reorder()` in the **forcats** package (Wickham, 2021):

---

[1]To focus on the new methodology, we analyze a tiny fraction of the nearly 33,000 songs analyzed by Pavlik. We COULD but won't analyze all 33,000 songs.

DOI: 10.1201/9780429288340-16

```
spotify <- spotify %>%
  select(artist, title, popularity) %>%
  mutate(artist = fct_reorder(artist, popularity, .fun = 'mean'))
```

You're encouraged to view the resulting `spotify` data in full. But even just a few snippets reveal a *hierarchical* or *grouped* data structure:

```
# First few rows
head(spotify, 3)
# A tibble: 3 x 3
  artist title        popularity
  <fct>  <chr>             <dbl>
1 Alok   On & On              79
2 Alok   All The Lies         56
3 Alok   Hear Me Now          75

# Number of songs
nrow(spotify)
[1] 350

# Number of artists
nlevels(spotify$artist)
[1] 44
```

Specifically, our total sample of 350 songs is comprised of multiple songs for each of 44 artists who were, in turn, sampled from the population of all artists that have songs on Spotify (Figure 16.1).

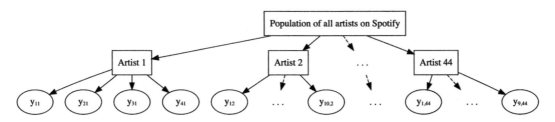

FIGURE 16.1: Hierarchical data on artists.

The `artist_means` data frame summarizes the number of songs observed on each artist along with the mean popularity of these songs. Shown below are the two artists with the lowest and highest mean popularity ratings. Keeping in mind that we've eliminated from our analysis the nearly 10% of Spotify songs which have a popularity rating of 0, each artist in our sample should be seen as a relative success story.

```
artist_means <- spotify %>%
  group_by(artist) %>%
  summarize(count = n(), popularity = mean(popularity))

artist_means %>%
```

```
  slice(1:2, 43:44)
# A tibble: 4 x 3
  artist         count popularity
  <fct>          <int>      <dbl>
1 Mia X              4       13.2
2 Chris Goldarg     10       16.4
3 Lil Skies          3       79.3
4 Camilo             9         81
```

Recalling Chapter 15, we will consider three different approaches to analyzing this Spotify popularity data:

- **Complete pooling**
  Ignore artists and lump all songs together (Figure 15.5).

- **No pooling**
  Separately analyze each artist and assume that one artist's data doesn't contain valuable information about another artist (Figure 15.6).

- **Partial pooling (via hierarchical models)**
  Acknowledge the grouping structure, so that even though artists differ in popularity, they might share valuable information about each other *and* about the broader population of artists (Figure 16.1).

As you might suspect, the first two approaches oversimplify the analysis. The good thing about trying them out is that it will remind us why we should care about hierarchical models in the first place.

◎ **Goals**

- *Build* a hierarchical model of variable Y with *no* predictors X.
- *Simulate* and analyze this hierarchical model using **rstanarm**.
- Utilize hierarchical models for *predicting* Y.

To focus on the big picture, we keep this first exploration of hierarchical models to the *Normal* setting, a special case of the broader generalized linear hierarchical models we'll study in Chapter 18.

---

❶ **Warning**

As our Bayesian modeling toolkit expands, so must our R tools and syntax. Be patient with yourself here – as you see more examples throughout this unit, the *patterns* in the syntax will become more familiar.

16.1   Complete pooled model

Carefully analyzing group structured data requires some new notation. In general, we'll use i and j subscripts to track two pieces of information about each song in our sample: j indicates a song's *artist* and i indicates the specific *song* within each artist. To begin, let n_j denote the number of songs we have for artist j where $j \in \{1, 2, \ldots, 44\}$. For example, the first two artists have $n_1 = 4$ and $n_2 = 10$ songs, respectively:

```
head(artist_means, 2)
# A tibble: 2 x 3
  artist          count popularity
  <fct>           <int>      <dbl>
1 Mia X               4       13.2
2 Chris Goldarg      10       16.4
```

Further, the number of songs per artist runs as low as 2 and as high as 40:

```
artist_means %>%
  summarize(min(count), max(count))
# A tibble: 1 x 2
  `min(count)` `max(count)`
         <int>        <int>
1            2           40
```

Summing the number of songs n_j across *all* artists produces our *total* sample size n:

$$n = \sum_{j=1}^{44} n_j = n_1 + n_2 + \cdots + n_{44} = 350.$$

Beyond the artists are the songs themselves. To distinguish between songs, we will let Y_{ij} (or $Y_{i,j}$) represent the ith song for artist j where $i \in \{1, 2, \ldots, n_j\}$ and $j \in \{1, 2, \ldots, 44\}$. Thus, we can think of our sample of 350 total songs as the *collection* of our smaller samples on each of 44 artists:

$$Y := ((Y_{11}, Y_{21}, \ldots, Y_{n_1,1}), (Y_{12}, Y_{22}, \ldots, Y_{n_2,2}), \ldots, (Y_{1,44}, Y_{2,44}, \ldots, Y_{n_{44},44})).$$

As a first approach to analyzing this data, we'll implement a **complete pooled model** of song popularity which *ignores* the data's grouped structure, lumping all songs together in one sample. A density plot of this pool illustrates that there's quite a bit of variability in popularity from song to song, with ratings ranging from roughly 10 to 95.

```
ggplot(spotify, aes(x = popularity)) +
  geom_density()
```

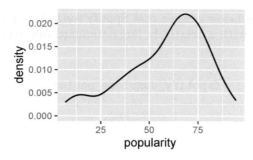

FIGURE 16.2: A density plot of the variability in popularity from song to song.

Though these ratings appear somewhat left skewed, we'll make one more simplification in our complete pooling analysis by assuming Normality. Specifically, we'll utilize the following **Normal-Normal complete pooled model** with a prior for μ that reflects our weak understanding that the average popularity rating is around 50 and, independently, a weakly informative prior for σ. These prior specifications can be confirmed using `prior_summary()`.

$$Y_{ij}|\mu,\sigma \sim N(\mu,\sigma^2)$$
$$\mu \sim N(50,52^2) \qquad (16.1)$$
$$\sigma \sim \text{Exp}(0.048)$$

Though we saw similar models in Chapters 5 and 9, it's important to review the meaning of parameters μ and σ before shaking things up. Since μ and σ are shared by *every* song Y_{ij}, we can think of them as **global parameters** which do not vary by artist j. Across *all* songs across *all* artists in the population:

- μ = global mean popularity; and
- σ = global standard deviation in popularity from song to song.

To simulate the corresponding posterior, note that the complete pooled model (16.1) is a simple Normal regression model in disguise. Specifically, if we substitute notation β_0 for global mean μ, (16.1) is an **intercept-only regression model** with no predictors X. With this in mind, we can simulate the complete pooled model using `stan_glm()` with the formula `popularity ~ 1` where the 1 specifies an "intercept-only" term:

```
spotify_complete_pooled <- stan_glm(
  popularity ~ 1,
  data = spotify, family = gaussian,
  prior_intercept = normal(50, 2.5, autoscale = TRUE),
  prior_aux = exponential(1, autoscale = TRUE),
  chains = 4, iter = 5000*2, seed = 84735)

# Get prior specifications
prior_summary(spotify_complete_pooled)
```

The posterior summaries below suggest that Spotify songs have a typical popularity rating, μ, around 58.39 points with a relatively large standard deviation from song to song, σ, around 20.67 points.

```
complete_summary <- tidy(spotify_complete_pooled,
                         effects = c("fixed", "aux"),
                         conf.int = TRUE, conf.level = 0.80)
complete_summary
# A tibble: 3 x 5
  term        estimate std.error conf.low conf.high
  <chr>          <dbl>     <dbl>    <dbl>     <dbl>
1 (Intercept)     58.4      1.10     57.0      59.8
2 sigma           20.7      0.776    19.7      21.7
3 mean_PPD        58.4      1.57     56.4      60.4
```

In light of these results, complete the quiz below.[2]

> **❷ Quiz Yourself!**
>
> Suppose the following three artists each release a new song on Spotify:
> - Mia X, the artist with the lowest mean popularity in our sample (13).
> - Beyoncé, an artist with nearly the highest mean popularity in our sample (70).
> - Mohsen Beats, a musical group that we *didn't* observe.
>
> Using the complete pooled model, what would be the approximate posterior predictive mean of each song's popularity?

Your intuition might have reasonably suggested that the predicted popularity for these three artists should be *different* – our data suggests that Beyoncé's song will be more popular than Mia X's. Yet, recall that by lumping all songs together, our complete pooled model *ignores* any artist-specific information. As a result, the posterior predicted popularity of any new song will be the *same* for every artist, those that are in our sample (Beyoncé and Mia X) and those that are not (Mohsen Beats). Further, the shared posterior predictive mean popularity will be roughly equivalent to the posterior expected value of the global mean, $E(\mu|y) \approx 58.39$. Plotting the posterior predictive popularity models (light blue) alongside the observed *mean* popularity levels (dark dots) for all 44 sampled artists illustrates the unfortunate consequences of this oversimplification – it treats all artists the same even though we know they're not.

```
set.seed(84735)
predictions_complete <- posterior_predict(spotify_complete_pooled,
                                           newdata = artist_means)

ppc_intervals(artist_means$popularity, yrep = predictions_complete,
              prob_outer = 0.80) +
  ggplot2::scale_x_continuous(labels = artist_means$artist,
                              breaks = 1:nrow(artist_means)) +
  xaxis_text(angle = 90, hjust = 1)
```

---

[2]Answer: The posterior predictive median will be around 58.39 for each artist.

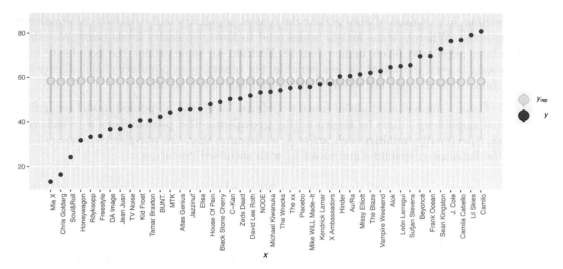

FIGURE 16.3: Posterior predictive intervals for artist song popularity, as calculated from a complete pooled model.

16.2   No pooled model

The complete pooled model ignored the grouped structure of our Spotify data – we have 44 unique artists in our sample and multiple songs per each. Next, consider a no pooled model which swings the opposite direction, separately analyzing the popularity rating of each individual artist. Figure 16.4 displays density plots of song popularity for each artist.

```
ggplot(spotify, aes(x = popularity, group = artist)) +
  geom_density()
```

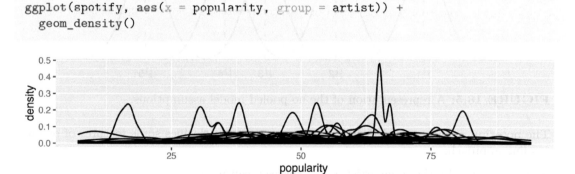

FIGURE 16.4: Density plots of the variability in popularity from song to song, by artist.

The punchline in this mess of lines is this: the *typical* popularity and *variability* in popularity differ by artist. Whereas some artists' songs tend to have popularity levels below 25, others tend to have popularity levels higher than 75. Further, for some artists, the popularity from song to song is very *consistent*, or less variable. For others, song popularity falls all over the map – some of their songs are wildly popular and others not so much.

Instead of assuming a shared *global* mean popularity level, the no pooled model recognizes that artist popularity differs by incorporating **group-specific parameters** μ_j which vary by artist. Specifically, **within** each artist j, we assume that the popularity of songs i are *Normally* distributed around some mean μ_j with standard deviation σ:

$$Y_{ij}|\mu_j,\sigma \sim N(\mu_j,\sigma^2) \tag{16.2}$$

Thus

- μ_j = mean song popularity for artist j; and
- σ = the standard deviation in popularity from song to song within each artist.

Figure 16.5 illustrates these parameters and the underlying model assumptions for just five of our 44 sampled artists. First, our Normal data model (16.2) assumes that the *mean* popularity μ_j varies between different artists j, i.e., some artists' songs tend to be more popular than other artists' songs. In contrast, it assumes that the *variability* in popularity from song to song, σ, is the *same* for each artist, and hence its lack of a j subscript. Though Figure 16.4 might bring the assumptions of Normality and equal variability into question, with such small sample sizes per artist, we can't put too much stock into the potential noise. The current simplifying assumptions are a reasonable place to start.

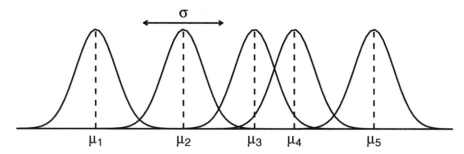

FIGURE 16.5: A representation of the no pooled model assumptions.

The notation of the no pooled data structure (16.2) also clarifies some features of our no pooling approach:

- Every song i by the same artist j, Y_{ij}, shares mean parameter μ_j.
- Songs by *different* artists j and k, Y_{ij} and Y_{ik}, have *different* mean parameters μ_j and μ_k. Thus, each artist gets a unique mean and one artist's mean doesn't tell us anything about another's.
- Whereas our model does *not* pool information about the players' mean popularity levels μ_j, by assuming that the artists share the σ standard deviation parameter, it *does* pool information from the artists to learn about σ. We *could* also assign unique standard deviation parameters to each artist, σ_j, but the shared standard deviation assumption simplifies things a great deal. In this spirit of simplicity, we'll continue to refer to (16.3) as a no pooled model, though "no" pooled would be more appropriate.

Finally, to complete the no pooled model, we utilize weakly informative Normal priors for the mean parameters μ_j (each centered at an average popularity rating of 50) and an Exponential prior for the standard deviation parameter σ. Since the 44 μ_j priors are uniquely tuned for the corresponding artist j, we do not list them here:

$$Y_{ij}|\mu_j, \sigma_j \sim N(\mu_j, \sigma^2)$$
$$\mu_j \sim N(50, s_j^2) \tag{16.3}$$
$$\sigma \sim \text{Exp}(0.048).$$

To simulate the no pooled model posteriors, we can make one call to `stan_glm()` by treating `artist` as a *predictor* in a regression model of `popularity` which lacks a global intercept term.[3] This setting is specified by the formula `popularity ~ artist - 1`:

```
spotify_no_pooled <- stan_glm(
  popularity ~ artist - 1,
  data = spotify, family = gaussian,
  prior = normal(50, 2.5, autoscale = TRUE),
  prior_aux = exponential(1, autoscale = TRUE),
  chains = 4, iter = 5000*2, seed = 84735)
```

Based on your understanding of the no pooling strategy, complete the quiz below.[4]

---

❷ Quiz Yourself!

Suppose the following three artists each release a new song on Spotify:

   a. Mia X, the artist with the lowest mean popularity in our sample (13).
   b. Beyoncé, an artist with nearly the highest mean popularity in our sample (70).
   c. Mohsen Beats, a musical group that we *didn't* observe.

Using the no pooled model, what will be the approximate posterior predictive mean of each song's popularity?

Since each artist gets a unique mean popularity parameter μ_j which is modeled using their unique song data, they each get a unique posterior predictive model for their next song's popularity. Having utilized weakly informative priors, it thus makes sense that Mia X's posterior predictive model should be centered near her observed sample mean of 13, whereas Beyoncé's should be near 70. In fact, Figure 16.6 confirms that the posterior predictive model for the popularity of *each* artist's next song is centered around the mean popularity of their observed sample songs.

---

[3]To make the connection to the no pooled model, let predictor X_{ij} be 1 for artist j and 0 otherwise. Then `spotify_no_pooled` models the mean in popularity Y_{ij}, for song i by artist j, by $\mu_1 X_{i1} + \mu_2 X_{i2} + \cdots + \mu_{44} X_{i,44} = \mu_j$, the mean popularity for artist j.

[4]Answer: a) Roughly 13; b) Roughly 70; c) NA. The no pooled model can't be used to predict the popularity of a Mohsen Beats song, since Mohsen Beats does not have any songs included in this sample.

```
# Simulate the posterior predictive models
set.seed(84735)
predictions_no <- posterior_predict(
  spotify_no_pooled, newdata = artist_means)

# Plot the posterior predictive intervals
ppc_intervals(artist_means$popularity, yrep = predictions_no,
              prob_outer = 0.80) +
  ggplot2::scale_x_continuous(labels = artist_means$artist,
                              breaks = 1:nrow(artist_means)) +
  xaxis_text(angle = 90, hjust = 1)
```

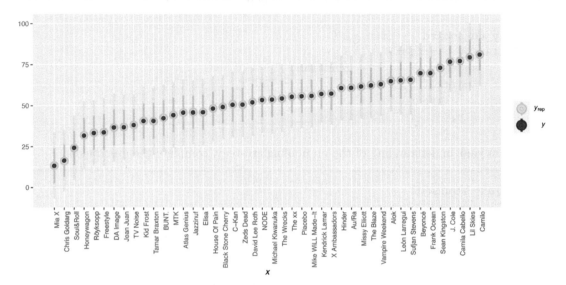

FIGURE 16.6: Posterior predictive intervals for artist song popularity, as calculated from a no pooled model.

If we didn't pause and reflect, this result might *seem* good. The no pooled model acknowledges that some artists tend to be more popular than others. However, there are two critical drawbacks. **First, the no pooled model ignores data on one artist when learning about the typical popularity of another.** This is especially problematic when it comes to artists for whom we have few data points. For example, our *low* posterior predictions for Mia X's next song were based on a measly 4 songs. The other artists' data suggests that these low ratings might just be a tough break – her next song might be more popular! Similarly, our *high* posterior predictions for Lil Skies' next song were based on only 3 songs. In light of the other artists' data, we might wonder whether this was beginner's luck that will be tough to maintain.

Second, the no pooled model cannot be generalized to artists outside our sample. Remember Mohsen Beats? Since the no pooled model is tailor-made to the artists in our sample, and Mohsen Beats isn't part of this sample, it *cannot* be used to predict the popularity of Mohsen Beats' next song. That is, just as the no pooled model assumes that no artist in our sample contains valuable information about the typical popularity of another artist in our sample, it assumes that our sampled artists tell us nothing about the typical popularity of artists that didn't make it into our sample. This also speaks to why we cannot account for the grouped structure of our data by including `artist` as a categorical *predictor* X. On top of violating the assumption of independence, doing so would allow us to learn about *only* the 44 artists in our sample.

16.3 Building the hierarchical model

16.3.1 The hierarchy

We hope we've re-convinced you that we must pay careful attention to hierarchically structured data. To this end, we'll finally concretize a hierarchical Normal model of Spotify popularity. There are a few layers to this model, two related to the hierarchically structured `spotify` data (Figure 16.1) and a third related to the prior models on our model parameters. A very general outline for this model is below, the layers merely numbered for reference:

Layer 1: $Y_{ij}|\mu_j, \sigma_y$ ~ model of how song popularity varies WITHIN artist j
Layer 2: $\mu_j|\mu, \sigma_\mu$ ~ model of how the typical popularity μ_j varies BETWEEN artists
Layer 3: $\mu, \sigma_y, \sigma_\mu$ ~ prior models for shared global parameters

$$(16.4)$$

Layer 1 deals with the smallest unit in our hierarchical data: individual songs within each artist. As with the *no* pooled model (16.2), this first layer is group or artist specific, and thus acknowledges that song popularity Y_{ij} depends in part on the artist. Specifically, **within** each artist j, we assume that the popularity of songs i are *Normally* distributed around some mean μ_j with standard deviation σ_y:

$$Y_{ij}|\mu_j, \sigma_y \sim N(\mu_j, \sigma_y^2)$$

Thus, the meanings of the Layer 1 parameters are the same as they were for the no pooled model:

- μ_j = mean song popularity for artist j; and
- σ_y = **within-group variability**, i.e., the standard deviation in popularity from song to song within each artist.

Moving forward, if we were to place separate priors on each artist mean popularity parameter μ_j, we'd be right back at the no pooled model (16.3). **Layer 2** of (16.4) is where the hierarchical model begins to distinguish itself. Instead of assuming that one artist's popularity level can't tell us about another, Layer 2 acknowledges that our 44 sampled artists are all drawn from the same broader population of Spotify artists. Within this population, popularity varies from artist to artist. We model this variability in mean popularity **between** artists by assuming that the individual mean popularity levels, μ_j, are *Normally* distributed around μ with standard deviation σ_μ,

$$\mu_j|\mu, \sigma_\mu \sim N(\mu, \sigma_\mu^2).$$

Thus, we can think of the two new parameters as follows:

- μ = the **global average** of mean song popularity μ_j across all artists j, i.e., the mean popularity rating for the most average artist; and
- σ_μ = **between-group variability**, i.e., the standard deviation in mean popularity μ_j from artist to artist.

A density plot of the *mean* popularity levels for our 44 artists indicates that this prior assumption is reasonable. The artists' observed sample means *do* appear to be roughly Normally distributed around some global mean:

```
ggplot(artist_means, aes(x = popularity)) +
  geom_density()
```

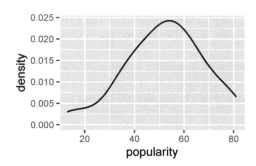

FIGURE 16.7: A density plot of the variability in mean song popularity from artist to artist.

Notation alert

- There's a difference between μ_j and μ. When a parameter has a subscript j, it refers to a feature of group j. When a parameter *doesn't* have a subscript j, it's the *global* counterpart, i.e., the same feature across all groups.

- Subscripts signal the group or layer of interest. For example, σ_y refers to the standard deviation of Y values within each group, whereas σ_μ refers to the standard deviation of means μ_j from group to group.

Though Layers 1 and 2 inform our understanding of the hierarchically structured data, remember that this is a Bayesian book. Thus, in the final **Layer 3** of our hierarchical model (16.4), we must specify prior models for the **global parameters** which describe and are shared across the entire population of Spotify artists: μ, σ_y, and σ_μ. As usual, the Normal model provides a reasonable prior for the mean parameter μ and the Exponential provides a reasonable prior for the standard deviation terms, σ_y and σ_μ.

Putting this all together, our final hierarchical model brings together the models of how individual song popularity Y_{ij} varies *within* artists (Layer 1) with the model of how mean popularity levels μ_j vary *between* artists (Layer 2) with our prior understanding of the entire population of artists (Layer 3). The weakly informative priors are specified here and confirmed below. Note that we again assume a baseline song popularity rating of 50.

$$
\begin{aligned}
Y_{ij}|\mu_j,\sigma_y &\sim N(\mu_j,\sigma_y^2) & &\text{model of individual songs within artist } j \\
\mu_j|\mu,\sigma_\mu &\overset{ind}{\sim} N(\mu,\sigma_\mu^2) & &\text{model of variability between artists} \\
\mu &\sim N(50,52^2) & &\text{prior models on global parameters} & (16.5) \\
\sigma_y &\sim \text{Exp}(0.048) \\
\sigma_\mu &\sim \text{Exp}(1)
\end{aligned}
$$

This type of model has a special name: one-way analysis of variance. We'll explore the motivation behind this moniker in Section 16.3.3.

One-way Analysis of Variance (ANOVA)

Suppose we have multiple data points on each of m groups. For each group $j \in \{1, 2, \ldots, m\}$, let Y_{ij} denote the ith observation and n_j denote the sample size. Then the one-way ANOVA model of Y_{ij} is:

$$
\begin{aligned}
Y_{ij}|\mu_j,\sigma_y &\sim N(\mu_j,\sigma_y^2) & &\text{model of individual observations within group } j \\
\mu_j|\mu,\sigma_\mu &\overset{ind}{\sim} N(\mu,\sigma_\mu^2) & &\text{model of how parameters vary between groups} \\
\mu &\sim N(m,s^2) & &\text{prior models on global parameters} \\
\sigma_y &\sim \text{Exp}(l_y) \\
\sigma_\mu &\sim \text{Exp}(l_\mu)
\end{aligned}
$$

$$(16.6)$$

16.3.2 Another way to think about it

In our hierarchical Spotify analysis, we've incorporated artist-specific mean parameters μ_j to indicate that the average song popularity can vary by artist j. Our hierarchical model (16.5) presents these artist-specific means μ_j as Normal deviations from the global mean song popularity μ with standard deviation σ_μ: $\mu_j \sim N(\mu, \sigma_\mu^2)$. *Equivalently*, we can think of the artist-specific means as random *tweaks* or *adjustments* b_j to μ,

$$\mu_j = \mu + b_j$$

where these tweaks are normal deviations from *0* with standard deviation σ_μ:

$$b_j \sim N(0, \sigma_\mu^2).$$

For example, suppose that the average popularity of *all* songs on Spotify is $\mu = 55$, whereas the average popularity for some artist j is $\mu_j = 65$. Then this artist's average popularity is 10 above the global average. That is, $b_j = 10$ and

$$\mu_j = \mu + b_j = 55 + 10 = 65.$$

In general, then, we can reframe Layers 1 and 2 of our hierarchical model (16.5) as follows:

$$Y_{ij}|\mu_j,\sigma_y \sim N(\mu_j,\sigma_y^2) \quad \text{with} \quad \mu_j = \mu + b_j$$

$$b_j|\sigma_\mu \stackrel{ind}{\sim} N(0,\sigma_\mu^2)$$

$$\mu \sim N(50,52^2) \tag{16.7}$$

$$\sigma_y \sim \text{Exp}(0.048)$$

$$\sigma_\mu \sim \text{Exp}(1)$$

Again, these two model formulations, (16.5) and (16.7), are equivalent. They're just two different ways to think about the hierarchical structure of our data. We'll use both throughout this book.

16.3.3 Within- vs between-group variability

Let's take in the model we've just built. The models we studied prior to Unit 4 allow us to explore *one* source of variability, that in the individual observations Y across an entire population. In contrast, the aptly named one-way "analysis of variance" models of hierarchical data enable us to decompose the variance in Y into *two* sources: (1) the variance of individual observations Y **within** any given group, σ_y^2; and (2) the variance of features **between** groups, σ_μ^2, i.e., from group to group. The *total* variance in Y is the sum of these two parts:

$$\text{Var}(Y_{ij}) = \sigma_y^2 + \sigma_\mu^2. \tag{16.8}$$

We can also think of this breakdown proportionately:

$$\frac{\sigma_y^2}{\sigma_\mu^2 + \sigma_y^2} = \begin{array}{l} \text{proportion of Var}(Y_{ij}) \text{ that can be explained by} \\ \text{differences in the observations within each group} \end{array}$$

$$\frac{\sigma_\mu^2}{\sigma_\mu^2 + \sigma_y^2} = \begin{array}{l} \text{proportion of Var}(Y_{ij}) \text{ that can be explained by} \\ \text{differences between groups} \end{array} \tag{16.9}$$

Or in the context of our Spotify analysis, we can quantify how much of the *total* variability in popularity across all songs and artists $(\text{Var}(Y_{ij}))$ can be explained by differences between the *songs within each artist* (σ_y^2) and differences *between the artists* themselves (σ_μ).

Figure 16.8 illustrates three scenarios for the *comparison* of these two sources of variability. At one extreme is the scenario in which the variability *between* groups is much greater than that *within* groups, $\sigma_\mu > \sigma_y$ (plot a). In this case, the majority of the variability in Y can be explained by differences between the groups. At the other extreme is the scenario in which the variability *between* groups is much less than that *within* groups, $\sigma_\mu < \sigma_y$ (plot c). This leads to very little distinction between the Y values for the two groups. As such, the majority of the variability in Y can be explained by differences between the observations within each group, not between the groups themselves. With 44 artists, it's tough to visually assess from the 44 density plots in Figure 16.4 where our Spotify data falls along this spectrum. However, our posterior analysis in Section 16.4 will reveal that our example is somewhere in the middle. The differences from song to song within any one artist are similar in scale to the differences between the artists themselves (à la plot b).

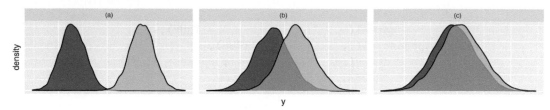

FIGURE 16.8: Simulated output for variable Y on two groups when $\sigma_y = 0.25\sigma_\mu$ (a), $\sigma_y = \sigma_\mu$ (b), and $\sigma_y = 4\sigma_\mu$ (c). Correspondingly, the within-group correlation among the Y values is 0.8 (a), 0.5 (b), and 0.2 (c).

Plot (a) illuminates another feature of grouped data: the multiple observations within any given group are *correlated*. The popularity of one Beyoncé song is related to the popularity of other Beyoncé songs. The popularity of one Mia X song is related to the popularity of other Mia X songs. In general, the one-way ANOVA model (16.6) assumes that, though observations on one group are independent of those on another, the **within-group correlation** for two observations i and k on the *same* group j is

$$\text{Cor}(Y_{ij}, Y_{kj}) = \frac{\sigma_\mu^2}{\sigma_\mu^2 + \sigma_y^2}. \tag{16.10}$$

Thus, the more *unique* the groups, with relatively large σ_μ, the greater the correlation within each group. For example, the within-group correlation in plot (a) of Figure 16.8 is 0.8, a figure close-ish to 1. In contrast, the more *similar* the groups, with relatively small σ_μ, the smaller the correlation within each group. For example, the within-group correlation in plot (c) is 0.2, a figure close-ish to 0. We'll return to these ideas once we simulate our model posterior, and thus can put some numbers to these metrics for our Spotify data.

16.4 Posterior analysis

16.4.1 Posterior simulation

Next step: posterior! Notice that our hierarchical Spotify model (16.5) has a total of 47 parameters: 44 artist-specific parameters (μ_j) and 3 global parameters (μ, σ_y, σ_μ). Exploring the posterior models of these whopping 47 parameters requires MCMC simulation. To this end, the `stan_glmer()` function operates very much like `stan_glm()`, with two small tweaks.

- To indicate that the `artist` variable defines the group structure of our data, as opposed to it being a predictor of `popularity`, the appropriate formula here is `popularity ~ (1 | artist)`.
- The prior for σ_μ is specified by `prior_covariance`. For this particular model, with only one set of artist-specific parameters μ_j, this is equivalent to an Exp(1) prior. (You will learn more about `prior_covariance` in Chapter 17.)

```
spotify_hierarchical <- stan_glmer(
  popularity ~ (1 | artist),
  data = spotify, family = gaussian,
```

```
  prior_intercept = normal(50, 2.5, autoscale = TRUE),
  prior_aux = exponential(1, autoscale = TRUE),
  prior_covariance = decov(reg = 1, conc = 1, shape = 1, scale = 1),
  chains = 4, iter = 5000*2, seed = 84735)

# Confirm the prior tunings
prior_summary(spotify_hierarchical)
```

MCMC diagnostics for all 47 parameters confirm that our simulation has stabilized:

```
mcmc_trace(spotify_hierarchical)
mcmc_dens_overlay(spotify_hierarchical)
mcmc_acf(spotify_hierarchical)
neff_ratio(spotify_hierarchical)
rhat(spotify_hierarchical)
```

Further, a quick posterior predictive check confirms that, though we can always do better, our hierarchical model (16.5) isn't *too* wrong when it comes to capturing variability in song popularity. A set of posterior simulated datasets of song popularity are consistent with the general features of the original popularity data.

```
pp_check(spotify_hierarchical) +
  xlab("popularity")
```

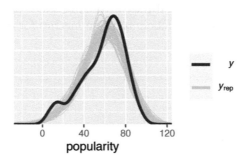

FIGURE 16.9: 100 posterior simulated datasets of song popularity (light blue) along with the actual observed popularity data (dark blue).

With this reassurance, let's dig into the posterior results. To begin, our spotify_hierarchical simulation contains Markov chains of length 20,000 for *each* of our 47 parameters. To get a sense for how this information is stored, check out the labels of the first and last few parameters:

```
# Store the simulation in a data frame
spotify_hierarchical_df <- as.data.frame(spotify_hierarchical)

# Check out the first 3 and last 3 parameter labels
spotify_hierarchical_df %>%
  colnames() %>%
```

```
    as.data.frame() %>%
    slice(1:3, 45:47)

1                                  (Intercept)
2                  b[(Intercept) artist:Mia_X]
3         b[(Intercept) artist:Chris_Goldarg]
4              b[(Intercept) artist:Camilo]
5                                        sigma
6 Sigma[artist:(Intercept),(Intercept)]
```

Due to the sheer number and type of parameters here, summarizing our posterior understanding of all of these parameters will take more care than it did for non-hierarchical models.
We'll take things one step at a time, putting in place the syntactical details we'll need for
all other hierarchical models in Unit 4, i.e., this won't be a waste of your time!

16.4.2 Posterior analysis of global parameters

Let's begin by examining the **global parameters** $(\mu, \sigma_y, \sigma_\mu)$ which are shared by all artists
within and beyond our particular sample. These parameters appear at the very top and very
bottom of the `stan_glmer()` output, and are labeled as follows:

- $\mu = $ (Intercept)
- $\sigma_y = $ sigma
- $\sigma_\mu^2 = $ Sigma[artist:(Intercept),(Intercept)][5]

To zero in on and obtain posterior summaries for μ, we can apply `tidy()` to the
`spotify_hierarchical` model with `effects = fixed`, where "fixed" is synonymous with
"non-varying" or "global." Per the results, there's an 80% chance that the *average* artist has
a mean popularity rating between 49.3 and 55.7.

```
tidy(spotify_hierarchical, effects = "fixed",
     conf.int = TRUE, conf.level = 0.80)
# A tibble: 1 x 5
  term          estimate std.error conf.low conf.high
  <chr>            <dbl>     <dbl>    <dbl>     <dbl>
1 (Intercept)       52.5      2.48     49.3      55.7
```

To call up the posterior medians for σ_y and σ_μ, we can specify `effects = "ran_pars"`, i.e.,
parameters related to randomness or variability:

```
tidy(spotify_hierarchical, effects = "ran_pars")
# A tibble: 2 x 3
  term                      group     estimate
  <chr>                     <chr>        <dbl>
1 sd_(Intercept).artist     artist        15.1
2 sd_Observation.Residual   Residual      14.0
```

---

[5]This is not a typo. The default output gives us information about the *standard deviation* within artists
(σ_y) but the *variance* between artists (σ_μ^2).

The posterior median of σ_y (`sd_Obervation.Residual`) suggests that, *within* any given artist, popularity ratings tend to vary by 14 points *from song to song*. The *between* standard deviation σ_μ (`sd_(Intercept).artist`) tends to be slightly higher at around 15.1. Thus, the *mean* popularity rating tends to vary by 15.1 points *from artist to artist*. By (16.10), these two·sources of variability suggest that the popularity levels among multiple songs *by the same artist* tend to have a moderate correlation near 0.54:

```
15.1^2 / (15.1^2 + 14.0^2)
[1] 0.5377
```

Thinking of this another way, 54% of the variability in song popularity is explained by differences between artists, whereas 46% is explained by differences among the songs within each artist:

```
14.0^2 / (15.1^2 + 14.0^2)
[1] 0.4623
```

16.4.3   Posterior analysis of group-specific parameters

Let's now turn our focus from the global features to the artist-specific features for our 44 sample artists. Recall that we can write the artist-specific mean popularity levels μ_j as *tweaks* to the global mean song popularity μ:

$$\mu_j = \mu + b_j.$$

Thus, b_j describes the *difference* between artist j's mean popularity and the global mean popularity. It's these tweaks that are simulated by `stan_glmer()` – each b_j has a corresponding Markov chain labeled `b[(Intercept) artist:j]`. We can obtain a `tidy()` posterior summary of all b_j terms using the argument `effects = "ran_vals"` (i.e., random artist-specific values):

```
artist_summary <- tidy(spotify_hierarchical, effects = "ran_vals",
                       conf.int = TRUE, conf.level = 0.80)

# Check out the results for the first & last 2 artists
artist_summary %>%
  select(level, conf.low, conf.high) %>%
  slice(1:2, 43:44)
# A tibble: 4 x 3
  level          conf.low conf.high
  <chr>             <dbl>     <dbl>
1 Mia_X             -40.7     -23.4
2 Chris_Goldarg     -39.3     -26.8
3 Lil_Skies          11.1      30.2
4 Camilo             19.4      32.4
```

Consider the 80% credible interval for `Camilo`'s b_j tweak: there's an 80% chance that `Camilo`'s mean popularity rating is between 19.4 and 32.4 *above* that of the average artist. Similarly, there's an 80% chance that Mia X's mean popularity rating is between 23.4 and 40.7 *below* that of the average artist. We can also *combine* our MCMC simulations for the global mean μ and artist tweaks b_j to directly simulate posterior models for the artist-specific means μ_j. For artist "j":

$$\mu_j = \mu + b_j = (\text{Intercept}) + \text{b}[(\text{Intercept}) \text{ artist:j}].$$

The **tidybayes** package provides some tools for this task. We'll take this one step at a time here, but combine these steps in later analyses. To begin, we use `spread_draws()` to extract the `(Intercept)` and `b[(Intercept) artist:j]` values for each artist in each iteration. We then sum these two terms to define `mu_j`. This produces an 880000-row data frame which contains 20000 MCMC samples of μ_j for each of the 44 artists j:

```
# Get MCMC chains for each mu_j
artist_chains <- spotify_hierarchical %>%
  spread_draws(`(Intercept)`, b[,artist]) %>%
  mutate(mu_j = `(Intercept)` + b)

# Check it out
artist_chains %>%
  select(artist, `(Intercept)`, b, mu_j) %>%
  head(4)
# A tibble: 4 x 4
# Groups:   artist [4]
  artist             `(Intercept)`     b   mu_j
  <chr>                      <dbl> <dbl>  <dbl>
1 artist:Alok                 51.1 16.3    67.4
2 artist:Atlas_Genius         51.1  8.46   59.5
3 artist:Au/Ra                51.1  4.18   55.3
4 artist:Beyoncé              51.1 15.0    66.1
```

For example, at the first set of chain values, `Alok`'s mean popularity (67.4) is 16.3 points *above* the average (51.1). Next, `mean_qi()` produces posterior summaries for each artist's mean popularity μ_j, including the posterior mean and an 80% credible interval:

```
# Get posterior summaries for mu_j
artist_summary_scaled <- artist_chains %>%
  select(-`(Intercept)`, -b) %>%
  mean_qi(.width = 0.80) %>%
  mutate(artist = fct_reorder(artist, mu_j))

# Check out the results
artist_summary_scaled %>%
  select(artist, mu_j, .lower, .upper) %>%
  head(4)
# A tibble: 4 x 4
  artist               mu_j .lower .upper
  <fct>               <dbl>  <dbl>  <dbl>
1 artist:Alok          64.3   60.3   68.3
2 artist:Atlas_Genius  46.9   38.8   55.0
```

```
3 artist:Au/Ra              59.5    52.1    66.7
4 artist:Beyoncé            69.1    65.6    72.7
```

For example, with 80% posterior certainty, we can say that Beyoncé's mean popularity rating μ_j is between 65.6 and 72.7. Plotting the 80% posterior credible intervals for all 44 artists illustrates the variability in our posterior understanding of their mean popularity levels μ_j:

```
ggplot(artist_summary_scaled,
       aes(x = artist, y = mu_j, ymin = .lower, ymax = .upper)) +
  geom_pointrange() +
  xaxis_text(angle = 90, hjust = 1)
```

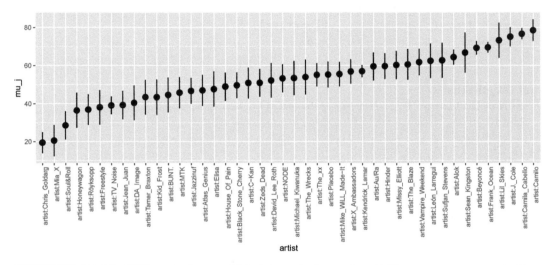

FIGURE 16.10: 80% posterior credible intervals for each artist's mean song popularity.

Not only do the μ_j posteriors vary in *location*, i.e., we expect some artists to be more popular than others, they vary in *scale* – some artists' 80% posterior credible intervals are much wider than others. Pause to think about *why* in the following quiz.[6]

❓Quiz Yourself!

At the more popular end of the spectrum, `Lil Skies` and `Frank Ocean` have similar posterior mean popularity levels. However, the 80% credible interval for Lil Skies is *much* wider than that for Frank Ocean. Explain why.

---

[6]Answer: Lil Skies has a much smaller sample size.

To sleuth out exactly why Lil Skies and Frank Ocean have similar posterior means but drastically different intervals, it's important to remember that not all is equal in our hierarchical data. Whereas our posterior understanding of Frank Ocean is based on 40 songs, the *most* of any artist in the dataset, we have only 3 songs for Lil Skies. Then we naturally have greater posterior certainty about Frank Ocean's popularity, and hence narrower intervals.

```
artist_means %>%
  filter(artist %in% c("Frank Ocean", "Lil Skies"))
# A tibble: 2 x 3
  artist        count popularity
  <fct>         <int>      <dbl>
1 Frank Ocean      40       69.8
2 Lil Skies         3       79.3
```

16.5 Posterior prediction

Our above analysis of the μ_j and μ parameters informed our understanding about general trends in popularity within and between artists. We now turn our attention from trends to specifics: predicting the popularity $Y_{\text{new},j}$ of the next *new* song released by artist j on Spotify. Though the process is different, we can utilize our hierarchical model (16.5) to make such a prediction for *any* artist, both those that did and didn't make it into our sample: How popular will the next song by Frank Ocean, an artist in our original sample, be? What about the next song by Mohsen Beats, an artist for whom we don't yet have any data? To connect this concept to hierarchical modeling, it's important to go through this process "by hand" before relying on the shortcut posterior_predict() function. We'll do that here.

First consider the **posterior prediction for an observed group** or artist, Frank Ocean, the $j = 39$th artist in our sample. The first layer of our hierarchical model (16.5) holds the key in this situation: it assumes that the popularity of individual Frank Ocean songs are Normally distributed around his own mean popularity level μ_j with standard deviation σ_y. Thus, to approximate the posterior predictive model for the popularity of Ocean's *next* song on Spotify, we can simulate a prediction from the Layer 1 model evaluated at each of the 20,000 MCMC parameter sets $\left\{\mu_j^{(i)}, \sigma_y^{(i)}\right\}$:

$$Y_{\text{new},j}^{(i)} | \mu_j, \sigma_y \sim N\left(\mu_j^{(i)}, \left(\sigma_y^{(i)}\right)^2\right). \tag{16.11}$$

The resulting predictions $\left\{Y_{\text{new},j}^{(1)}, Y_{\text{new},j}^{(2)}, \ldots, Y_{\text{new},j}^{(20000)}\right\}$ and corresponding posterior predictive model will reflect two sources of variability, and hence uncertainty, in the popularity of Ocean's next song:

- **within-group sampling variability** in Y, i.e., not all of Ocean's songs are equally popular; and

- **posterior variability** in the model parameters μ_j and σ_y, i.e., the underlying mean and variability in popularity across Ocean's songs are unknown and can, themselves, vary.

To construct this set of predictions, we first obtain the Markov chain for μ_j (mu_ocean) by summing the global mean μ (Intercept) and the artist adjustment (b[(Intercept) artist:Frank_Ocean]). We then simulate a prediction (y_ocean) from the Layer 1 model at each MCMC parameter set:

```
# Simulate Ocean's posterior predictive model
set.seed(84735)
ocean_chains <- spotify_hierarchical_df %>%
  rename(b = `b[(Intercept) artist:Frank_Ocean]`) %>%
  select(`(Intercept)`, b, sigma) %>%
  mutate(mu_ocean = `(Intercept)` + b,
         y_ocean = rnorm(20000, mean = mu_ocean, sd = sigma))

# Check it out
head(ocean_chains, 3)
  (Intercept)     b sigma mu_ocean y_ocean
1       51.08 16.62 14.64    67.70   77.47
2       51.82 19.84 13.28    71.66   70.01
3       52.40 21.09 15.21    73.49   85.36
```

We summarize Ocean's posterior predictive model using mean_qi() below. There's an 80% posterior chance that Ocean's next song will enjoy a popularity rating $Y_{new,j}$ between 51.3 and 87.6. This range is much *wider* than the 80% credible interval for Ocean's μ_j parameter (66.6, 72.2). Naturally, we can be much more certain about Ocean's underlying *mean* song popularity than in the popularity of any *single* Ocean song.

```
# Posterior summary of Y_new,j
ocean_chains %>%
  mean_qi(y_ocean, .width = 0.80)
# A tibble: 1 x 6
  y_ocean .lower .upper .width .point .interval
    <dbl>  <dbl>  <dbl>  <dbl> <chr>  <chr>
1    69.4   51.3   87.6    0.8 mean   qi

# Posterior summary of mu_j
artist_summary_scaled %>%
  filter(artist == "artist:Frank_Ocean")
# A tibble: 1 x 7
  artist     mu_j .lower .upper .width .point .interval
  <fct>     <dbl>  <dbl>  <dbl>  <dbl> <chr>  <chr>
1 artist:F~  69.4   66.6   72.2    0.8 mean   qi
```

Next consider **posterior prediction for a yet unobserved group**. No observed songs for Mohsen Beats means that we do *not* have any information about their mean popularity μ_j, and thus can't take the same approach as we did for Ocean. What we *do* know is this: (1) Mohsen Beats is an artist within the broader population of artists, (2) mean popularity levels among these artists are Normally distributed around some global mean μ with between-artist standard deviation σ_μ (Layer 2 of (16.5)), and (3) our 44 sampled artists have informed our posterior understanding of this broader population. Then to approximate the posterior predictive model for the popularity of Mohsen Beats' next song, we can simulate 20,000 predictions $\left\{ Y_{\text{new,mohsen}}^{(1)}, Y_{\text{new,mohsen}}^{(2)}, \ldots, Y_{\text{new,mohsen}}^{(20000)} \right\}$ through a two-step process:

- **Step 1:** Simulate a potential *mean* popularity level μ_{mohsen} for Mohsen Beats by drawing from the Layer 2 model evaluated at each MCMC parameter set $\left\{ \mu^{(i)}, \sigma_\mu^{(i)} \right\}$:

$$\mu_{\text{mohsen}}^{(i)} | \mu, \sigma_\mu \sim N\left(\mu^{(i)}, \left(\sigma_\mu^{(i)} \right)^2 \right).$$

- **Step 2:** Simulate a prediction of song popularity $Y_{\text{new,mohsen}}$ from the Layer 1 model evaluated at each MCMC parameter set $\left\{ \mu_{\text{mohsen}}^{(i)}, \sigma_y^{(i)} \right\}$:

$$Y_{\text{new,mohsen}}^{(i)} | \mu_{\text{mohsen}}, \sigma_y \sim N\left(\mu_{\text{mohsen}}^{(i)}, \left(\sigma_y^{(i)} \right)^2 \right).$$

The additional step in our Mohsen Beats posterior prediction process reflects a *third* source of variability. When predicting song popularity for a new group, we must account for:

- **within-group sampling variability** in Y, i.e., not all of Mohsen Beats' *songs* are equally popular;

- **between-group sampling variability** in μ_j, i.e., not all *artists* are equally popular; and

- **posterior variability** in the global model parameters $(\sigma_y, \mu, \sigma_\mu)$.

We implement these ideas to simulate 20,000 posterior predictions for Mohsen Beats below. Knowing *nothing* about Mohsen Beats other than that they are an artist, we're able to predict with 80% posterior certainty that their next song will have a popularity rating somewhere between 25.72 and 78.82:

```
set.seed(84735)
mohsen_chains <- spotify_hierarchical_df %>%
  mutate(sigma_mu = sqrt(`Sigma[artist:(Intercept),(Intercept)]`),
         mu_mohsen = rnorm(20000, `(Intercept)`, sigma_mu),
         y_mohsen = rnorm(20000, mu_mohsen, sigma))

# Posterior predictive summaries
mohsen_chains %>%
  mean_qi(y_mohsen, .width = 0.80)
# A tibble: 1 x 6
  y_mohsen .lower .upper .width .point .interval
     <dbl>  <dbl>  <dbl>  <dbl> <chr>  <chr>
1     52.3   25.7   78.8    0.8 mean   qi
```

Finally, below we replicate these "by hand" simulations for Frank Ocean and Mohsen Beats using the `posterior_predict()` shortcut function. Plots of the resulting posterior predictive models illustrate two key features of our understanding about our two artists (Figure 16.11). First, we anticipate that Ocean's next song will have a higher probability rating. This makes sense – since Ocean's observed sample songs were popular, we'd expect his next song to be more popular than that of an artist for whom we have no information. Second, we have greater posterior certainty about Ocean's next song. This again makes sense – since we actually have some data for Ocean (and don't for Mohsen Beats), we *should* be more confident in our ability to predict his next song's popularity. We'll continue our hierarchical posterior prediction exploration in the next section, paying special attention to how the results compare with those from no pooled and complete pooled models.

```
set.seed(84735)
prediction_shortcut <- posterior_predict(
  spotify_hierarchical,
  newdata = data.frame(artist = c("Frank Ocean", "Mohsen Beats")))

# Posterior predictive model plots
mcmc_areas(prediction_shortcut, prob = 0.8) +
  ggplot2::scale_y_discrete(labels = c("Frank Ocean", "Mohsen Beats"))
```

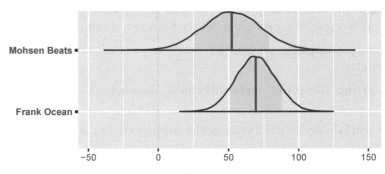

FIGURE 16.11: Posterior predictive models for the popularity of the next songs by Mohsen Beats and Frank Ocean.

16.6 Shrinkage & the bias-variance trade-off

In the previous section, we constructed posterior predictive models of song popularity for just one observed artist, Frank Ocean. We now apply these techniques to all 44 artists in our sample and plot the resulting 80% posterior prediction intervals and means (Figure 16.12).

```
set.seed(84735)
predictions_hierarchical <- posterior_predict(spotify_hierarchical,
                                              newdata = artist_means)

# Posterior predictive plots
ppc_intervals(artist_means$popularity, yrep = predictions_hierarchical,
              prob_outer = 0.80) +
  ggplot2::scale_x_continuous(labels = artist_means$artist,
                              breaks = 1:nrow(artist_means)) +
  xaxis_text(angle = 90, hjust = 1) +
  geom_hline(yintercept = 58.4, linetype = "dashed")
```

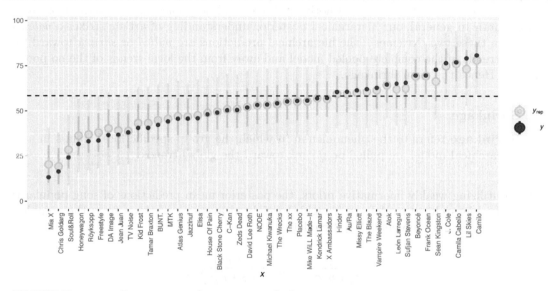

FIGURE 16.12: Posterior predictive intervals for artist song popularity, as calculated from a hierarchical model. The horizontal dashed line represents the average popularity across all songs.

In this plot we can observe a phenomenon called **shrinkage**.

❓**Quiz Yourself!**

What might "shrinkage" mean in this Spotify example and why might it occur?

To understand shrinkage, we first need to remember where we've been. In Section 16.1 we started our Spotify analysis with a complete pooled model which entirely ignored the fact that our data is grouped by artist (16.1). As a result, its posterior mean predictions of song popularity were the same for each artist (Figure 16.3). When using weakly informative priors, this shared prediction is *roughly* equivalent to the global mean popularity level across all $n = 350$ songs in the `spotify` sample, ignoring artist:

$$\overline{y}_{\text{global}} = \frac{1}{n} \sum_{\text{all } i,j} y_{ij}.$$

We swung the other direction in Section 16.2, using a no pooled model which separately analyzed each artist (16.3). As such, when using weakly informative priors, the no pooled posterior predictive means were *roughly* equivalent to the sample mean popularity levels for each artist (Figure 16.6). For any artist j, this sample mean is calculated by averaging the popularity levels y_{ij} of each song i across the artist's n_j songs:

$$\overline{y}_j = \frac{1}{n_j} \sum_{i=1}^{n_j} y_{ij}.$$

Figure 16.12 contrasts the hierarchical model posterior mean predictions with the complete pooled model predictions (dashed horizontal line) and no pooled model predictions (dark blue dots). In general, our hierarchical posterior understanding of artists strikes a *balance* between these two extremes – the hierarchical predictions are *pulled* or *shrunk* toward the global trends of the complete pooled model and away from the local trends of the no pooled model. Hence the term shrinkage.

Shrinkage

Shrinkage refers to the phenomenon in which the group-specific local trends in a hierarchical model are pulled or *shrunk* toward the global trends.

When utilizing weakly informative priors, the posterior mean predictions of song popularity from the hierarchical model are (roughly) weighted averages of those from the complete pooled ($\overline{y}_{\text{global}}$) and no pooled ($\overline{y}_j$) models:

$$\frac{\sigma_y^2}{\sigma_y^2 + n_j \sigma_\mu^2} \overline{y}_{\text{global}} + \frac{n_j \sigma_\mu^2}{\sigma_y^2 + n_j \sigma_\mu^2} \overline{y}_j.$$

In posterior predictions for artist j, the *weights* given to the global and local means depend upon how much data we have on artist j (n_j) as well as the comparison of the *within*-group and *between*-group variability in song popularity (σ_y and σ_μ). These weights highlight a couple of scenarios in which individualism fades, i.e., our hierarchical posterior predictions shrink away from the group-specific means $\overline{y}_j$ and toward the global mean $\overline{y}_{\text{global}}$:

- Shrinkage increases as the number of observations on group j, n_j, decreases. That is, we rely more and more on global trends to understand a group for which we have little data.

- Shrinkage increases when the variability within groups, σ_y, is large in comparison to the variability between groups, σ_μ. That is, we rely more and more on global trends to understand a group when there is little distinction in the patterns from one group to the next (see Figure 16.8).

We can see these dynamics at play in Figure 16.12 – some artists shrunk more toward the global mean popularity levels than others. The artists that shrunk the *most* are those with smaller sample sizes n_j and popularity levels at the extremes of the spectrum. Consider two of the most popular artists: `Camila Cabello` and `Lil Skies`. Though Cabello's observed mean popularity is slightly lower than that of `Lil Skies`, it's based on a *much* bigger sample size:

```
artist_means %>%
  filter(artist %in% c("Camila Cabello", "Lil Skies"))
# A tibble: 2 x 3
  artist           count popularity
  <fct>            <int>      <dbl>
1 Camila Cabello      38       77.1
2 Lil Skies            3       79.3
```

As such, Lil Skies' posterior predictive mean popularity shrunk closer to the global mean – the data on the other artists in our sample suggests that `Lil Skies` might have beginner's luck, and thus that their next song will likely be below their current three-song average. This makes sense and is a compelling feature of hierarchical models. Striking a balance between complete and no pooling, hierarchical models allow us to:

1. **generalize** the observations on our sampled groups to the broader population; while
2. **borrowing strength** or information from all sampled groups when learning about any individual sampled group.

Hierarchical models also offer balance in the bias-variance trade-off we discussed in Chapter 11. Before elaborating, take a quick quiz.

❓ Quiz Yourself!

Consider the three modeling approaches in this chapter: no pooled, complete pooled, and hierarchical models.

1. Suppose different researchers were to conduct their own Spotify analyses using different samples of artists and songs. Which approach would be the most **variable** from sample to sample? The *least* variable?
2. Which approach would tend to be the most **biased** in estimating artists' mean popularity levels? That is, which model's posterior mean estimates will tend to be the furthest from the observed means? Which would tend to be the *least* biased?

The extremes presented by the complete pooled and no pooled models correspond to the extremes of the bias-variance trade-off. Consider the complete pooled model. By pooling all cases together, this model is very rigid and won't vary much if based on a different sample of Spotify artists. BUT it also tends to be overly simple and miss the nuances in artists' mean popularity (Figure 16.3). Thus, complete pooled models tend to have **higher bias and lower variance**.

No pooled models have the opposite problem. With the built-in flexibility to detect group-specific trends, they tend to have less bias than complete pooled models. BUT, since they're tailored to the artists in our sample, if we sampled a different set of Spotify artists our no pooled models could change quite a bit, and thus produce unstable conclusions. Thus, no pooled models tend to have **lower bias and higher variance**.

Hierarchical models offer a balanced alternative. Unlike complete pooled models, hierarchical models take group-specific trends into account, and thus will be less biased. And unlike no pooled models, hierarchical models take global trends into account, and thus will be less variable. Hierarchical models!

16.7   Not everything is hierarchical

Now that we've seen the power of hierarchical models, you might question whether we should have been using them all along. The answer is no. To determine *when* hierarchical models are appropriate, we must understand how our data was collected – did it include some natural grouping structure? Answering this question often comes down to understanding the role of a categorical variable within a dataset. For example, in the no pooled model of song `popularity`, we essentially treated the categorical `artist` variable as a *predictor*. In the hierarchical model, we treated `artist` as a *grouping variable*.

> **Grouping variable vs predictor**
>
> Suppose our dataset includes a categorical variable X for which we have multiple observations per category. To determine whether X encodes some *grouping* structure or is a potential *predictor* of our variable Y, the following distinctions can help:
>
> - If the observed data on X covers *all* categories of interest, it's likely *not* a grouping variable, but rather a potential predictor.
>
> - If the observed categories on X are merely a random sample from many of interest, it *is* a potential grouping variable.

Let's apply this thinking to the categorical `artist` variable. Our sample data included multiple observations for a mere random sample of 44 among *thousands* of artists on Spotify. Hence, treating `artist` as a predictor (as in the no pooled model) would limit our understanding to only these artists. In contrast, treating it as a grouping variable (as in the hierarchical model) allows us to not only learn about the 44 artists in our data, but the broader population of artists from which they were sampled.

Check out another example. Reconsider the `bikes` data from Chapter 9 which, on each of 500 days, records the number of `rides` taken by Capital Bikeshare members and whether the day fell on a `weekend`:

```
data(bikes)
bikes %>%
  select(rides, weekend) %>%
  head(3)
```

```
  rides weekend
1   654    TRUE
2  1229   FALSE
3  1454   FALSE
```

> **❷ Quiz Yourself!**
>
> In a model of `rides`, is the `weekend` variable a potential *grouping variable* or *predictor*?

There are only two possible categories for the `weekend` variable: each day either falls on a weekend (`TRUE`) or it doesn't (`FALSE`). Our dataset covers both instances:

```
bikes %>%
  group_by(weekend) %>%
  tally()
# A tibble: 2 x 2
  weekend     n
  <lgl>   <int>
1 FALSE     359
2 TRUE      141
```

Since the observed `weekend` values cover all categories of interest, it's *not* a grouping variable. Rather, it's a potential *predictor* that can help us explore whether ridership is different on weekends vs weekdays.

Consider yet another example. To address disparities in vocabulary levels among children from households with different income levels, the Abdul Latif Jameel Poverty Action Lab (J-PAL) evaluated the effectiveness of a digital vocabulary learning program, the *Big Word Club* (Kalil et al., 2020). To do so, they enrolled 818 students across 47 different schools in a vocabulary study. Data on the students' vocabulary levels *after* completing this program (`score_a2`) was obtained through the Inter-university Consortium for Political and Social Research (ICPSR) and stored in the `big_word_club` dataset in the **bayesrules** package. We'll keep only the students for whom we have a vocabulary score:

```
data(big_word_club)
big_word_club <- big_word_club %>%
  select(score_a2, school_id) %>%
  na.omit()

big_word_club %>%
  slice(1:2, 602:603)
  score_a2 school_id
1       28         1
2       25         1
3       31        47
4       34        47
```

❓ **Quiz Yourself!**

In a model of `score_a2`, is the `school_id` variable a potential *grouping variable* or *predictor*?

In this setting, `school_id` is a grouping variable. Consider two clues: we observed *multiple* students within each of 47 schools and these 47 schools are merely a small sample from the *thousands* of schools out there. In this case, treating `school_id` as a *predictor* of `score_a2` (à la no pooling) would only allow us to learn about our small sample of schools, whereas treating `school_id` as a *grouping variable* in a *hierarchical* model of `score_a2` would allow us to extend our conclusions to the broader population of all schools.

16.8 Chapter summary

In Chapter 16 we explored how to model hierarchical data. Suppose we have multiple data points on each of m groups. For each group $j \in \{1, 2, \ldots, m\}$, let Y_{ij} denote the ith observation and n_j denote the sample size. Thus, our data on variable Y is the *collection* of smaller samples on each of the m groups:

$$Y := ((Y_{11}, Y_{21}, \ldots, Y_{n_1,1}), (Y_{12}, Y_{22}, \ldots, Y_{n_2,2}), \ldots, (Y_{1,m}, Y_{2,m}, \ldots, Y_{n_m,m})).$$

And the hierarchical **one-way analysis of variance** model of Y consists of three layers:

$$
\begin{aligned}
Y_{ij}|\mu_j, \sigma_y &\sim N(\mu_j, \sigma_y^2) & &\text{model of individual observations within group } j \\
\mu_j|\mu, \sigma_\mu &\overset{ind}{\sim} N(\mu, \sigma_\mu^2) & &\text{model of how parameters vary between groups} \\
\mu &\sim N(m, s^2) & &\text{prior models on global parameters} \\
\sigma_y &\sim \text{Exp}(l_y) \\
\sigma_\mu &\sim \text{Exp}(l_\mu).
\end{aligned}
\tag{16.12}
$$

Equivalently, we can decompose the μ_j terms into $\mu_j = \mu + b_j$ and model the variability in b_j:

$$
\begin{aligned}
Y_{ij}|\mu_j, \sigma_y &\sim N(\mu_j, \sigma_y^2) \quad \text{with } \mu_j = \mu + b_j \\
b_j|\mu, \sigma_\mu &\overset{ind}{\sim} N(0, \sigma_\mu^2) \\
\mu &\sim N(m, s^2) \\
\sigma_y &\sim \text{Exp}(l_y) \\
\sigma_\mu &\sim \text{Exp}(l_\mu).
\end{aligned}
$$

Here's a list of cool things about this model:

- it acknowledges that, though observations on one group are independent of those on another, observations *within* any given group are correlated;
- group-varying or group-specific parameters μ_j provide insight into group-specific trends;
- global parameters $(\mu, \sigma_y, \sigma_\mu)$ provide insight into the broader population;
- to learn about any one group, this model borrows strength from the information on other groups, thereby *shrinking* group-specific local phenomena toward the global trends; and
- this model is less variable than the no pooling approach and less biased than the complete pooling approach.

16.9 Exercises

16.9.1 Conceptual exercises

Exercise 16.1 (Shrinkage). The plot below illustrates the distribution of critic ratings for 7 coffee shops. Suppose we were to model coffee shop ratings using three different approaches: complete pooled, no pooled, and hierarchical. For each model, sketch what the posterior mean ratings for the 7 coffee shops might look like on the plot below.

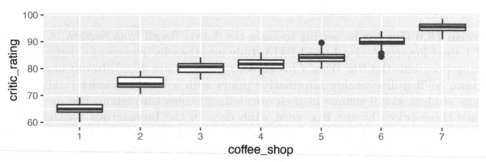

Exercise 16.2 (Grouping variable or predictor?).

   a.   The `climbers_sub` data in the `bayesrules` package contains outcomes for 2076 climbers that have sought to summit a Himalayan mountain peak. In a model of climber `success`, is `expedition_id` a potential predictor or a grouping variable? Explain.

   b.   In a model of climber `success`, is `season` a potential predictor or a grouping variable? Explain.

   c.   The `coffee_ratings` data in the `bayesrules` package contains ratings for 1339 different coffee batches. In a model of coffee ratings (`total_cup_points`), is `processing_method` a potential predictor or a grouping variable? Explain.

   d.   In a model of coffee ratings (`total_cup_points`), is `farm_name` a potential predictor or a grouping variable? Explain.

Exercise 16.3 (Speed typing: interpret the coefficients). Alicia loves typing. To share the appreciation, she invites four friends to each take 20 speed-typing tests. Let Y_{ij} be the time it takes friend j to complete test i.

   a.   In modeling Y_{ij}, explain why it's important to account for the grouping structure introduced by observing each friend multiple times.

   b.   Suppose we were to model the outcomes Y_{ij} by (16.5). Interpret the meaning of all model coefficients in terms of what they might illuminate about typing speeds: $(\mu_j, \mu, \sigma_y, \sigma_\mu)$.

Exercise 16.4 (Speed typing: sketch the data). In the spirit of Figure 16.8, sketch what density plots of your four friends' typing speed outcomes Y_{ij} might look like under each scenario below.

   a.   The overall results of the 20 timed tests are remarkably similar among the four friends.

b.   Each person is quite consistent in their typing times, but there are big differences from person to person – some tend to type much faster than others.

c.   Within the subjects, there doesn't appear to be much correlation in typing time from test to test.

Exercise 16.5 (Speed typing: connecting to the model). For each scenario in the above exercise, indicate the corresponding relationship between σ_y and σ_μ: $\sigma_y < \sigma_\mu$, $\sigma_y \approx \sigma_\mu$, or $\sigma_y > \sigma_\mu$.

16.9.2   Applied exercises

Exercise 16.6 (Big words: getting to know the data). Recall from Section 16.7 the Abdul Latif Jameel Poverty Action Lab (J-PAL) study into the effectiveness of a digital vocabulary learning program, the *Big Word Club* (BWC) (Kalil et al., 2020). In our analysis of this program, we'll utilize weakly informative priors with a baseline understanding that the average student saw 0 change in their vocabulary scores throughout the program. We'll balance these priors by the `big_word_club` data in the **bayesrules** package. For each student participant, `big_word_club` includes a `school_id` and the *percentage change* in vocabulary scores over the course of the study period (`score_pct_change`). We keep only the students that participated in the BWC program (`treat == 1`), and thus eliminate the control group.

```
data("big_word_club")
big_word_club <- big_word_club %>%
  filter(treat == 1) %>%
  select(school_id, score_pct_change) %>%
  na.omit()
```

a)   How many schools participated in the *Big Word Club*?

b)   What's the range in the number of student participants per school?

c)   On average, at which school did students exhibit the greatest improvement in vocabulary? The least?

d)   Construct and discuss a plot which illustrates the variability in `score_pct_change` within and between schools.

Exercise 16.7 (Big words: setting up the model). In the next exercises you will explore a hierarchical one-way ANOVA model (16.12) of Y_{ij}, the percentage change in vocabulary scores, for student i in school j.

a)   Why is a hierarchical model, vs a complete or no pooled model, appropriate in our analysis of the BWC program?

b)   Compare and contrast the meanings of model parameters μ and μ_j in the context of this vocabulary study.

c)   Compare and contrast the meanings of model parameters σ_y and σ_μ in the context of this vocabulary study.

Exercise 16.8 (Big words: simulating the model).

a)   Simulate the hierarchical posterior model of parameters $(\mu_j, \mu, \sigma_y, \sigma_\mu)$ using 4 chains, each of length 10000.

b)   Construct and discuss Markov chain trace, density, and autocorrelation plots.

c)   Construct and discuss a `pp_check()` of the chain output.

Exercise 16.9 (Big words: global parameters). In this exercise, we'll explore the global parameters of our BWC model: $(\mu, \sigma_y, \sigma_\mu)$.

a)   Construct and interpret a 95% credible interval for μ.

b)   Is there ample evidence that, on average, student vocabulary levels improve throughout the vocabulary program? Explain.

c)   Which appears to be larger: the variability in vocabulary performance *between* or *within* schools? Provide posterior evidence and explain the implication of this result in the context of the analysis.

Exercise 16.10 (Big words: focusing on schools). Next, let's dig into the school-specific means, μ_j.

a)   Construct and discuss a plot of the 80% posterior credible intervals for the average percent change in vocabulary score for all schools in the study, μ_j.

b)   Construct and interpret the 80% posterior credible interval for μ_{10}.

c)   Is there ample evidence that, on average, vocabulary scores at School 10 improved by more than 5% throughout the duration of the vocabulary program? Provide posterior evidence.

Exercise 16.11 (Big words: predicting vocab levels). Suppose we *continue* the vocabulary study at each of Schools 6 and 17 (which participated in the current study) and Bayes Prep, a school which is new to the study. In this exercise you'll make predictions about $Y_{new,j}$, the vocabulary performance of a student that's *new* to the study from each of these three schools j.

a)   *Without* using the `posterior_predict()` shortcut function, simulate posterior predictive models of $Y_{new,j}$ for School 6 and Bayes Prep. Display the first 6 posterior predictions for both schools.

b)   Using your simulations from part (a), construct, interpret, and compare the 80% posterior predictive intervals of $Y_{new,j}$ for School 6 and Bayes Prep.

c)   Using `posterior_predict()` this time, simulate posterior predictive models of $Y_{new,j}$ for each of School 6, School 17, and Bayes Prep. Illustrate your simulation results using `mcmc_areas()` and discuss your findings.

d)   Finally, construct, plot, and discuss the 80% posterior prediction intervals for all schools in the original study.

Exercise 16.12 (Big words: shrinkage). Re-examine the posterior predictive plots from Exercise 16.11. Would you say that there is a little or a lot of shrinkage in this analysis? Provide evidence and explain why you think this is the case.

Exercise 16.13 (Open exercise: voices). The `voices` data in the **bayesrules** package contains the results of an experiment in which subjects participated in role-playing dialog under various conditions. Of interest to the researchers was the subjects' voice pitches (in Hz). In this open-ended exercise, complete a hierarchical analysis of voice pitch, without any predictors.

17

(Normal) Hierarchical Models with Predictors

In Chapter 15 we convinced you that there's a *need* for hierarchical models. In Chapter 16 we built our *first* hierarchical model – a Normal hierarchical model of Y with *no* predictors X. Here we'll take the next natural step by building a Normal hierarchical regression model of Y *with* predictors X. Going full circle, let's return to the Cherry Blossom 10 mile running race analysis that was featured in Chapter 15. Our goal is to better understand variability in running times: To what extent do some people run faster than others? How are running times associated with age, and to what extent does this differ from person to person?

In answering these questions, we'll utilize the `cherry_blossom_sample` data from the **bayesrules** package, shortened to `running` here. This data records multiple `net` running times (in minutes) for each of 36 different runners in their 50s and 60s that entered the 10-mile race in multiple years.

```
# Load packages
library(bayesrules)
library(tidyverse)
library(rstanarm)
library(bayesplot)
library(tidybayes)
library(broom.mixed)

# Load data
data(cherry_blossom_sample)
running <- cherry_blossom_sample
```

But it turns out that the `running` data is missing some `net` race times. Since functions such as `prediction_summary()`, `add_fitted_draws()`, and `add_predicted_draws()` require *complete* information on each race, we'll omit the rows with incomplete observations. In doing so, it's important to use `na.omit()` *after* selecting our variables of interest so that we don't throw out observations that have complete information on these variables just because they have incomplete information on variables we don't care about.

```
# Remove NAs
running <- running %>%
  select(runner, age, net) %>%
  na.omit()
```

With multiple observations per runner, this data is hierarchical or **grouped**. To acknowledge this grouping structure, let Y_{ij} denote the net running time and X_{ij} the age for runner j in their ith race. In modeling Y_{ij} by X_{ij}, Chapter 15 previewed that it would be a mistake to ignore the data's grouped structure: a complete pooling approach ignores the fact that we

DOI: 10.1201/9780429288340-17

have multiple *dependent* observations on each runner, and a no pooling approach assumes that data on one runner can't tell us anything about another runner (thus that our data also can't tell us anything about the general running population). Instead, our analysis of the relationship between running times and age will combine two Bayesian modeling paradigms we've developed throughout the book:

- regression models of Y by predictors X when our data does *not* have a group structure (Chapters 9 through 14); and
- hierarchical models of Y which account for group-structured data but do *not* use predictors X (Chapter 16).

◎ **Goals**

- Build hierarchical *regression* models of response variable Y by predictors X.
- Evaluate and compare hierarchical and non-hierarchical models.
- Use hierarchical models for posterior prediction.

17.1   First steps: Complete pooling

To explore the association between running times (Y_{ij}) and age (X_{ij}), let's start by *ignoring* the data's grouped structure. This isn't the right approach, but it provides a good point of comparison and a building block to a better model. To this end, the **complete pooled regression model** of Y_{ij} from Chapter 15 assumes an age-specific linear relationship $\mu_i = \beta_0 + \beta_1 X_{ij}$ with weakly informative priors:

$$
\begin{aligned}
Y_{ij}|\beta_0,\beta_1,\sigma &\sim N(\mu_i,\sigma^2) \quad \text{where } \mu_i = \beta_0 + \beta_1 X_{ij} \\
\beta_{0c} &\sim N(0,35^2) \\
\beta_1 &\sim N(0,15^2) \\
\sigma &\sim \text{Exp}(0.072)
\end{aligned}
\tag{17.1}
$$

This model depends upon three **global parameters**: intercept β_0, age coefficient β_1, and variability from the regression model σ. Our posterior simulation results for these parameters from Section 15.1, stored in `complete_pooled_model`, are summarized below.

```
# Posterior summary statistics
model_summary <- tidy(complete_pooled_model,
                      conf.int = TRUE, conf.level = 0.80)
model_summary
# A tibble: 2 x 5
  term          estimate std.error conf.low conf.high
  <chr>            <dbl>     <dbl>    <dbl>     <dbl>
1 (Intercept)      75.2      24.6     43.7      106.
2 age               0.268     0.446   -0.302      0.842
```

The vibe of this complete pooled model is captured by Figure 17.1: it lumps together all race results and describes the relationship between running time and age by a common global model. Lumped together in this way, a scatterplot of `net` running times versus `age` exhibit a weak relationship with a posterior median model of 75.2 + 0.268 age.

```
# Posterior median model
B0 <- model_summary$estimate[1]
B1 <- model_summary$estimate[2]
ggplot(running, aes(x = age, y = net)) +
  geom_point() +
  geom_abline(aes(intercept = B0, slope = B1))
```

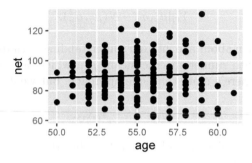

FIGURE 17.1: A scatterplot of net running times versus age along with the posterior median model from the complete pooled model.

This posterior median estimate of the age coefficient β_1 suggests that running times tend to increase by a mere 0.27 minutes for each year in age. Further, with an 80% posterior credible interval for β_1 which straddles 0, (-0.3, 0.84), our complete pooled regression model suggests there's *not* a significant relationship between running time and age. Our intuition (and personal experience) says this is wrong — as adults age they tend to slow down. This intuition is correct.

17.2 Hierarchical model with varying intercepts

17.2.1 Model building

Chapter 15 revealed that it would indeed be a mistake to stop our runner analysis with the complete pooled regression model (17.1). Thus, our next goal is to incorporate the data's grouped structure *while* maintaining our age predictor X_{ij}. Essentially, we want to combine the regression principles from the complete pooled regression model with the grouping principles from the simple Normal hierarchical model *without* a predictor from Chapter 16:

$$
\begin{aligned}
Y_{ij}|\mu_j,\sigma_y &\sim N(\mu_j,\sigma_y^2) \quad &\text{model of running times WITHIN runner } j \\
\mu_j|\mu,\sigma_\mu &\stackrel{ind}{\sim} N(\mu,\sigma_\mu^2) \quad &\text{model of how typical running times vary BETWEEN runners} \\
\mu,\sigma_y,\sigma_\mu &\sim \dots \quad &\text{prior models on global parameters}
\end{aligned}
$$

(17.2)

This hierarchical regression model will build from (17.2) and unfold in three similar layers.

17.2.1.1   Layer 1: Variability within runner

The first layer of the simple Normal hierarchical model (17.2) assumes that each runner's net running times Y_{ij} vary normally around their own mean time μ_j, with no consideration of their age X_{ij}:

$$
Y_{ij}|\mu_j,\sigma_y \sim N(\mu_j,\sigma_y^2).
$$

To incorporate information about age into our understanding of running times within runners, we can replace the μ_j with runner-specific means μ_{ij}, which depend upon the runner's age in their ith race, X_{ij}. There's more than one approach, but we'll start with the following:

$$
\mu_{ij} = \beta_{0j} + \beta_1 X_{ij}.
$$

Thus, the first layer of our hierarchical model describes the relationship between net times and age **within** each runner j by:

$$
Y_{ij}|\beta_{0j},\beta_1,\sigma_y \sim N(\mu_{ij},\sigma_y^2) \quad \text{where} \quad \mu_{ij} = \beta_{0j} + \beta_1 X_{ij}. \tag{17.3}
$$

For each runner j, (17.3) assumes that their running times are Normally distributed around an age- and runner-specific mean, $\beta_{0j} + \beta_1 X_{ij}$, with standard deviation σ_y. This model depends upon three parameters: β_{0j}, β_1, and σ_y. Paying special attention to subscripts, only β_{0j} depends upon j, and thus is runner- or **group-specific**:

- β_{0j} = intercept of the regression model for runner j.

The other parameters are **global**, or shared across all runners j:

- β_1 = global age coefficient, i.e., the typical change in a runner's net time per one year increase in age; and
- σ_y = **within-group variability** around the mean regression model $\beta_{0j} + \beta_1 X_{ij}$, and hence a measure of the *strength* of the relationship between an individual runner's time and their age.

Putting this together, the first layer of our hierarchical model (17.3) assumes that relationships between running time and age *randomly* vary from runner to runner, having *different* intercepts β_{0j} but a shared age coefficient β_1. Figure 17.2 illustrates these assumptions and helps us translate them within our context: though some runners tend to be faster or slower than others, runners experience roughly the same increase in running times as they age.

17.2.1.2   Layer 2: Variability between runners

As with the simple hierarchical model (17.2), the first layer of our hierarchical regression model (17.3) captured the relationship between running time and age *within* runners. It's in the next layer that we must capture how the relationships between running time and age vary from runner to runner, i.e., **between** runners.

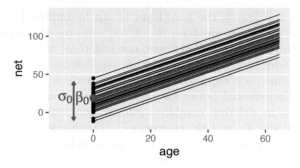

FIGURE 17.2: Hypothetical mean regression models, $\beta_{0j} + \beta_1 X$, for our 36 sampled runners under (17.3). The black dots represent the runner-specific intercepts β_{0j} which vary normally around β_0 with standard deviation σ_0.

❓ **Quiz Yourself!**

Which of our current model parameters, $(\beta_{0j}, \beta_1, \sigma_y)$, will we model in the next layer of the hierarchical model? What is a reasonable structure for this model?

Since it's the only regression feature that we're assuming *can* vary from runner to runner, the next layer will model variability in the intercept parameters β_{0j}. It's important to recognize here that our 36 sample runners are drawn from the same broader population of runners. Thus, instead of taking a *no pooling* approach, which assigns each β_{0j} a unique prior, and hence assumes that one runner j can't tell us about another, these intercepts should *share* a prior. To this end we'll assume that the runner-specific intercept parameters, and hence baseline running speeds, vary *normally* around some mean β_0 with standard deviation σ_0:

$$\beta_{0j} | \beta_0, \sigma_0 \overset{ind}{\sim} N(\beta_0, \sigma_0^2). \tag{17.4}$$

Figure 17.2 adds context to this layer of the hierarchical model which depends upon two new parameters:

- β_0 = the **global average intercept** across all runners, i.e., the *average* runner's baseline speed; and
- σ_0 = **between-group variability** in intercepts β_{0j}, i.e., the extent to which baseline speeds vary from runner to runner.

17.2.1.3 Layer 3: Global priors

We've now completed the first two layers of our hierarchical regression model which reflect the relationships of running time and age, within and between runners. As with any Bayesian model, the last step is to specify our priors.

❓ **Quiz Yourself!**

For which model parameters must we specify priors in the final layer of our hierarchical regression model?

Building from (17.3) and (17.4), the final layer of our hierarchical regression model must specify priors for the global parameters: β_0, β_1, σ_y, and σ_0. It's these global parameters that describe the entire population of runners, not just those in our sample. As usual, we'll utilize independent Normal priors for regression coefficients β_0 and β_1, where our prior understanding of baseline β_0 is expressed through the centered intercept β_{0c}. Further, we'll utilize independent Exponential priors for standard deviation terms σ_y and σ_0. The final hierarchical regression model thus combines information about the relationship between running times Y_{ij} and age X_{ij} **within** runners (17.3) with information about how baseline speeds β_{0j} vary **between** runners (17.4) with our prior understanding of the broader **global** population of runners. As embodied by Figure 17.2, this particular model is often referred to as a **hierarchical random intercepts model**:

$$
\begin{aligned}
Y_{ij}|\beta_{0j},\beta_1,\sigma_y &\sim N(\mu_{ij},\sigma_y^2) \quad \text{with} \quad \mu_{ij} = \beta_{0j} + \beta_1 X_{ij} \quad &&\text{(regression model}\\
& &&\text{WITHIN runner } j)\\
\beta_{0j}|\beta_0,\sigma_0 &\overset{ind}{\sim} N(\beta_0,\sigma_0^2) &&\text{(variability in baseline speeds}\\
& &&\text{BETWEEN runners)}\\
\beta_{0c} &\sim N(m_0,s_0^2) &&\text{(priors on global parameters)}\\
\beta_1 &\sim N(m_1,s_1^2)\\
\sigma_y &\sim \text{Exp}(l_y)\\
\sigma_0 &\sim \text{Exp}(l_0).
\end{aligned}
$$

$$(17.5)$$

The complete set of model assumptions is summarized below.

Normal hierarchical regression assumptions

Let Y_{ij} and X_{ij} denote observations for the ith observation in group j. The appropriateness of the Bayesian Normal hierarchical regression model (17.5) of Y_{ij} by X_{ij} depends upon the following assumptions.

- **Structure of the data**
  Conditioned on predictor X_{ij}, the outcomes Y_{ij} on any one group j are *independent* of those on another group k, Y_{ik}. However, different data points *within* the same group j, Y_{ij} and Y_{hj}, are *correlated*.
- **Structure of the relationship**
  Within any group j, the typical outcome of Y_{ij} (μ_{ij}) can be written as a **linear function** of predictor X_{ij}.
- **Structure of the variability within groups**
  Within any group j and at any predictor value X_{ij}, the observed values of Y_{ij} will vary **normally** around mean μ_{ij} with consistent standard deviation σ_y.
- **Structure of the variability between groups**
  The group-specific baselines or intercepts, β_{0j}, vary **normally** around a global intercept β_0 with standard deviation σ_0.

Connecting the hierarchical and complete pooled models

The complete pooled model (17.1) is a *special case* of (17.5). These two models are equivalent when $\sigma_0 = 0$, i.e., when the intercepts do *not* differ from group to group.

17.2.2   Another way to think about it

Recall from Chapter 16 that there are two ways to think about group-specific parameters. First, in our hierarchical regression model of running times (17.5), we incorporated runner-specific intercept parameters β_{0j} to indicate that the baseline speed can vary from runner to runner j. We thought of these runner-specific intercepts as normal deviations from some global intercept β_0 with standard deviation σ_0: $\beta_{0j} \sim N(\beta_0, \sigma_0^2)$. *Equivalently*, we can think of the runner-specific intercepts as random *tweaks* or *adjustments* b_{0j} to β_0,

$$\beta_{0j} = \beta_0 + b_{0j}$$

where these tweaks are normal deviations from *0* with standard deviation σ_0:

$$b_{0j} \sim N(0, \sigma_0^2).$$

Consider an example. Suppose that some runner j has a baseline running speed of $\beta_{0j} = 24$ minutes, whereas the average baseline speed across *all* runners is $\beta_0 = 19$ minutes. Thus, at any age, runner j tends to run 5 minutes slower than average. That is, $b_{0j} = 5$ and

$$\beta_{0j} = \beta_0 + b_{0j} = 19 + 5 = 24.$$

In general, then, we can reframe Layers 1 and 2 of our hierarchical model (17.5) as follows:

$$
\begin{aligned}
Y_{ij}|\beta_{0j}, \beta_1, \sigma_y &\sim N(\mu_{ij}, \sigma_y^2) \quad \text{with} \quad \mu_{ij} = (\beta_0 + b_{0j}) + \beta_1 X_{ij} \\
b_{0j}|\sigma_0 &\stackrel{ind}{\sim} N(0, \sigma_0^2) \\
\beta_{0c} &\sim N(m_0, s_0^2) \\
\beta_1 &\sim N(m_1, s_1^2) \\
\sigma_y &\sim \text{Exp}(l_y) \\
\sigma_0 &\sim \text{Exp}(l_0).
\end{aligned}
\tag{17.6}
$$

17.2.3   Tuning the prior

With the hierarchical regression model framework in place (17.5), let's tune the priors to match our prior understanding in this running context. Pretending that we didn't already see data in Chapter 15, our prior understanding is as follows:

- The typical runner in this age group runs somewhere between an 8-minute mile and a 12-minute mile during a 10-mile race, and thus has a net time somewhere between 80 and 120 minutes for the entire race. As such we'll set the prior model for the centered global intercept to $\beta_{0c} \sim N(100, 10^2)$. (This centered intercept is much easier to think about than the raw intercept β_0, the typical net time for a 0-year-old runner!)
- We're pretty sure that the typical runner's net time in the 10-mile race will, on average, *increase* over time. We're not very sure about the rate of this increase, but think it's likely between 0.5 and 4.5 minutes per year. Thus, we'll set our prior for the global age coefficient to $\beta_1 \sim N(2.5, 1^2)$.

- Beyond the typical net time for the typical runner, we do not have a clear prior understanding of the variability between runners (σ_0), nor of the degree to which a runner's
net times might fluctuate from their regression trend (σ_y). Thus, we'll utilize weakly
informative priors on these standard deviation parameters.

Our final tuning of the hierarchical random intercepts model follows, where the priors on σ_y
and σ_0 are assigned by the `stan_glmer()` simulation below:

$$Y_{ij}|\beta_{0j},\beta_1,\sigma_y \sim N(\mu_{ij},\sigma_y^2) \quad \text{with} \quad \mu_{ij} = \beta_{0j} + \beta_1 X_{ij}$$

$$\beta_{0j}|\beta_0,\sigma_0 \stackrel{ind}{\sim} N(\beta_0,\sigma_0^2)$$

$$\beta_{0c} \sim N(100,10^2)$$

$$\beta_1 \sim N(2.5,1^2) \tag{17.7}$$

$$\sigma_y \sim \text{Exp}(0.072)$$

$$\sigma_0 \sim \text{Exp}(1).$$

To get a sense for the combined meaning of our prior models, we simulate 20,000 prior
parameter sets using `stan_glmer()` with the following special arguments:

- We specify the model of `net` times by `age` by the formula `net ~ age + (1 | runner)`.
This essentially combines a non-hierarchical regression formula (`net ~ age`) with that
for a hierarchical model with no predictor (`net ~ (1 | runner)`).
- We specify `prior_PD = TRUE` to indicate that we wish to simulate parameters from the
prior, not posterior, models.

```
running_model_1_prior <- stan_glmer(
  net ~ age + (1 | runner),
  data = running, family = gaussian,
  prior_intercept = normal(100, 10),
  prior = normal(2.5, 1),
  prior_aux = exponential(1, autoscale = TRUE),
  prior_covariance = decov(reg = 1, conc = 1, shape = 1, scale = 1),
  chains = 4, iter = 5000*2, seed = 84735,
  prior_PD = TRUE)
```

The simulation results describe 20,000 prior plausible scenarios for the relationship between
running time and age, within and between our 36 runners. Though we encourage you to
plot many more on your own, we show just 4 prior plausible scenarios of what the mean
regression models, $\beta_{0j} + \beta_1 X$, might look like for our 36 runners (Figure 17.3 left). The
variety across these prior scenarios reflects our general uncertainty about running. Though
each scenario is consistent with our sense that runners likely slow down over time, the rate of
increase ranges quite a bit. Further, in examining the distances between the runner-specific
regression lines, some scenarios reflect a plausibility that there's little difference between
runners, whereas others suggest that some runners might be much faster than others.

Finally, we also simulate 100 datasets of race outcomes from the prior model, across a variety
of runners and ages. The 100 density plots in Figure 17.3 (right) reflect the distribution of
the `net` times in these simulated datasets. There is, again, quite a range in these simulations.
Though some span ridiculous outcomes (e.g., negative `net` running times), the variety
in the simulations and the general set of values they cover, adequately reflect our prior
understanding and uncertainty. For example, since a 25- to 30-minute mile is a good walking

(not running) pace, the upper values near 250-300 minutes for the entire 10-mile race seem reasonable.

```
set.seed(84735)
running %>%
  add_fitted_draws(running_model_1_prior, n = 4) %>%
  ggplot(aes(x = age, y = net)) +
    geom_line(aes(y = .value, group = paste(runner, .draw))) +
    facet_wrap(~ .draw)

running %>%
  add_predicted_draws(running_model_1_prior, n = 100) %>%
  ggplot(aes(x = net)) +
    geom_density(aes(x = .prediction, group = .draw)) +
    xlim(-100,300)
```

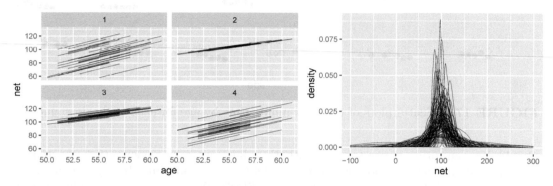

FIGURE 17.3: Simulated scenarios under the prior models of the hierarchical regression model (17.7). At left are 4 prior plausible sets of 36 runner-specific relationships between running time and age, $\beta_{0j} + \beta_1 X$. At right are density plots of 100 prior plausible sets of net running time data.

17.2.4 Posterior simulation & analysis

After all of the buildup, let's counter our prior understanding of the relationship between running time and age with some data. We plot the running times by age for each of our 36 runners below. This reaffirms our model-building process and some of our prior hunches. Most runners' times *do* tend to increase with age, and there is variability between the runners themselves – some tend to be faster than others.

```
ggplot(running, aes(x = age, y = net)) +
  geom_point() +
  facet_wrap(~ runner)
```

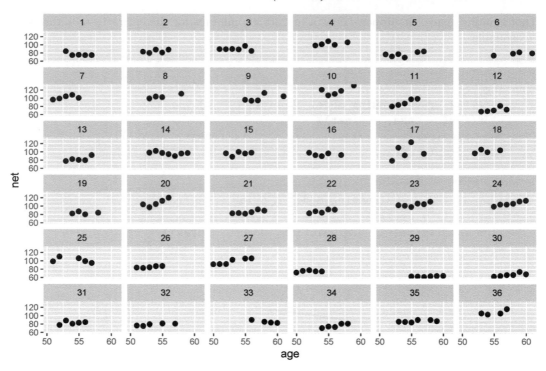

FIGURE 17.4: Scatterplots of the net running time by age for each of the 36 runners.

Combining our prior understanding with this data, we take a syntactical shortcut to simulate the posterior random intercepts model (17.5) of `net` times by `age`: we `update()` the `running_model_1_prior` with `prior_PD = FALSE`. We encourage you to follow this up with a check of the prior tunings as well as some Markov chain diagnostics:

```
# Simulate the posterior
running_model_1 <- update(running_model_1_prior, prior_PD = FALSE)

# Check the prior specifications
prior_summary(running_model_1)

# Markov chain diagnostics
mcmc_trace(running_model_1)
mcmc_dens_overlay(running_model_1)
mcmc_acf(running_model_1)
neff_ratio(running_model_1)
rhat(running_model_1)
```

There are a whopping 40 parameters in our model: 36 runner-specific intercept parameters (β_{0j}) in addition to 4 global parameters $(\beta_0, \beta_1, \sigma_y, \sigma_0)$. These are labeled as follows in the `stan_glmer()` simulation results:

- `(Intercept)` $= \beta_0$
- `age` $= \beta_1$
- `b[(Intercept) runner:j]` $= b_{0j} = \beta_{0j} - \beta_0$, the *difference* between runner j's baseline speed and the average baseline speed

- `sigma` $= \sigma_y$
- `Sigma[runner:(Intercept),(Intercept)]` $= \sigma_0^2$

We'll keep our focus on the big themes here, first those related to the relationship between running time and age for the *typical* runner, and then those related to the *variability* from this average.

17.2.4.1   Posterior analysis of the global relationship

To begin, consider the **global relationship** between running time and age for the *typical* runner:

$$\beta_0 + \beta_1 X.$$

Posterior summaries for β_0 and β_1, which are `fixed` across runners, are shown below.

```
tidy_summary_1 <- tidy(running_model_1, effects = "fixed",
                       conf.int = TRUE, conf.level = 0.80)
tidy_summary_1
# A tibble: 2 x 5
  term        estimate std.error conf.low conf.high
  <chr>          <dbl>     <dbl>    <dbl>     <dbl>
1 (Intercept)     19.1      11.9     3.74      34.5
2 age              1.30     0.213    1.02       1.58
```

Accordingly, there's an 80% chance that the *typical* runner tends to slow down somewhere between 1.02 and 1.58 minutes per year. The fact that this range is entirely and comfortably above 0 provides significant evidence that the *typical* runner tends to slow down with age. This assertion is visually supported by the 200 posterior plausible global model lines below, superimposed with their posterior median, all of which exhibit positive associations between running time and age. In plotting these model lines, note that we use `add_fitted_draws()` with `re_formula = NA` to specify that we are interested in the *global*, not group-specific, model of running times:

```
B0 <- tidy_summary_1$estimate[1]
B1 <- tidy_summary_1$estimate[2]
running %>%
  add_fitted_draws(running_model_1, n = 200, re_formula = NA) %>%
  ggplot(aes(x = age, y = net)) +
    geom_line(aes(y = .value, group = .draw), alpha = 0.1) +
    geom_abline(intercept = B0, slope = B1, color = "blue") +
    lims(y = c(75, 110))
```

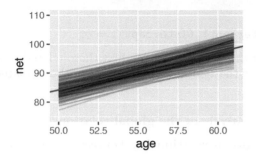

FIGURE 17.5: 200 posterior plausible global model lines, $\beta_0 + \beta_1 X$, for the relationship between running time and age.

Statistical vs practical significance

When we use the term "significant" in the discussion above, we're referring to *statistical* significance. This merely indicates that a positive relationship *exists*. In this case, we also believe that a per-year slowdown of 1.02 to 1.58 minutes is of a *meaningful magnitude* that has real-world implications. That is, this positive relationship is also *practically* significant.

Don't let the details distract you from the important punchline here! By incorporating the group structure of our data, our hierarchical random intercepts model has what the complete pooled model (17.1) lacked: the power to detect a significant relationship between running time and age. Our discussion in Chapter 15 revealed *why* this happens in our `running` analysis: pooling all runners' data together masks the fact that most *individual* runners slow down with age.

17.2.4.2 Posterior analysis of group-specific relationships

In our next step, let's examine what the hierarchical random intercepts model reveals about the **runner-specific relationships** between net running time and age,

$$\beta_{0j} + \beta_1 X_{ij} = (\beta_0 + b_{0j}) + \beta_1 X_{ij}.$$

We'll do so by combining what we learned about the global age parameter β_1 above, with information on the runner-specific intercept terms β_{0j}. The latter will require the specialized syntax we built up in Chapter 16, and thus some patience. First, the b[(Intercept) runner:j] chains correspond to the *difference* in the runner-specific and global intercepts b_{0j}. Thus, we obtain MCMC chains for each $\beta_{0j} = \beta_0 + b_{0j}$ by adding the (Intercept) chain to the b[(Intercept) runner:j] chains via spread_draws() and mutate(). We then use median_qi() to obtain posterior summaries of the β_{0j} chain for each runner j:

```
# Posterior summaries of runner-specific intercepts
runner_summaries_1 <- running_model_1 %>%
  spread_draws(`(Intercept)`, b[,runner]) %>%
  mutate(runner_intercept = `(Intercept)` + b) %>%
  select(-`(Intercept)`, -b) %>%
  median_qi(.width = 0.80) %>%
  select(runner, runner_intercept, .lower, .upper)
```

Consider the results for runners 4 and 5. With a posterior median intercept of 30.8 minutes vs 6.7 minutes, runner 4 seems to have a slower baseline speed than runner 5. Thus, at any shared age, we would expect runner 4 to run roughly 24.1 minutes slower than runner 5 (30.8 − 6.7):

```
runner_summaries_1 %>%
  filter(runner %in% c("runner:4", "runner:5"))
# A tibble: 2 x 4
  runner    runner_intercept .lower .upper
  <chr>                <dbl>  <dbl>  <dbl>
1 runner:4              30.8   15.3   46.3
2 runner:5              6.66  -8.42   21.9
```

These observations are echoed in the plots below, which display 100 posterior plausible models of `net` time by `age` for runners 4 and 5:

```
# 100 posterior plausible models for runners 4 & 5
running %>%
  filter(runner %in% c("4", "5")) %>%
  add_fitted_draws(running_model_1, n = 100) %>%
  ggplot(aes(x = age, y = net)) +
    geom_line(
      aes(y = .value, group = paste(runner, .draw), color = runner),
      alpha = 0.1) +
    geom_point(aes(color = runner))
```

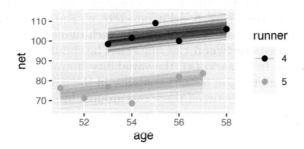

FIGURE 17.6: 100 posterior plausible models of running time by age, $\beta_{0j} + \beta_1 X$, for subjects $j \in \{4, 5\}$.

We can similarly explore the models for all 36 runners, $\beta_{0j} + \beta_1 X_{ij}$. For a quick comparison, the runner-specific posterior *median* models are plotted below and superimposed with the posterior median global model, $\beta_0 + \beta_1 X_{ij}$. This drives home the point that the global model represents the relationship between running time and age for the most *average* runner. The individual runner models vary around this global average, some with faster baseline speeds ($\beta_{0j} < \beta_0$) and some with slower baseline speeds ($\beta_{0j} > \beta_0$).

```
# Plot runner-specific models with the global model
ggplot(running, aes(y = net, x = age, group = runner)) +
  geom_abline(data = runner_summaries_1, color = "gray",
              aes(intercept = runner_intercept, slope = B1)) +
  geom_abline(intercept = B0, slope = B1, color = "blue") +
  lims(x = c(50, 61), y = c(50, 135))
```

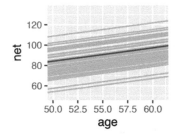

FIGURE 17.7: The posterior median models for our 36 runners j as calculated from the hierarchical random intercepts model (gray), with the posterior median global model (blue).

17.2.4.3 Posterior analysis of within- and between-group variability

All of this talk about the variability in runner-specific models brings us to a posterior consideration of the final remaining model parameters, σ_y and σ_0. Whereas σ_y measures the variability from the mean regression model **within** each runner, σ_0 measures the variability in baseline running speeds **between** the runners. The simulated datasets in Figure 17.8 provide some intuition. In scenario (a), the variability from the mean model *within* both groups (σ_y) is quite small relative to the variability in the models *between* the groups (σ_0), leading to a great distinction between these two groups. In scenario (b), σ_y is larger than σ_0, leading to little distinction between the groups.

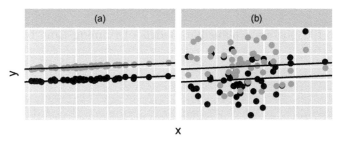

FIGURE 17.8: Simulated output for the relationship between response variable Y and predictor X when $\sigma_y < \sigma_0$ (a) and $\sigma_y > \sigma_0$ (b).

Posterior `tidy()` summaries for our variance parameters suggest that the running analysis is more like scenario (a) than scenario (b). For a given runner j, we estimate that their observed running time at any age will deviate from *their* mean regression model by roughly 5.25 minutes (σ_y). By the authors' assessment (none of us professional runners!), this deviation is rather small in the context of a long 10-mile race, suggesting a rather strong relationship between running times and age *within runners*. In contrast, we expect that baseline speeds vary by roughly 13.3 minutes from runner to runner (σ_0).

```
tidy_sigma <- tidy(running_model_1, effects = "ran_pars")
tidy_sigma
# A tibble: 2 x 3
  term                     group     estimate
  <chr>                    <chr>        <dbl>
1 sd_(Intercept).runner    runner        13.3
2 sd_Observation.Residual  Residual      5.25
```

Comparatively then, the posterior results suggest that $\sigma_y < \sigma_0$ – there's greater variability in the models *between* runners than variability from the model *within* runners. Think about this another way. As with the simple hierarchical model in Section 16.3.3, we can decompose the *total* variability in race times across all runners and races into that explained by the variability between runners and that explained by the variability within each runner (16.8):

$$\mathrm{Var}(Y_{ij}) = \sigma_0^2 + \sigma_y^2.$$

Thus, *proportionally* (16.9), differences between runners account for roughly 86.62% (the majority!) of the total variability in racing times, with fluctuations among individual races within runners explaining the other 13.38%:

```
sigma_0 <- tidy_sigma[1,3]
sigma_y <- tidy_sigma[2,3]
sigma_0^2 / (sigma_0^2 + sigma_y^2)
  estimate
1   0.8662
sigma_y^2 / (sigma_0^2 + sigma_y^2)
  estimate
1   0.1338
```

17.3 Hierarchical model with varying intercepts & slopes

Since you've gotten this far in the book, you know that once in a while it's important to stand back from the details and ask: can we do even better? A plot of the data for just four of our runners, Figure 17.9, suggests that the hierarchical random intercepts model (17.5) might oversimplify reality. Though this model recognizes that some runners tend to be faster than others, it assumes that the *change* in running time with age (β_1) is the *same* for each runner. In reality, whereas some runners *do* slow down at similar rates (e.g., runners 4 and 5), some slow down quicker (e.g., runner 20) and some barely at all (e.g., runner 29).

```
# Plot runner-specific models in the data
running %>%
  filter(runner %in% c("4", "5", "20", "29")) %>%
  ggplot(., aes(x = age, y = net)) +
    geom_point() +
    geom_smooth(method = "lm", se = FALSE) +
    facet_grid(~ runner)
```

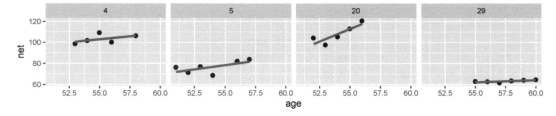

FIGURE 17.9: Scatterplots and observed trends in running times vs age for four subjects.

A snapshot of the observed trends for all 36 runners provides a more complete picture of just how much the change in net time with age might vary by runner:

```
ggplot(running, aes(x = age, y = net, group = runner)) +
  geom_smooth(method = "lm", se = FALSE, size = 0.5)
```

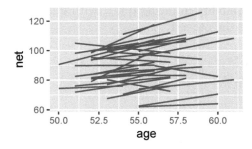

FIGURE 17.10: Observed trends in running time versus age for the 36 subjects (gray) along with the posterior median model (blue).

---

❷ Quiz Yourself!

How can we modify the random intercepts model (17.5) to recognize that the rate at which running time changes with age might vary from runner to runner?

17.3.1   Model building

Inspired by Figure 17.10, our goal is to build a model which recognizes that in the relationship between running time and age, *both* the intercept (i.e., baseline speed) and slope (i.e., rate at which speed changes with age) might vary from runner to runner. To this end, we can replace the global age coefficient β_1 in (17.5) by a runner-specific coefficient β_{1j}. Thus, the model of the relationship between running time and age **within** each runner j becomes:

$$Y_{ij}|\beta_{0j},\beta_{1j},\sigma_y \sim N(\mu_{ij},\sigma_y^2) \quad \text{where } \mu_{ij} = \beta_{0j} + \beta_{1j}X_{ij}. \tag{17.8}$$

Accordingly, just as we assumed in (17.5) that the *runner-specific* intercepts β_{0j} are Normally distributed around some *global* intercept β_0 with standard deviation σ_0, we now also assume that the *runner-specific* age coefficients β_{1j} are Normally distributed around some *global* age coefficient β_1 with standard deviation σ_1:

$$\beta_{0j}|\beta_0,\sigma_0 \sim N(\beta_0,\sigma_0^2) \quad \text{and} \quad \beta_{1j}|\beta_1,\sigma_1 \sim N(\beta_1,\sigma_1^2) \tag{17.9}$$

But these priors aren't yet complete – β_{0j} and β_{1j} work *together* to describe the model for runner j, and thus are *correlated*. Let $\rho \in [-1,1]$ represent the correlation between β_{0j} and β_{1j}. To reflect this correlation, we represent the *joint* Normal model of β_{0j} and β_{1j} by

$$\begin{pmatrix} \beta_{0j} \\ \beta_{1j} \end{pmatrix} \Big| \beta_0,\beta_1,\sigma_0,\sigma_1 \;\sim\; N\left(\begin{pmatrix} \beta_0 \\ \beta_1 \end{pmatrix},\, \Sigma\right) \tag{17.10}$$

where (β_0,β_1) is the joint mean and Σ is the 2x2 **covariance matrix** which encodes the variability and correlation amongst β_{0j} and β_{1j}:

$$\Sigma = \begin{pmatrix} \sigma_0^2 & \rho\sigma_0\sigma_1 \\ \rho\sigma_0\sigma_1 & \sigma_1^2 \end{pmatrix}. \tag{17.11}$$

Though this notation looks overwhelming, it simply indicates that β_{0j} and β_{1j} are *both* marginally Normal (17.9) and have correlation ρ. The correlation ρ between the runner-specific intercepts and slopes, β_{0j} and β_{1j}, isn't just a tedious mathematical detail. It's an interesting feature of the hierarchical model! Figure 17.11 provides some insight. Plot (a) illustrates the scenario in which there's a strong *negative* correlation between β_{0j} and β_{1j} – models that start out lower (with small β_{0j}) tend to increase at a more rapid rate (with higher β_{1j}). In plot (c) there's a strong *positive* correlation between β_{0j} and β_{1j} – models that start out higher (with larger β_{0j}) also tend to increase at a more rapid rate (with higher β_{1j}). In between these two extremes, plot (b) illustrates the scenario in which there's no correlation between the group-specific intercepts and slopes.

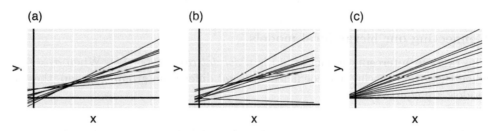

FIGURE 17.11: Simulated output for the relationship between response variable Y and predictor X when $\rho < 0$ (a), $\rho = 0$ (b), and $\rho > 0$ (c).

Consider the implications of this correlation in the context of our running analysis.[1]

❓ Quiz Yourself!

1. In our running example, what would it mean for β_{0j} and β_{1j} to be *negatively* correlated?
   a. Runners that start out *slower* (i.e., with a higher baseline), also tend to slow down at a more rapid rate.
   b. The rate at which runners slow down over time isn't associated with how fast they start out.
   c. Runners that start out *faster* (i.e., with a lower baseline), tend to slow down at a more rapid rate.
2. Similarly, what would it mean for β_{0j} and β_{1j} to be *positively* correlated?

The completed hierarchical model (17.12) pulls together (17.8) and (17.10) with priors for the global parameters. For reasons you might imagine, this is often referred to as a **hierarchical random intercepts and slopes model**:

$$Y_{ij}|\beta_{0j}, \beta_{1j}, \sigma_y \; \sim N(\mu_{ij}, \sigma_y^2) \quad \text{where} \quad \mu_{ij} = \beta_{0j} + \beta_{1j}X_{ij}$$

$$\begin{pmatrix} \beta_{0j} \\ \beta_{1j} \end{pmatrix} | \beta_0, \beta_1, \sigma_0, \sigma_1 \; \sim N\left(\begin{pmatrix} \beta_0 \\ \beta_1 \end{pmatrix}, \Sigma \right)$$

$$\begin{aligned}
\beta_{0c} &\sim N(100, 10^2) \\
\beta_1 &\sim N(2.5, 1^2) \\
\sigma_y &\sim \text{Exp}(0.072) \\
\Sigma &\sim \text{(decomposition of covariance)}.
\end{aligned}$$

(17.12)

Equivalently, we can re-express the random intercepts and slopes as random tweaks to the global intercept and slope: $\mu_{ij} = (\beta_0 + b_{0j}) + (\beta_1 + b_{1j})X_{ij}$ with

$$\begin{pmatrix} b_{0j} \\ b_{1j} \end{pmatrix} | \sigma_0, \sigma_1 \; \sim N\left(\begin{pmatrix} 0 \\ 0 \end{pmatrix}, \Sigma \right).$$

Connecting our hierarchical models

The random intercepts model (17.5) is a *special case* of (17.12). When $\sigma_1 = 0$, i.e., when the group-specific age coefficients do *not* differ from group to group, these two models are equivalent.

Most of the pieces in this model are familiar. For global parameters β_0 and β_1 we use the tuned Normal priors from (17.7). For σ_y we use a weakly informative prior. Yet there is one big new piece. We need a *joint* prior model to express our understanding of how the *combined* σ_0, σ_1, and ρ parameters define covariance matrix Σ (17.11). At the writing of this book, the `stan_glmer()` function allows users to define this prior through a **decomposition of covariance**, or `decov()`, model. Very generally speaking, this model *decomposes* our prior model for the covariance matrix into prior information about three separate pieces:

---

[1]Answers: 1 = c; 2 = a

1. the correlation between the group-specific intercepts and slopes, ρ (Figure 17.11);

2. the combined degree to which the intercepts and slopes vary by group, $\sigma_0^2 + \sigma_1^2$; and[2]

3. the relative proportion of this variability between groups that's due to differing intercepts vs differing slopes,

$$\pi_0 = \frac{\sigma_0^2}{\sigma_0^2 + \sigma_1^2} \quad \text{vs} \quad \pi_1 = \frac{\sigma_1^2}{\sigma_0^2 + \sigma_1^2}.$$

Figure 17.12 provides some context on this third piece, displaying a few scenarios for the relationship between π_0 and π_1. In general, π_0 and π_1 always sum to 1, and thus have a push-and-pull relationship. For example, when $\pi_0 \approx 1$ and $\pi_1 \approx 0$, the variability in intercepts (σ_0^2) is large in comparison to the variability in slopes (σ_1^2). Thus, the majority of the variability between group-specific models is explained by differences in *intercepts* (plot a). In contrast, when $\pi_0 \approx 0$ and $\pi_1 \approx 1$, the majority of the variability between group-specific models is explained by differences in *slopes* (plot c). In between these extremes, when π_0 and π_1 are both approximately 0.5, roughly half of the variability between groups can be explained by differing intercepts and the other half by differing slopes (plot b).

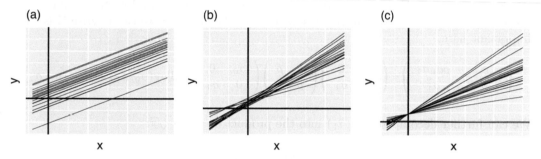

FIGURE 17.12: Simulated output for the relationship between response variable Y and predictor X when $\pi_0 = 1$ and $\pi_1 = 0$ (a), $\pi_0 = \pi_1 = 0.5$ (b), and $\pi_0 = 0$ and $\pi_1 = 1$ (c).

In our analysis, we'll utilize the weakly informative default setting for the hierarchical random intercepts and slopes model: `decov(reg = 1, conc = 1, shape = 1, scale = 1)` in **rstanarm** notation. This makes the following prior assumptions regarding the three pieces above:

1. the correlation ρ is equally likely to be anywhere between -1 and 1;
2. we have weakly informative prior information about the total degree to which the intercepts and slopes vary by runner; and
3. the relative proportion of the variability between runners that's due to differing intercepts is equally likely to be anywhere between 0 and 1, i.e., we're not at all sure if there's more, less, or the same level of variability in the baseline speeds from runner to runner, β_{0j}, than in the rate at which their speeds change over time, β_{1j}.

We'll utilize these default assumptions for the covariance prior in this book. Beyond the defaults, specifying and tuning the decomposition of covariance prior requires the acquisition

---

[2]Technically, this is the combined variability in group-specific intercepts and slopes when assuming they are uncorrelated.

of two new probability models. We present more *optional* detail in the next section and refer the curious reader to Gabry and Goodrich (2020a) for a more mathematical treatment that scales up to models beyond those considered here.

17.3.2   Optional: The decomposition of covariance model

Let's take a closer look at the decomposition of covariance model. To begin, we define the three pieces that are important to specifying our prior understanding of the random intercepts and slopes covariance matrix Σ (17.11), and thus the $(\sigma_0, \sigma_1, \rho)$ parameters by which it's defined. These pieces are numbered in accordance to their corresponding interpretations above:

$$R = \begin{pmatrix} 1 & \rho \\ \rho & 1 \end{pmatrix} \tag{1}$$

$$\tau = \sqrt{\sigma_0^2 + \sigma_1^2} \tag{2}$$

$$\pi = \begin{pmatrix} \pi_0 \\ \pi_1 \end{pmatrix} = \begin{pmatrix} \frac{\sigma_0^2}{\sigma_0^2 + \sigma_1^2} \\ \frac{\sigma_1^2}{\sigma_0^2 + \sigma_1^2} \end{pmatrix} \tag{3}$$

We can decompose Σ into a product which depends on R, τ, and π. If you know some linear algebra, you can confirm this result, though the fact that we can rewrite Σ is what's important here:

$$\begin{pmatrix} \sigma_0^2 & \rho\sigma_0\sigma_1 \\ \rho\sigma_0\sigma_1 & \sigma_1^2 \end{pmatrix} = \begin{pmatrix} \sigma_0 & 0 \\ 0 & \sigma_1 \end{pmatrix} \begin{pmatrix} 1 & \rho \\ \rho & 1 \end{pmatrix} \begin{pmatrix} \sigma_0 & 0 \\ 0 & \sigma_1 \end{pmatrix} = \text{diag}(\sigma_0, \sigma_1)\, R\, \text{diag}(\sigma_0, \sigma_1).$$

We can further decompose (σ_0, σ_1) into the product of τ and $\sqrt{\pi}$:

$$\begin{pmatrix} \sigma_0 \\ \sigma_1 \end{pmatrix} = \tau\sqrt{\pi}.$$

And since we can rewrite Σ using R, τ, and π, we can also express our prior *understanding* of Σ by our combined prior understanding of these three pieces. This joint prior, which we simply expressed above by

$$\Sigma \sim (\text{decomposition of covariance})$$

is actually defined by three individual priors:

$$R \sim \text{LKJ}(\eta)$$
$$\tau \sim \text{Gamma}(s, r)$$
$$\pi \sim \text{Dirichlet}(2, \delta)$$

Let's begin with the "LKJ" prior model on the correlation matrix R with **regularization hyperparameter** $\eta > 0$. In our model (17.12), R depends only on the correlation ρ between the group-specific intercepts β_{0j} and slopes β_{1j}. Thus, the LKJ prior model simplifies to a prior model on ρ with pdf

$$f(\rho) = \left[2^{1-2\eta} \frac{\Gamma(2\eta)}{\Gamma(\eta)\Gamma(\eta)} \right] (1 - \rho^2)^{\eta-1} \quad \text{for } \rho \in [-1, 1].$$

Figure 17.13 displays the LKJ pdf under a variety of regularization parameters η, illustrating the important comparison of η to 1:

- Setting $\eta < 1$ indicates a prior understanding that the group-specific intercepts and slopes are most likely strongly correlated, though we're not sure if this correlation is negative or positive.
- When $\eta = 1$, the LKJ model is uniform from -1 to 1, indicating that the correlation between the intercepts and slopes is equally likely to be anywhere in this range – we're not really sure.
- Setting $\eta > 1$ indicates a prior understanding that the group-specific intercepts and slopes are most likely weakly correlated ($\rho \approx 0$). The greater η is, the tighter and tighter the LKJ hugs values of ρ near 0.

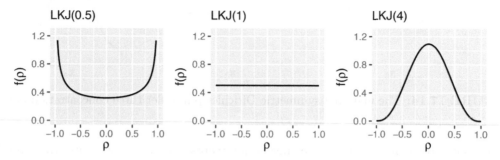

FIGURE 17.13: The LKJ pdf under three regularization parameters.

Next, for the total standard deviation in the intercepts and slopes, $\tau = \sqrt{\sigma_0^2 + \sigma_1^2}$, we utilize the "usual" Gamma prior (or its Exponential special case). Finally, consider the prior for the π_0 and π_1 parameters. Recall that π_0 and π_1 measure the relative proportion of the variability between groups that's due to differing intercepts vs differing slopes, respectively. Thus, π_0 and π_1 are both restricted to values between 0 and 1 and must sum to 1:

$$\pi_0 + \pi_1 = \frac{\sigma_0^2}{\sigma_0^2 + \sigma_1^2} + \frac{\sigma_1^2}{\sigma_0^2 + \sigma_1^2} = 1.$$

Accordingly, we can utilize a joint **symmetric Dirichlet**$(2, \delta)$ **prior** with **concentration hyperparameter** $\delta > 0$ for (π_0, π_1). The symmetric Dirichlet pdf defines the relative prior plausibility of valid (π_0, π_1) pairs:

$$f(\pi_0, \pi_1) = \frac{\Gamma(2\delta)}{\Gamma(\delta)\Gamma(\delta)} (\pi_0 \pi_1)^{\delta - 1} \quad \text{for} \quad \pi_0, \pi_1 \in [0, 1] \quad \text{and} \quad \pi_0 + \pi_1 - 1.$$

In fact, in the special case when we have only two group-specific parameters, β_{0j} and β_{1j}, the symmetric Dirichlet model for (π_0, π_1) is equivalently expressed by:

$$\pi_0 \sim \text{Beta}(\delta, \delta) \quad \text{and} \quad \pi_1 = 1 - \pi_0.$$

Figure 17.14 displays the marginal symmetric Dirichlet pdf, i.e., Beta pdf, for π_0 under a variety of concentration parameters δ, illustrating the important comparison of δ to 1:

- Setting $\delta < 1$ places more prior weight on proportions π_0 near 0 or 1. This indicates a prior understanding that relatively little ($\pi_0 \approx 0$) or relatively much ($\pi_0 \approx 1$) of the variability between groups is explained by differences in intercepts rather than differences in slopes (Figure 17.12 plots a and c).

- When $\delta = 1$, the marginal pdf on π_0 is uniform from 0 to 1, indicating that the variability in intercepts explains anywhere between 0 and all of the variability between groups – we're not really sure.
- Setting $\delta > 1$ indicates a prior understanding that roughly half of the variability between groups is explained by differences in intercepts and the other half by differences in slopes (Figure 17.12 plot b). The greater δ is, the tighter and tighter this prior hugs values of π_0 near 0.5.

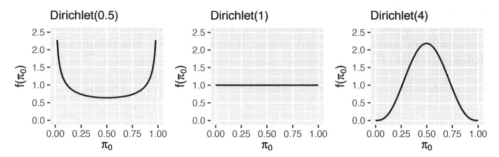

FIGURE 17.14: The marginal symmetric Dirichlet pdf under three concentration parameters.

When our models have both group-specific intercepts and slopes, we'll use the following default decomposition of variance priors which indicate general uncertainty about the correlation between group-specific intercepts and slopes, the overall variability in group-specific model, and the relative degree to which this variability is explained by differing intercepts vs differing slopes:

$$R \sim \text{LKJ}(1)$$
$$\tau \sim \text{Gamma}(1,1), \text{ i.e., } \text{Exp}(1)$$
$$\pi \sim \text{Dirichlet}(2,1)$$

In **rstanarm** notation, this prior is expressed by decov(reg = 1, conc = 1, shape = 1, scale = 1). To tune this prior, you can change the LKJ regularization parameter η, the Gamma shape and scale parameters, and the Dirichlet concentration parameter δ.

17.3.3   Posterior simulation & analysis

Finally, let's simulate the posterior of our hierarchical random intercepts and slopes model of running times (17.12). This requires one minor tweak to our stan_glmer() call: instead of using the formula net ~ age + (1 | runner) we use net ~ age + (age | runner).

```
running_model_2 <- stan_glmer(
  net ~ age + (age | runner),
  data = running, family = gaussian,
  prior_intercept = normal(100, 10),
  prior = normal(2.5, 1),
  prior_aux = exponential(1, autoscale = TRUE),
  prior_covariance = decov(reg = 1, conc = 1, shape = 1, scale = 1),
  chains = 4, iter = 5000*2, seed = 84735, adapt_delta = 0.99999
)
```

```
# Confirm the prior model specifications
prior_summary(running_model_2)
```

> **❶ Warning**
>
> **This simulation will be slooooooooow.** Notice the additional argument in our
> `stan_glmer()` syntax: `adapt_delta = 0.99999`. Simply put, `adapt_delta` is a tun-
> ing parameter for the underlying MCMC algorithm, the technical details of which
> are outside the scope of this book. Prior to this example, we've been running our
> simulations using the default `adapt_delta` value of 0.95. However, in this particular
> example, the default produces a warning: `There were 1 divergent transitions`
> `after warmup`. This warning indicates that the MCMC algorithm had a tough time
> exploring the posterior plausible range of our parameter values. When encountering
> this issue, one strategy is to increase `adapt_delta` to some value closer to 1. Doing
> so produces a *much slower*, but more stable, simulation. For more technical details on
> divergent transitions and how to address them, we recommend Modrak (2019).

17.3.3.1 Posterior analysis of the global and group-specific parameters

Remember thinking that the 40 parameters in the random intercepts model was a lot? This
new model has 78 parameters: 36 runner-specific intercept parameters β_{0j}, 36 runner-specific
age coefficients β_{1j}, and 6 global parameters $(\beta_0, \beta_1, \sigma_y, \sigma_0, \sigma_1, \rho)$. Let's examine these piece
by piece, starting with the global model of the relationship between running time and age,

$$\beta_0 + \beta_1 X.$$

The results here for the random intercepts and slopes model (17.12) are quite similar to
those for the random intercepts model (17.5): the posterior median model is $18.5 + 1.32$ age.

```
# Quick summary of global regression parameters
tidy(running_model_2, effects = "fixed", conf.int = TRUE, conf.level = 0.80)
# A tibble: 2 x 5
  term        estimate std.error conf.low conf.high
  <chr>          <dbl>     <dbl>    <dbl>     <dbl>
1 (Intercept)    18.5      11.6     3.61      33.6
2 age             1.32      0.217   1.04       1.59
```

Since the global mean model $\beta_0 + \beta_1 X$ captures the relationship between running time and
age for the average runner, we shouldn't be surprised that our two hierarchical models
produced similar assessments. Where these two models start to differ is in their assessments
of the runner-specific relationships. Obtaining the MCMC chains for the runner-specific
intercepts and slopes gets quite technical. We encourage you to pick through the code below,
line by line. Here are some important details to pick up on:

- `spread_draws()` uses `b[term, runner]` to grab the chains for all runner-specific parameters. As usual now, these chains correspond to b_{0j} and b_{1j}, the *differences* between the
  runner-specific vs global intercepts and age coefficients.

- `pivot_wider()` creates separate columns for each of the b_{0j} and b_{1j} chains and names these `b_(Intercept)` and `b_age`.
- `mutate()` obtains the runner-specific intercepts $\beta_{0j} = \beta_0 + b_{0j}$, named `runner_intercept`, by summing the global `(Intercept)` and runner-specific adjustments `b_(Intercept)`. The runner-specific β_{1j} coefficients, `runner_age`, are created similarly.

```
# Get MCMC chains for the runner-specific intercepts & slopes
runner_chains_2 <- running_model_2 %>%
  spread_draws(`(Intercept)`, b[term, runner], `age`) %>%
  pivot_wider(names_from = term, names_glue = "b_{term}",
              values_from = b) %>%
  mutate(runner_intercept = `(Intercept)` + `b_(Intercept)`,
         runner_age = age + b_age)
```

From these chains, we can obtain the posterior medians for each runner-specific intercept and age coefficient. Since we're only obtaining posterior medians here, we use `summarize()` in combination with `group_by()` instead of using the `median_qi()` function:

```
# Posterior medians of runner-specific models
runner_summaries_2 <- runner_chains_2 %>%
  group_by(runner) %>%
  summarize(runner_intercept = median(runner_intercept),
            runner_age = median(runner_age))

# Check it out
head(runner_summaries_2, 3)
# A tibble: 3 x 3
  runner     runner_intercept runner_age
  <chr>                 <dbl>      <dbl>
1 runner:1               18.6       1.06
2 runner:10              18.5       1.75
3 runner:11              18.5       1.32
```

Figure 17.15 plots the posterior median models for all 36 runners.

```
ggplot(running, aes(y = net, x = age, group = runner)) +
  geom_abline(data = runner_summaries_2, color = "gray",
              aes(intercept = runner_intercept, slope = runner_age)) +
  lims(x = c(50, 61), y = c(50, 135))
```

Hmph. Are you surprised? *We* were slightly surprised. The slopes do differ, but not as drastically as we expected. But then we remembered – **shrinkage**! Consider sample runners 1 and 10. Their posteriors suggest that, on average, runner 10's running time increases by just 1.06 minute per year, whereas runner 1's increases by 1.75 minutes per year:

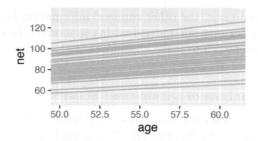

FIGURE 17.15: The posterior median models for the 36 runners, as calculated from the hierarchical random intercepts and slopes model.

```
runner_summaries_2 %>%
  filter(runner %in% c("runner:1", "runner:10"))
# A tibble: 2 x 3
  runner    runner_intercept runner_age
  <chr>                <dbl>      <dbl>
1 runner:1              18.6       1.06
2 runner:10             18.5       1.75
```

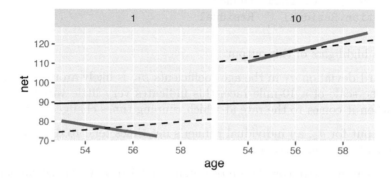

FIGURE 17.16: For runners 1 and 10, the posterior median relationships between running time and age from the hierarchical random intercepts and slopes model (dashed) are contrasted by the observed no pooled models (blue) and the complete pooled model (black).

Figure 17.16 contrasts these posterior median models for runners 1 and 10 (dashed lines) by the complete pooled posterior models (black) and no pooled posterior models (blue). As usual, the hierarchical models strike a balance between these two extremes. Like the no pooled models, the hierarchical models *do* vary between the two runners. Yet the difference is not as stark. The hierarchical models are drawn away from the no pooled models and *toward* the complete pooled models. Though this shrinkage is subtle for runner 10, the association between running time and age switches from negative to positive for runner 1. This is to be expected. Unlike the no pooled approach, which models runner-specific relationships using only runner-specific data, our hierarchical model assumes that one runner's behavior can tell us about another's. Further, we have very few data points on each runner – at most 7 races. With so few observations, the other runners' information has ample influence on our posterior understanding for any one individual (as it should). In the case of runner 1, the other 35 runners' data is enough to make us think that this runner, too, will eventually slow down.

17.3.3.2   Posterior analysis of within- and between-group variability

Stepping back, we should also ask ourselves: *Is it worth it?* Incorporating the random runner-specific age coefficients introduced 37 parameters into our model of running time by age. Yet at least visually, there doesn't appear to be much variation among the slopes of the runner-specific models. For a numerical assessment of this variation, we can examine the posterior trends in σ_1 (sd_age.runner). While we're at it, we'll also check out σ_0 (sd_(Intercept).runner), ρ (cor_(Intercept).age.runner), and σ_y (sd_Observation.Residual):

```
tidy(running_model_2, effects = "ran_pars")
# A tibble: 4 x 3
  term                            group      estimate
  <chr>                           <chr>         <dbl>
1 sd_(Intercept).runner           runner         1.34
2 sd_age.runner                   runner         0.251
3 cor_(Intercept).age.runner      runner       -0.0955
4 sd_Observation.Residual         Residual       5.17
```

Consider some highlights of this output:

- The standard deviation σ_1 in the age coefficients β_{1j} is likely around 0.251 minutes per year. On the scale of a 10-mile race, this indicates very little variability between the runners when it comes to the rate at which running times change with age.

- Per the output for σ_y, an individual runner's net times tend to deviate from their own mean model by roughly 5.17 minutes.

- There's a weak negative correlation of roughly -0.0955 between the runner-specific β_{0j} and β_{1j} parameters. Thus, it seems that, *ever so slightly*, runners that start off faster tend to slow down at a faster rate.

17.4 Model evaluation & selection

We have now built a sequence of three models of running time by age: a complete pooled model (17.1), a hierarchical random intercepts model (17.5), and a hierarchical random intercepts and slopes model (17.12). Figure 17.17 provides a reminder of the general assumptions behind these three models.

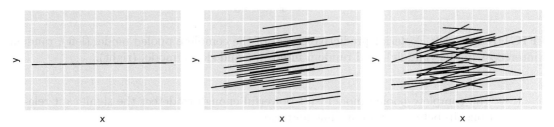

FIGURE 17.17: The idea behind three different approaches to modeling Y by X when utilizing group-structured data: complete pooling (left), hierarchical random intercepts (middle), and hierarchical random intercepts and slopes (right).

So which one should we use? To answer this question, we can compare our three models using the framework of Chapters 10 and 11, and asking these questions: (1) How *fair* is each model? (2) How *wrong* is each model? (3) How *accurate* are each model's posterior predictions? Consider question (1). The context and data collection procedure is the same for each model. Since the data has been anonymized and runners are aware that race results will be public, we think this data collection process is fair. Further, though the models produce slightly different conclusions about the relationship between running time and age (e.g., the hierarchical models conclude this relationship is *significant*), none of these conclusions seem poised to have a negative impact on society or individuals. Thus, our three models are equally *fair*.

Next, consider question (2). **Posterior predictive checks** suggest that the complete pooled model comparatively underestimates the variability in running times – datasets of running time simulated from the complete pooled posterior tend to exhibit a slightly narrower range than the running times we actually observed. Thus, the complete pooled model is *more wrong* than the hierarchical models.

```
pp_check(complete_pooled_model) +
  labs(x = "net", title = "complete pooled model")
pp_check(running_model_1) +
  labs(x = "net", title = "running model 1")
pp_check(running_model_2) +
  labs(x = "net", title = "running model 2")
```

In fact, we *know* that the complete pooled model is wrong. By ignoring the data's grouped structure, it incorrectly assumes that each race observation is independent of the others. Depending upon the trade-offs, we might live with this wrong but simplifying assumption in some analyses. Yet at least two signs point to this being a mistake for our running analysis.

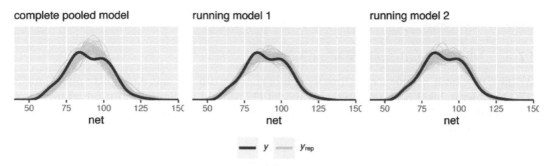

FIGURE 17.18: Posterior predictive checks of the complete pooled model (left), random intercepts model (middle), and random intercepts and slopes model (right).

1. The complete pooled model isn't powerful enough to detect the significant relationship between running time and age.
2. Not only have we seen *visual* evidence that some runners tend to be significantly faster or slower than others, the posterior prediction summaries in Section 17.2.4 suggest that there's significant variability between runners (σ_0).

In light of this discussion, let's drop the complete pooled model from consideration. In choosing between `running_model_1` and `running_model_2`, consider question (3): what's the *predictive accuracy* of these models? Recall some approaches to answering this question from Chapter 11: posterior prediction summaries and ELPD. To begin, we use the `prediction_summary()` function from the **bayesrules** package to compare how well these two models predict the running outcomes of the 36 runners that were part of our sample.

```
# Calculate prediction summaries
set.seed(84735)
prediction_summary(model = running_model_1, data = running)
    mae mae_scaled within_50 within_95
1 2.626      0.456    0.6865     0.973
prediction_summary(model = running_model_2, data = running)
    mae mae_scaled within_50 within_95
1 2.53      0.4424    0.7027     0.973
```

By all metrics, `running_model_1` and `running_model_2` produce similarly accurate posterior predictions. For both models, the *observed* net running times tend to be 2.63 and 2.53 minutes, or 0.46 and 0.44 standard deviations, from their posterior mean *predictions*. The posterior predictive models also have similar coverage in terms of the percent of observed running times that fall within their 50% and 95% prediction intervals.

Thinking beyond our own sample of runners, we could also utilize `prediction_summary_cv()` to obtain cross-validated metrics of posterior predictive accuracy. The *idea* is the same as for non-hierarchical models, but the details change to reflect the grouped structure of our data. To explore how well our models predict the running behavior of runners that *weren't* included in our sample data, we divide the *runners*, not the individual race outcomes, into distinct folds. For example, for a 10-fold cross validation with 36 runners, each fold would include data on 3 or 4 of the sample runners. Thus, we would train each of 10 models using data on 32 or 33 of our sample runners and test it on the other 3 or 4. We include code for the curious but do not run it here.

```
prediction_summary_cv(model = running_model_1, data = running,
                      k = 10, group = "runner")
```

> **❶ Warning**
>
> For hierarchical models, the `prediction_summary_cv()` function divides *groups*, not individual outcomes Y, into distinct folds. Thus, if we have 10 groups, 10-fold cross-validation will build each training model using data on 9 groups and test it on the 10th group. This approach makes sense when we want to assess how well our model generalizes to new *groups* outside our sample for which we have *no* data. But this isn't always our goal. Instead, we might want to assess how well our model predicts the new *outcomes* of groups for which we have at least *some* data. For example, instead of evaluating how well our model predicts the **net** times for new *runners*, we might wish to evaluate how well it predicts the *next* **net** time of a runner for whom we have data on past races. In short, `prediction_summary_cv()` is *not* a catch-all. For a discussion of the various approaches to cross-validation for hierarchical models, as well as how these can be implemented from scratch, we recommend Vehtari (2019).

Finally, consider one last comparison of our two hierarchical models: the cross-validated **expected log-predictive densities (ELPD)**. The estimated ELPD for `running_model_1` is lower (worse) than, though within two standard errors of, the `running_model_2` ELPD. Hence, by this metric, there is *not* a significant difference in the posterior predictive accuracy of our two hierarchical models.

```
# Calculate ELPD for the 2 models
elpd_hierarchical_1 <- loo(running_model_1)
elpd_hierarchical_2 <- loo(running_model_2)
```

```
# Compare the ELPD
loo_compare(elpd_hierarchical_1, elpd_hierarchical_2)
                elpd_diff se_diff
running_model_2  0.0       0.0
running_model_1 -1.6       1.2
```

After reflecting upon our model evaluation, we're ready to make a final determination: we choose `running_model_1`. The choice of `running_model_1` over the `complete_pooled_model` was pretty clear: the latter was wrong and didn't have the power to detect a relationship between running time and age. The choice of `running_model_1` over `running_model_2` comes down to this: the complexity introduced by the additional random age coefficients in `running_model_2` produced little apparent change or benefit. Thus, the additional complexity simply isn't worth it (at least not to us).

17.5   Posterior prediction

Finally, let's use our preferred model, `running_model_1`, to make some posterior predictions. Suppose we want to predict the running time that three different runners will achieve when they're 61 years old: runner 1, runner 10, and Miles. Though Miles' running prowess is a mystery, we observed runners 1 and 10 in our sample. Should their trends continue, we expect that runner 10's time will be slower than that of runner 1 when they're both 61:

```
# Plot runner-specific trends for runners 1 & 10
running %>%
  filter(runner %in% c("1", "10")) %>%
  ggplot(., aes(x = age, y = net)) +
    geom_point() +
    facet_grid(~ runner) +
    lims(x = c(54, 61))
```

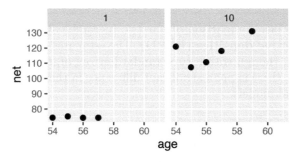

FIGURE 17.19: The observed net running times by age for runners 1 and 10.

In general, let $Y_{new,j}$ denote a *new* observation on an *observed* runner j, specifically runner j's running time at age 61. As in Chapter 16, we can approximate the posterior predictive model for $Y_{new,j}$ by simulating a prediction from the first layer of (17.3), that which describes the variability in race times Y_{ij}, evaluated at each of the 20,000 parameter sets $\left\{\beta_{0j}^{(i)}, \beta_1^{(i)}, \sigma_y^{(i)}\right\}$ in our MCMC simulation:

$$Y_{new,j}^{(i)} | \beta_{0j}, \beta_1, \sigma_y \sim N\left(\mu_{ij}^{(i)}, \left(\sigma_y^{(i)}\right)^2\right) \quad \text{where} \quad \mu_{ij}^{(i)} = \beta_{0j}^{(i)} + \beta_1^{(i)} \cdot 61.$$

The resulting posterior predictive model will reflect two sources of uncertainty in runner j's race time: the **within-group sampling variability** σ_y (we can't perfectly predict runner j's time from their mean model); and **posterior variability** in β_{0j}, β_1, and σ_y (the parameters defining runner j's relationship between running time and age are unknown and random). Since we don't have any data on the baseline speed for our new runner, Miles, there's a third source of uncertainty in predicting his race time: **between-group sampling variability** σ_0 (baseline speeds vary from runner to runner). Though we recommend doing these simulations "by hand" to connect with the concepts of posterior prediction (as we did in Chapter 16), we'll use the `posterior_predict()` shortcut function to simulate the posterior predictive models for our three runners:

```
# Simulate posterior predictive models for the 3 runners
set.seed(84735)
predict_next_race <- posterior_predict(
  running_model_1,
    newdata = data.frame(runner = c("1", "Miles", "10"),
                         age = c(61, 61, 61)))
```

These posterior predictive models are plotted in Figure 17.20. As anticipated from their previous trends, our posterior expectation is that runner 10 will have a slower time than runner 1 when they're 61 years old. Our posterior predictive model for Miles' net time is somewhere in between these two extremes. The posterior median prediction is just under 100 minutes, similar to what we'd get if we plugged an age of 61 into the *global* posterior median model for the average runner:

```
B0 + B1 * 61
[1] 98.54
```

That is, without any information about Miles, our default assumption is that he's an average runner. Our *uncertainty* in this assumption is reflected by the relatively wide posterior predictive model. Naturally, having observed data on runners 1 and 10, we're more certain about how fast they will be when they're 61. But Miles is a wild card – he *could* be really fast or really slow!

```
# Posterior predictive model plots
mcmc_areas(predict_next_race, prob = 0.8) +
  ggplot2::scale_y_discrete(labels = c("runner 1", "Miles", "runner 10"))
```

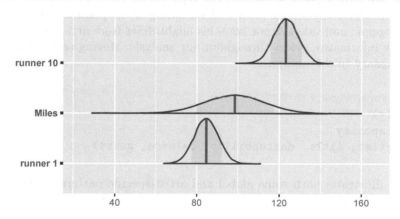

FIGURE 17.20: Posterior predictive models for the net running times at age 61 for sample runners 1 and 10, as well as Miles, a runner that wasn't in our original sample.

17.6   Details: Longitudinal data

The **running** data on **net** times by **age** is **longitudinal**. We observe each runner over time, where this time (or aging) is of primary interest. Though our hierarchical models of this relationship account for the correlation in running times within any runner, they make a simplifying assumption about this correlation: it's the *same* across all ages. In contrast, you might imagine that observations at one age are more strongly correlated with observations at similar ages. For example, a runner's net time at age 60 is likely more strongly correlated with their net time at age 59 than at age 50. It's beyond the scope of this book, but we *can* adjust the structure of our hierarchical models to account for a longitudinal correlation structure. We encourage the interested reader to check out the **bayeslongitudinal** R package (Carreño and Cuervo, 2017) and the foundational paper by Laird and Ware (1982).

17.7   Example: Danceability

Let's implement our hierarchical regression tools in a different context, by revisiting our `spotify` data from Chapter 16. There we studied a hierarchical model of song popularity, accounting for the fact that we had grouped data with multiple songs per sampled artist. Here we'll switch our focus to a song's *danceability* and how this might be explained by two features: its *genre* and *valence* or mood. The danceability and valence of a song are both measured on a scale from 0 (low) to 100 (high). Thus, lower valence scores are assigned to negative / sad / angry songs and higher scores to positive / happy / euphoric songs.

Starting out, we have a vague sense that the *typical* song has a danceability rating around 50, yet we don't have any strong prior understanding of the possible relationship between danceability, genre, and valence, nor how this might differ from artist to artist. Thus, we'll utilize weakly informative priors throughout our analysis. Moving on from the priors, load the data necessary to this analysis:

```
# Import and wrangle the data
data(spotify)
spotify <- spotify %>%
  select(artist, title, danceability, valence, genre)
```

Figure 17.21 illustrates both some global and artist-specific patterns in the relationships among these variables:

```
ggplot(spotify, aes(y = danceability, x = genre)) +
  geom_boxplot()
ggplot(spotify, aes(y = danceability, x = valence)) +
  geom_point()
ggplot(spotify, aes(y = danceability, x = valence, group = artist)) +
  geom_smooth(method = "lm", se = FALSE, size = 0.5)
```

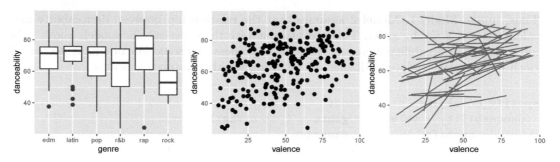

FIGURE 17.21: The relationship of danceability with genre (left) and valence (middle) for individual songs. The relationship of danceability with valence by artist (right).

When pooling all songs together, notice that `rock` songs tend to be the least danceable and `rap` songs the most (by a slight margin). Further, danceability seems to have a weak but positive association with valence. Makes sense. Sad songs tend not to inspire dance. Taking this all with a grain of salt, recall that the global models might mask what's actually going on. To this end, the artist-specific models in the final plot paint a more detailed picture. The two key themes to pick up on here are that (1) some artists' songs tend to be more danceable than others, and (2) the association between danceability and valence might differ among artists, though it's typically positive.

To model these relationships, let's define some notation. For song i within artist j, let Y_{ij} denote danceability and X_{ij1} denote valence. Next, note that there are 6 different genres, `edm` being the baseline or reference. Thus, we must define 5 additional predictors, one for each non-`edm` genre. Specifically, let $(X_{ij2}, X_{ij3}, \ldots, X_{ij6})$ be indicators of whether a song falls in the `latin`, `pop`, `r&b`, `rap`, and `rock` genres, respectively. For example,

$$X_{ij2} = \begin{cases} 1 & \text{Latin genre} \\ 0 & \text{otherwise.} \end{cases}$$

Thus, for an `edm` song, all genre indicators are 0. We'll consider two possible models of danceability by valence and genre. The first layer of **Model 1** assumes that $Y_{ij}|(\beta_{0j}, \beta_1, \beta_2, \ldots, \beta_6, \sigma_y) \sim N(\mu_{ij}, \sigma_y^2)$ with

$$\mu_{ij} = \beta_{0j} + \beta_1 X_{ij1} + \beta_2 X_{ij2} + \cdots + \beta_6 X_{ij6}.$$

The global coefficients $(\beta_1, \beta_2, \ldots, \beta_6)$ reflect an assumption that the relationships between danceability, valence, and genre are similar for each artist. Yet the artist-specific intercepts β_{0j} assume that, when holding constant a song's valence and genre, some artists' songs tend to be more danceable than other artists' songs.

The first layer of **Model 2** incorporates additional artist-specific valence coefficients, assuming $Y_{ij}|(\beta_{0j}, \beta_{1j}, \beta_2, \ldots, \beta_6, \sigma_y) \sim N(\mu_{ij}, \sigma_y^2)$ with

$$\mu_{ij} = \beta_{0j} + \beta_{1j} X_{ij1} + \beta_2 X_{ij2} + \cdots + \beta_6 X_{ij6}.$$

Thus, unlike Model 1, Model 2 assumes that the relationship between danceability and valence might *differ* by artist. This is consistent with the artist-specific models above – for most artists, danceability increased with the happiness of a song. For others, it decreased. Both models are simulated utilizing the following weakly informative priors:

```
spotify_model_1 <- stan_glmer(
  danceability ~ valence + genre + (1 | artist),
  data = spotify, family = gaussian,
  prior_intercept = normal(50, 2.5, autoscale = TRUE),
  prior = normal(0, 2.5, autoscale = TRUE),
  prior_aux = exponential(1, autoscale = TRUE),
  prior_covariance = decov(reg = 1, conc = 1, shape = 1, scale = 1),
  chains = 4, iter = 5000*2, seed = 84735)

spotify_model_2 <- stan_glmer(
  danceability ~ valence + genre + (valence | artist),
  data = spotify, family = gaussian,
  prior_intercept = normal(50, 2.5, autoscale = TRUE),
  prior = normal(0, 2.5, autoscale = TRUE),
  prior_aux = exponential(1, autoscale = TRUE),
  prior_covariance = decov(reg = 1, conc = 1, shape = 1, scale = 1),
  chains = 4, iter = 5000*2, seed = 84735)

# Check out the prior specifications
prior_summary(spotify_model_1)
prior_summary(spotify_model_2)
```

Posterior predictive checks of these two models are similar, both models producing posterior simulated datasets of song danceability that are consistent with the main features in the original song data.

```
pp_check(spotify_model_1) +
  xlab("danceability")
pp_check(spotify_model_2) +
  xlab("danceability")
```

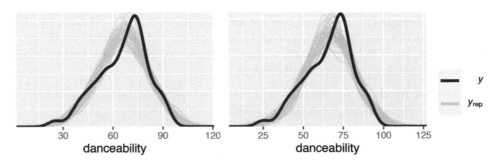

FIGURE 17.22: Posterior predictive checks of Spotify Model 1 (left) and Model 2 (right).

Yet upon a comparison of their ELPDs, we think it's best to go forward with Model 1. The quality of the two models do not significantly differ, and Model 1 is substantially simpler. Without more data per artist, it's difficult to know if the artist-specific valence coefficients are insignificant due to the fact that the relationship between danceability and valence *doesn't* vary by artist, or if we simply don't have enough data per artist to determine that it does.

```
# Calculate ELPD for the 2 models
elpd_spotify_1 <- loo(spotify_model_1)
elpd_spotify_2 <- loo(spotify_model_2)

# Compare the ELPD
loo_compare(elpd_spotify_1, elpd_spotify_2)
                elpd_diff se_diff
spotify_model_2  0.0       0.0
spotify_model_1 -5.3       3.0
```

Digging into Model 1, first consider posterior summaries for the global model parameters $(\beta_0, \beta_1, \ldots, \beta_6)$. For the average artist in any genre, we'd expect danceability to increase by between 2.16 and 3 points for every *10*-point increase on the valence scale – a statistically significant but fairly marginal bump. Among genres, it appears that when controlling for valence, only `rock` is significantly less danceable than `edm`. Its 80% credible interval is the only one to lie entirely above or below 0, suggesting that for the average artist, the typical danceability of a `rock` song is between 1.42 and 12 points lower than that of an `edm` song with the same valence.

```
tidy(spotify_model_1, effects = "fixed",
     conf.int = TRUE, conf.level = 0.80)
# A tibble: 7 x 5
  term          estimate std.error conf.low conf.high
  <chr>            <dbl>     <dbl>    <dbl>     <dbl>
1 (Intercept)     53.3       3.02    49.4      57.1
2 valence          0.258     0.0330   0.216     0.300
3 genrelatin       0.362     3.58    -4.15      5.02
4 genrepop         0.466     2.67    -2.96      3.89
5 genrer&b        -2.43      2.85    -6.05      1.22
6 genrerap         1.68      3.15    -2.29      5.68
7 genrerock       -6.73      4.14   -12.0      -1.42
```

In interpreting these summaries, keep in mind that the genre coefficients directly compare each genre to `edm` alone and not, say, `rock` to `r&b`. In contrast, `mcmc_areas()` offers a useful visual comparison of all genre posteriors. Other than the `rock` coefficient, 0 is a fairly posterior plausible value for the other genre coefficients, reaffirming that these genres aren't significantly more or less danceable than `edm`. There's also quite a bit of overlap in the posteriors. As such, though there's evidence that some of these genres are more danceable than others (e.g., `rap` vs `r&b`), the difference isn't substantial.

```
# Plot the posterior models of the genre coefficients
mcmc_areas(spotify_model_1, pars = vars(starts_with("genre")), prob = 0.8) +
  geom_vline(xintercept = 0)
```

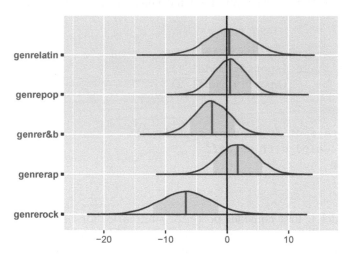

FIGURE 17.23: Posterior models for the genre-related coefficients in Spotify Model 1.

Finally, consider some posterior results for two artists in our sample: Missy Elliott and Camilo. The below `tidy()` summary compares the typical danceability levels of these artists to the average artist, $b_{0j} = \beta_{0j} - \beta_0$. When controlling for valence and genre, Elliott's songs tend to be significantly more danceable than the average artist's, whereas Camilo's tend to be less danceable. For example, there's an 80% posterior chance that Elliott's typical song danceability is between 2.78 and 14.8 points higher than average.

```
tidy(spotify_model_1, effects = "ran_vals",
    conf.int = TRUE, conf.level = 0.80) %>%
  filter(level %in% c("Camilo", "Missy_Elliott")) %>%
  select(level, estimate, conf.low, conf.high)
# A tibble: 2 x 4
  level          estimate conf.low conf.high
  <chr>             <dbl>    <dbl>     <dbl>
1 Camilo            -2.00    -7.29      3.44
2 Missy_Elliott      8.69     2.78      14.8
```

To predict the danceability of their *next* songs, our hierarchical regression model takes into consideration the artists' typical danceability levels *as well as* the song's valence and genre. Suppose that both artists' next songs have a valence score of 80, but true to their genres, Elliot's is a rap song and Camilo's is in the Latin genre. Figure 17.24 plots both artists' posterior predictive models along with that of Mohsen Beats, a `rock` artist that wasn't in our sample but also releases a song with a valence level of 80. As we'd expect, the danceability of Elliott's song is likely to be the highest among these three. Further, though Camilo's typical danceability is lower than average, we expect Mohsen Beats's song to be even less danceable since it's of the least danceable genre.

```
# Simulate posterior predictive models for the 3 artists
set.seed(84735)
predict_next_song <- posterior_predict(
  spotify_model_1,
  newdata = data.frame(
    artist = c("Camilo", "Mohsen Beats", "Missy Elliott"),
    valence = c(80, 60, 90), genre = c("latin", "rock", "rap")))

# Posterior predictive model plots
mcmc_areas(predict_next_song, prob = 0.8) +
 ggplot2::scale_y_discrete(
   labels = c("Camilo", "Mohsen Beats", "Missy Elliott"))
```

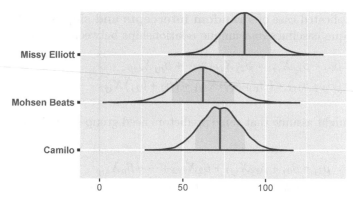

FIGURE 17.24: Posterior predictive models for the popularity of the next songs by Missy Elliott, Mohsen Beats, and Camilo.

One last thing

If you're paying close attention, you might notice a flaw in our model. It produces posterior predictions of danceability that exceed the maximum danceability level of 100. This is a consequence of using the Normal to model a response variable with a limited range (here 0 - 100). The Normal assumption was "good enough" here, but we might consider replacing it with, say, a Beta model.

17.8   Chapter summary

In Chapter 17 we explored how to build a hierarchical model of Y by predictors X when working with group structured or hierarchical data. Suppose we have multiple data points on each of m groups. For each group $j \in \{1, 2, \ldots, m\}$, let Y_{ij} and $(X_{ij1}, X_{ij2}, \ldots, X_{ijp})$ denote the ith set of observed data on response variable Y and p different predictors X. Then a Normal hierarchical regression model of Y vs X consists of three layers: it combines information about the relationship between Y and X **within** groups, with information about how these relationships vary **between** groups, with our prior understanding of the broader

global population. Letting β_j denote a set of group-specific parameters and $(\beta, \sigma_y, \sigma)$ a set of global parameters:

$$
\begin{aligned}
Y_{ij} | \beta_j, \sigma_y &\sim N(\mu_{ij}, \sigma_y^2) && \text{regression model within group } j \\
\beta_j | \beta, \sigma &\sim N(\beta, \sigma^2) && \text{variability in regression parameters between groups} \\
\beta, \sigma_y, \sigma, \dots &\sim \cdots && \text{prior models on global parameters}
\end{aligned}
$$

Where we have some choices to make is in the definition of the regression mean μ_{ij}. In the simplest case of a **random intercepts model**, we assume that groups might have unique baselines β_{0j}, yet share a common relationship between Y and X:

$$
\begin{aligned}
\mu_{ij} &= \beta_{0j} + \beta_1 X_{ij1} + \beta_2 X_{ij2} + \cdots + \beta_p X_{ijp} \\
&= (\beta_0 + b_{0j}) + \beta_1 X_{ij1} + \beta_2 X_{ij2} + \cdots + \beta_p X_{ijp}.
\end{aligned}
$$

In the most complicated case of a **random intercepts and slopes model**, we assume that groups have unique baselines *and* unique relationships between Y and each X:

$$
\begin{aligned}
\mu_{ij} &= \beta_{0j} + \beta_{1j} X_{ij1} + \beta_{2j} X_{ij2} + \cdots + \beta_{pj} X_{ijp} \\
&= (\beta_0 + b_{0j}) + (\beta_1 + b_{1j}) X_{ij1} + (\beta_2 + b_{2j}) X_{ij2} + \cdots + (\beta_p + b_{pj}) X_{ijp}.
\end{aligned}
$$

In between, we might assume that some predictors need group-specific coefficients and others don't:

$$
\begin{aligned}
\mu_{ij} &= \beta_{0j} + \beta_{1j} X_{ij1} + \beta_2 X_{ij2} + \cdots + \beta_p X_{ijp} \\
&= (\beta_0 + b_{0j}) + (\beta_1 + b_{1j}) X_{ij1} + \beta_2 X_{ij2} + \cdots + \beta_p X_{ijp}.
\end{aligned}
$$

17.9   Exercises

17.9.1   Conceptual exercises

Exercise 17.1 (Translating assumptions into model notation). To test the relationship between reaction times and sleep deprivation, researchers enlisted 3 people in a 10-day study. Let Y_{ij} denote the reaction time (in ms) to a given stimulus and X_{ij} the number of days of sleep deprivation for the ith observation on subject j. For each set of assumptions below, use mathematical notation to represent an appropriate Bayesian hierarchical model of Y_{ij} vs X_{ij}.

    a.   Not only do some people tend to react more quickly than others, sleep deprivation might impact some people's reaction times more than others.

    b.   Though some people tend to react more quickly than others, the impact of sleep deprivation on reaction time is the same for all.

    c.   Nobody has inherently faster reaction times, though sleep deprivation might impact some people's reaction times more than others.

Exercise 17.2 (Sketch the assumption: Part 1). Continuing with the sleep study, suppose we model the relationship between reaction time Y_{ij} and days of sleep deprivation X_{ij} using the random intercepts model (17.5).

a)   Sketch a plot of data that we might see if $\sigma_y > \sigma_0$.
b)   Explain what $\sigma_y > \sigma_0$ would mean in the context of the sleep study.
c)   Repeat part a assuming $\sigma_y < \sigma_0$.
d)   Repeat part b assuming $\sigma_y < \sigma_0$.

Exercise 17.3 (Sketch the assumption: Part 2). Suppose instead that we model the relationship between reaction time Y_{ij} and days of sleep deprivation X_{ij} using the random intercepts and slopes model (17.12) (with different priors).

a)   Sketch a plot of subject-specific trends that we might see if the correlation parameter were *positive* $\rho > 0$.
b)   Explain what $\rho > 0$ would mean in the context of the sleep study.
c)   Repeat part a for $\rho < 0$.
d)   Repeat part b for $\rho < 0$.

Exercise 17.4 (Making meaning of models). To study the relationship between weight and height among pug puppies, you collect data on 10 different litters, each containing 4 to 6 puppies born to the same mother. Let Y_{ij} and X_{ij} denote the weight and height, respectively, for puppy i in litter j.

a)   Write out formal model notation for model 1, a *random intercepts* model of Y_{ij} vs X_{ij}.
b)   Write out formal model notation for model 2, a *random intercepts and slopes* model of Y_{ij} vs X_{ij}.
c)   Summarize the key differences in the assumptions behind models 1 and 2. Root this discussion in the puppy context.

Exercise 17.5 (Translating models to code). Suppose we had `weight` and `height` data for the puppy study. Write out appropriate `stan_glmer()` model code for models 1 and 2 from Exercise 17.4.

17.9.2   Applied exercises

Exercise 17.6 (Sleep: setting up the model). In the above conceptual exercises, you considered a sleep study of the relationship between reaction time and the number of days of sleep deprivation. You will explore this relationship in more depth here. To this end, suppose researchers tell us that on a typical day, the average person should have a reaction time of roughly 250ms to the stimulus used in the sleep study. Beyond this baseline, we'll balance weakly informative priors with the `sleepstudy` data from the **lme4** package to better understand reaction times. Specifically, consider two possible models as expressed by `stan_glmer()` syntax:

model	formula
1	`Reaction ~ Days + (1 | Subject)`
2	`Reaction ~ Days + (Days | Subject)`

a)   What's the grouping variable in the `sleepstudy` data and why is it important to incorporate this grouping structure into our analysis?

b)   Use formal notation to define the hierarchical regression structure of models 1 and 2. (You will tune the priors in the next exercise.)

c)   Summarize the key differences between models 1 and 2. Root this discussion in the sleep study.

d)   Using the `sleepstudy` data, construct and discuss a plot that helps you explore which model is more appropriate: 1 or 2.

Exercise 17.7 (Sleep: simulating the model). Continuing with the sleep analysis, let's simulate and dig into the hierarchical posteriors.

a)   Simulate the posteriors of models 1 and 2. Remember to use a baseline reaction time of 250ms, and weakly informative priors otherwise.

b)   For model 2, report the global posterior median regression model of `Reaction` time.

c)   For model 2, construct and interpret 80% credible intervals for the `Days` regression coefficient.

d)   For model 2, calculate and interpret the posterior medians of σ_y, σ_0, σ_1, and ρ.

Exercise 17.8 (Sleep: group-specific inference). Next, let's dig into what Model 2 indicates about the individuals that participated in the sleep study.

a)   Use your posterior simulation to identify the person for whom reaction time changes the *least* with sleep deprivation. Write out their posterior median regression model.

b)   Repeat part a, this time for the person for whom reaction time changes the *most* with sleep deprivation.

c)   Use your posterior simulation to identify the person that has the slowest baseline reaction time. Write out their posterior median regression model.

d)   Repeat part c, this time for the person that has the fastest baseline reaction time.

e)   Simulate, plot, and discuss the posterior predictive model of reaction time after 5 days of sleep deprivation for two subjects: you and Subject 308. You're encouraged to try this from scratch before relying on the `posterior_predict()` shortcut.

Exercise 17.9 (Sleep: Which model?).

a)   Evaluate the two models of reaction time: Are they wrong? Are they fair? How accurate are their posterior predictions?

b)   Which of the two models do you prefer and what does this indicate about the relationship between reaction time and sleep deprivation? Justify your answer with posterior evidence.

Exercise 17.10 (Voices: setting up the model). Does one's voice pitch change depending on attitude? To address this question, Winter and Grawunder (2012) conducted a study in which each subject participated in various role-playing dialogs. These dialogs spanned different contexts (e.g., asking for a favor) and were approached with different attitudes (polite vs informal). In the next exercises you'll explore a hierarchical regression analysis of Y_{ij}, the average voice pitch in subject j's ith dialog session (measured in Hz), by X_{ij}, whether or not the dialog was polite (vs informal). Beyond a baseline understanding that the typical voice pitch is around 200 Hz, you should utilize weakly informative priors.

a) Using formal notation, define the hierarchical regression model of Y_{ij} vs X_{ij1}. In doing so, assume that baseline voice pitch differs from subject to subject, but that the impact of attitude on voice pitch is similar among all subjects.

b) Compare and contrast the meanings of model parameters β_{0j} and β_0 in the context of this voice pitch study. NOTE: Remember that X_{ij} is a categorical indicator variable.

c) Compare and contrast the meanings of model parameters σ_y and σ_0 in the context of this voice pitch study.

Exercise 17.11 (Voices: check out some data). To balance our weakly informative priors for the model of pitch by attitude, check out some data.

a) Load the **voices** data from the **bayesrules** package. How many study subjects are included in this sample? In how many dialogs did each subject participate?

b) Construct and discuss a plot which illustrates the relationship between voice pitch and attitude both within and between subjects.

Exercise 17.12 (Voices: simulating the model). Continuing with the voice pitch analysis, in this exercise you will simulate and dig into the hierarchical posterior of your model parameters.

a) Simulate the hierarchical posterior model of voice pitch by attitude. Construct and discuss trace plots, density plots, autocorrelation plots, and a `pp_check()` of the chain output.

b) Construct and interpret a 95% credible interval for β_0.

c) Construct and interpret a 95% credible interval for β_1.

d) Is there ample evidence that, for the average subject, voice pitch differs depending on attitude (polite vs informal)? Explain.

Exercise 17.13 (Voices: focusing on the individual). Continuing with the voice pitch analysis, in this exercise you will focus on specific subjects.

a) Report the global posterior median model of the relationship between voice pitch and attitude.

b) Report and contrast the posterior median models for two subjects in our data: A and F.

c) Using `posterior_predict()`, simulate posterior predictive models of voice pitch in a *new* polite dialog for three different subjects: A, F, and *you*. Illustrate your simulation results using `mcmc_areas()` and discuss your findings.

17.9.3 Open-ended exercises

Exercise 17.14 (Open exercise: coffee). What makes some coffee beans more winning than others? In this open-ended exercise, construct, interpret, and evaluate a hierarchical regression model of the professional rating or "total cup points" awarded to a batch of coffee beans. Do so using the `coffee_ratings_small` data from the **bayesrules** package which contains information for multiple batches of coffee beans from each of 27 farms. *You* get to choose which predictors to use.

Exercise 17.15 (Sleep: different priors). In our earlier sleep analysis, we utilized weakly informative priors. Pretending that you haven't already seen the data, specify a model of `Reaction` time by `Days` of sleep deprivation using priors that you tune yourself. Use prior simulation to illustrate your prior understanding.

18

Non-Normal Hierarchical Regression & Classification

A master chef becomes a master chef by mastering the basic elements of cooking, from flavor to texture to smell. When cooking then, they can combine these elements without relying on rigid step-by-step cookbook directions. Similarly, in building statistical models, Bayesian *or* frequentist, there is no rule book to follow. Rather, it's important to familiarize ourselves with some basic modeling building blocks and develop the ability to use these in different combinations to suit the task at hand. With this, in Chapter 18 you will practice cooking up new models from the ingredients you already have. To focus on the new concepts in this chapter, we'll utilize weakly informative priors throughout. Please review Chapters 12 and 13 for a refresher on tuning prior models in the Poisson and logistic regression settings, respectively. The same ideas apply here.

> ◎ **Goals**
>
> Expand our **generalized hierarchical regression model** toolkit by combining
> - hierarchical regression techniques (Chapter 17) with
> - Poisson and Negative Binomial regression models for count response variables Y (Chapter 12) and logistic regression models for binary categorical response variables Y (Chapter 13).

```
# Load packages
library(bayesrules)
library(tidyverse)
library(bayesplot)
library(rstanarm)
library(tidybayes)
library(broom.mixed)
library(janitor)
```

18.1 Hierarchical logistic regression

Whether for the thrill of thin air, a challenge, or the outdoors, mountain climbers set out to summit great heights in the majestic Nepali Himalaya. Success is not guaranteed – poor weather, faulty equipment, injury, or simply bad luck, mean that not all climbers reach their destination. This raises some questions. What's the probability that a mountain climber makes it to the top? What factors might contribute to a higher success rate? Beyond a *vague* sense that the typical climber might have a 50/50 chance at success, we'll balance our weakly informative prior understanding of these questions with the `climbers_sub` data in

DOI: 10.1201/9780429288340-18

the **bayesrules** package, a mere *subset* of data made available by The Himalayan Database (2020) and distributed through the **#tidytuesday** project (R for Data Science, 2020b):

```
# Import, rename, & clean data
data(climbers_sub)
climbers <- climbers_sub %>%
  select(expedition_id, member_id, success, year, season,
         age, expedition_role, oxygen_used)
```

This dataset includes the outcomes for 2076 climbers, dating back to 1978. Among them, only 38.87% successfully summited their peak:

```
nrow(climbers)
[1] 2076
climbers %>%
  tabyl(success)
 success    n percent
   FALSE 1269  0.6113
    TRUE  807  0.3887
```

---

❷ Quiz Yourself!

As you might imagine given its placement in this book, the `climbers` data has an underlying grouping structure. Identify which of the following variables encodes that grouping structure: `expedition_id`, `member_id`, `season`, `expedition_role`, or `oxygen_used`.

Since `member_id` is essentially a row (or climber) identifier and we only have one observation per climber, this is *not* a grouping variable. Further, though `season`, `expedition_role`, and `oxygen_used` each have categorical levels which we observe more than once, these are potential *predictors* of `success`, not grouping variables.[1] This leaves `expedition_id` – this *is* a grouping variable. The `climbers` data spans 200 different *expeditions*:

```
# Size per expedition
climbers_per_expedition <- climbers %>%
  group_by(expedition_id) %>%
  summarize(count = n())

# Number of expeditions
nrow(climbers_per_expedition)
[1] 200
```

---

[1] For example, the observed `season` categories (`Autumn`, `Spring`, `Summer`, `Winter`) are a fixed and complete set of options, not a random sample of categories from a broader population of seasons.

Each expedition consists of multiple climbers. For example, our first three expeditions set out with 5, 6, and 12 climbers, respectively:

```
climbers_per_expedition %>%
  head(3)
# A tibble: 3 x 2
  expedition_id count
  <chr>         <int>
1 AMAD03107         5
2 AMAD03327         6
3 AMAD05338        12
```

It would be a *mistake* to ignore this grouping structure and otherwise assume that our individual climber outcomes are independent. Since each expedition works as a *team*, the success or failure of one climber in that expedition depends in part on the success or failure of another. Further, all members of an expedition start out with the same destination, with the same leaders, and under the same weather conditions, and thus are subject to the same external factors of success. Beyond it being the right thing to do then, accounting for the data's grouping structure can also illuminate the degree to which these factors introduce variability in the success rates *between* expeditions. To this end, notice that more than 75 of our 200 expeditions had a 0% success rate – i.e., *no* climber in these expeditions successfully summited their peak. In contrast, nearly 20 expeditions had a 100% success rate. In between these extremes, there's quite a bit of variability in expedition success rates.

```
# Calculate the success rate for each exhibition
expedition_success <- climbers %>%
  group_by(expedition_id) %>%
  summarize(success_rate = mean(success))

# Plot the success rates across exhibitions
ggplot(expedition_success, aes(x = success_rate)) +
  geom_histogram(color = "white")
```

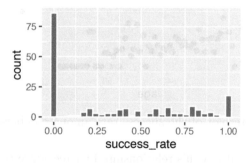

FIGURE 18.1: A histogram of the success rates for the 200 climbing expeditions.

18.1.1   Model building & simulation

Reflecting the grouped nature of our data, let Y_{ij} denote whether or not climber i in expedition j successfully summits their peak:

$$Y_{ij} = \begin{cases} 1 & \text{yes} \\ 0 & \text{no} \end{cases}$$

There are several potential predictors of climber success in our dataset. We'll consider only two: a climber's age and whether they received supplemental *oxygen* in order to breathe more easily at high elevation. As such, define:

$$X_{ij1} = \text{age of climber } i \text{ in expedition } j$$
$$X_{ij2} = \text{whether or not climber } i \text{ in expedition } j \text{ received oxygen.}$$

By calculating the proportion of success at each age and oxygen use combination, we get a sense of how these factors are related to climber success (albeit a wobbly sense given the small sample sizes of some combinations). In short, it appears that climber success decreases with age and drastically increases with the use of oxygen:

```
# Calculate the success rate by age and oxygen use
data_by_age_oxygen <- climbers %>%
  group_by(age, oxygen_used) %>%
  summarize(success_rate = mean(success))

# Plot this relationship
ggplot(data_by_age_oxygen, aes(x = age, y = success_rate,
                               color = oxygen_used)) +
  geom_point()
```

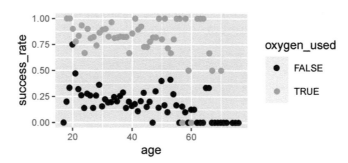

FIGURE 18.2: A scatterplot of the success rate among climbers by age and oxygen use.

In building a Bayesian model of this relationship, first recognize that the Bernoulli model is reasonable for our binary response variable Y_{ij}. Letting π_{ij} be the *probability* that climber i in expedition j successfully summits their peak, i.e., that $Y_{ij} = 1$,

$$Y_{ij}|\pi_{ij} \sim \text{Bern}(\pi_{ij}).$$

Way back in Chapter 13, we explored a **complete pooling** approach to expanding this simple model into a **logistic regression model** of Y by a set of predictors X:

$$Y_{ij}|\beta_0,\beta_1,\beta_2 \overset{ind}{\sim} \text{Bern}(\pi_{ij}) \quad \text{with} \quad \log\left(\frac{\pi_{ij}}{1-\pi_{ij}}\right) = \beta_0 + \beta_1 X_{ij1} + \beta_2 X_{ij2}$$

$$\beta_{0c} \sim N\left(m_0, s_0^2\right)$$
$$\beta_1 \sim N\left(m_1, s_1^2\right)$$
$$\beta_2 \sim N\left(m_2, s_2^2\right).$$

This is a great start, BUT it doesn't account for the grouping structure of our data. Instead, consider the following hierarchical alternative with independent, weakly informative priors tuned below by `stan_glmer()` and with a prior model for β_0 expressed through the centered intercept β_{0c}. After all, it makes more sense to think about the baseline success rate among the *typical* climber, β_{0c}, than among 0-year-old climbers that don't use oxygen, β_0. To this end, we started our analysis with a weak understanding that the typical climber has a 0.5 probability of success, or a log(odds of success) = 0.

$$
\begin{aligned}
Y_{ij}|\beta_{0j},\beta_1,\beta_2 &\sim \text{Bern}(\pi_{ij}) &\text{(model within expedition } j\text{)}\\
\text{with} \quad \log\left(\tfrac{\pi_{ij}}{1-\pi_{ij}}\right) &= \beta_{0j} + \beta_1 X_{ij1} + \beta_2 X_{ij2}\\
\beta_{0j}|\beta_0,\sigma_0 &\overset{ind}{\sim} N(\beta_0,\sigma_0^2) &\text{(variability between expeditions)}\\
\beta_{0c} &\sim N\left(0, 2.5^2\right) &\text{(global priors)}\\
\beta_1 &\sim N\left(0, 0.24^2\right)\\
\beta_2 &\sim N\left(0, 5.51^2\right)\\
\sigma_0 &\sim \text{Exp}(1).
\end{aligned}
$$

$$(18.1)$$

Equivalently, we can reframe this **random intercepts logistic regression model** by expressing the expedition-specific intercepts as *tweaks* to the global intercept,

$$\log\left(\frac{\pi_{ij}}{1-\pi_{ij}}\right) = (\beta_0 + b_{0j}) + \beta_1 X_{ij1} + \beta_2 X_{ij2}$$

where $b_{0j}|\sigma_0 \overset{ind}{\sim} N(0,\sigma_0^2)$. Consider the meaning of, and assumptions behind, the model parameters:

- The **expedition-specific** intercepts β_{0j} describe the underlying success rates, as measured by the log(odds of success), for each expedition j. These acknowledge that some expeditions are inherently more successful than others.

- The expedition-specific intercepts β_{0j} are assumed to be Normally distributed around some **global** intercept β_0 with standard deviation σ_0. Thus, β_0 describes the *typical* baseline success rate across all expeditions, and σ_0 captures the **between-group variability** in success rates from expedition to expedition.

- β_1 describes the **global** relationship between success and age when controlling for oxygen use. Similarly, β_2 describes the global relationship between success and oxygen use when controlling for age.

Putting this all together, our random intercepts logistic regression model (18.1) makes the simplifying (but we think reasonable) assumption that expeditions might have *unique* intercepts β_{0j} but share *common* regression parameters β_1 and β_2. In plain language, though the *underlying success rates* might differ from expedition to expedition, being younger or using oxygen aren't more beneficial in one expedition than in another.

To simulate the model posterior, the `stan_glmer()` code below combines the best of two worlds: `family = binomial` specifies that ours is a *logistic* regression model (à la Chapter 13) and the `(1 | expedition_id)` term in the model formula incorporates our hierarchical grouping structure (à la Chapter 17):

```
climb_model <- stan_glmer(
  success ~ age + oxygen_used + (1 | expedition_id),
  data = climbers, family = binomial,
  prior_intercept = normal(0, 2.5, autoscale = TRUE),
  prior = normal(0, 2.5, autoscale = TRUE),
  prior_covariance = decov(reg = 1, conc = 1, shape = 1, scale = 1),
  chains = 4, iter = 5000*2, seed = 84735
)
```

You're encouraged to follow this simulation with a confirmation of the prior specifications and some MCMC diagnostics:

```
# Confirm prior specifications
prior_summary(climb_model)

# MCMC diagnostics
mcmc_trace(climb_model, size = 0.1)
mcmc_dens_overlay(climb_model)
mcmc_acf(climb_model)
neff_ratio(climb_model)
rhat(climb_model)
```

Whereas these diagnostics confirm that our MCMC simulation is on the right track, a **posterior predictive check** indicates that our *model* is on the right track. From each of 100 posterior simulated datasets, we record the proportion of climbers that were successful using the `success_rate()` function. These success rates range from roughly 37% to 41%, in a tight window around the actual observed 38.9% success rate in the `climbers` data.

```
# Define success rate function
success_rate <- function(x){mean(x == 1)}

# Posterior predictive check
pp_check(climb_model, nreps = 100,
         plotfun = "stat", stat = "success_rate") +
  xlab("success rate")
```

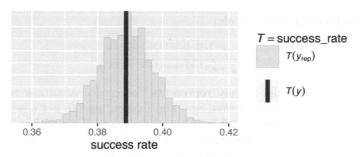

FIGURE 18.3: A posterior predictive check of the hierarchical logistic regression model of climbing success. The histogram displays the proportion of climbers that were successful in each of 100 posterior simulated datasets. The vertical line represents the observed proportion of climbers that were successful in the climbers data.

18.1.2   Posterior analysis

In our posterior analysis of mountain climber success, let's focus on the global. Beyond being comforted by the fact that we're correctly accounting for the grouping structure of our data, we aren't interested in any *particular* expedition. With that, below are some posterior summaries for our global regression parameters β_0, β_1, and β_2:

```
tidy(climb_model, effects = "fixed", conf.int = TRUE, conf.level = 0.80)
# A tibble: 3 x 5
  term           estimate std.error conf.low conf.high
  <chr>             <dbl>     <dbl>    <dbl>     <dbl>
1 (Intercept)      -1.42     0.479    -2.04    -0.822
2 age              -0.0474   0.00910  -0.0594  -0.0358
3 oxygen_usedTRUE   5.79     0.478     5.20     6.43
```

To begin, notice that the 80% posterior credible interval for **age** coefficient β_1 is comfortably below 0. Thus, we have *significant* posterior evidence that, when controlling for whether or not a climber uses oxygen, the likelihood of success decreases with age. More specifically, translating the information in β_1 from the *log*(odds) to the *odds* scale, there's an 80% chance that the *odds* of successfully summiting drop somewhere between 3.5% and 5.8% for every extra year in age: $(e^{-0.0594}, e^{-0.0358}) = (0.942, 0.965)$. Similarly, the 80% posterior credible interval for the **oxygen_usedTRUE** coefficient β_2 provides *significant* posterior evidence that, when controlling for age, the use of oxygen dramatically increases a climber's likelihood of summiting the peak. There's an 80% chance that the use of oxygen could correspond to anywhere between a 182-fold increase to a 617-fold increase in the odds of success: $(e^{5.2}, e^{6.43}) = (182, 617)$. Oxygen please!

Combining our observations on β_1 and β_2, the posterior median model of the relationship between climbers' log(odds of success) and their age (X_1) and oxygen use (X_2) is

$$\log\left(\frac{\pi}{1-\pi}\right) = -1.42 - 0.0474X_1 + 5.79X_2.$$

Or, on the probability of success scale:

$$\pi = \frac{e^{-1.42-0.0474X_1+5.79X_2}}{1 + e^{-1.42-0.0474X_1+5.79X_2}}.$$

This posterior median model merely represents the *center* among a *range* of posterior plausible relationships between success, age, and oxygen use. To get a sense for this range, Figure 18.4 plots 100 posterior plausible alternative models. Both with oxygen and without, the probability of success decreases with age. Further, at any given age, the probability of success is *drastically* higher when climbers use oxygen. However, our posterior certainty in these trends varies quite a bit by age. We have *much* less certainty about the success rate for older climbers on oxygen than for younger climbers on oxygen, for whom the success rate is uniformly high. Similarly, but less drastically, we have less certainty about the success rate for younger climbers who don't use oxygen than for older climbers who don't use oxygen, for whom the success rate is uniformly low.

```
climbers %>%
  add_fitted_draws(climb_model, n = 100, re_formula = NA) %>%
  ggplot(aes(x = age, y = success, color = oxygen_used)) +
    geom_line(aes(y = .value, group = paste(oxygen_used, .draw)),
              alpha = 0.1) +
    labs(y = "probability of success")
```

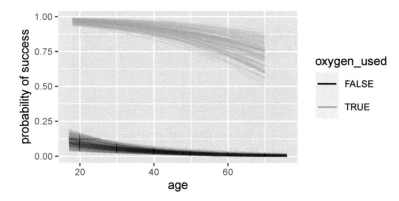

FIGURE 18.4: 100 posterior plausible models for the probability of climbing success by age and oxygen use.

18.1.3   Posterior classification

Suppose four climbers set out on a new expedition. Two are 20 years old and two are 60 years old. Among both age pairs, one climber plans to use oxygen and the other does not:

```
# New expedition
new_expedition <- data.frame(
  age = c(20, 20, 60, 60), oxygen_used = c(FALSE, TRUE, FALSE, TRUE),
  expedition_id = rep("new", 4))
new_expedition
  age oxygen_used expedition_id
```

```
1   20      FALSE           new
2   20      TRUE            new
3   60      FALSE           new
4   60      TRUE            new
```

Naturally, they each want to know the probability that they'll reach their summit. For a reminder on how to simulate a posterior predictive model for a *new* group from scratch, please review Section 16.5. Here we jump straight to the `posterior_predict()` shortcut function to simulate 20,000 0-or-1 posterior predictions for each of our 4 new climbers:

```
# Posterior predictions of binary outcome
set.seed(84735)
binary_prediction <- posterior_predict(climb_model, newdata = new_expedition)
```

```
# First 3 prediction sets
head(binary_prediction, 3)
      1 2 3 4
[1,]  0 1 0 1
[2,]  0 1 0 1
[3,]  0 1 0 1
```

For each climber, the probability of success is approximated by the observed proportion of success among their 20,000 posterior predictions. Since these probabilities incorporate uncertainty in the baseline success rate of the new expedition, they are more moderate than the global trends observed in Figure 18.4:

```
# Summarize the posterior predictions of Y
colMeans(binary_prediction)
     1      2      3      4
0.2788 0.8040 0.1440 0.6515
```

These posterior predictions provide more insight into the connections between age, oxygen, and success. For example, our posterior prediction is that climber 1, who is 20 years old and does *not* plan to use oxygen, has a 27.88% chance of summiting the peak. This probability is naturally lower than for climber 2, who is also 20 but *does* plan to use oxygen. It's also higher than the posterior prediction of success for climber 3, who also doesn't plan to use oxygen but is 60 years old. Overall, the posterior prediction of success is *highest* for climber 2, who is younger and plans to use oxygen, and *lowest* for climber 3, who is older and doesn't plan to use oxygen.

In Chapter 13 we discussed the option of turning such posterior *probability* predictions into **posterior classifications** of binary outcomes: yes or no, do we anticipate that the climber will succeed or not? If we used a simple 0.5 posterior probability cut-off to make this determination, we would recommend that climbers 1 and 3 *not* join the expedition (at least, not without oxygen) and give climbers 2 and 4 the go ahead. Yet in this particular context, we should probably leave it up to individual climbers to interpret their own results and make their own yes-or-no decisions about whether to continue on their expedition. For example, a 65.16% chance of success might be worth the hassle and risk to some but not to others.

18.1.4   Model evaluation

To conclude our climbing analysis, let's ask: Is our hierarchical logistic model a *good* model? Long story short, the answer is *yes*. First, our model is **fair**. The data we used are part of public record and we do not foresee our analysis having any negative impact on individuals or society. (Again, boring answers to the question of fairness are the best kind.) Second, our posterior predictive check in Figure 18.3 demonstrated that our model **doesn't seem too wrong** – our posterior simulated success rates hover around the observed success rate in our data. Finally, to the question of **posterior classification accuracy**, we can compare our posterior classifications of success to the actual outcomes for the 2076 climbers in our dataset. By default, let's start out with a **probability cut-off** of 0.5 – if a climber's *probability* of success exceeds 0.5, we'll predict that they'll succeed. We implement and evaluate this classification rule using `classification_summary()` below.

```
set.seed(84735)
classification_summary(data = climbers, model = climb_model, cutoff = 0.5)
```

```
$confusion_matrix
      y    0   1
FALSE 1174  95
 TRUE   77 730
```

```
$accuracy_rates

sensitivity       0.9046
specificity       0.9251
overall_accuracy  0.9171
```

Overall, under this classification rule, our model successfully predicts the outcomes for 91.71% of our climbers. This is pretty fantastic given that we're only utilizing information on the climbers' ages and oxygen use, among many possible other considerations (e.g., destination, season, etc.). Yet given the consequences of misclassification in this particular context (e.g., risk of injury), we should prioritize **specificity**, our ability to anticipate when a climber might *not* succeed. To this end, our model correctly predicted only 92.51% of the climbing failures. To increase this rate, we can change the probability cut-off in our classification rule.

❓ Quiz Yourself!

What cut-off can we utilize to achieve a specificity of at least 95% while also maintaining the highest possible sensitivity?

In general, to increase specificity, we can *increase* the probability cut-off, thereby making it more *difficult* to predict "success." After some trial and error, it seems that cut-offs of roughly 0.65 or higher will achieve a desired 95% specificity level. This switch to 0.65 naturally decreases the sensitivity of our posterior classifications, from 90.46% to 81.54%, and thus our ability to detect when a climber *will* be successful. We think the added caution is worth it.

```
set.seed(84735)
classification_summary(data = climbers, model = climb_model, cutoff = 0.65)

$confusion_matrix
      y    0    1
  FALSE 1214   55
   TRUE  149  658

$accuracy_rates

sensitivity      0.8154
specificity      0.9567
overall_accuracy 0.9017
```

18.2 Hierarchical Poisson & Negative Binomial regression

Vacation rental services, such as AirBnB, offer travelers alternatives to hotel rooms. The AirBnB inventory and price range are wide, leading us to ask: why do some listings have more reviews (hence presumably more guests) than others? Beyond a *vague* baseline understanding that the typical listing has around 20 reviews, we're unsure of the dynamics in the AirBnB market, and thus will otherwise utilize weakly informative priors. We'll balance these priors by the `airbnb` data in the `bayesrules` package, originally collated by Trinh and Ameri (2016) and distributed by Legler and Roback (2021). This dataset contains information on 1561 listings across 43 Chicago neighborhoods, and hence *multiple listings per neighborhood*:

```
# Load data
data(airbnb)

# Number of listings
nrow(airbnb)
[1] 1561

# Number of neighborhoods
airbnb %>%
  summarize(nlevels(neighborhood))
  nlevels(neighborhood)
1                    43
```

We're not interested in these neighborhoods in particular. Rather, we'd like to use these data on these neighborhoods to learn about the broader AirBnB market. However, since listings within the same neighborhood share many features (e.g., similar location, amenities, public transit), it would be a mistake to assume their independence. Thus, to reflect and study the variability in the number of AirBnB `reviews` *between* and *within* neighborhoods, we'll incorporate the `neighborhood` grouping structure in our analysis.

474   18 Non-Normal Hierarchical Regression & Classification

18.2.1   Model building & simulation

In exploring why the number of `reviews` varies from AirBnB listing to listing, we'll consider two possible factors: its overall user `rating` (on a 1 to 5 scale) and the privacy allotted by its `room_type`, i.e., whether the renter gets an entire private unit, gets a private room within a shared unit, or shares a room. More specifically, for listing i in neighborhood j, we will model response variable

$$Y_{ij} = \text{number of reviews}$$

by visitor rating X_{ij1} and room type, where an entire private unit is the reference level and we have indicator variables for the two other room types:

$$X_{ij2} = \begin{cases} 1 & \text{private room} \\ 0 & \text{otherwise} \end{cases} \quad \text{and} \quad X_{ij3} = \begin{cases} 1 & \text{shared room} \\ 0 & \text{otherwise} \end{cases}$$

Figure 18.5 displays the trends in the number of reviews as well as their relationship with a listing's rating and room type. In examining the variability in Y_{ij} alone, note that the majority of listings have fewer than 20 reviews, though there's a long right skew. Further, the volume of reviews tends to increase with ratings and privacy levels.

```
ggplot(airbnb, aes(x = reviews)) +
  geom_histogram(color = "white", breaks = seq(0, 200, by = 10))
ggplot(airbnb, aes(y = reviews, x = rating)) +
  geom_jitter()
ggplot(airbnb, aes(y = reviews, x = room_type)) +
  geom_violin()
```

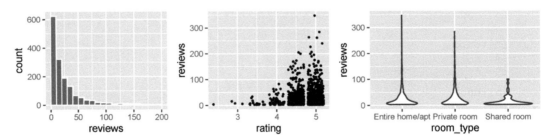

FIGURE 18.5: Plots of the number of reviews received across AirBnB listings (left) as well as the relationship of the number of reviews with a listing's rating (middle) and room type (right).

We can further break down these dynamics within each neighborhood. We show just three here to conserve precious space: Albany Park (a residential neighborhood in northern Chicago), East Garfield Park (a residential neighborhood in central Chicago), and The Loop (a commercial district and tourist destination). In general, notice that Albany Park listings tend to have fewer reviews, no matter their rating or room type.

```
airbnb %>%
  filter(neighborhood %in%
          c("Albany Park", "East Garfield Park", "The Loop")) %>%
  ggplot(aes(y = reviews, x = rating, color = room_type)) +
    geom_jitter() +
    facet_wrap(~ neighborhood)
```

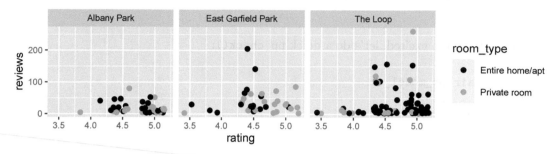

FIGURE 18.6: A scatterplot of an AirBnB listing's number of reviews by its rating and room type, for three neighborhoods.

In building a regression model for the number of reviews, the first step is to consider reasonable probability models for data Y_{ij}. Since the Y_{ij} values are *non-negative skewed counts*, a **Poisson** model is a good starting point. Specifically, letting *rate* λ_{ij} denote the expected number of reviews received by listing i in neighborhood j,

$$Y_{ij}|\lambda_{ij} \sim \text{Pois}(\lambda_{ij}).$$

The **hierarchical Poisson regression model** below builds this out to incorporate (1) the rating and room type predictors (X_{ij1}, X_{ij2}, X_{ij3}) and (2) the `airbnb` data's grouped structure. Beyond a general understanding that the typical listing has around 20 reviews (hence $\log(20) \approx 3$ logged reviews), this model utilizes independent, weakly informative priors tuned by `stan_glmer()`:

$$
\begin{aligned}
Y_{ij}|\beta_{0j},\beta_1,\beta_2,\beta_3 &\sim \text{Pois}(\lambda_{ij}) \quad \text{with} \quad \log(\lambda_{ij}) = \beta_{0j} + \beta_1 X_{ij1} + \beta_2 X_{ij2} + \beta_3 X_{ij3} \\
\beta_{0j}|\beta_0,\sigma_0 &\stackrel{ind}{\sim} N(\beta_0,\sigma_0^2) \\
\beta_{0c} &\sim N\left(3, 2.5^2\right) \\
\beta_1 &\sim N\left(0, 7.37^2\right) \\
\beta_2 &\sim N\left(0, 5.04^2\right) \\
\beta_3 &\sim N\left(0, 14.19^2\right) \\
\sigma_0 &\sim \text{Exp}(1).
\end{aligned}
\tag{18.2}
$$

Taking a closer look, this model assumes that neighborhoods might have unique intercepts β_{0j} but share common regression parameters (β_1,β_2,β_3). In plain language: though some neighborhoods might be more popular AirBnB destinations than others (hence their listings tend to have more reviews), the *relationship* of reviews with rating and room type is the same for each neighborhood. For instance, ratings aren't more influential to reviews in one neighborhood than in another. This assumption greatly simplifies our analysis while still accounting for the grouping structure in the data. To simulate the posterior, we specify our `family = poisson` data structure and incorporate the neighborhood-level grouping structure through (1 | neighborhood):

```
airbnb_model_1 <- stan_glmer(
  reviews ~ rating + room_type + (1 | neighborhood),
  data = airbnb, family = poisson,
  prior_intercept = normal(3, 2.5, autoscale = TRUE),
  prior = normal(0, 2.5, autoscale = TRUE),
  prior_covariance = decov(reg = 1, conc = 1, shape = 1, scale = 1),
  chains = 4, iter = 5000*2, seed = 84735
)
```

Before getting too excited, let's do a quick `pp_check()`:

```
pp_check(airbnb_model_1) +
  xlim(0, 200) +
  xlab("reviews")
```

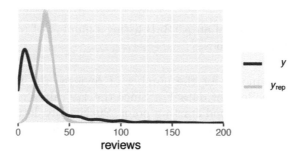

FIGURE 18.7: A posterior predictive check of the AirBnB Poisson regression model.

Figure 18.7 indicates that our hierarchical Poisson regression model significantly underestimates the variability in reviews from listing to listing, while overestimating the typical number of reviews. We've been here before! Recall from Chapter 12 that an underlying Poisson regression assumption is that, at any set of predictor values, the *average* number of reviews is equal to the *variance* in reviews:

$$E(Y_{ij}) = \text{Var}(Y_{ij}) = \lambda_{ij}.$$

The `pp_check()` calls this assumption into question. To address the apparent **overdispersion** in the Y_{ij} values, we swap out the Poisson model in (18.2) for the more flexible Negative Binomial model, picking up the additional reciprocal dispersion parameter $r > 0$:

$$
\begin{aligned}
Y_{ij}|\beta_{0j}, \beta_1, \beta_2, \beta_3, r &\sim \text{NegBin}(\mu_{ij}, r) \quad \text{with} \quad \log(\mu_{ij}) = \beta_{0j} + \beta_1 X_{ij1} + \beta_2 X_{ij2} + \beta_3 X_{ij3} \\
\beta_{0j}|\beta_0, \sigma_0 &\stackrel{ind}{\sim} N(\beta_0, \sigma_0^2) \\
\beta_{0c} &\sim N\left(3, 2.5^2\right) \\
\beta_1 &\sim N\left(0, 7.37^2\right) \\
\beta_2 &\sim N\left(0, 5.04^2\right) \\
\beta_3 &\sim N\left(0, 14.19^2\right) \\
r &\sim \text{Exp}(1) \\
\sigma_0 &\sim \text{Exp}(1)
\end{aligned}
$$

$$(18.3)$$

Equivalently, we can express the random intercepts as tweaks to the global intercept,

$$\log(\mu_{ij}) = (\beta_0 + b_{0j}) + \beta_1 X_{ij1} + \beta_2 X_{ij2} + \beta_3 X_{ij3}$$

where $b_{0j}|\sigma_0 \overset{ind}{\sim} N(0, \sigma_0^2)$. To simulate the posterior of the **hierarchical Negative Binomial regression model**, we can swap out `family = poisson` for `family = neg_binomial_2`:

```
airbnb_model_2 <- stan_glmer(
  reviews ~ rating + room_type + (1 | neighborhood),
  data = airbnb, family = neg_binomial_2,
  prior_intercept = normal(3, 2.5, autoscale = TRUE),
  prior = normal(0, 2.5, autoscale = TRUE),
  prior_aux = exponential(1, autoscale = TRUE),
  prior_covariance = decov(reg = 1, conc = 1, shape = 1, scale = 1),
  chains = 4, iter = 5000*2, seed = 84735
)
```

Though not perfect, the Negative Binomial model does a *much* better job of capturing the behavior in reviews from listing to listing:

```
pp_check(airbnb_model_2) +
  xlim(0, 200) +
  xlab("reviews")
```

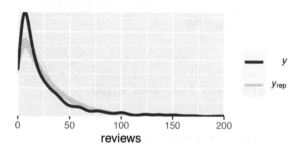

FIGURE 18.8: A posterior predictive check of the Negative Binomial regression model of AirBnB reviews.

18.2.2 Posterior analysis

Let's now make meaning of our posterior simulation results, beginning with the global relationship of reviews with ratings and room type, $\log(\lambda_{ij}) = \beta_0 + \beta_1 X_{ij1} + \beta_2 X_{ij2} + \beta_3 X_{ij3}$, or:

$$\lambda_{ij} = e^{\beta_0 + \beta_1 X_{ij1} + \beta_2 X_{ij2} + \beta_3 X_{ij3}}.$$

Below are posterior summaries of our global parameters $(\beta_0, \beta_1, \beta_2, \beta_3)$. The posterior model of β_1 reflects a significant and substantive positive association between reviews and rating. When controlling for room type, there's an 80% chance that the volume of reviews increases somewhere between 1.17 and 1.45 *times*, or 17 and 45 percent, for every extra point in rating: $(e^{0.154}, e^{0.371}) = (1.17, 1.45)$. In contrast, the posterior model of β_3 illustrates that shared rooms are negatively associated with reviews. When controlling for ratings, there's an 80% chance that the volume of reviews for shared room listings is somewhere between 52 and 76 percent as high as for listings that are entirely private: $(e^{-0.659}, e^{-0.275}) = (0.52, 0.76)$.

```
tidy(airbnb_model_2, effects = "fixed", conf.int = TRUE, conf.level = 0.80)
# A tibble: 4 x 5
    term              estimate std.error conf.low conf.high
    <chr>                <dbl>     <dbl>    <dbl>     <dbl>
1 (Intercept)           1.99     0.405   1.48e+0     2.52
2 rating                0.265    0.0852  1.54e-1     0.371
3 room_typePriva~       0.0681   0.0520  9.54e-4     0.136
4 room_typeShare~      -0.471    0.149  -6.59e-1    -0.275
```

Let's also peak at some **neighborhood-specific** AirBnB models,

$$\lambda_{ij} = e^{\beta_{0j} + \beta_1 X_{ij1} + \beta_2 X_{ij2} + \beta_3 X_{ij3}}$$

where β_{0j}, the baseline review rate, varies from neighborhood to neighborhood. We'll again focus on just three neighborhoods: Albany Park, East Garfield Park, and The Loop. The below posterior summaries evaluate the *differences* between these neighborhoods' baselines and the global intercept, $b_{0j} = \beta_{0j} - \beta_0$:

```
tidy(airbnb_model_2, effects = "ran_vals",
     conf.int = TRUE, conf.level = 0.80) %>%
  select(level, estimate, conf.low, conf.high) %>%
  filter(level %in% c("Albany_Park", "East_Garfield_Park", "The_Loop"))
# A tibble: 3 x 4
    level               estimate conf.low conf.high
    <chr>                  <dbl>    <dbl>     <dbl>
1 Albany_Park            -0.234   -0.427   -0.0548
2 East_Garfield_Park      0.205    0.0385   0.397
3 The_Loop                0.0101  -0.0993   0.124
```

Note that AirBnB listings in Albany Park have atypically few reviews, those in East Garfield Park have atypically large numbers of reviews, and those in The Loop do not significantly differ from the average. Though not dramatic, these differences from neighborhood to neighborhood play out in posterior predictions. For example, below we simulate posterior predictive models of the number of reviews for three listings that each have a 5 rating and each offer privacy, yet they're in three different neighborhoods. Given the differing review baselines in these neighborhoods, we anticipate that the Albany Park listing will have fewer reviews than the East Garfield Park listing, though the predictive ranges are quite wide:

```
# Posterior predictions of reviews
set.seed(84735)
predicted_reviews <- posterior_predict(
  airbnb_model_2,
  newdata = data.frame(
    rating = rep(5, 3),
    room_type = rep("Entire home/apt", 3),
    neighborhood = c("Albany Park", "East Garfield Park", "The Loop")))
mcmc_areas(predicted_reviews, prob = 0.8) +
  ggplot2::scale_y_discrete(
    labels = c("Albany Park", "East Garfield Park", "The Loop")) +
  xlim(0, 150) +
  xlab("reviews")
```

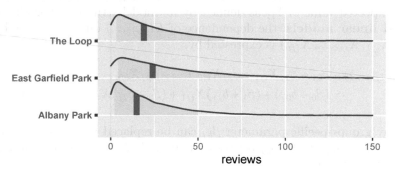

FIGURE 18.9: Posterior predictive models for the number of reviews for AirBnB listings in three different neighborhoods.

18.2.3 Model evaluation

To wrap things up, let's formally evaluate the quality of our listing analysis. First, to the question of whether or not our model is **fair**. Kinda. In answering our research question about AirBnB, we tried to learn something from the data available to us. What was available was information about the AirBnB market in *Chicago*. Thus, we'd be hesitant to use our analysis for anything more than general conclusions about the broader market. Second, our posterior predictive check in Figure 18.8 indicated that **our model isn't TOO wrong** – at least it's much better than taking a Poisson approach! Finally, **our posterior predictions of review counts are not very accurate**. The *observed* number of reviews for a listing tends to be 18 reviews, or 0.68 standard deviations, from its posterior mean prediction. This level of error is quite large on the scale of our data – though the number of reviews per listing ranges from roughly 0 to 200, most listings have below 20 reviews (Figure 18.5). To *improve* our prediction of the number of reviews a listing receives, we might incorporate additional predictors into our analysis.

```
set.seed(84735)
prediction_summary(model = airbnb_model_2, data = airbnb)
  mae mae_scaled within_50 within_95
1 17.7      0.678    0.5183    0.9571
```

18.3   Chapter summary

In Chapter 18 we combined the Unit 3 principles of generalized linear models with the Unit 4 principles of hierarchical modeling, thereby expanding our hierarchical toolkit to include Normal, Poisson, Negative Binomial, and logistic regression. In general, let $E(Y_{ij}|\ldots)$ denote the expected value of Y_{ij} as defined by its model structure. For all **hierarchical generalized linear models**, the dependence of $E(Y_{ij}|\ldots)$ on a linear combination of p predictors $(X_{ij1}, X_{ij2}, \ldots, X_{ijp})$ is expressed by

$$g\left(E(Y_{ij}|\ldots)\right) = \beta_{0j} + \beta_{1j}X_{ij1} + \beta_{2j}X_{ij2} + \cdots + \beta_{pj}X_{ijp}$$
$$= (\beta_0 + b_{0j}) + (\beta_1 + b_{1j})X_{ij1} + (\beta_2 + b_{2j})X_{ij2} + \cdots + (\beta_p + b_{pj})X_{ijp}$$

where (1) any group-specific parameter β_{kj} can be replaced by a global parameter β_k in cases of little variability in the β_{kj} between groups and (2) the **link function** $g(\cdot)$ depends upon the model structure:

model	link function
$Y_{ij}|\ldots \sim N(\mu_{ij}, \sigma_y^2)$	$g(\mu_{ij}) = \mu_{ij}$
$Y_{ij}|\ldots \sim \text{Pois}(\lambda_{ij})$	$g(\lambda_{ij}) = \log(\lambda_{ij})$
$Y_{ij}|\ldots \sim \text{NegBinom}(\mu_{ij}, r)$	$g(\mu_{ij}) = \log(\mu_{ij})$
$Y_{ij}|\ldots \sim \text{Bern}(\pi_{ij})$	$g(\pi_{ij}) = \log\left(\frac{\pi_{ij}}{1-\pi_{ij}}\right)$

18.4   Exercises

18.4.1   Applied & conceptual exercises

Exercise 18.1 (We know how to do a lot of stuff). For each model scenario, specify an appropriate structure for the data model, note whether the model is hierarchical, and if so, identify the grouping variable. Though you do not need to simulate the models, be sure to justify your selections using the data provided. To learn more about these datasets, type `?name_of_dataset` into the console.

a) Using the `coffee_ratings` data in R, researchers wish to model whether a batch of coffee beans is of the Robusta `species` based on its `flavor`.
b) Using the `trees` data in R, researchers wish to model a tree's `Height` by its `Girth`.
c) Using the `radon` data in the `rstanarm` package, researchers wish to model a home's `log_radon` levels by its `log_uranium` levels.

d)   Using the `roaches` data in the `rstanarm` package, researchers wish to model the number of roaches in an urban apartment by whether or not the apartment has received a pest control `treatment`.

Exercise 18.2 (Book banning: setting up the model). People have both failed and succeeded at getting books banned, and hence ideas censored, from public libraries and education. In the following exercises, you'll explore whether certain book characteristics can help predict whether or not a book challenge is successful. To do so, you'll balance *weakly informative priors* with the `book_banning` data in the **bayesrules** package. This data, collected by Fast and Hegland (2011) and presented by Legler and Roback (2021), includes features and outcomes for 931 book challenges made in the US between 2000 and 2010. Let Y_{ij} denote the outcome of the ith book challenge in state j, i.e., whether or not the book was `removed`. You'll consider three potential predictors of outcome: whether the reasons for the challenge include concern about `violent` material (X_{ij1}), `antifamily` material (X_{ij2}), or the use of inappropriate `language` (X_{ij3}).

a)   In your book banning analysis, you'll use the `state` in which the book challenge was made as a grouping variable. Explain why it's reasonable (and important) to assume that the book banning outcomes within any given state are *dependent*.

b)   Write out an appropriate hierarchical regression model of Y_{ij} by $(X_{ij1}, X_{ij2}, X_{ij3})$ using formal notation. Assume each state has its own intercept, but that the states share the same predictor coefficients.

c)   Dig into the `book_banning` data. What state had the most book challenges? The least?

d)   Which state has the greatest book removal rate? The smallest?

e)   Visualize and discuss the relationships between the book challenge outcome and the three predictors.

Exercise 18.3 (Book banning: simulating the model). Next, let's simulate and dig into the posterior model of the book banning parameters.

a)   Simulate the posterior of your hierarchical book banning model. Construct trace, density, and autocorrelation plots of the chain output.

b)   Report the posterior median global model. Interpret each number in this model.

c)   Are each of violence, anti-family material, and inappropriate language significantly linked to the outcome of a book challenge (when controlling for the others)? Explain.

d)   What combination of book features is most commonly associated with a book that's banned? With a book that's not banned? Explain.

Exercise 18.4 (Book banning: will the book be banned?).

a)   How accurate are your model's posterior predictions of whether a book will be banned? Provide evidence.

b)   Interpret the posterior medians of $b_{0j} = \beta_{0j} - \beta_0$ for two states j: Kansas (KS) and Oregon (OR).

c)   Suppose a book is challenged in both Kansas and Oregon for its language, but not for violence or anti-family material. Construct and compare posterior models for whether or not the book will be banned in these two states.

Exercise 18.5 (Basketball!). The Women's National Basketball Association (WNBA) is a professional basketball league in North America. Founded in the mid-1990s, the league has grown to include twelve teams. Though teams might have up to 12 players, only five play at a time. Players then rotate in and out throughout the game, with the best combination of players typically *starting* the game. The goal of our analysis is to better understand why some players get more game time than other players. Beyond the general understanding that the typical player gets 400 minutes of game time throughout the season, we'll balance *weakly informative priors* with the `basketball` data in the **bayesrules** package. This data includes information about various players from the 2019 season including:

variable	meaning
total_minutes	the total number of minutes played throughout the season
games_played	the number of games played throughout the season
starter	whether or not the player started in more than half of the games that they played
avg_points	the average number of points scored per game
team	team name

a) How many players are in the dataset?
b) How many teams are represented in the dataset?
c) Construct and discuss a plot of `total_minutes` vs `avg_points` and `starter`. Is this what you would expect to see?
d) Construct and discuss a plot of `total_minutes` vs `games_played` and `starter`. NOTE: Incorporating `games_played` into our analysis provides an important point of comparison for the total number of minutes played.

Exercise 18.6 (Basketball! Build a model).

a) Using formal notation, write out a hierarchical Poisson regression model of `total_minutes` by `avg_points`, `starter`, and `games_played`. Use `team` as the grouping variable.
b) Explain why the Poisson might be a reasonable model structure here, and why it might *not* be.
c) Explain why it's important to include `team` as a grouping variable.
d) Simulate this model using weakly informative priors and perform a `pp_check()`.

Exercise 18.7 (Basketball! Normal or Poisson or Negative Binomial?). Above, you analyzed a player's total number of minutes played using a Poisson hierarchical regression model. Redo this analysis using a *Normal* and a *Negative Binomial* hierarchical model. Evaluate and compare the three models, and explain which you prefer.

Exercise 18.8 (More basketball!). Utilize your final chosen model (Poisson, Normal, or Negative Binomial) to explore the relationship between the total number of minutes played by a player and their average points per game in more depth.

(a) Summarize your key findings. Some things to consider along the way: Can you interpret every model parameter (both global and team-specific)? Can you summarize the key trends? Which trends are significant? How good is your model?

(b)   Predict the total number of minutes that a player will get throughout a season if they play in 30 games, they start each game, and they score an average of 15 points per game.

18.4.2   Open-ended exercises

Exercise 18.9 (Open exercise: basketball analysis with multiple predictors). In this open-ended exercise, complete an analysis of the number of games started by WNBA players using multiple predictors of your choosing.

Exercise 18.10 (Open exercise: more climbing). In Chapter 18, you analyzed the relationship of a climber's success with their age and oxygen use. In this open-ended exercise, continue your climbing analysis by considering other possible predictors. These might include any combination of personal attributes (`age`, `oxygen_used`, `injured`), time attributes (`year`, `season`), or attributes of the climb itself (`highpoint_metres`, `height_metres`).

19

Adding More Layers

Throughout this book, we've laid the foundations for Bayesian thinking and modeling. But in the broader scheme of things, we've just scratched the surface. This last chapter marks the end of this book, not the end of the Bayesian modeling toolkit. There's so much more we wish we could share, but one book can't cover it all. (Perhaps a sequel?! *Bayes Rules 2! The Bayesianing* or *Bayes Rules 2! Happy Bayes Are Here Again!*) Hopefully *Bayes Rules!* has piqued your curiosity and you feel equipped to further your Bayesian explorations. We conclude here by nudging our hierarchical models one step further, to address two questions.

> ◎ **Goals**
> - We've utilized **individual-level predictors** to better understand the trends among individuals within groups. How can we also utilize **group-level predictors** to better understand the trends among the groups themselves?
> - What happens when we have *more than one grouping variable*?

We'll explore these questions through two case studies. To focus on the new concepts in this chapter, we'll also utilize weakly informative priors throughout. For a more expansive treatment, we recommend Legler and Roback (2021) or Gelman and Hill (2006). Though these resources utilize a frequentist framework, if you've read this far, you have the skills to consider their work through a Bayesian lens.

```
# Load packages
library(bayesrules)
library(tidyverse)
library(bayesplot)
library(rstanarm)
library(janitor)
library(tidybayes)
library(broom.mixed)
```

19.1 Group-level predictors

In Chapter 18 we explored how the number of reviews varies from AirBnB listing to listing. We might also ask: what makes some AirBnB listings more expensive than others? We have a weak prior understanding here that the typical listing costs around $100 per night. Beyond this baseline, we'll supplement a weakly informative prior understanding of the AirBnB market by the `airbnb` dataset in the **bayesrules** package. Recall that `airbnb` contains

DOI: 10.1201/9780429288340-19

information on 1561 listings across 43 Chicago neighborhoods, and hence *multiple listings per neighborhood*:

```
data(airbnb)

# Number of listings
nrow(airbnb)
[1] 1561

# Number of neighborhoods & other summaries
airbnb %>%
  summarize(nlevels(neighborhood), min(price), max(price))
  nlevels(neighborhood) min(price) max(price)
1                    43         10       1000
```

The observed listing prices, ranging from \$10 to \$1000 per night, are highly skewed. Thus, to facilitate our eventual *modeling* of what makes some listings more expensive than others, we'll work with the symmetric *logged* prices. (Trust us for now and we'll provide further justification below!)

```
ggplot(airbnb, aes(x = price)) +
  geom_histogram(color = "white", breaks = seq(0, 500, by = 20))
ggplot(airbnb, aes(x = log(price))) +
  geom_histogram(color = "white", binwidth = 0.5)
```

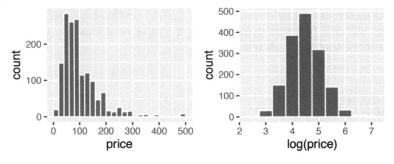

FIGURE 19.1: Histograms of AirBnB listing prices in dollars (left) and log(dollars) (right).

19.1.1 A model using only individual-level predictors

As you might expect, an AirBnB listing's price is positively associated with the number of `bedrooms` it has, its overall user `rating` (on a 1 to 5 scale), and the privacy allotted by its `room_type`, i.e., whether the renter gets an entire private unit, gets a private room within a shared unit, or shares a room:

```
ggplot(airbnb, aes(y = log(price), x = bedrooms)) +
  geom_jitter()
ggplot(airbnb, aes(y = log(price), x = rating)) +
  geom_jitter()
```

```
ggplot(airbnb, aes(y = log(price), x = room_type)) +
  geom_boxplot()
```

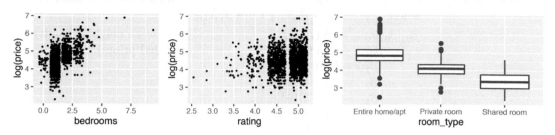

FIGURE 19.2: Plots of the log(price) of AirBnB listings by the number of bedrooms (left), user rating (middle), and room type (right).

Further, though we're not interested in the particular Chicago neighborhoods in the `airbnb` data (rather we want to use this data to learn about the broader market), we shouldn't simply ignore them either. In fact, boxplots of the listing prices in each neighborhood hint at correlation within neighborhoods. As is true with real estate in general, AirBnB listings tend to be less expensive in some neighborhoods (e.g., 13 and 41) and more expensive in others (e.g., 21 and 36):

```
ggplot(airbnb, aes(y = log(price), x = neighborhood)) +
  geom_boxplot() +
  scale_x_discrete(labels = c(1:44))
```

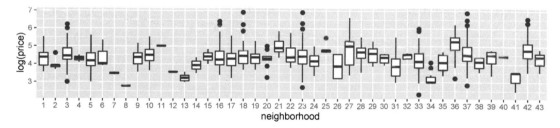

FIGURE 19.3: Boxplots of the log(price) for AirBnB listings, by neighborhood.

To this end, we can build a **hierarchical model** of AirBnB prices by a listing's number of bedrooms, rating, and room type *while accounting for* the neighborhood grouping structure. For listing i in neighborhood j, let Y_{ij} denote the price, X_{ij1} the number of bedrooms, X_{ij2} the rating,

$$X_{ij3} = \begin{cases} 1 & \text{private room} \\ 0 & \text{otherwise} \end{cases} \quad \text{and} \quad X_{ij4} = \begin{cases} 1 & \text{shared room} \\ 0 & \text{otherwise}. \end{cases}$$

Given the symmetric, continuous nature of the logged prices, we'll implement the following **Normal hierarchical regression model** of $log(Y)$ by (X_1, X_2, X_3, X_4):

$$
\begin{aligned}
log(Y_{ij})|\beta_{0j}, \beta_1, \dots, \beta_4, \sigma_y &\sim N(\mu_{ij}, \sigma_y^2) && \text{(within neighborhood } j\text{)} \\
\text{where } \mu_{ij} &= \beta_{0j} + \beta_1 X_{ij1} + \cdots + \beta_4 X_{ij4} \\
\beta_{0j}|\beta_0, \sigma_0 &\overset{ind}{\sim} N(\beta_0, \sigma_0^2) && \text{(between neighborhoods)} \\
\beta_{0c} &\sim N(4.6, 1.6^2) && \text{(global priors)} \\
\beta_1 &\sim N(0, 2.01^2) \\
\beta_2 &\sim N(0, 4.66^2) \\
\beta_3 &\sim N(0, 3.19^2) \\
\beta_4 &\sim N(0, 8.98^2) \\
\sigma_y &\sim \text{Exp}(1.6) \\
\sigma_0 &\sim \text{Exp}(1)
\end{aligned}
$$

$$\text{(19.1)}$$

Notice that (19.1) allows for random neighborhood-specific intercepts yet assumes shared predictor coefficients. That is, we assume that the baseline listing prices might vary from neighborhood to neighborhood, but that the listing features have the same association with price in each neighborhood. Further, beyond our vague understanding that the typical listing has a nightly price of $100 (hence a log(price) of roughly 4.6), our weakly informative priors are tuned by `stan_glmer()`. The corresponding posterior is simulated below with a `prior_summary()` which confirms the prior specifications in (19.1).

```
airbnb_model_1 <- stan_glmer(
  log(price) ~ bedrooms + rating + room_type + (1 | neighborhood),
  data = airbnb, family = gaussian,
  prior_intercept = normal(4.6, 2.5, autoscale = TRUE),
  prior = normal(0, 2.5, autoscale = TRUE),
  prior_aux = exponential(1, autoscale = TRUE),
  prior_covariance = decov(reg = 1, conc = 1, shape = 1, scale = 1),
  chains = 4, iter = 5000*2, seed = 84735
)

prior_summary(airbnb_model_1)
```

The `pp_check()` in Figure 19.4 (left) reassures us that ours is a reasonable model – 100 posterior simulated datasets of logged listing prices have features similar to the original logged listing prices. It's not because we're brilliant, but because we tried other things first. Our first approach was to model *price*, instead of *logged* price. Yet a `pp_check()` confirmed that this original model didn't capture the skewed nature in prices (Figure 19.4 right).

```
pp_check(airbnb_model_1) +
  labs(title = "airbnb_model_1 of log(price)") +
  xlab("log(price)")
```

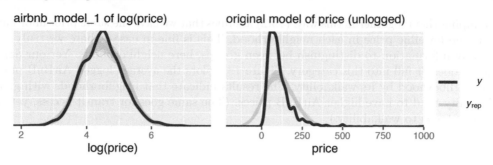

FIGURE 19.4: Posterior predictive checks of the Normal models of logged AirBnB listing prices (left) and raw, unlogged listing prices (right).

19.1.2 Incorporating group-level predictors

If this were Chapter 17, we might stop our analysis with this model – it's pretty good! Yet we wouldn't be maximizing the information in our `airbnb` data. In addition to features on the *individual listings*...

```
airbnb %>%
  select(price, neighborhood, bedrooms, rating, room_type) %>%
  head(3)
  price neighborhood bedrooms rating     room_type
1    85  Albany Park        1    5.0   Private room
2    35  Albany Park        1    5.0   Private room
3   175  Albany Park        2    4.5 Entire home/apt
```

we have features of the broader *neighborhoods* themselves, such as ratings for walkability and access to public transit (on 0 to 100 scales). These features are shared by each listing in the neighborhood. For example, all listings in `Albany Park` have a `walk_score` of 87 and a `transit_score` of 62:

```
airbnb %>%
  select(price, neighborhood, walk_score, transit_score) %>%
  head(3)
  price neighborhood walk_score transit_score
1    85  Albany Park         87            62
2    35  Albany Park         87            62
3   175  Albany Park         87            62
```

Putting this together, `airbnb` includes **individual-level predictors** (e.g., rating) and **group-level predictors** (e.g., walkability). The latter are *ignored* by our current model (19.1). Consider the neighborhood-specific intercepts β_{0j}. The original model (19.1) uses the *same prior* for each β_{0j}, assuming that the baseline logged prices in neighborhoods are Normally distributed around some mean logged price β_0 with standard deviation σ_0:

$$\beta_{0j}|\beta_0,\sigma_0 \overset{ind}{\sim} N(\beta_0,\sigma_0^2). \tag{19.2}$$

By lumping them together in this way, (19.2) assumes that we have the *same* prior information about the baseline price in each neighborhood. This is fine in cases where we truly don't have any information to distinguish between groups, here neighborhoods. Yet our `airbnb` analysis doesn't fall into this category. Figure 19.5 plots the average logged AirBnB price in each neighborhood by its walkability. The results indicate that neighborhoods with greater walkability tend to have higher AirBnB prices. (The same goes for transit access, yet we'll limit our focus to walkability.)

```
# Calculate mean log(price) by neighborhood
nbhd_features <- airbnb %>%
  group_by(neighborhood, walk_score) %>%
  summarize(mean_log_price = mean(log(price)), n_listings = n()) %>%
  ungroup()

# Plot mean log(price) vs walkability
ggplot(nbhd_features, aes(y = mean_log_price, x = walk_score)) +
  geom_point() +
  geom_smooth(method = "lm", se = FALSE)
```

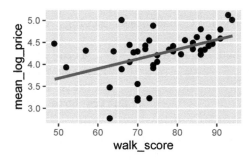

FIGURE 19.5: A scatterplot of the average logged AirBnB listing price by the walkability score for each neighborhood.

What's more, the average logged price by neighborhood appears to be *linearly* associated with walkability. Incorporating this observation into our model of β_{0j}, (19.2), isn't much different than incorporating individual-level predictors X into the first layer of the hierarchical model. First, let U_j denote the walkability of neighborhood j. Notice a couple of things about this notation:

- Though U_j is ultimately a predictor of price, we utilize "U" instead of "X" to distinguish it as a neighborhood-level, not listing-level predictor.
- Since all listings i in neighborhood j share the same walkability value, U_j needs only a j subscript.

Next, we can replace the global trend in β_{0j}, β_0, with a neighborhood-specific linear trend μ_{0j} informed by walkability U_j:

$$\mu_{0j} = \gamma_0 + \gamma_1 U_j. \tag{19.3}$$

This switch introduces two new model parameters which describe the linear trend between the baseline listing price and walkability of a neighborhood:

- intercept γ_0 technically measures the average logged price we'd expect for neighborhoods with 0 walkability (though no such neighborhood exists); and

- slope γ_1 measures the expected change in a neighborhood's typical logged price with each extra point in walkability score.

Our final AirBnB model thus expands upon (19.1) by incorporating the group-level regression model (19.3) along with prior models for the new group-level regression parameters. Given the large number of model parameters, we do not write out the independent and weakly informative priors here. These can be obtained using `prior_summary()` below.

$$
\begin{aligned}
\log(Y_{ij}) | \beta_{0j}, \beta_1, \ldots, \beta_4, \sigma_y &\sim N(\mu_j, \sigma_y^2) &&\text{with } \mu_j = \beta_{0j} + \beta_1 X_{ij1} + \cdots + \beta_4 X_{ij4} \\
\beta_{0j} | \gamma_0, \gamma_1, \sigma_0 &\stackrel{ind}{\sim} N(\mu_{0j}, \sigma_0^2) &&\text{with } \mu_{0j} = \gamma_0 + \gamma_1 U_j \\
\beta_1, \ldots, \beta_4, \gamma_0, \gamma_1, \sigma_0, \sigma_y &\sim \text{ some priors}
\end{aligned}
\tag{19.4}
$$

The new notation here might make this model appear more complicated or different than it is. First, consider what this model implies about the expected price of an AirBnB listing. For neighborhoods j that were *included* in our `airbnb` data, the expected log price of a listing i is defined by

$$
\beta_{0j} + \beta_1 X_{ij1} + \beta_2 X_{ij2} + \beta_3 X_{ij3} + \beta_4 X_{ij4}
$$

where intercepts β_{0j} vary from neighborhood to neighborhood *according to their walkability*. *Beyond* the included neighborhoods, the expected log price for an AirBnB listing is defined by replacing β_{0j} with its walkability-dependent mean $\gamma_0 + \gamma_1 U_j$:

$$
(\gamma_0 + \gamma_1 U_j) + \beta_1 X_{ij1} + \beta_2 X_{ij2} + \beta_3 X_{ij3} + \beta_4 X_{ij4}.
$$

The parentheses here emphasize the structure of (19.4): the baseline price, $\gamma_0 + \gamma_1 U_j$, varies from neighborhood to neighborhood depending upon the neighborhood walkability predictor, U_j. Removing them emphasizes another point: walkability U_j is really just like every other predictor, except that all listings in the same neighborhood share a U_j value.

Finally, consider the **within-group** and **between-group** variability parameters, σ_y and σ_0. Since the first layers of both models utilize the same regression structure of price *within* neighborhoods, σ_y has the same meaning in our original model (19.1) and new model (19.4): σ_y measures the unexplained variability in listing prices *within* any neighborhood, given the listings' `bedrooms`, `rating`, and `room_type`. Yet, by altering our model of how the typical logged prices vary *between* neighborhoods, the meaning of σ_0 has changed:

- in (19.1), σ_0 reflects the unexplained variability in baseline prices β_{0j} from neighborhood to neighborhood;
- in (19.4), σ_0 reflects the unexplained variability in baseline prices β_{0j} from neighborhood to neighborhood, *after taking the neighborhoods' walkability into account*.

Incorporating group-level predictors

Consider grouped data, where individual i in group j has response variable Y_{ij} and **individual-level predictor** X_{ij}. Further, let U_j denote a **group-level predictor**, the values of which are shared by every individual in group j. The underlying structure of a hierarchical model of Y_{ij} which includes *both* individual- and group-level predictors is given by (19.5):

$$\text{model of } Y_{ij} \text{ within group } j: \beta_{0j} + \beta_1 X_{ij}$$
$$\text{model of } \beta_{0j} \text{ between groups: } \gamma_0 + \gamma_1 U_j \qquad (19.5)$$
$$\beta_1, \gamma_0, \gamma_1, \dots \sim \text{ some priors.}$$

The first layer of (19.5) reflects the relationship between *individual* Y_{ij} and X_{ij} values, with intercepts β_{0j} that vary by group j. The next layer reflects how our prior understanding of the *group* parameters β_{0j} might be informed by the *group*-level predictor U_j. Pulling these two layers together, the expected relationship between Y_{ij} and X_{ij} is

$$(\gamma_0 + \gamma_1 U_j) + \beta_1 X_{ij}$$

where intercept $(\gamma_0 + \gamma_1 U_j)$ depends upon a group's U_j value.

19.1.3   Posterior simulation & global analysis

To simulate the posteriors of our hierarchical model (19.4), we need only plunk the group-level `walk_score` predictor directly into the `stan_glmer()` formula. Since all listings in the same neighborhood share the same `walk_score`, it is automatically recognized as a group-level predictor.

```
airbnb_model_2 <- stan_glmer(
  log(price) ~ walk_score + bedrooms + rating + room_type +
    (1 | neighborhood),
  data = airbnb, family = gaussian,
  prior_intercept = normal(4.6, 2.5, autoscale = TRUE),
  prior = normal(0, 2.5, autoscale = TRUE),
  prior_aux = exponential(1, autoscale = TRUE),
  prior_covariance = decov(reg = 1, conc = 1, shape = 1, scale = 1),
  chains = 4, iter = 5000*2, seed = 84735
)

# Don't forget to check the prior specifications!
prior_summary(airbnb_model_2)
```

Let's dig into and compare the global median models associated with `airbnb_model_2` and `airbnb_model_1`, our original model which did *not* include the neighborhood-level walkability predictor (19.1). We combine the summaries of the regression parameters for ease of comparison, yet encourage you to also check out the separate `model_1_mean` and `model_2_mean` summaries:

```
# Get relationship summaries for both models
model_1_mean <- tidy(airbnb_model_1, effects = "fixed")
model_2_mean <- tidy(airbnb_model_2, effects = "fixed")

# Combine the summaries for both models
combined_summaries <- model_1_mean %>%
  right_join(., model_2_mean, by = "term",
             suffix = c("_model_1", "_model_2")) %>%
  select(-starts_with("std.error"))

combined_summaries
# A tibble: 6 x 3
  term                estimate_model_1 estimate_model_2
  <chr>                          <dbl>            <dbl>
1 (Intercept)                     3.21             1.92
2 bedrooms                       0.265            0.265
3 rating                         0.220            0.221
4 room_typePrivate r~           -0.538           -0.537
5 room_typeShared ro~           -1.06            -1.06
6 walk_score                        NA           0.0166
```

Based on these `tidy()` summaries, the posterior median models of log(price) for the two models are:

$$\text{model 1:} \qquad 3.21 + 0.265X_{ij1} + 0.22X_{ij2} - 0.538X_{ij3} - 1.06X_{ij4}$$
$$\text{model 2:} \quad (1.92 + 0.0166U_j) + 0.265X_{ij1} + 0.221X_{ij2} - 0.537X_{ij3} - 1.06X_{ij4}.$$

With the exception of the intercept terms, the posterior median models are nearly indistinguishable. This makes sense. Including the group-level walkability predictor in `airbnb_model_2` essentially replaces the original global intercept β_0 in `airbnb_model_1` with $\gamma_0 + \gamma_1 U_j$, *without* tweaking the individual-level X predictors. We can also *interpret* `airbnb_model_2`'s posterior median coefficients as usual, both listing- and neighborhood-level, while being mindful of the logged scale of the price response variable. For example, when controlling for the other model predictors, the typical *logged* price for a shared room is roughly 1.06 less than that for an entire private home. More meaningfully, the typical *price* for a shared room is roughly *one-third* of that for an entire private home ($e^{-1.06} \approx 0.35$). Further, for every extra *10* points in a neighborhood's walkability rating, we expect the typical price of listings in that neighborhood to increase by roughly 18 percent ($e^{10*0.0166} \approx 1.18$).

In a similar spirit, let's obtain and compare posterior summaries for the standard deviation parameters, σ_y (`sd_Observation.Residual`) and σ_0 (`sd_(Intercept).neighborhood`):

```
# Get variance summaries for both models
model_1_var <- tidy(airbnb_model_1, effects = "ran_pars")
model_2_var <- tidy(airbnb_model_2, effects = "ran_pars")

# Combine the summaries for both models
model_1_var %>%
  right_join(., model_2_var, by = "term",
             suffix = c("_model_1", "_model_2")) %>%
```

```
   select(-starts_with("group"))
# A tibble: 2 x 3
  term                  estimate_model_1 estimate_model_2
  <chr>                          <dbl>           <dbl>
1 sd_(Intercept).nei~            0.279           0.202
2 sd_Observation.Res~            0.365           0.366
```

The posterior medians of the **within-group variability** parameter σ_y are nearly indistinguishable for our two models: 0.365 vs 0.366. This suggests that incorporating the neighborhood-level walkability predictor didn't improve our understanding of the variability in individual listing prices *within* neighborhoods, i.e., why some listings are more expensive than others in the same neighborhood. Makes sense! Since all listings within a neighborhood share the same walkability value U_j, including this information in `airbnb_model_2` doesn't help us distinguish between listings in the same neighborhood.

In contrast, the posterior median of the **between-group variability** parameter σ_0 is *notably* smaller in `airbnb_model_2` than in `airbnb_model_1`: 0.202 vs 0.279. Recall that σ_0 reflects our uncertainty about neighborhood baseline prices β_{0j}. Thus, the drop in σ_0 from `airbnb_model_1` to `airbnb_model_2` indicates that, by including the neighborhood-level walkability predictor, we have increased our certainty about the neighborhood-level β_{0j} parameter. Or, in words, walkability helps explain why some neighborhoods tend to have more expensive listings than others. This, too, makes sense!

An observation about group-level predictors

Including a group-level predictor tends to increase our certainty about **between-group** trends (how groups differ from one another) while not improving our certainty about **within-group** trends (how individual observations within the same group differ from one another).

19.1.4 Posterior group-level analysis

In a final consideration of the impact of the walkability predictor on our AirBnB price model, let's dig into the neighborhood-level trends. Consider two of the 43 neighborhoods: Edgewater and Pullman. The typical AirBnB listing prices in these two neighborhoods are nearly equivalent. *However*, Edgewater is much more walkable than Pullman (89 vs 49), something we've seen to be a desirable feature in the AirBnB market:

```
nbhd_features %>%
  filter(neighborhood %in% c("Edgewater", "Pullman"))
# A tibble: 2 x 4
  neighborhood walk_score mean_log_price n_listings
  <fct>              <int>          <dbl>      <int>
1 Edgewater             89           4.47         35
2 Pullman               49           4.47          5
```

Our two different models, (19.1) and (19.4), formulate different baseline prices β_{0j} for these two neighborhoods. Letting b_{0j} denote a neighborhood j adjustment:

$$\begin{aligned} \text{model 1:} \quad \beta_{0j} &= \qquad\qquad \beta_0 + b_{0j} \\ \text{model 2:} \quad \beta_{0j} &= \quad \gamma_0 + \gamma_1 U_j + b_{0j}. \end{aligned} \tag{19.6}$$

To calculate the posterior median intercepts β_{0j} for both neighborhoods in both models, we can utilize the posterior median values of $(\beta_0, \gamma_0, \gamma_1)$, (3.21, 1.92, 0.0166), from the earlier `tidy(..., effects = "fixed")` summaries:

```
combined_summaries %>%
  filter(term %in% c("(Intercept)", "walk_score"))
# A tibble: 2 x 3
  term        estimate_model_1 estimate_model_2
  <chr>                  <dbl>            <dbl>
1 (Intercept)             3.21             1.92
2 walk_score                NA           0.0166
```

Further, we can obtain the neighborhood adjustments b_{0j} for both models from the `tidy(..., effects = "ran_vals")` summaries below:

```
# Get neighborhood summaries from both models
model_1_nbhd <- tidy(airbnb_model_1, effects = "ran_vals")
model_2_nbhd <- tidy(airbnb_model_2, effects = "ran_vals")

# Combine the summaries for both models
model_1_nbhd %>%
  right_join(., model_2_nbhd, by = "level",
             suffix = c("_model_1", "_model_2")) %>%
  select(-starts_with(c("group", "term", "std.error"))) %>%
  filter(level %in% c("Edgewater", "Pullman"))
# A tibble: 2 x 3
  level      estimate_model_1 estimate_model_2
  <chr>                 <dbl>            <dbl>
1 Edgewater            0.0705           -0.109
2 Pullman              0.0640            0.320
```

Then plugging into (19.6), the posterior median intercepts β_{0j} for both neighborhoods in both models are as follows:

neighborhood	model 1 intercept	model 2 intercept
Edgewater	$3.21 + 0.0705 = 3.2805$	$1.92 + 0.0166*89 - 0.109 = 3.2884$
Pullman	$3.21 + 0.064 = 3.274$	$1.92 + 0.0166*49 + 0.32 = 3.0534$

There are some cool and intuitive things to notice in this table:

- In `airbnb_model_1`, the two neighborhoods have nearly identical intercepts. This isn't surprising – `airbnb_model_1` ignores the fact that Edgewater is much more walkable than Pullman. Thus, since the typical prices are so similar in the two neighborhoods, so too are their intercepts.

- In `airbnb_model_2`, Pullman's intercept is much lower than Edgewater's. This also isn't surprising – `airbnb_model_2` takes into account that neighborhood prices are positively associated with walkability ($1.92 + 0.0166*U_j$). Since Pullman's walkability is so much lower than Edgewater's, so too is its intercept.

Looking beyond Pullman and Edgewater, Figure 19.6 plots the pairs of `airbnb_model_1` intercepts (open circles) and `airbnb_model_2` intercepts (closed circles) for all 43 sample neighborhoods. Like the observed average log(prices) in these neighborhoods (Figure 19.5), the `airbnb_model_2` intercepts are positively associated with walkability. The posterior median model of this association is captured by $\gamma_0 + \gamma_1 U_j \approx 1.92 + 0.0166 U_j$.

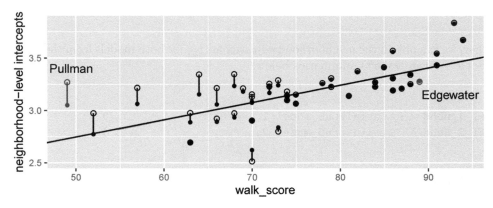

FIGURE 19.6: For each of the 43 neighborhoods in `airbnb`, the posterior median neighborhood-level intercepts from `airbnb_model_1` (open circles) and `airbnb_model_2` (closed circles) are plotted versus neighborhood walkability. Vertical lines connect the neighborhood intercept pairs. The sloped line represents the `airbnb_model_2` posterior median model of log(price) by walkability, $\gamma_0 + \gamma_1 U_j \approx 1.92 + 0.0166 U_j$.

The *comparison* between the two models' intercepts is also notable here. As our numerical calculations above confirm, Edgewater's intercepts are quite similar in the two models. (It's tough to even visually distinguish between them!) Since its `airbnb_model_1` intercept was already so close to the price vs walkability trend, incorporating the walkability predictor in `airbnb_model_2` didn't do much to change our mind about Edgewater. In contrast, Pullman's `airbnb_model_1` intercept implied a much higher baseline price than we would expect for a neighborhood with such low walkability. Upon incorporating walkability, `airbnb_model_2` thus **pulled** Pullman's intercept down, closer to the trend.

We've been here before. Hierarchical models **pool** information across all groups, allowing what we learn about some groups to improve our understanding of others. As evidenced by the `airbnb_model_2` neighborhood intercepts that are pulled toward the trend with walkability, this pooling is intensified by incorporating a group-level predictor and is especially pronounced for neighborhoods that either (1) have `airbnb_model_1` intercepts that fall *far* from the trend or (2) have small sample sizes. For example, Pullman falls into both categories. Not only is its `airbnb_model_1` intercept quite far above the trend with walkability, our `airbnb` data included only 5 listings in Pullman (contrasted by 35 in Edgewater):

```
nbhd_features %>%
  filter(neighborhood %in% c("Edgewater", "Pullman"))
# A tibble: 2 x 4
  neighborhood walk_score mean_log_price n_listings
  <fct>             <int>          <dbl>      <int>
1 Edgewater            89           4.47         35
2 Pullman              49           4.47          5
```

In this case, the pooled information from the other neighborhoods regarding the relationship between prices and walkability has a lot of sway in our posterior understanding about Pullman.

Another observation about group-level predictors

Utilizing group-level predictors increases our ability to pool information across groups, thereby especially enhancing our understanding of groups with small sample sizes.

19.1.5   We're just scratching the surface!

The model we're considering here just scratches the surface. We can go deeper by connecting it to other themes we've considered throughout Units 3 and 4. To name a few:

- we can incorporate *more than one* group-level predictor;
- group-level and individual-level predictors might *interact*; and
- group-level predictors might help us better understand *group-specific "slopes"* or regression parameters, not just group-specific intercepts.

19.2   Incorporating two (or more!) grouping variables

19.2.1   Data with two grouping variables

In Chapter 18 we used the `climbers` data to model the success of a mountain climber in summiting a peak, by their age and use of oxygen:

```
# Import, rename, & clean data
data(climbers_sub)
climbers <- climbers_sub %>%
  select(peak_name, expedition_id, member_id, success,
         year, season, age, expedition_role, oxygen_used)
```

In doing so, we were mindful of the fact that climbers were **grouped** into different expeditions, the success of one climber in an exhibition being directly tied to the success of others. For example, 5 climbers participated in the "AMAD03107" expedition:

```
# Summarize expeditions
expeditions <- climbers %>%
  group_by(peak_name, expedition_id) %>%
  summarize(n_climbers = n())

head(expeditions, 2)
# A tibble: 2 x 3
# Groups:   peak_name [1]
  peak_name   expedition_id n_climbers
  <fct>       <chr>              <int>
1 Ama Dablam  AMAD03107              5
2 Ama Dablam  AMAD03327              6
```

But that's not all! If you look more closely, you'll notice *another* grouping factor in the data: the *mountain peak* being summited. For example, our dataset includes 27 different expeditions with a total of 210 different climbers that set out to summit the Ama Dablam peak:

```
# Summarize peaks
peaks <- expeditions %>%
  group_by(peak_name) %>%
  summarize(n_expeditions = n(), n_climbers = sum(n_climbers))

head(peaks, 2)
# A tibble: 2 x 3
  peak_name    n_expeditions n_climbers
  <fct>                <int>      <int>
1 Ama Dablam              27        210
2 Annapurna I              6         62
```

Altogether, the `climbers` dataset includes information about 2076 individual climbers, grouped together in 200 expeditions, to 46 different peaks:

```
# Number of climbers
nrow(climbers)
[1] 2076
# Number of expeditions
nrow(expeditions)
[1] 200
# Number of peaks
nrow(peaks)
[1] 46
```

Further, these groupings are **nested**: the data consists of climbers **within** expeditions and expeditions **within** peaks. That is, a given climber does not set out on every expedition nor does a given expedition set out to summit every peak. Figure 19.7 captures a simplified version of this nested structure in pictures, assuming only 2 climbers within each of 6 expeditions and 2 expeditions within each of 3 peaks.

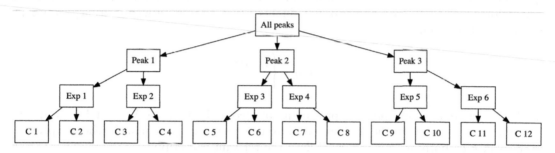

FIGURE 19.7: In the nested group structure, climbers (C) are nested within expeditions (Exp), which are nested within peaks.

19.2.2 Building a model with two grouping variables

Just as we shouldn't ignore the fact that the climbers are grouped by expedition, we shouldn't ignore the fact that expeditions are grouped by the peak they try to summit. After all, due to different elevations, steepness, etc., some peaks are easier to summit than others. Thus, the success of expeditions that pursue the same peak are inherently related. At the easier end of the climbing spectrum, 3 of the 46 sample peaks had a success rate of 1 – all sampled climbers that set out for those 3 peaks were successful. At the tougher end, 20 peaks had a success rate of 0 – *none* of the climbers that set out for those peaks were successful.

```
climbers %>%
  group_by(peak_name) %>%
  summarize(p_success = mean(success)) %>%
  ggplot(., aes(x = p_success)) +
    geom_histogram(color = "white")
```

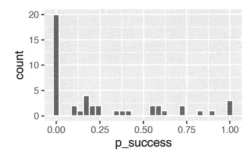

FIGURE 19.8: A histogram of the climbing success rates at 46 different mountain peaks.

Now, we don't really care about the particular 46 peaks represented in the `climbers` dataset. These are just a sample from a vast world of mountain climbing. Thus, to incorporate it into our analysis of climber success, we'll include peak name as a **grouping variable**, not a predictor. This second grouping variable, in addition to expedition group, requires a new subscript. Let Y_{ijk} denote whether or not climber i that sets out with expedition j to summit peak k is successful,

$$Y_{ijk} = \begin{cases} 1 & \text{yes} \\ 0 & \text{no} \end{cases}$$

and π_{ijk} denote the corresponding probability of success. Further, let X_{ijk1} and X_{ijk2} denote the climber's age and whether they received oxygen, respectively. In models where expedition j or peak k or both are ignored, we'll drop the relevant subscripts.

Given the binary nature of response variable Y, we can utilize **hierarchical logistic regression** for its analysis. Consider two approaches to this task. Like our approach in Chapter 18, **Model 1 assumes that baseline success rates vary by expedition** j, and thus incorporates expedition-specific intercepts β_{0j}. In past chapters, we learned that we can think of β_{0j} as a random tweak or adjustment b_{0j} to some global intercept β_0:

$$\begin{aligned} \log\left(\tfrac{\pi_{ij}}{1-\pi_{ij}}\right) &= \beta_{0j} &+\beta_1 X_{ij1} + \beta_2 X_{ij2} \\ &= (\beta_0 + b_{0j}) &+\beta_1 X_{ij1} + \beta_2 X_{ij2} \end{aligned}$$

Accordingly, we'll specify Model 1 as follows, where the random tweaks b_{0j} are assumed to be Normally distributed around 0, and thus the random intercepts β_{0j} Normally distributed around β_0, with standard deviation σ_b. Further, the weakly informative priors are tuned by `stan_glmer()` below, where we again utilize a baseline prior assumption that the typical climber has a 0.5 probability, or 0 log(odds), of success:

$$\begin{aligned} Y_{ij}|\beta_0,\beta_1,\beta_2,b_{0j} &\sim \text{Bern}(\pi_{ij}) && \text{(within expedition } j) \\ \text{with } \log\left(\tfrac{\pi_{ij}}{1-\pi_{ij}}\right) &= (\beta_0 + b_{0j}) + \beta_1 X_{ij1} + \beta_2 X_{ij2} \\ b_{0j}|\sigma_b &\overset{ind}{\sim} N(0,\sigma_b^2) && \text{(between expeditions)} \\ \beta_{0c} &\sim N(0,2.5^2) && \text{(global priors)} \\ \beta_1 &\sim N(0,0.24^2) \\ \beta_2 &\sim N(0,5.51^2) \\ \sigma_b &\sim \text{Exp}(1). \end{aligned} \qquad (19.7)$$

Next, let's acknowledge our second grouping factor. **Model 2 assumes that baseline success rates vary by expedition j AND peak k**, thereby incorporating expedition- and peak-specific intercepts β_{0jk}. Following our approach to Model 1, we obtain these β_{0jk} intercepts by adjusting the global intercept β_0 by expedition-tweak b_{0j} and peak-tweak p_{0k}:

$$\begin{aligned}
\log\left(\frac{\pi_{ijk}}{1-\pi_{ijk}}\right) &= & \beta_{0jk} && +\beta_1 X_{ijk1} + \beta_2 X_{ijk2} \\
&= & (\beta_0 + b_{0j} + p_{0k}) && +\beta_1 X_{ijk1} + \beta_2 X_{ijk2}.
\end{aligned}$$

Thus, for expedition j and peak k, we've **decomposed** the intercept β_{0jk} into three pieces:

- β_0 = the global baseline success rate across all climbers, expeditions, and peaks
- b_{0j} = an adjustment to β_0 for climbers in expedition j
- p_{0k} = an adjustment to β_0 for expeditions that try to summit peak k.

The complete **Model 2** specification follows, where the independent weakly informative priors are specified by `stan_glmer()` below:

$$Y_{ijk}|\beta_0,\beta_1,\beta_2,b_{0j},p_{0k} \sim \text{Bern}(\pi_{ijk})$$

$$\text{with } \log\left(\frac{\pi_{ijk}}{1-\pi_{ijk}}\right) = (\beta_0 + b_{0j} + p_{0k}) + \beta_1 X_{ijk1} + \beta_2 X_{ijk2}$$

$$b_{0j}|\sigma_b \overset{ind}{\sim} N(0,\sigma_b^2)$$

$$p_{0k}|\sigma_p \overset{ind}{\sim} N(0,\sigma_p^2)$$

$$\beta_{0c} \sim N(0,2.5^2)$$

$$\beta_1 \sim N(0,0.24^2) \qquad\qquad (19.8)$$

$$\beta_2 \sim N(0,5.51^2)$$

$$\sigma_b \sim \text{Exp}(1)$$

$$\sigma_p \sim \text{Exp}(1).$$

Note that the two **between** variance parameters are interpreted as follows:

- σ_b = variability in success rates from expedition to expedition *within a peak*; and
- σ_p = variability in success rates from peak to peak.

This is quite a philosophical leap! We'll put some specificity into the details by simulating the model posteriors in the next section.

19.2.3 Simulating models with two grouping variables

The posteriors corresponding to Models 1 and 2, (19.7) and (19.8), are simulated below. Notice that to incorporate the additional peak-related grouping variable in Model 2, we merely add `(1 | peak_name)` to the model formula:

```
climb_model_1 <- stan_glmer(
  success ~ age + oxygen_used + (1 | expedition_id),
  data = climbers, family = binomial,
  prior_intercept = normal(0, 2.5, autoscale = TRUE),
  prior = normal(0, 2.5, autoscale = TRUE),
  prior_covariance = decov(reg = 1, conc = 1, shape = 1, scale = 1),
```

```
  chains = 4, iter = 5000*2, seed = 84735
)

climb_model_2 <- stan_glmer(
  success ~ age + oxygen_used + (1 | expedition_id) + (1 | peak_name),
  data = climbers, family = binomial,
  prior_intercept = normal(0, 2.5, autoscale = TRUE),
  prior = normal(0, 2.5, autoscale = TRUE),
  prior_covariance = decov(reg = 1, conc = 1, shape = 1, scale = 1),
  chains = 4, iter = 5000*2, seed = 84735
)
```

The `tidy()` summaries below illuminate the posterior models of the regression coefficients $(\beta_0, \beta_1, \beta_2)$ for both Models 1 and 2. In general, these two models lead to similar conclusions about the expected relationship between *climber* success with age and use of oxygen: aging doesn't help, but oxygen does. For example, by the posterior median estimates of β_1 and β_2 from `climb_model_2`, the odds of success are roughly cut in half for every extra 15 years in age ($e^{15*-0.0475} = 0.49$) and increase nearly 500-fold with the use of oxygen ($e^{6.19} = 486.65$).

```
# Get trend summaries for both models
climb_model_1_mean <- tidy(climb_model_1, effects = "fixed")
climb_model_2_mean <- tidy(climb_model_2, effects = "fixed")

# Combine the summaries for both models
climb_model_1_mean %>%
  right_join(., climb_model_2_mean, by ="term",
             suffix = c("_model_1", "_model_2")) %>%
  select(-starts_with("std.error"))
# A tibble: 3 x 3
  term            estimate_model_1 estimate_model_2
  <chr>                      <dbl>            <dbl>
1 (Intercept)                -1.42            -1.55
2 age                      -0.0474          -0.0475
3 oxygen_usedTRUE             5.79             6.19
```

What *does* change quite a bit from `climb_model_1` to `climb_model_2` is our understanding of what contributes to the overall *variability* in success rates. Both models acknowledge that *some* variability can be accounted for by the age and oxygen use among climbers within an expedition. Yet individual climbers do *not* account for all variability in success. To this end, `climb_model_1` attributes the remaining variability to differences between expeditions – some expeditions are set up to be more successful than others. Doubling down on this idea, `climb_model_2` assumes that the remaining variability might be explained by *both* the inherent differences between expeditions and those between peaks – some peaks are easier to climb than others. The posterior medians for these variability sources are summarized below, where `sd_(Intercept).expedition_id` corresponds to the standard deviation in success rates between expeditions (σ_b) and `sd_(Intercept).peak_name` to the standard deviation between peaks (σ_p):

```
# Get variance summaries for both models
climb_model_1_var <- tidy(climb_model_1, effects = "ran_pars")
climb_model_2_var <- tidy(climb_model_2, effects = "ran_pars")

# Combine the summaries for both models
climb_model_1_var %>%
  right_join(., climb_model_2_var, by = "term",
             suffix =c("_model_1", "_model_2")) %>%
  select(-starts_with("group"))
# A tibble: 2 x 3
  term                   estimate_model_1 estimate_model_2
  <chr>                         <dbl>            <dbl>
1 sd_(Intercept).exp~            3.65             3.10
2 sd_(Intercept).pea~            NA               1.85
```

Starting with `climb_model_1`, the variability in success rates from expedition to expedition has a posterior median value of 3.65. This variability is *redistributed* in `climb_model_2` to both peaks *and* expeditions within peaks. There are a few patterns to pick up on here:

- The variability in success rates from peak to peak, σ_p, is *smaller* than that from expedition to expedition within any given peak, σ_b. This suggests that there are greater differences between expeditions on the same peak than between the peaks themselves.

- The posterior median of σ_b drops from `climb_model_1` to `climb_model_2`. This makes sense for two reasons. First, some of the expedition-related variability in `climb_model_1` is being redistributed and attributed to peaks in `climb_model_2`. Second, σ_b measures the variability in success across *all* expeditions in `climb_model_1`, but the variability across expeditions *within the same peak* in `climb_model_2` – naturally, the outcomes of expeditions on the same peak are more consistent than the outcomes of expeditions across all peaks.

19.2.4   Examining the group-specific parameters

Finally, let's examine what `climb_model_2` indicates about the success rates π_{ijk} across different peaks k, expeditions j, and climbers i:

$$\log\left(\frac{\pi_{ijk}}{1 - \pi_{ijk}}\right) = (\beta_0 + b_{0j} + p_{0k}) + \beta_1 X_{ijk1} + \beta_2 X_{ijk2}. \qquad (19.9)$$

Earlier we observed the posterior properties of the global parameters $(\beta_0, \beta_1, \beta_2)$:

```
# Global regression parameters
climb_model_2_mean %>%
  select(term, estimate)
# A tibble: 3 x 2
  term              estimate
  <chr>                <dbl>
1 (Intercept)          -1.55
2 age                  -0.0475
3 oxygen_usedTRUE       6.19
```

Further, for each of the 200 sample expeditions *and* 46 sample peaks, `group_levels_2` provides a tidy posterior summary of the associated b_{0j} and p_{0k} adjustments to the global baseline success rate β_0:

```
# Group-level terms
group_levels_2 <- tidy(climb_model_2, effects = "ran_vals") %>%
  select(level, group, estimate)
```

For example, expeditions to the Ama Dablam peak have a *higher than average* success rate, with a *positive* peak tweak of $p_{0k} = 2.92$. In contrast, expeditions to Annapurna I have a lower than average success rate, with a *negative* peak tweak of $p_{0k} = $ -2.04:

```
group_levels_2 %>%
  filter(group == "peak_name") %>%
  head(2)
# A tibble: 2 x 3
  level       group      estimate
  <chr>       <chr>         <dbl>
1 Ama_Dablam  peak_name      2.92
2 Annapurna_I peak_name     -2.04
```

Further, among the various *expeditions* that tried to summit Ama Dablam, both `AMAD03107` and `AMAD03327` had higher than average success rates, and thus positive expedition tweaks b_{0j} (0.00575 and 3.32, respectively):

```
group_levels_2 %>%
  filter(group == "expedition_id") %>%
  head(2)
# A tibble: 2 x 3
  level     group          estimate
  <chr>     <chr>             <dbl>
1 AMAD03107 expedition_id   0.00575
2 AMAD03327 expedition_id      3.32
```

We can combine this global, peak-specific, and expedition-specific information to model the success rates for three different groups of climbers. In cases where the group's expedition or destination peak falls outside the observed groups in our `climbers` data, the corresponding tweak is set to 0 – i.e., in the face of the unknown, we assume *average* behavior for the new expedition or peak:

- **Group a** climbers join expedition `AMAD03327` to Ama Dablam, and thus have positive expedition and peak tweaks, $b_{0j} = 3.32$ and $p_{0j} = 2.92$;
- **Group b** climbers join a *new* expedition to Ama Dablam, and thus have a neutral expedition tweak and a positive peak tweak, $b_{0j} = 0$ and $p_{0j} = 2.92$; and

- **Group c** climbers join a *new* expedition to Mount Pants Le Pants, a peak not included in our `climbers` data, and thus have neutral expedition and peak tweaks, $b_{0j} = p_{0j} = 0$.

Plugging these expedition and peak tweaks, along with the posterior medians for $(\beta_0, \beta_1, \beta_2)$, into (19.9) reveals the posterior median models of success for the three groups of climbers:

$$
\begin{aligned}
\text{Group a:} \quad & \log\left(\tfrac{\pi}{1-\pi}\right) = & (-1.55 + 3.32 + 2.92) & \quad -0.0475 X_1 + 6.19 X_2 \\
\text{Group b:} \quad & \log\left(\tfrac{\pi}{1-\pi}\right) = & (-1.55 + 0 + 2.92) & \quad -0.0475 X_1 + 6.19 X_2 \\
\text{Group c:} \quad & \log\left(\tfrac{\pi}{1-\pi}\right) = & (-1.55 + 0 + 0) & \quad -0.0475 X_1 + 6.19 X_2
\end{aligned}
$$

Since `AMAD03327` has a higher-than-average success rate among expeditions to Ama Dablam, notice that climbers in Group a are "rewarded" with a positive expedition tweak. Similarly, since `Ama Dablam` has a higher-than-average success rate among the *peaks*, the climbers in Groups a and b receive a positive peak tweak. As a result, climbers in Group c on a *new* expedition to a *new* peak have the lowest expected success rate and climbers in Group a the highest.

19.2.5 We're just scratching the surface!

The two-grouping structure we're considering here just scratches the surface. We can expand on the theme. For example:

- we might have *more than two* grouping variables;
- we might incorporate these grouping variables for group-level "slopes" or regression parameters, not just group-level intercepts; and
- our grouping variables might have a *crossed* or *non-nested structure* in which levels of "grouping variable 1" might occur with multiple different levels of "grouping variable 2" (unlike the expedition teams which pursue one, not multiple, peaks).

19.3 Exercises

19.3.1 Conceptual exercises

Exercise 19.1 (Individual- vs group-level predictors: Part I). In the Chapter 18 exercises, you utilized the `book_banning` data to model whether or not a book was `removed` while accounting for the grouping structure in `state`, i.e., there being multiple book challenges per state. Indicate whether each variable below is a potential book-level or state-level predictor of `removed`. Support your claim with evidence.

    a)   `language`
    b)   `political_value_index`
    c)   `hs_grad_rate`
    d)   `antifamily`

Exercise 19.2 (Individual- vs group-level predictors: Part II). In Chapter 19, you utilized the `climbers_sub` data to model whether or not a mountain climber had `success`, while accounting for the grouping structure in `expedition_id` and `peak_id`. Indicate whether each variable below is a potential climber-level, expedition-level, or peak-level predictor of `success`. Support your claim with evidence.

a) `height_metres`
b) `age`
c) `count`
d) `expedition_role`
e) `first_ascent_year`

Exercise 19.3 (Two groups: Part I). To study the occurrence of widget defects, researchers enlisted 3 different workers at each of 4 different factories into a study. Each worker produced 5 widgets and researchers recorded the number of defects in each widget.

a) There are two grouping variables in this study. Name them.
b) In the spirit of Figure 19.7, draw a diagram which illustrates the grouping structure of the resulting study data.
c) Is the study data "nested"? Explain.

Exercise 19.4 (Two groups: Part II). Continuing with the widget study, let Y_{ijk} be the number of defects for the ith widget made by worker j at factory k. Suppose the following is a reasonable model of Y_{ijk}:

$$Y_{ijk}|\beta_0, b_{0j}, p_{0k} \sim N(\mu_{ijk}, \sigma_y^2) \quad \text{with} \quad \mu_{ijk} = \beta_0 + b_{0j} + f_{0k}$$
$$b_{0j}|\sigma_b \stackrel{ind}{\sim} N(0, \sigma_b^2)$$
$$f_{0k}|\sigma_f \stackrel{ind}{\sim} N(0, \sigma_f^2)$$
$$\beta_{0c}, \sigma_b, \sigma_f \sim \text{some independent priors.}$$

a) Explain the meaning of the β_0 term in this context.
b) Explain the meaning of the b_{0j} and f_{0k} terms in this context.
c) Suppose that the variance parameters $(\sigma_y, \sigma_b, \sigma_f)$ have posterior median values $(2, 10, 1)$. Compare and contrast these values in the context of widget manufacturing.

19.3.2 Applied exercises

Exercise 19.5 (Spotify: double the groups). Do happier songs tend to be more popular? To answer this question, we'll model `popularity` by `valence` using the `spotify` data in the `bayesrules` package. In doing so, we'll utilize weakly informative priors with a general understanding that the typical artist has a popularity rating of around 50. *For illustrative purposes only*, we'll restrict our attention to just 6 artists:

```
data(spotify)
spotify_small <- spotify %>%
  filter(artist %in% c("Beyoncé", "Camila Cabello", "Camilo",
                       "Frank Ocean", "J. Cole", "Kendrick Lamar")) %>%
  select(artist, album_id, popularity, valence)
```

a) In previous models of `popularity`, we've only acknowledged the grouping structure imposed by the `artist`. Yet there's a second grouping variable in the `spotify` data. What is it?

b) In the spirit of Figure 19.7, draw a diagram which illustrates the grouping structure of the `spotify` data.

c) Challenge: plot the relationship between `popularity` and `valence`, while indicating the two grouping variables.

Exercise 19.6 (Spotify: two models). In this exercise you will compare two Normal hierarchical regression models of `popularity` by `valence`. For simplicity, utilize random intercepts but not random slopes throughout.

a) Define and use careful notation to write out a model of `popularity` by `valence` which accounts for the `artist` grouping variable but *ignores* the other grouping variable.

b) Simulate this model and perform a `pp_check()`. Use this to explain the consequence of ignoring the other grouping variable.

c) Define and use careful notation to write out an appropriate model of `popularity` by `valence` which accounts for *both* grouping variables.

d) Simulate this model and perform a `pp_check()`. How do the `pp_check()` results compare to those for your first model?

Exercise 19.7 (Spotify: digging in). Let's dig into your model that accounts for both grouping variables in the `spotify` data.

a) Write out the posterior median model of the relationship between `popularity` and `valence` for songs in the following groups:
   - Albums by artists not included in the `spotify_small` dataset
   - A new album by Kendrick Lamar
   - Kendrick Lamar's "good kid, m.A.A.d city" album (`album_id` 748dZDqSZy6aPXKcI9H80u)

b) Compare the posterior median models from part a. What do they tell us about the relevant artists and albums?

c) Which of the 6 sample artists gets the highest "bump" or tweak in their baseline popularity?

d) Which sample album gets the highest "bump" or tweak in its baseline popularity? And which artist made this album?

Exercise 19.8 (Spotify: understanding variability). Your Spotify model has three variance parameters. Construct, interpret, and compare posterior summaries of these three parameters. For example, what do they tell you about the music industry: is there more variability in the popularity from song to song within the same album, from album to album within the same artist, or from artist to artist?

Exercise 19.9 (Big words: incorporating group-level predictors). In the Chapter 16 exercises, you utilized the `big_word_club` data to explore the effectiveness of a digital vocabulary learning program, the *Big Word Club* (BWC) (Kalil et al., 2020). You will build upon this analysis here, modeling the percent change in students' vocabulary scores throughout the study period (`score_pct_change`) while accounting for there being multiple student participants per `school_id`.

```
# Load & process the data
data("big_word_club")
bwc <- big_word_club %>%
  mutate(treat = as.factor(treat))
```

a) Consider five potential predictors of `score_pct_change`: `treat` (whether or not the student participated in the BWC or served as a control), `age_months`, `private_school`, `esl_observed`, `free_reduced_lunch`. Explain whether each is a student-level or school-level predictor.

b) Use careful notation to specify a hierarchical regression model of `score_pct_change` by these five predictors. Utilize weakly informative priors with a baseline understanding that the typical student might see 0 improvement in their vocabulary score.

c) Simulate the model posterior and perform a `pp_check()`.

Exercise 19.10 (Big words: interpretation).

a) Discuss your conclusions from the output in `tidy(..., effects = "fixed", conf.int = TRUE)`.

b) Discuss your conclusions from the output in `tidy(..., effects = "ran_pars", conf.int = TRUE)`.

Exercise 19.11 (Big words: models by school). Write out the posterior median models of `score_pct_change` for students in each of the following schools.

a) School "30," which participated in the vocabulary study

b) School "47," which participated in the vocabulary study

c) "Manz Elementary," a public school at which 95% of students receive free or reduced lunch, and which did not participate in the vocabulary study

d) "South Elementary," a private school at which 10% of students receive free or reduced lunch, and which did not participate in the vocabulary study

Exercise 19.12 (Big words: wrap it up).

a) Reflecting on your work above, what school features are associated with greater vocabulary improvement among its students?

b) Reflecting on your work above, what student features are associated with greater vocabulary improvement?

19.4   Goodbye!

Goodbye, dear readers. We hope that after working through this book, you feel empowered to go forth and do some Bayes things.

Bibliography

Antonio, N., de Almeida, A., and Nunes, L. (2019). Hotel booking demand datasets. *Data in Brief*, 22:41–49.

Bachynski, K. (2019). *No Game for Boys to Play: The History of Youth Football and the Origins of a Public Health Crisis*. UNC Press Books.

Baumer, B. S., Garcia, R. L., Kim, A. Y., Kinnaird, K. M., and Ott, M. Q. (2020). Integrating data science ethics into an undergraduate major. *arXiv preprint arXiv:2001.07649*.

Baumer, B. S., Horton, N., and Kaplan, D. (2021). *mdsr: Complement to Modern Data Science with R*. R package version 0.2.4.

Bechdel, A. (1986). *Dykes to Watch Out For*. Firebrand Books.

Belenky, G., Wesensten, N. J., Thorne, D. R., Thomas, M. L., Sing, H. C., Redmond, D. P., Russo, M. B., and Balkin, T. J. (2003). Patterns of performance degradation and restoration during sleep restriction and subsequent recovery: A sleep dose-response study. *Journal of Sleep Research*, 12:1–12.

Benjamin, R. (2019). *Race After Technology: Abolitionist Tools for the New Jim Code*. John Wiley & Sons.

Berger, J. O. (1984). *The Likelihood Principle (lecture notes-monograph series)*. Institute of Mathematical Statistics.

Birds Canada (2018). `https://www.birdscanada.org/`.

Blackwell, D. (1969). *Basic Statistics*. McGraw Hill.

Blangiardo, M. and Cameletti, M. (2015). *Spatial and Spatio-Temporal Bayesian models with R - INLA*. Wiley.

Blitzstein, J. and Hwang, J. (2019). *Introduction to Probability*. Chapman & Hall / CRC Texts in Statistical Science, second edition.

Bolker, B. and Robinson, D. (2021). *broom.mixed: Tidying Methods for Mixed Models*. R package version 0.2.7.

Brooks, S., Gelman, A., Jones, G., and Meng, X.-L. (2011). *Handbook of Markov Chain Monte Carlo*. CRC Press.

Cards Against Humanity (2017). Pulse of the Nation. `https://thepulseofthenation.com/`.

Carreño, E. J. C. and Cuervo, E. C. (2017). *bayeslongitudinal: Adjust Longitudinal Regression Models Using Bayesian Methodology*. R package version 0.1.0.

Dastin, J. (2018). Amazon scraps secret AI recruiting tool that showed bias against women. `https://www.reuters.com/article/us-amazon-com-jobs-automation-insight/amazon-scraps-secret-ai-recruiting-tool-that-showed-bias-against-women-idUSKCN1MK08G`.

DOI: 10.1201/9780429288340-19

D'Ignazio, C. and Klein, L. F. (2020). *Data Feminism.* MIT Press.

Dogucu, M., Johnson, A., and Ott, M. (2021). *bayesrules: Datasets and Supplemental Functions from the Bayes Rules! Book.* R package version 0.0.2.

Dua, D. and Graff, C. (2017). UCI Machine Learning Repository. `https://archive.ics.uci.edu/ml`.

Eckhardt, R. (1987). Stan Ulam, John Von Neumann, and the Monte Carlo method. *Los Alamos Science Special Issue.*

El-Gamal, M. A. and Grether, D. M. (1995). Are people Bayesian? Uncovering behavioral strategies. *Journal of the American Statistical Association,* 90(432):1137–1145.

Eubanks, V. (2018). *Automating Inequality: How High-Tech Tools Profile, Police, and Punish the Poor.* St. Martin's Press.

Fanaee-T, H. and Gama, J. (2014). Event labeling combining ensemble detectors and background knowledge. *Progress in Artificial Intelligence,* 2:113–127.

Fast, S. and Hegland, T. (2011). Book challenges: A statistical examination. *Project for Statistics 316-Advanced Statistical Modeling, St. Olaf College.*

Firke, S. (2021). *janitor: Simple Tools for Examining and Cleaning Dirty Data.* R package version 2.1.0.

Gabry, J. and Goodrich, B. (2020a). Estimating generalized linear models with group-specific terms with rstanarm. `https://mc-stan.org/rstanarm/articles/glmer.html`.

Gabry, J. and Goodrich, B. (2020b). Prior distributions for rstanarm models. `https://mc-stan.org/rstanarm/articles/priors.html`.

Gabry, J. and Goodrich, B. (2020c). *rstanarm: Bayesian Applied Regression Modeling via Stan.* R package version 2.21.1.

Gabry, J., Simpson, D., Vehtari, A., Betancourt, M., and Gelman, A. (2019). Visualization in Bayesian workflow. *J. R. Stat. Soc. A,* 182:389–402.

Gebru, T., Morgenstern, J., Vecchione, B., Vaughan, J. W., Wallach, H., Daumé III, H., and Crawford, K. (2018). Datasheets for datasets. *arXiv preprint arXiv:1803.09010.*

Gelman, A. and Hill, J. (2006). *Data Analysis Using Regression and Multilevel/Hierarchical Models.* Cambridge University Press.

Goodman, S. (2011). A dirty dozen: twelve p-value misconceptions. *Seminars in Hematology,* 45:135–140.

Gorman, K. B., Williams, T. D., and Fraser, W. R. (2014). Ecological sexual dimorphism and environmental variability within a community of Antarctic penguins (Genus Pygoscelis). *PLoS ONE,* 9(3)(e90081).

Guo, J., Gabry, J., Goodrich, B., and Weber, S. (2020). *rstan: R Interface to Stan.* R package version 2.21.2.

Hadavas, C. (2020). How automation bias encourages the use of flawed algorithms. `https://slate.com/technology/2020/03/ice-lawsuit-hijacked-algorithm.html`.

Harmon, A. (2019). As cameras track Detroit's residents, a debate ensues over racial bias. *New York Times.*

Horst, A., Hill, A., and Gorman, K. (2020). *palmerpenguins: Palmer Archipelago (Antarctica) Penguin Data*. R package version 0.1.0.

Kalil, A., Mayer, S., and Oreopoulos, P. (2020). Closing the word gap with Big Word Club: Evaluating the impact of a tech-based early childhood vocabulary program. *Ann Arbor, MI: Inter-university Consortium for Political and Social Research [distributor]*.

Kay, M. (2021). *tidybayes: Tidy Data and Geoms for Bayesian Models*. R package version 3.0.1.

Kim, A. Y., Ismay, C., and Chunn, J. (2020). *fivethirtyeight: Data and Code Behind the Stories and Interactives at FiveThirtyEight*. R package version 0.6.1.

Laird, N. and Ware, J. (1982). Random-effects models for longitudinal data. *Biometrics*, 38 4:963–974.

Legler, J. and Roback, P. (2021). *Beyond Multiple Linear Regression: Applied Generalized Linear Models and Multilevel Models in R*. Chapman and Hall/CRC.

Lock, R. H., Lock, P. F., Morgan, K. L., Lock, E. F., and Lock, D. F. (2016). *Statistics: Unlocking the Power of Data*. John Wiley & Sons.

Lum, K., Price, M., Guberek, T., and Ball, P. (2010). Measuring elusive populations with Bayesian model averaging for multiple systems estimation: A case study on lethal violations in Casanare, 1998-2007. *Statistics, Politics and Policy*, 1(1).

Mbuvha, R. and Marwala, T. (2020). Bayesian inference of COVID-19 spreading rates in South Africa. *PLOS ONE*, 15(8):1–16.

McElreath, R. (2019). Statistical Rethinking winter 2019 lecture 12. `https://www.youtube.com/watch?v=hRJtKCIDTwc`.

McGrayne, S. (2012). *The theory that would not die - How Bayes' Rule cracked the Enigma code, hunted down Russian submarines and emerged triumphant from two centuries of controversy*. Yale University Press.

Meyer, D., Dimitriadou, E., Hornik, K., Weingessel, A., and Leisch, F. (2021). *e1071: Misc Functions of the Department of Statistics, Probability Theory Group (Formerly: E1071), TU Wien*. R package version 1.7-9.

Milgram, S. (1963). Behavioral study of obedience. *Journal of Abnormal and Social Psychology*, 67:371–378.

Mitchell, M., Wu, S., Zaldivar, A., Barnes, P., Vasserman, L., Hutchinson, B., Spitzer, E., Raji, I. D., and Gebru, T. (2019). Model cards for model reporting. *Proccedings of the Conference on Fairness, Accountability, and Transparency*.

Modrak, M. (2019). Divergent transitions – a primer. `https://discourse.mc-stan.org/t/divergent-transitions-a-primer/17099`.

MuseumofModernArt (2020). MoMA – collection. *GitHub repository*.

Noble, S. U. (2018). *Algorithms of Oppression: How Search Engines Reinforce Racism*. NYU Press.

Pavlik, K. (2019). Understanding classifying genres using Spotify audio features. `https://www.kaylinpavlik.com/classifying-songs-genres/`.

R for Data Science (2018). Christmas bird counts. `https://github.com/rfordatascience/tidytuesday/tree/master/data/2019/2019-06-18`.

R for Data Science (2020a). Coffee ratings. `https://github.com/rfordatascience/tid ytuesday/blob/master/data/2020/2020-07-07`.

R for Data Science (2020b). Himalayan climbing expeditions. `https://github.com/rford atascience/tidytuesday/tree/master/data/2020/2020-09-22`.

R for Data Science (2020c). Hotels. `https://github.com/rfordatascience/tidytuesd ay/blob/master/data/2020/2020-02-11`.

R for Data Science (2020d). Spotify songs. `https://github.com/rfordatascience/tid ytuesday/blob/master/data/2020/2020-01-21/readme.md`.

Raji, I. D., Smart, A., White, R. N., Mitchell, M., Gebru, T., Hutchinson, B., Smith-Loud, J., Theron, D., and Barnes, P. (2020). Closing the AI accountability gap: Defining an end-to-end framework for internal algorithmic auditing. In *Proceedings of the 2020 Conference on Fairness, Accountability, and Transparency*, pages 33–44.

Roberts, S. (2020). How to think like an epidemiologist. *New York Times*.

Shu, K., Sliva, A., Wang, S., Tang, J., and Liu, H. (2017). Fake news detection on social media: A data mining perspective. *ACM SIGKDD Explorations Newsletter*, 19(1):22–36.

Singh, R., Meier, T. B., Kuplicki, R., Savitz, J., Mukai, I., Cavanagh, L., Allen, T., Teague, T. K., Nerio, C., Polanski, D., et al. (2014). Relationship of collegiate football experience and concussion with hippocampal volume and cognitive outcomes. *Journal of the American Medical Association*, 311(18):1883–1888.

Stan development team (2019). Stan user's guide. `https://mc-stan.org/docs/2_25/stan-users-guide/index.html`.

The Himalayan Database (2020). `https://www.himalayandatabase.com/`.

Trinh, L. and Ameri, P. (2016). AirBnB price determinants: A multilevel modeling approach. *Project for Statistics 316-Advanced Statistical Modeling, St. Olaf College.*

Vats, D. and Knudson, C. (2018). Revisiting the Gelman-Rubin diagnostic. *arXiv preprint arXiv:1812.09384.*

Vehtari, A. (2019). Cross-validation for hierarchical models. `https://avehtari.github.io/modelselection/rats_kcv.html`.

Vehtari, A., Gelman, A., Simpson, D., Carpenter, B., and Bürkner, P.-C. (2021). Rank-normalization, folding, and localization: An improved $\widehat{R}$ for assessing convergence of MCMC. *Bayesian Analysis*, 16:667–718.

Warbelow, S., Avant, C., and Kutney, C. (2019). 2019 State Equality Index. *Human Rights Campaign Foundation.*

Wasserstein, R. L. (2016). The ASA's statement on p-values: Context, process, and purpose. *The American Statistician*, 70:129–133.

Wickham, H. (2016). *ggplot2: Elegant Graphics for Data Analysis*. Springer-Verlag New York.

Wickham, H. (2021). *forcats: Tools for Working with Categorical Variables (Factors)*. R package version 0.5.1.

Wickham, H., François, R., Henry, L., and Müller, K. (2021). *dplyr: A Grammar of Data Manipulation*. R package version 1.0.6.

Williams, G. J. (2011). *Data Mining with Rattle and R: The Art of Excavating Data for Knowledge Discovery*. Use R! Springer.

Winter, B. and Grawunder, S. (2012). The phonetic profile of Korean formal and informal speech registers. *Journal of Phonetics*, 40:808–815.

Index